Advanced Exercise Physiology: Essential Concepts and Applications

Jonathan K. Ehrman, PhD

Dennis J. Kerrigan, PhD

Steven J. Keteyian, PhD

Henry Ford Health System

Human Kinetics

Library of Congress Cataloging-in-Publication Data

Names: Ehrman, Jonathan K., 1962- author. | Kerrigan, Dennis J., 1974-
  author. | Keteyian, Steven J, author.
Title: Advanced exercise physiology : essential concepts and applications /
  Jonathan K. Ehrman, Dennis J. Kerrigan, Steven J. Keteyian.
Description: Champaign, IL : Human Kinetics, [2018] | Includes
  bibliographical references and index.
Identifiers: LCCN 2017007456 | ISBN 9781492505716 (print)
Subjects: | MESH: Exercise--physiology
Classification: LCC QP301 | NLM QT 256 | DDC 612/.044--dc23 LC record available at https://lccn.loc.
  gov/2017007456

ISBN: 978-1-4925-0571-6 (print)

The web addresses cited in this text were current as of May 2017, unless otherwise noted.

Senior Acquisitions Editor: Amy M. Tocco; **Developmental Editor:** Kevin Matz; **Managing Editors:** Stephanie M. Ebersohl, Carly S. O'Connor, Anna Lan Seaman, and Caitlin Husted; **Copyeditor:** Amanda M. Eastin-Allen; **Indexer:** Andrea Hepner; **Permissions Manager:** Dalene Reeder; **Senior Graphic Designer:** Angela K. Snyder; **Graphic Designer:** Whitney Milburn; **Cover Designer:** Keri Evans; **Photographs:** © Human Kinetics, unless otherwise noted; **Photo Production Manager:** Jason Allen; **Senior Art Manager:** Kelly Hendren; **Illustrations:** © Human Kinetics, unless otherwise noted; **Printer:** Sheridan Books

Printed in the United States of America   10 9 8 7 6 5 4 3 2 1

The paper in this book is certified under a sustainable forestry program.

Human Kinetics
Website: www.HumanKinetics.com

United States: Human Kinetics
P.O. Box 5076
Champaign, IL 61825-5076
800-747-4457
e-mail: info@hkusa.com

Canada: Human Kinetics
475 Devonshire Road Unit 100
Windsor, ON N8Y 2L5
800-465-7301 (in Canada only)
e-mail: info@hkcanada.com

Europe: Human Kinetics
107 Bradford Road
Stanningley
Leeds LS28 6AT, United Kingdom
+44 (0) 113 255 5665
e-mail: hk@hkeurope.com

For information about Human Kinetics' coverage in other areas of the world,
please visit our website: www.HumanKinetics.com   E6460

Contents

Preface

The goal of this advanced exercise physiology text is to provide a concise teaching resource that can be presented in its entirety in an advanced-level (i.e., master's) course over a typical 15-week college semester. In part to meet this goal we removed some of the material that is covered in the undergraduate setting because such knowledge is a prerequisite for entry into graduate programs.

Chapters 1 through 5 provide the background for exercise physiology by taking students through the body's various organ systems. Each of these chapters begins with a brief review of the basics as they relate to exercise physiology and then provides a detailed overview of each chapter topic at the graduate level. These chapters also examine the effects of aging on physiological and psychological parameters of physical activity and exercise ability. Chapters 6 and 7 discuss exercise training and testing methods and principles as well as exercise training adaptations and responses.

The remaining chapters cover a variety of specific topics, including a full review of body composition assessment techniques (chapter 8); the effects of various environments on the physiology of exercise and athletic performance (Chapter 9); and the relationship between physical activity, fitness, and regular exercise and public health (Chapter 10). Concern-

eBook
available at
HumanKinetics.com

ing public health, much data shows that the lack of exercise and physical activity plays an important role in the health of individuals. Conversely, a plethora of information describing the positive effects of exercise and regular physical activity on health is now available to the general public. Finally, chapter 11 examines two other important and emerging concepts in the field of exercise physiology: genomics and pharmacology.

Special features in most chapters include practical material to address the clinical and human performance aspects of exercise physiology. Other features include information on the rich history and research aspects of exercise physiology. Finally, each chapter ends with a bulleted list that summarizes the major concepts, a full reference list, a list of review questions, and other study tools such as suggested readings.

Additionally, the following ancillaries are available to instructors who have adopted the text:

- A test package containing over 250 multiple-choice, true-or-false, and fill-in-the-blank questions, which can be used for building quizzes or to supplement existing exams

- An image bank containing key figures and tables from the text, which can be used to create PowerPoint or other classroom presentations

Instructors can access these materials at www.HumanKinetics.com/AdvancedExercisePhysiology.

Acknowledgments

I acknowledge the many blessings I have received from God, my family, and my coworkers throughout the years that have allowed me to be in the position to contribute to this text. I am grateful for my education at Central Michigan University, Wayne State University, and The Ohio State University, and I have enjoyed the delightful benefits of a fantastic work environment, led by Dr. Steven Keteyian, that has surrounded me with numerous clinical exercise physiologists and other allied health-care professionals who have greatly enriched my knowledge-base and brought me many, many wonderful opportunities. Finally, my family has been a great source of love and joy throughout the past 25 years. I thank my wife, Janel, and children, Joshua, Jacob, Jared, and Johanna, for many, many years of encouragement, affection, and pure enjoyment.

—Jonathan K. Ehrman

It is important to recognize that this work would not be possible without the guidance, encouragement, and support of many teachers, friends, and colleagues, both past and present. Specifically, I would like to recognize my mentors and co-authors on this book, Drs. Jonathan Ehrman and Steven Keteyian.

Most importantly, my inspiration for my work is my family. To my best friend, Robin, who has supported my efforts each step along the way, I am blessed to have you as my wife. To my children, Veronica, Liam, and Emma, you each bring me joy in your own ways and push me to be the best I can be. To my parents, Dennis and Jane, I am grateful for the love of learning and value of hard work that you taught me.

—Dennis J. Kerrigan

I believe strongly that it is important to recognize those that have contributed to advancing one's intellectual and professional growth and development. To that end, I would like to acknowledge Sidney Goldstein, MD; Henry Kim, MD; the late Donald J. Magilligan, MD; James Scott; the late Robert Shepard, PhD; and W. Douglass Weaver, MD. Additionally, I would be remiss if I did not recognize and thank my coworkers and students, who, for nearly 40 years, have shared with me a passion for the use of exercise testing and training in the care of patients with clinically manifest disease. Finally, the enjoyment and appreciation I experience from my professional work pales when compared to the love, joy, and gratefulness I hold for my family: Courtland Keteyian; Stephanie, Jake, and Ian Stacy; Jacob Keteyian; Aram Keteyian; and, without end, my wife Lynette Laidler-Keteyian. Thank you and many blessings.

—Steven J. Keteyian

Chapter 1

Exercise Metabolism

We thank Micah Zuhl, PhD, for his contributions to this chapter.

Metabolism is the sum of all chemical reactions occurring in a living organism, which take place in an organized and highly regulated fashion. Potential energy is stored in the form of carbohydrate, fat, and protein, which are the major energy substrates obtained from the diet. During cellular metabolism, the energy released from the breakdown of food molecular bonds is used to synthesize a high-energy phosphate molecule, adenosine triphosphate (ATP), which is sometimes termed the *energy currency* of the cell. The energy that is released from further breakdown of ATP is used for a wealth of metabolic processes in the body, namely the myosin–actin cross-bridge cycling in skeletal muscle. During exercise, the energy demand of the contracting muscle must be met by changes in the metabolic rate. This is achieved through coordination and complex regulation of ATP synthesis. The goal of this chapter is to understand the complexity of each metabolic pathway and how exercise activates, regulates, and adapts each system.

FOUNDATIONS OF METABOLISM

Metabolism is regulated by the principles of bioenergetics, which is the process of converting substrates to energy through a series of enzymatic steps, or pathways. In humans, metabolic reactions result in free energy used for work (muscle contraction) and heat release; however, it is very difficult to measure both work energy and heat production individually. Therefore, it has been established that all metabolic reactions (work + heat) eventually result in heat production, which can be measured as kilocalories (kcal). A kilocalorie is the basic energy unit and is calculated as the amount of heat required to raise 1 kg of water 1 °C from 14.5 °C to 15.5 °C.

The rate of energy production is controlled by the amount of available substrate along with enzyme activity. Substrates are the molecules that, when combined, form a specific end product. Enzymes assist the action of substrate to product (or vice versa) by speeding up the rate of reaction. Metabolic pathways involve multiple reactions, each catalyzed by an enzyme. However, these systems have specific enzymes that turn on the entire pathway of reactions. Regulation of a rate-controlling enzyme essentially controls the entire biochemical pathway. The key regulatory enzymes are discussed in detail later in the chapter.

Skeletal muscle has sensitive biochemical controls that activate metabolic pathways based on ATP demand and produce energy through one of three pathways:

1. ATP-PCr (phosphagen-creatine system)
2. Glycolysis (aerobic and nonaerobic)
3. Oxidative phosphorylation

ATP is supplied through the ATP-PCr and glycolytic systems during short high-intensity contractions, whereas oxidative phosphorylation plays a larger role in energy production during low- to moderate-intensity exercise. However, these systems do not work independently of each other, and all contribute variable amounts of ATP during each exercise condition. In addition, the capacity of each system is based on skeletal muscle fiber type and exercise training (e.g., resistance vs. aerobic) (36). The activation, regulation, and interaction of these systems during exercise are the primary foci of this chapter. Further, we discuss how these systems adapt to various types of exercise training.

ENERGY TRANSFER

Before in-depth discussion about regulation of ATP production via metabolic pathways can begin, additional background on energy transfer must be reviewed. The basis of metabolism relies on two laws of bioenergetics: Energy cannot be created or destroyed but rather changed from one form to another, and energy transfer proceeds in the direction of increased entropy and free energy release. In this context, the stored chemical energy from ingested nutrients (i.e., carbohydrate, fat, and protein) must be transferred to mechanical energy to perform muscle contractions. In addition, the rate of energy release or energy storage must be regulated by ATP demand and supported by the direction of each reaction (product to substrate or vice versa).

Entropy (S), a measure of disorder or randomness, governs the directionality of all reactions (the second law of bioenergetics). In the simplest context of thermodynamics, all reactions in the universe proceed toward increased entropy. **Enthalpy** (ΔH), another common measurement in bioenergetics, represents total energy transfer. Enthalpy can be separated into the energy used for heat release and the energy used to perform cellular work, termed *free energy* (sometimes called *free enthalpy*). When the change or transfer of enthalpy is negative ($-\Delta H$), heat and free energy are being released from a reaction. When the change in enthalpy is positive ($+\Delta H$), heat and free energy are being absorbed. Free energy (ΔG) is quantified as the difference between enthalpy and entropy, or total energy minus disorder:

$$\Delta G = \Delta H - T\Delta S \qquad \text{(Equation 1.1)}$$

where T is the absolute temperature.

A negative ΔG, which indicates free energy release, may occur in several ways based on the variables in equation 1.2. Both $T\Delta S$ and ΔH are negative, but the absolute value of ΔH is higher, resulting in a negative ΔG. Another scenario is that ΔH is positive and $T\Delta S$ is positive, but the absolute value of ΔH is lower than $T\Delta S$. Finally, ΔH is negative and $T\Delta S$ is positive. What is interesting is that in all scenarios energy transfer will occur only if there is an increase in entropy (net positive $T\Delta S$) and a net negative ΔG.

Scenario 1: $-\Delta G = -\Delta H - -T\Delta S$
(absolute ΔH must be higher than $T\Delta S$)
Scenario 2: $-\Delta G = \Delta H - T\Delta S$
(absolute ΔH must be lower than $T\Delta S$)
Scenario 3: $-\Delta G = -\Delta H - T\Delta S$
(negative minus a positive) $\qquad$ (Equation 1.2)

Equilibrium Constant

As previously discussed, reactions move in the direction of increase in entropy. When the movement of reactant to product reaches maximal entropy, it is called the *equilibrium constant* (K_{eq}). When this occurs, the energy released from the reaction is complete. The K_{eq} determines the free energy release from a reaction, which is diagrammed in equation 1.3, where the movement of reactants (A) to product (B) indicates the ratio of concentrations of products and reactants. A larger K_{eq} demonstrates that more products are being formed and greater free energy is being released.

$$K_{eq} = \text{reactants (A)} <> \text{products (B)}$$
$$\text{or } \dot{K}_{eq} = \frac{[\text{products B}]}{[\text{reactants A}]} \qquad \text{(Equation 1.3)}$$

Understanding the K_{eq} allows us to introduce the standard free energy change equation, where free energy release is based on how large or small the K_{eq} is. In this equation, R represents the gas constant (1.987 cal $\cdot$ mol^{-1} $\cdot$ K^{-1}), T is the standard temperature ($25°C$ or 298 K), and ln K_{eq} is the natural log of the K_{eq} under these standard laboratory conditions. $\Delta G°'$ represents the standard free energy change that occurs when reactants are converted to products in a reaction under standard conditions (R and T) at pH 7.

$$\Delta G°' = -RT \ln K_{eq} \qquad \text{(Equation 1.4)}$$

We know that a negative $\Delta G°'$ represents free energy release and, based on equation 1.4, a higher K_{eq} equates to a more negative $\Delta G°'$. Conversely, a lower K_{eq} equals a much higher $\Delta G°'$ because the natural log of numbers smaller than 1 have a negative value. For example, if the ratio of products to reactants is 0.005, then the $\Delta G°'$ will be roughly $+3.31$ kcal/mol: $-592.12 \times \ln(.005)$. Further, if the K_{eq} is 5.00, then the $\Delta G°'$ will equal -9.52 kcal/mol, which would be a tremendous amount of free energy release! The most common unit of measure for free energy release is the kilocalorie (definition stated previously); 1 kcal = 4,184 joules (J; the international unit of energy measurement). It is very important to understand that the standard free energy change ($\Delta G^{\circ\prime}$) is a theoretical number because it is based on standard conditions (gas, temperature, and pH constant) that exist only in a controlled laboratory environment. During exercise, muscle temperature and pH change and greatly influence the amount of free energy change during each reaction. A perfect example of this is that the contractile force of skeletal muscle is greatly reduced when the free energy release from ATP breakdown is lowered due to a decrease in muscle pH and temperature levels (72).

In a biological system, if the free energy is released from a reaction, it is called an *exergonic reaction* and will occur spontaneously ($-\Delta G$). Conversely, if the free energy is absorbed or if the reaction favors an increase in reactants, it is called an *endergonic reaction* and requires energy input ($+\Delta G$) from an exergonic reaction to proceed in the direction of the products. An **equilibrium** reaction indicates that the reaction is in balance (reactants = products), and free energy change does not occur (no change in G).

A common example of the interplay between an exergonic reaction and an endergonic reaction is demonstrated in the creatine kinase reaction. This two-part reaction utilizes the free energy release from the breakdown of creatine phosphate (CrP) to fuel ATP synthesis. This is called a *coupled reaction* because the exergonic reaction provides energy input to drive the endergonic reaction toward increased product.

High-Energy Phosphates

ATP is the energy currency of the cell and is vital for free energy transfer in humans. It is a complex molecule that comprises adenosine (adenine group and a 5-carbon ribose sugar group) and three phosphate groups linked in a series from carbon 5 of ribose. The enzyme ATPase catalyzes the hydrolysis (i.e., the splitting of a molecule using water) of ATP and yields adenosine diphosphate (ADP), (inorganic phosphate) (Pi), and a hydrogen ion (H^+) (equation 1.5).

$$\text{ATPase}$$
$$\text{ATP} + H_2O \longleftrightarrow \text{ADP} + \text{Pi} + \text{H} + \Delta G^{\circ\prime} = \text{-7.3 kcal}$$

(Equation 1.5)

The standard free energy release ($\Delta G^{\circ\prime}$) when ATP is broken down is -7.3 kcal $\cdot$ mol^{-1} (54). In comparison,

the $\Delta G^{\circ\prime}$ of CrP is -10.3 kcal $\cdot$ mol^{-1}, which means that more free energy release occurs from the breakdown of CrP than from that of ATP. Why, then, does the cell prefer to utilize ATP as the major energy for cellular work and not CrP? It takes an equal amount of energy input to synthesize these energy phosphates. In this context, to generate ATP, equation 1.5 must be reversed from right to left (the reaction is endergonic in this direction) and requires 7.3 kcal of input. In addition, the synthesis of CrP requires 10.3 kcal of energy input. ATP is the more favorable molecule because it provides modest energy release and requires only modest energy input for synthesis. Metabolism is highly regulated and responsive to immediate changes in ATP. When exercise begins and ATP usage increases, the cell responds by activating systems of ATP production. This is the only way humans can sustain exercise for any length of time. This leads us into our discussion of the energy systems.

Oxidative–Reduction Reactions

Oxidative and reduction reactions are common in metabolism and involve the transfer of electrons. A substance is reduced when it gains an electron, and it is oxidized when it transfers an electron. In metabolism, the hydrogen atom (sometimes referred to as a *proton*) commonly is used as an indicator for oxidative–reduction reactions. Although the passing of an electron is not always evident in metabolic reactions, the hydrogen atom can be followed because it contains one proton and one electron. A molecule that accepts a proton is referred to as *reduced,* and one that transfers (or loses) a proton is referred to as *oxidized.* These reactions are catalyzed by dehydrogenases. The most common oxidative reduction reactions in metabolism are the transfer of protons to and from nicotinamide adenine dinucleotide (NAD) and flavin adenine dinucleotide (FAD).

$$\text{NAD} + \text{H} \longleftrightarrow \text{NADH}$$
$$\text{FAD} + 2\,\text{H} \longleftrightarrow \text{FADH}_2 \qquad \text{(Equation 1.6)}$$

Later we explain that the end result of aerobic metabolism is the transfer of electrons to oxygen, which makes oxygen the strongest oxidizer. In addition, the importance of the ratio between NADH and NAD (called the *redox state*) is discussed.

ENERGY SUBSTRATES FOR METABOLISM

The major nutrients, or substrates, that are used at rest and during exercise are stored carbohydrate (glycogen), circulating blood glucose, and stored lipids (fat). Energy from these substrates is harnessed upon their entry into metabolic pathways and is used to generate ATP.

Under resting conditions, ATP is adequately supplied through the breakdown of fats and carbohydrate. As the intensity of exercise progresses from low to high, a shift occurs toward greater utilization of carbohydrate and less utilization of fat. Protein or amino acids assist in key enzymatic steps of metabolism but contribute minimally to ATP synthesis. However, under extreme exercise situations (e.g., long-duration aerobic exercise), proteins can be broken down for energy. This section reviews the basic characteristics of carbohydrate, lipids, and protein to assist in further understanding exercise metabolism.

Carbohydrate

Carbohydrate is a compound made from the combination of carbon, hydrogen, and oxygen atoms in the formula $C_nH_{2n}O_n$. Carbohydrate is one of the human body's main fuel sources for energy, providing approximately 4 kcal of energy $\cdot$ g^{-1} when catabolized. At rest, the body derives about half of its energy from carbohydrate; the remaining fuel comes from fats. After feeding, most carbohydrate enters the bloodstream via the gastrointestinal tract as a 6-carbon glucose molecule, which is transported to all tissues. Glucose uptake into skeletal muscle or liver has two fates: entry into metabolism for ATP synthesis (i.e., glycolysis) or storage in the cell cytoplasm as glycogen (68). Skeletal muscle glycogen is broken down and used for ATP synthesis during intense exercise, whereas liver glycogen is broken down to glucose and released into circulation for blood glucose maintenance. Roughly 2,000 kcal (500 g) is stored as skeletal muscle glycogen, whereas the liver stores 500 kcal (125 g; table 1.1). This macronutrient is classified into three main categories—monosaccharides, oligosaccharides, and polysaccharides—based on the number of simple sugars bonded together within the compound.

Monosaccharides

Monosaccharides—the basic unit of a carbohydrate—are the building blocks for disaccharides and polysaccharides. A monosaccharide is a 3- to 7-carbon molecule categorized by the number of carbon atoms. These molecules are referred to as triose (3C), tetrose (4C), pentose (5C), hexose (6C), and heptose (7C). Fructose, galactose, and glucose are all examples of hexose monosaccharides. In terms of energy metabolism, glucose is the preferred carbohydrate. This hexose comprises 6 carbon atoms, 12 hydrogen atoms, and 6 oxygen atoms ($C_6H_{12}O_6$). Glucose is obtained through diet and can be synthesized in the human body. In terms of dietary intake, a complex carbohydrate is digested and catabolized into glucose. Glucose molecules in the small intestine are then absorbed into the blood. The absorbed glucose will meet one of three fates: immediate use as a substrate in the energy pathway, storage in liver and muscle as glycogen, or conversion to fat. In contrast, the synthesis of glucose in the body occurs mainly in the liver through a process called *gluconeogenesis*. Utilization of carbon residues from amino acids, glycerol, pyruvate, and lactate allow for this process to occur. The glucose molecules created by gluconeogenesis are used immediately for energy when the body's glucose supplies are depleted.

Oligosaccharides

Oligosaccharides are formed when 2 to 10 monosaccharides covalently bond together. Disaccharides, a main oligosaccharide, form when only 2 monosaccharides combine. Three major disaccharides include maltose (glucose + glucose), sucrose (glucose + fructose), and lactose (glucose + galactose). Disaccharides must be broken down to glucose for absorption and for energy metabolism.

Polysaccharides

Polysaccharides, meaning "many sugars," are long chains of 10 or more monosaccharides linked together. Cellulose, starch, and glycogen are common polysaccharides. Cellulose is an important structural component in plants, whereas starch and glycogen are important in terms of storage. In plants, starch is the storage form of carbohydrate and comprises amylose (a straight chain of monosaccharides) and amylopectin (a highly branched structure of monosaccharides). Glycogen—the storage form of carbohydrate in animals—is a highly branched structure that comprises glucose molecules linked together. When blood glucose supply exceeds normal limits (~5 mmol $\cdot$ L^{-1} or ~90 mg $\cdot$ dL^{-1}), the monosaccharides combine to form this large polysaccharide in the muscles and liver through a process called *glucogenesis*. Glycogen's branched structure, consisting of several reducing ends, allows for a rapid breakdown back into glucose. This process occurs by cleaving the monosaccharide units one at a time. The glycogen stored in the muscles is a major source of glucose during exercise. Liver breakdown of glycogen to glucose is used to maintain blood glucose levels during prolonged exercise or starvation. This process is termed *glycogenolysis.* When glycogen stores are depleted in both the muscles and the liver, carbohydrate energy production is maintained through gluconeogenesis.

Lipids

Lipids are macronutrients that comprise the same atoms as carbohydrate but differ in their molecular structure. When comparing potential energy stores in the body, there is a larger storage of fats than of carbohydr-ate because fats are insoluble (i.e., do not attract water when stored), whereas carbohydrate is soluble (i.e., stored with water, thus increasing the weight of the potential energy). Fat catabolism allows for more energy per gram than does carbohydrate catabolism (about 9 kcal · g⁻¹), making it ideal for prolonged, low- to moderate-intensity exercise. Humans store substantially more energy in the form of fats—roughly 75,000 kcal for a 65-kg lean adult male (~73,000 kcal adipose tissue and ~1,500 kcal intramuscular; table 1.1).

Several types of lipids exist in the body. These include fatty acids, triacylglycerols, phospholipids, and steroids. Phospholipids and steroids are not major contributors to energy metabolism. Phospholipids form the lipid bilayer of the cellular membrane, providing structural support. Steroids, derived from cholesterol, are also in the cellular membrane and are important components of hormones.

Fatty Acids

Fatty acids are found in several types of lipids such as triacylglycerols. They comprise a long carbon chain, each bounded by hydrogen, and a carboxyl group at one end. Fatty acids are categorized based on the type of bonds formed between the carbon atoms. Saturated fatty acids contain single bonds between each of the carbons. Unsaturated fatty acids are joined together by double bonds; they can be further broken down to monounsaturated fatty acids (one double bond) and polyunsaturated fatty acids (more than one double bond). Fatty acids are the main lipid fuel source, providing energy to the active muscles. The common fatty acids used in energy metabolism include palmitic acid (16-carbon saturated fatty acid), stearic acid (18-carbon saturated fatty acid), and oleic acid (18-carbon monounsaturated fatty acid).

Triacylglycerols

Triacylglycerols are the storage form of lipids in the body. Triacylglycerols, also called *triglycerides,* comprise a glycerol (an alcohol) backbone with three chemically bonded fatty acids, each varying in length and type. The synthesis of triacylglycerols, called *esterification,* occurs in adipocytes and muscle tissue, where several reactions take place in order to attach three fatty acids to a single glycerol backbone. Synthesis and storage happen in the cytoplasm of adipocytes (fat cells) and are vital in terms of long-term energy storage.

Triacylglycerols are catabolized to provide the energy needed to sustain the muscle activity. Lipolysis—the process that catabolizes a triacylglycerol down to its

Table 1.1 Body Stores of Fuels and Energy

	g	kcal
Carbohydrate		
Liver glycogen	110	451
Muscle glycogen	500	2,050
Glucose in body fluids	15	62
Fat		
Subcutaneous and visceral	7,800	73,320
Intramuscular	161	1,513
Total Fat	8,586	77,396

Note: These estimates are based on a body weight of 65 kg with 12% body fat.

Reprinted, by permission, from L.W. Kenney, J.H. Wilmore, and D.L. Costill, 2015, *Physiology of sport and exercise,* 6th ed. (Champaign, IL: Human Kinetics), 53.

functional components—occurs through the use of the enzyme lipase. Following lipase activity in adipose tissue, fatty acid transportation occurs. Fatty acids are carried to active muscles by plasma proteins called *albumin,* creating a structure called a *free fatty acid.* Intramuscular lipolysis also occurs, releasing fatty acids to be catabolized for energy. Fatty acids in skeletal muscle and from adipose tissue are then modified to enter the muscle cell's mitochondria. These modified fatty acids are broken down two carbons at a time through a series of reactions called beta-oxidation. Beta-oxidation results in the byproduct acetyl coenzyme A (acetyl-CoA), which is able to enter the energy pathway system. The glycerol from lipolysis is circulated to the liver for processing or incorporated into the energy pathway to yield a small amount of energy.

Protein

Proteins vary from carbohydrate and lipids because their molecular makeup includes nitrogen. Proteins are polypeptide chains created from a combination of the 20 different amino acids. Proteins yield about 4 kcal of energy $\cdot$ g^{-1} (similar to carbohydrate); however, proteins are not considered a predominant source of energy. Proteins mainly are stored in the skeletal muscle and provide up to 10% of the energy needed during prolonged exercise, conditions of carbohydrate and lipid depletion, or starvation. Proteins must be modified before they are used as an energy substrate because the human body is unable to process nitrogen. Proteins have three fates once their nitrogen is removed by deamination or transamination: They are converted into glucose (gluconeogenesis), they are converted into a fatty acid (lipogenesis), or the carbon skeleton of the amino acids enters the energy pathway system.

ENERGY SYSTEMS

The three major metabolic pathways that generate ATP for the contracting muscle are phosphocreatine, glycolysis/glycogenolysis, and oxidative phosphorylation. Advanced exercise physiology students should have a basic understanding of the structure of each system, so the main focus is on regulation (please see table 1.2 for all key regulatory enzymes).

ATP-PCr

The ATP-PCr system is considered substrate metabolism and is the simplest of the energy systems. Skeletal muscle

Table 1.2 Regulatory Enzymes

Enzyme	Activator	Inhibitor
ATP-PCr	ADP	PCr
PFK	Decrease ATP/ADP AMP Epinephrine Pi	Increase ATP/ADP Isocitrate dehydrogenase citrate
PK	Decrease ATP/ADP	PCr Increase ATP/ADP
PDH	Ca^{++} Decrease ATP/ADP Decrease NADH/NAD Decrease acetyl-CoA/CoA Pyruvate	Increase ATP/ADP Increase NADH/NAD
Glycogen phosphorylase	Ca^{++} Epinephrine Pi AMP	
Citrate synthase	Decrease ATP/ADP Decrease ATP/ADP	Increase ATP/ADP Increase NADH/NAD
Isocitrate dehydrogenase	Decrease ATP/ADP Decrease ATP/ADP Increase Ca^{++}	Increase ATP/ADP Increase NADH/NAD
ETC	Increase ADP	
Hormone-sensitive lipase	Epinephrine	

Important Energy Ratios

Metabolism is regulated by substrate changes, which result in activation or inhibition of various enzymes. The ratios of various substrates to products (or oxidants to reducers) are the key to activating key regulatory enzyme activity. These include ATP/ADP, NADH/NAD, and acetyl-CoA/CoA. For example, a decline in the ATP/ADP ratio indicates that ADP levels are increasing and that ATP levels possibly are decreasing. This will activate phosphofructokinase (PFK), which is a key glycolytic pathway enzyme, and results in the acceleration of ATP synthesis. The NADH/NAD ratio commonly is called the *redox ratio*; a decline in this ratio indicates greater NAD or decreased NADH. This causes activation of pyruvate dehydrogenase (PDH) reaction, which is an important enzyme for aerobic metabolism of carbohydrate.

stores a very small amount of ATP ($\sim$6-8 mmol $\cdot$ kg^{-1} of wet wt) along with a high-energy phosphate molecule called CrP in the cell cytosol. At the onset of intense muscle contraction, stored ATP is hydrolyzed to ADP, Pi, and a proton (H$^+$). Simultaneously, stored CrP donates a phosphate group to ADP to form ATP through the creatine kinase reaction, which is considered the most immediate means for regenerating ATP. This system is activated at the onset of exercise and during intense muscle contraction (e.g., sprinting, resistance exercise) and supports rapid ATP demand for 5 to 15 s. Interestingly, ATP levels are preserved during intense muscle work, whereas PCr stores deplete rapidly ($\sim$3 mmol $\cdot$ kg^{-1} of wet wt) (61).

The ATP-PCr energy system consists of four reactions and is activated immediately when muscle contraction begins. It is capable of responding to the initial demand for ATP while the other energy pathways upregulate. This is possible because the major substrates are ATP and phosphocreatine, which are both stored in the cytosol of cell.

Reaction 1: ATP + H$_2$O $\longleftrightarrow$ ADP + Pi + H$^+$

Reaction 2: PCr + ADP + H$^+$ $\longleftrightarrow$ ATP + Cr

Reaction 3: ADP + ADP + H$^+$ $\longleftrightarrow$ ATP + AMP

Reaction 4: AMP + H$_2$O + H$^+$ $\longrightarrow$ IMP + NH$_4$

(Equation 1.7)

Note: AMP-adenosine monophosphate; IMP – inosine monophosphate; Cr – creatine

Let's begin our discussion with reaction 1, which is the ATPase reaction. A small amount of ATP is stored in skeletal muscle and is used immediately upon muscle contraction. The breakdown of ATP increases concentrations of the substrates ADP, Pi, and H$^+$, and the ADP

and H$^+$ are used as substrates for reaction 2. The creatine kinase reaction (reaction 2) is an equilibrium reaction and is dependent on stored levels of phosphocreatine, which is approximately 28 mmol $\cdot$ kg^{-1} of wet weight in a healthy human. PCr donates a phosphate group to ADP, forming ATP and consuming a proton (H$^+$). As soon as ADP levels increase due to ATP breakdown, the reaction moves to the right as substrate levels (i.e., ADP) increase. This increases ATP production and further supports muscle contraction. The creatine kinase reaction will continue until PCr levels are nearly depleted; this occurs in as little as 10 s.

Simultaneously, the adenylate kinase (reaction 3) is activated as ADP levels continue to increase from ATP breakdown. This reaction forms additional ATP by donating a phosphate from one ADP to another and results in ATP and AMP (19). The adenylate kinase reaction provides only a provisional amount of ATP but is very sensitive to minor shifts in ATP levels and increasing ADP (19). The major contribution of this reaction is an increase in levels of AMP, which is a major signaling molecule for activation of glycogenolysis (28,47,70). This is discussed further in the Glycolysis/ Glycogenolysis section.

The AMP deaminase reaction removes AMP and forms IMP and ammonia (NH$_4$). The importance of this reaction is the regulation of AMP, which is tightly controlled because of the allosteric nature of the molecule. In other words, AMP has regulatory effects on other areas of metabolism, which are discussed soon. The AMP deaminase reaction is not reversible and forms ammonia, which is removed by the liver and kidneys.

To summarize, the activation of the ATP-PCr pathway is determined by the slight change in substrate levels. A small change in ATP levels results in immediate activation of reactions 2 through 4. Bangsbo et al. (3) demonstrated that ADP levels increase proportionally to the intensity of muscle contraction in addition to a parallel decrease in muscle PCr levels.

Glycolysis/Glycogenolysis

Glycolysis is the process of harvesting the energy from the breakdown of the 6-carbon glucose molecule to generate ATP. Plasma glucose may enter into the pathway, or stored glycogen, which is termed *glycogenolysis*. Glucose 6-phosphate is the central entry molecule for both glucose and glycogen; however, one can see from figure 1.1 that glycogenolysis requires one additional reaction compared with glycolysis. The pathway yields only two or three ATP per glucose or glycogen molecule, respectively, but the reactions occur rapidly, supporting ATP supply during intense muscle contraction. During low- to moderate-intensity exercise, glucose is the major substrate into the glycolytic pathway, and glycogen utilization is greater during higher intensity exercise (64, 16). This pathway is capable of functioning in the presence (i.e., aerobic metabolism) or absence (i.e., anaerobic metabolism) of oxygen. The goal of metabolism is to meet the ATP demand of the working muscle. If ATP cannot be supplied from oxidative sources (i.e., oxygen), then glycolysis/glycogenolysis can support ATP synthesis in the absence of oxygen. This important concept is discussed further later in the chapter. The glycolytic pathway is the only nutrient-dependent energy system that can generate ATP in both the absence and presence of oxygen. For this reason, we discuss nonaerobic (oxygen is not present in the reaction) and aerobic (oxygen is present in the reaction) glycolysis/glycogenolysis separately.

Nonaerobic Glycolysis/ Glycogenolysis

Nonaerobic glycolysis/glycogenolysis—the breakdown of glucose or glycogen in the absence of oxygen—occurs in the cytosol of the muscle cell. Both molecules enter the pathway through glucose 6-phosphate and yield two pyruvate, two or three ATP, water, and two NADH. The two pyruvate molecules are further reduced to two lactate molecules (figure 1.1).

Glucose enters the pathway via blood glucose transport into the skeletal muscle and then conversion to glucose 6-phosphate through the hexokinase reaction. This reaction requires energy input from ATP

Performance and Sprinting Training

Sprint training is a common component of training programs for both power and endurance athletes. It consists of near-maximal aerobic efforts lasting 10 to 30 s (sometimes longer) followed by a short recovery (similar to high-intensity interval training). Nonaerobic energy systems account for roughly 95% of energy metabolism during a maximal 10-s sprint, and the percentage decreases as the duration of the sprint increases (e.g., 73% during a 30-s sprint). For this reason, it is common for sprint athletes to engage in shorter sprints, whereas endurance athletes perform longer (e.g., >30 s) sprints.

Several metabolic benefits result from sprint training, including adaptations to the ATP-PCr, glycolytic, and oxidative systems. Creatine kinase and adenylate kinase enzyme activities have been shown to increase by 36% and 20%, respectively, after 5-s maximal intervals among sprint athletes (19,67). Interestingly, these adaptations were not seen when intervals were increased to 30 s or when longer intervals were combined with shorter bouts. Glycolytic adaptations such as increased PFK, PK, LDH, and glycogen phosphorylase have been shown to increase after longer sprint intervals (e.g., > 10 s). Shorter intervals (< 30 s) have been shown to decrease TCA enzyme levels (19). Furthermore, the combination of short and long intervals produced no change in oxidative enzymes (22). However, performing longer intervals (> 30 s) or multiple sets (> 10) has been shown to enhance citrate synthase and succinate dehydrogenase levels (41). In summary, metabolic adaptations to sprint training are dictated by the length of the sprint or the number of intervals. Therefore, an aerobic athlete may benefit by performing longer intervals (> 30 s), whereas a sprinter may benefit from shorter bouts (< 10 s) (55).

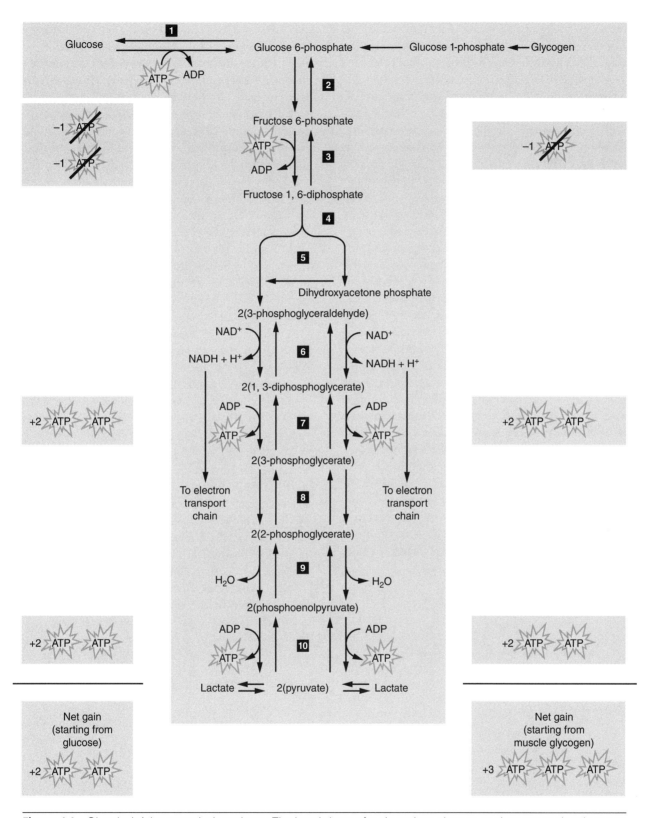

Figure 1.1 Glycolysis/glycogenolysis pathway. The breakdown of a six-carbon glucose or glycogen molecule to two three-carbon pyruvate molecules and either 2 or 3 ATP. The pathway is 10 reactions for glucose and requires 2 ATP input. The breakdown of glycogen is 9 reactions and bypasses the hexokinase reaction requiring 1 ATP in the input phase.

breakdown. Glycogen is stored in the cell cytosol and can therefore enter the pathway without the need for transport. Glycogen must be broken down to glucose 1-phosphate through activation of the phosphorylase enzyme. Figure 1.2 shows that an inorganic phosphate is a substrate in the reaction and that it comes from ATP breakdown, making the reaction more exergonic. In addition, phosphorylase is activated by increasing levels of AMP (28) produced by the adenylate kinase reaction, providing a link from the ATP-PCr system (previously discussed). Further activators of phosphorylase are intracellular calcium and epinephrine, which both increase as intensity of exercise increases. Given all the factors stated previously (increasing Pi, AMP levels, increasing Ca^{++}, and epinephrine), it is easy to understand that glycogenolysis is more active during high-intensity exercise when Pi, AMP, Ca^{++}, and epinephrine levels are also high (figure 1.2). The importance of the phosphorylase enzyme can be demonstrated when examining patients with McArdle disease, who have a phosphorylase deficiency. It consistently has been shown that these patients have reduced exercise tolerance (15,57). In addition, phosphorylase inhibitors are beginning to be used as treatment for type 2 diabetes to help lower plasma glucose levels; however, a major side effect is reduced muscle function (2).

An advantage of glycogenolysis is that it bypasses the hexokinase reaction and uses one fewer ATP molecule. This is the reason why nonaerobic glycolysis yields two ATP and glycogenolysis results in three ATP. Once glucose 1-phosphate is formed, it is quickly converted to glucose 6-phosphate through the phosphoglucomutase reaction.

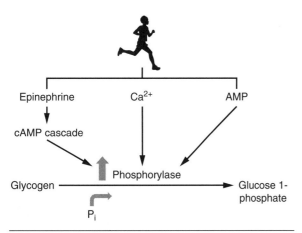

Figure 1.2 Glycogen phosphorylase reaction and regulators. High intensity or prolonged exercise upregulates key activators of the phosphorylase enzyme resulting in greater glycogen utilization and sparing blood glucose.

Only the glycolytic reactions that are of utmost importance to our understanding of how the pathway is regulated are discussed in detail. Most of the reactions in figure 1.2 are equilibrium reactions. Once substrate levels increase, equilibrium reactions will move toward product formation. A major regulator of the glycolytic pathway is the PFK reaction. This reaction converts fructose 6-phosphate to fructose 1,6-bisphosphate and uses the energy input from ATP to make the reaction exergonic. The activation of PFK results in substrate flux and drives all the subsequent reactions toward product formation. PFK is activated by ADP, AMP, and Pi, which all increase when ATP turnover is high (35). Stable levels of ATP inhibit the enzyme, and some in vitro evidence has shown that the lowering of pH (increase in H^+) is also an inhibitor. However, work by Spriet (61) has shown that PFK is able to match ATP demand despite muscle pH of 6.4 to 6.5.

The next step is the splitting of the fructose 1,6-bisphosphate molecule into two 3-carbon glyceraldehyde 3-phosphate (G3P) molecules. The reactions that occur prior to the formation of G3P are considered the energy input phase of glycolysis/glycogenolysis, and all reactions that occur after are considered the energy output phase. Keep in mind that each reaction occurs twice (two G3P molecules). The G3P dehydrogenase reaction is another very important step in the glycolytic pathway in which the addition of a phosphate to G3P forms 1,3-bisphosphoglycerate. In addition, the electron carrier NAD is also a substrate and harnesses an electron and proton (H^+) that are released from the reaction. This reduces NAD to $NADH_2$. The importance of this reaction is detailed later during our discussion of lactate formation.

The phosphoglycerate kinase reaction is the first ATP-producing reaction in the pathway and results from the transfer of a phosphate from 1,3-biphosphoglycerate to ADP. The pyruvate kinase reaction is the second ATP-producing reaction in the pathway as a phosphate is transferred from phosphoenolpyruvate to ADP. The products of the pyruvate kinase (PK) reaction are ATP and pyruvate. The PK reaction is the final reaction in glycolysis and is a key regulatory enzyme in the glycolytic pathway, similar to PFK. It is inhibited by ATP and PCr and is activated by rising ADP, which is a substrate in the reaction. Once can see that glycolysis/glycogenolysis is regulated at the beginning (i.e., PFK) and the end (i.e., PK) of the pathway.

If we track the products of a single 6-carbon glucose or glycogen molecule through the PK reaction, there will be four ATP, two $NADH_2$, and two pyruvate. Keep in mind that glycolysis uses two ATP (hexokinase, PFK), whereas glycogenolysis requires one ATP (PFK), which results in a net of two and three ATP, respectively.

Lactate Production

Before beginning the discussion on lactate production, we refer back to the G3P dehydrogenase reaction (reaction #6 in figure 1.1). Remember, NAD is a substrate and accepts a proton and electron from the reaction and is reduced to NADH. During low- to moderate-intensity exercise, NADH produced from glycolysis/glycogenolysis shuttles its protons and electrons into the mitochondria (electron transport chain) and is oxidized back to NAD for continued use in the G3P dehydrogenase reaction. Two shuttle systems—malate-aspartate and glycerol-phosphate—support the oxidation of NADH (47). The malate-aspartate shuttle transfers protons and electrons from glycolysis-produced NADH (cytosolic) to mitochondrial NAD. The glycerol-phosphate shuttle transfers cytosolic $NADH_2$ to mitochondrial FAD (figure 1.3). The mitochondria are impenetrable to $NADH_2$ movement across the membrane in the absence of these shuttles (49).

In addition, during low- to moderate-intensity exercise, pyruvate is transported into the mitochondria and enters the PDH complex (discussed later). However, if the ATP demand of the contracting muscle increase (i.e., vigorous exercise) and the glycolytic rate accelerates, large amounts of $NADH_2$ may accumulate in the cell cytosol. If the $NADH_2$ are incapable of being oxidized back to NAD via the shuttling mechanism, then pyruvate will accept the protons from $NADH_2$ and be reduced to lactate by the lactate dehydrogenase enzyme (LDH).

$$\text{Pyruvate} + \text{NADH} \longleftrightarrow \text{lactate} + \text{NAD} \text{ (Equation 1.8)}$$

The reaction occurs in the cytosol and supports glycolysis/glycogenolysis because it frees up NAD to move back up into the G3P dehydrogenase reaction. Without the regeneration of NAD, glycolysis would slow down tremendously. Skeletal muscle lactate levels are around 1 mmol · kg^{-1} at rest and may reach levels of 10 to 20 mmol · kg^{-1} during maximal exercise (29,34). The generation of lactate allows for glycolysis to continue and for ATP production to continue at a high rate.

Exercise training enhances the capacity of the skeletal muscle to oxidize NADH back to NAD, and it may be a result of increased levels of the malate-aspartate shuttle or the glycerol-phosphate shuttle (58,59). This may indicate that the two shuttles are important bottle-necks in the pathway. However, another hypothesis has been proposed by Brooks et al. (9) and is very important to include in our discussion. The supposition is that most pyruvate produced in the cytosol is reduced to lactate, thus freeing up NAD. The lactate molecules are then transported into the mitochondria and converted back to pyruvate via the reverse direction of LDH reaction. This proposal is still highly controversial in the field, and data from other labs have not consistently been able to locate LDH or a lactate shuttle mechanism in skeletal muscle mitochondria (74). It may be that all mechanisms play a role in supporting the oxidation of $NADH_2$ back to NAD.

The accumulating lactate in the muscle must be removed. It would be disadvantageous for the LDH reaction to move toward pyruvate formation in the cell cytosol because it would require NAD, which is needed for the G3P dehydrogenase reaction. Skeletal muscle lactate is transported into the bloodstream via the monocarboxylate transporters, which are proteins that span the sarcolemma and T tubule. The monocarboxylate transporters transport one lactate and one hydrogen molecule into blood circulation. This is the reason why we can measure blood lactate levels. Once in the blood, lactate is used and removed by the heart, liver, and brain. In addition, it has been shown that skeletal muscle (mainly type I slow oxidative fibers) express monocarboxylate transporters and may play a role in lactate clearance and removal (23,56).

In summary, as the ATP demand and the glycolytic rate increase, pyruvate will enter the LDH reaction, forming lactate and NAD. We previously discussed the end products of glycolysis/glycogenolysis. Under nonaerobic conditions, the two pyruvates are converted to two lactate molecules. The nonaerobic glycolytic system may sustain energy supply for several minutes due to the high speed of ATP production. Although only minimal ATP is produced, the reactions occur at very fast rates. Is lactate production due to limited oxygen availability or to the fact that ATP demand exceeds the rate of supply from aerobic metabolism? The answer to this important question is discussed in the Metabolic Regulation During Exercise section.

Aerobic Glycolysis/ Glycogenolysis

The aerobic metabolism of glucose or glycogen is the complete oxidation of a carbohydrate molecule. The end products of glycolysis/glycogenolysis have been discussed in detail. If $NADH_2$ is able to shuttle its protons and electrons into the mitochondria, then lactate production will be minimal. If not reduced to lactate, pyruvate will enter the mitochondria through the mitochondria pyruvate carrier (8), which has been shown to be required for mitochondrial uptake (8).

Pyruvate enters the PDH complex, which is located in the mitochondrial matrix (figure 1.3), and catalyzes the conversion of pyruvate to acetyl-CoA. The PDH complex cleaves a carbon from the 3-carbon pyruvate, yielding the 2-carbon acetyl-CoA. The carbon from the reaction is released as carbon dioxide (CO_2). The total yield from the PDH reaction for one glucose molecule (remember, we have two pyruvate) is two acetyl-CoA, two $NADH_2$, and two CO_2.

What are the fates of each byproduct of the PDH complex? The $NADH_2$ enter the electron transport chain, the CO_2 are blown off through the respiratory system, and the acetyl-CoA enter the tricarboxylic acid (TCA) cycle (figure 1.3). Acetyl-CoA is a very important molecule in metabolism because it is the entry point into the TCA cycle for the major nutrients (i.e., carbohydrate, fat, and protein). The PDH complex is inhibited by high levels of acetyl-CoA/CoA and ATP/ADP and low levels of NADH/NAD, which results in slowing of the glycolytic rate. If acetyl-CoA levels are stable and NADH and ATP are adequate, then there exists no need for energy contribution from glycolysis/glycogenolysis.

The conversion of pyruvate to acetyl-CoA in the PDH reaction is nonreversible and signals carbohydrate entry into aerobic metabolism. The components of aerobic

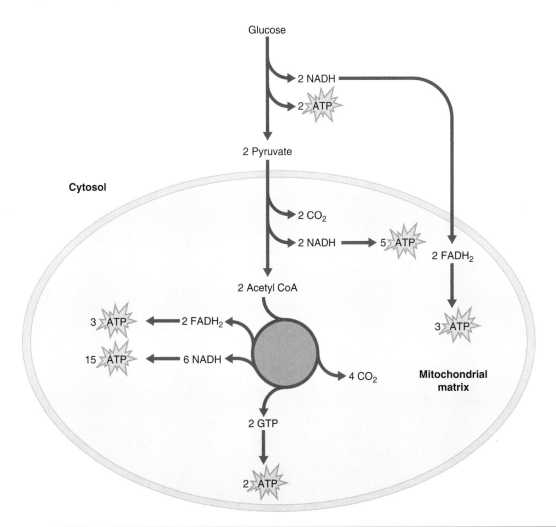

Figure 1.3 Byproducts of complete glucose oxidation. Key steps in this process are: NADH transport through the glycerol-phosphate shuttle yielding 2 $FADH_2$; Pyruvate entry into the PDH reaction; and Acetyl CoA formation and entry into the TCA cycle.

Reprinted, by permission, from V. Mougios, 2006, *Exercise biochemistry* (Champaign, IL: Human Kinetics), 161.

metabolism (TCA and electron transport chain [ETC]) are discussed in the next section. Keep in mind that PDH reaction occurs twice for each carbohydrate molecule, which means two acetyl-CoA and two turns of the TCA. Figures 1.1 and 1.2 diagram the end products of glycolysis/glycogenolysis and the PDH reaction, respectively.

Oxidative Phosphorylation

Oxidative phosphorylation is the third energy system that we discuss. This pathway is aerobic and depends on the presence of oxygen to generate ATP from the breakdown of both carbohydrate and fat. All of the reactions occur in the mitochondria, a double membrane specialized organelle (figure 1.4) that is located in the cell cytosol adjacent to the contractile units. The rate of energy production is much slower compared with the ATP-PCr and glycolytic systems, but the amount of total ATP is much greater. For this reason, the pathway is more active during low- to moderate-intensity aerobic exercise of longer duration.

Oxidative phosphorylation is a complex system that uses the stored energy from electron and proton transfer to phosphorylate ADP to ATP. Oxygen is required in this pathway because it is the final electron acceptor and essentially pulls electrons and protons through the pathway. We separate our discussion into three areas: the TCA cycle, ETC, and ATP synthase.

TCA Cycle

The TCA cycle consists of nine reactions that result in the decarboxylation (i.e., removal of CO_2) of acetyl-CoA and—most important—the production of NADH and $FADH_2$. The 2-carbon acetyl-CoA enters the pathway at the citrate synthase reaction and is combined with the 4-carbon oxaloacetate, which produces the 6-carbon citrate molecule. In reactions 4 and 5, NAD is a substrate and is reduced to NADH; CO_2 is released as a byproduct in both reactions. Further, in reaction 6, a guanosine diphosphate (GDP) molecule is phosphorylated by the addition of an inorganic phosphate and yields guanosine triphosphate (GTP), which is the equivalent of ATP. Last, in reactions 7 and 8, $FADH_2$ and NADH are produced, respectively. The sum of the byproducts from one turn of the TCA cycle (or one acetyl-CoA) is detailed as follows.

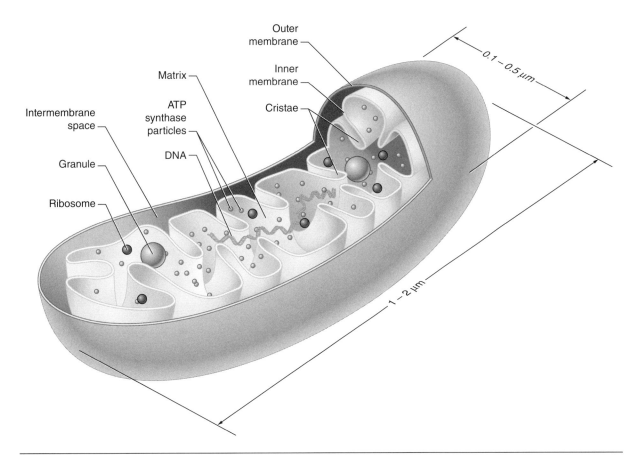

Figure 1.4   The structure and components of skeletal muscle mitochondria.

Important to understand is that the 2-carbon acetyl-CoA molecule is constantly combined with the 4-carbon oxaloacetate molecule, but the two carbons of acetyl-CoA are removed via CO_2 (reactions 4 and 5), setting for the addition of the next acetyl-CoA. The cycle is occurring constantly but not necessarily as circular events. Keep in mind that molecules may enter or leave the pathway and that reactions may occur spontaneously.

2 CO_2

3 NADH

1 GTP (ATP)

1 $FADH_2$

The TCA cycle is regulated by the citrate synthase, isocitrate dehydrogenase, and α-ketoglutarate dehydrogenase reactions. High levels of ATP/ADP inhibit citrate synthase, whereas a high ratio of NADH/NAD inhibits isocitrate dehydrogenase and α-ketoglutarate dehydrogenase. In addition, both isocitrate dehydrogenase and α-ketoglutarate dehydrogenase are activated by an increase in mitochondria calcium levels. In this context, during exercise the ATP/ADP and NADH/NAD ratios are slightly reduced, which activates each of these enzymes and accelerates acetyl-CoA into the pathway. Also, mitochondria calcium levels increase during prolonged exercise, slightly contributing to further activation of the pathway.

Electron Transport Chain (ETC)

The ETC consists of four complexes and the ATP synthase complex (discussed in the next section) that are located on the mitochondria inner membrane (figure 1.5). The pathway is a series of reactions that oxidize NADH and $FADH_2$ back to NAD and FAD, with the end being the reduction of oxygen to water. The purpose of the ETC is to create potential energy that can be used

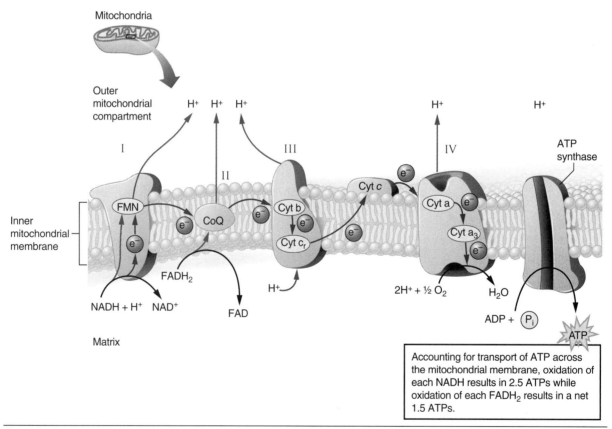

Accounting for transport of ATP across the mitochondrial membrane, oxidation of each NADH results in 2.5 ATPs while oxidation of each $FADH_2$ results in a net 1.5 ATPs.

Figure 1.5 Electron transport chain is a series of reactions that transports electrons from NADH and $FADH_2$ to oxygen forming water. At complexes I, III, and IV protons are pumped into the outer membrane space creating a potential energy for ATP synthesis through the ATP synthase complex.

to phosphorylate ADP to ATP; this is why it is called *oxidative phosphorylation* (oxidation of electron carriers and phosphorylation of ADP).

Potential energy is created through the formation of an electrical gradient by pumping protons into the inner mitochondrial space. How does this occur? The electron and proton carriers, NADH and $FADH_2$, deliver two electrons and protons to the ETC. The electrons travel down the chain, but the protons are pumped into the inner mitochondrial membrane. Protons are pumped at complexes I, III, and IV. If we look carefully, NADH delivers its electrons and protons at complex I, whereas $FADH_2$ delivers its products at complex II, indicating that fewer protons are being pumped per $FADH_2$. The number of protons pumped remains unknown and may vary. Some estimate that four protons are pumped at complex I, two at complex III, and four at complex IV (43,44).

It is very important to understand that each successive complex of the ETC has an increasing affinity for electrons. NADH has a very low affinity for electrons and is a great donor (low redox potential). Oxygen has the greatest affinity for electrons and the highest redox potential (1). Why are the intermediary steps required, or why doesn't oxygen simply accept the electrons from NADH? The reason is because the free energy drop would be too severe and all the energy would be released as heat. Using each intermediary complex, the energy transfer is gradual (passing the high-energy electrons slowly to oxygen) and allows for energy to be harnessed for work (1). Oxygen is the final electron acceptor in complex IV, where two protons and electrons are combined with oxygen to form water (H_2O).

$$2H^+ + 2e^- + \tfrac{1}{2} O_2 \rightarrow H_2O \qquad \text{(Equation 1.9)}$$

ATP Synthase

The ATP synthase is a protein complex located on the inner mitochondrial membrane in close proximity to the ETC complexes. The energy gradient created by proton transfer in the ETC is used to phosphorylate ADP to ATP. The protons travel down the ATP synthase protein toward the matrix and are used to reverse the ATP hydrolysis reaction, thus forming ATP. The $\Delta G^{\circ\prime}$ of ATP is –7.5 kcal, so this amount of energy must be input to reverse the reaction (equation 1.5) or move it toward ATP synthesis.

The synthesis of one ATP molecule is thought to require use of four protons in the ATP synthase protein.

Therefore, we can estimate the amount of ATP generated from NADH and $FADH_2$. Remember that NADH delivers its products at complex I, resulting in 10 protons being pumped (4 complex I, 2 complex III, and 4 complex IV). $FADH_2$ arrives at complex II and results in 6 protons being pumped. Based on this information, NADH produces 2.5 ATP (10/4) and $FADH_2$ produces 1.5 ATP (6/4). These values will be important when we discuss ATP totals from the metabolic pathways.

The regulation of oxidative phosphorylation is based on the concentrations of the substrate ADP. If ADP levels decline only slightly, the pathway is activated. In addition, as ATP demand increases, so does oxygen consumption, indicating greater activation of oxidative energy production.

Lipid Metabolism

This section examines the processes of lipid metabolism.

Lipolysis

Lipolysis is the breakdown of a triacylglycerol to three fatty acids and glycerol. Hormone-sensitive lipase catalyzes the reaction; it is activated by epinephrine, glucagon, and growth hormones and is inhibited by increasing insulin levels (discussed further in chapter 5). Once the free fatty acids are cleaved from the adipose cell, they must be circulated and transported into the contracting muscle and mitochondria.

Because fatty acids are insoluble in aqueous media, they are carried in the blood by the protein albumin, which provides transport to the muscle. Uptake across the skeletal muscle sarcolemma is accomplished by the transport proteins fatty acid translocase (FAT/CD36) and fatty acid binding protein (FABP) (14,26,48,66). Interestingly, an increase in skeletal muscle FAT/CD36 and FABP content results in greater fat burning and is related to weight loss (5,60). Once transported into the cell cytosol, the fatty acid is raised to a higher potential energy by the addition of a coenzyme A, forming a fatty acyl-CoA. This occurs in the mitochondria outer membrane and is catalyzed by the acyl-CoA synthase enzyme. This reaction requires energy input from two ATP, similar to the energy input phase of glycolysis. The fatty acyl-CoA is then shuttled into the mitochondrial matrix through the carnitine shuttle, which is a family of enzymes that performs two functions: combines the fatty acyl with carnitine (which allows for movement across the inner membrane) and catalyzes the reverse reaction (which releases the fatty acyl-CoA into the matrix). The fatty acyl-CoA is now ready for entry into beta-oxidation.

It is important to mention that lipolysis occurs in adipose storage sites throughout the body. In addition, skeletal muscle contains lipid droplets called *intramuscular triglycerides* (IMTGs) that are mobilized during exercise. What site is preferred during exercise? Researchers have been attempting to answer this question. Romijn et al. (53) demonstrated that at 25% $\dot{V}O_2$max free fatty acids from adipose tissue release are the major site for lipolysis; however, at an intensity of 65% $\dot{V}O_2$max, IMTG becomes the major fat source to the contracting muscle. Conversely, it has been shown that IMTG levels do not change during exercise, decline during recovery, and then return to pre-exercise levels after 30 h of recovery (36). This indicates that IMTG may be used during recovery but replenished. It is safe to assume that both adipose storage sites and IMTG are utilized during exercise.

Beta-Oxidation

Beta-oxidation is the breakdown of the fatty acyl-CoA to actyl-CoA, NADH, and FADH$_2$. It consists of four reactions that result in the cleaving of two carbons from the fatty acyl-CoA molecule forming one acetyl-CoA (two carbons), one NADH, and one FADH$_2$. Most fatty acyl-CoA molecules have an even number of carbons; the most common is palmitate, which has 16 carbons. After the first round of beta-oxidation, 14 carbons of the palmitate molecule will remain; after the second round, 12 carbons will remain, and so on (figure 1.6). The molecule will be completely oxidized after seven rounds of beta-oxidation. (Round 8 is not needed because a 2-carbon acetyl-CoA will be left over.) The total byproducts from the complete oxidation of a palmitate fatty acid are listed.

8 acetyl-CoA

7 NADH

7 FADH$_2$

Protein Metabolism

Protein degradation and metabolism provide minimal energy during exercise under conditions of proper nutrient storage levels. However, during prolonged exercise without adequate caloric intake or in situations of starvation, protein is broken down to glucose and used to generate ATP. This is termed *gluconeogenesis,* and it

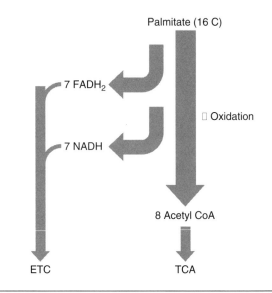

Figure 1.6 The breakdown of a 16 carbon palmitate molecule through β-oxidation. The yield is 7 FADH$_2$, 7 NADH, and 8 acetyl-CoA molecules.

occurs in the liver. It is estimated that proteins provide 5% to 10% of energy during prolonged exercise. In addition, amino acid breakdown from stored pools of amino acid contributes to the synthesis of TCA cycle intermediary products (e.g., α-ketoglutarate).

During prolonged exercise conditions, activation of gluconeogenesis results in the reversal of the glycolytic pathway. This is mediated by the conversion of the amino acids pyruvate, malate, and oxaloacetate to phosphophenolpyruvate, yielding glucose. The glucose that is produced is transferred into the bloodstream to preserve plasma glucose and to support skeletal muscle glycolysis. Furthermore, during starvation, muscle protein is broken down rapidly to amino acids and transported to the liver for glucose synthesis, which is used mainly by the brain and nervous system.

Proteins are a large part of the muscle mass, but their building blocks—amino acids—are present in small amounts. These free amino acids can be converted to intermediaries of metabolic pathways. Glutamine, glutamate, and alanine are considered the main amino acids that are utilized for this purpose. Common TCA intermediaries that are generated include α-ketoglutarate, oxaloacetate, and fumarate.

The rest of this chapter focuses on carbohydrate and lipid metabolism. However, it is key to understand the importance of proper nutrition during exercise to prevent the breakdown of skeletal muscle protein.

TOTAL YIELD FROM METABOLISM

Understanding the ATP yield from metabolic pathways allows us to better compare carbohydrate versus fat oxidation and further supports our later discussion about why carbohydrate breakdown is more advantageous during higher intensity exercise. The total amount of ATP from each pathway is reviewed in detail in the following sections.

ATP Totals

We previously discussed that nonaerobic glycolysis yields two ATP, whereas glycogenolysis produces three ATP. Now let's discuss the number of ATP generated from the complete oxidation of one glucose molecule. (Remember the ATP equivalents for NADH and $FADH_2$?) We separate the discussion into three components: glycolysis, PDH reaction, and oxidative phosphorylation.

Glycolysis total = 5 ATP

2 ATP produced = 2 ATP

2 NADH produced and converted to 2 $FADH_2$ = 3 ATP (2×1.5) (shuttled into the mitochondria and assuming the glycerol phosphate shuttle)

PDH total = 5 ATP

2 NADH = 5 ATP (2×2.5)

Oxidative phosphorylation total = 20 ATP

2 acetyl-CoA (2 turns of the TCA)

6 NADH = 15 ATP (6×2.5)

2 $FADH_2$ = 3 ATP (2×1.5)

2 GTP = 2 ATP

If we add each component together, the total ATP yield from the breakdown of a glucose molecule is 30 ATP (31 for glycogen). Can you calculate the total if the malate-aspartate shuttle is used for cytosolic NADH?

Let's calculate the total ATP generated from the breakdown of one complete palmitate fatty acid molecule. Remember, we consume two ATP initially for the formation of the fatty acyl-CoA molecule. We need to calculate only oxidative phosphorylation totals because fat can be broken down only in the presence of oxygen.

Oxidative phosphorylation total = 108 ATP

8 acetyl-CoA (8 turns of the TCA)

24 NADH = 60 ATP (24×2.5)

8 $FADH_2$ = 12 ATP (12×1.5)

8 GTP = 8 ATP

7 NADH = 17.5 ATP (7×2.5)

7 $FADH_2$ = 10.5 ATP (7×1.5)

The total ATP produced from the breakdown of one fatty acyl-CoA molecule yields 108 ATP (106 ATP in reality because we must subtract the two ATP that were used as energy input).

The complete breakdown of a fatty acid molecule generates a far greater amount of ATP compared with a glucose/glycogen molecule. Why then do we prefer to use glucose/glycogen during higher intensity exercise? If we compare ATP production per oxygen consumption, our interpretation changes because fat requires more oxygen (equation 1.6) for complete oxidation. Carbohydrate yields approximately 5.1 ATP per oxygen molecule consumed, whereas fat generates approximately 4.6 ATP per oxygen molecule consumed. In addition, the glycolytic pathway activates immediately as small changes in substrates move the reactions, whereas beta-oxidation takes time for activation.

Complete oxidation of glucose

$C_6H_{12}O_6 + 6 O_2 \rightarrow$ 30 ATP + 6 CO_2 + 6 H_2O

30 or 31 ATP/6 O_2 = ~5 ATP

Complete oxidation of palmitate

$C_{16}H_{32}O_2 + 23 O_2 \rightarrow$ 106 ATP + 16 CO_2 + 16 H_2O

106 ATP/23 O_2 = ~4.6 ATP

(Equation 1.10)

CO_2 Production

Metabolic CO_2 is produced from two sites: the PDH reaction and the TCA cycle. Based on our understanding of the metabolism, glucose oxidation produces CO_2 at both sites, whereas fat oxidation produces CO_2 at only one (TCA). The complete oxidation of 6-carbon glucose produces only 6 CO_2, whereas a 16-carbon palmitate produces 16 CO_2. Palmitate generates more CO_2 because it requires eight turns of the TCA (two CO_2 per turn), and glucose results in two turns. However, if we view CO_2 production as a ratio to oxygen consumption, then glucose produces much more CO_2. This ratio commonly is referred to as the *respiratory exchange ratio* (RER) ($\dot{V}CO_2/\dot{V}O_2$). When measuring expired gases, an RER value of 0.70 indicates 100% metabolism coming from fat oxidation, whereas an RER value of 1.00 represents

---

Clinical Perspective: Metabolic Disease

Type 2 diabetes mellitus affects nearly 9% of the global population over age 18 years, which is nearly 560 million people worldwide. It is a condition of impaired insulin sensitivity, commonly called *insulin resistance.* This results in abnormal insulin-stimulated glucose uptake into skeletal muscle and elevated levels of glucose and possibly insulin. Acute exercise stimulates glucose uptake through insulin-independent mechanisms (in the absence of insulin) during muscle contraction and has been established as a preferred treatment for type 2 diabetes mellitus (65). Acute exercise also sensitizes the skeletal muscle to insulin and results in enhanced glucose uptake after exercise. In fact, the benefits of one bout of acute exercise on glucose uptake can last for up to 48 h in people with diabetes (33). For this reason, it is recommended that diabetics do not go more than 2 d between bouts of aerobic exercise in order to maintain insulin sensitivity (18).

Exercise and insulin stimulate glucose uptake through regulation of glucose transporter 4 (GLUT4), which is located in the cell cytosol (65). Upon activation by either insulin or exercise, GLUT4 migrates to the cell membrane, where it facilitates glucose transport into the cell. Interestingly, exercise and insulin activate GLUT4 through different signaling mechanisms (73).

An acute 60-min bout of aerobic exercise at 55% intensity has been shown to increase GLUT4 protein expression among individuals with type 2 diabetes (32). Chronic training at low to moderate intensity (55%–65%) increases GLUT4 expression and has been the traditional exercise recommendation for diabetic patients (18). However, high-intensity interval training recently has become more common among diabetic patients. Just 2 wk of performing 60-s cycling sprints resulted in enhanced GLUT4 expression and lower plasma glucose levels along with enhanced levels of TCA and ETC enzymes among type 2 diabetics (40). In summary, exercise is an important and proven treatment for type 2 diabetes mellitus. Immediate benefits are gained from one acute training bout, and chronic exercise provides added benefits.

100% carbohydrate metabolism. Resting RER typically is closer to 0.85 as both nutrients contribute to resting energy expenditure.

RER for glucose = $6\,CO_2/6\,O_2 = 1.00$

RER for palmitate = $16\,CO_2/23\,O_2 = 0.70$

(Equation 1.11)

Where CO_2 is produced and the RER during exercise are very important concepts that provide information about exercise intensity and nutrient oxidation. These concepts are discussed in chapter 7.

METABOLIC REGULATION DURING EXERCISE

The foundational metabolism knowledge has been discussed in depth, and now we are prepared to discuss how each of these systems interact and are regulated during exercise. We begin our discussion by analyzing metabolism under resting conditions, during short bouts of intense exercise, during steady-state conditions, and finally at maximal exercise (i.e., $\dot{V}O_2max$). Many factors influence metabolism, such as skeletal muscle fiber types and the type of exercise training (e.g., aerobic vs. resistance). These are discussed in minor detail. Further, we must constantly remember that the goal of exercise metabolism is to meet the ATP demand of the contracting muscle.

Resting Metabolism

Oxidative phosphorylation is the major source for energy production under resting conditions. Both fat and carbohydrate are broken down at rest to support basic metabolic processes such as organ function and food digestion. The energy source is heavily influenced by substrate availability despite no change in the metabolic demand. For example, consuming a large amount of carbohydrate will inhibit fatty acid oxidation and shift toward more plasma glucose oxidation with little change in metabolic rate.

Early experiments demonstrated that the key regulatory molecule is acetyl-CoA, which is the entry molecule for both fatty acid and glucose oxidation (50). An increase in acetyl-CoA and NADH levels from fatty acids inhibits the PDH reaction, thus directing metabolism toward greater fat utilization (24). In addition, in vitro work has shown that the increase in citrate (from the TCA) inhibits PFK, again resulting in greater fatty acid oxidation. The inhibition of PFK and PDH results in an increased accumulation of G6P in the muscle, which inactivates hexokinase and reduces glucose uptake into the muscle. This mechanism preserves plasma glucose levels, which is a major homeostatic variable. This regulation has been termed the *glucose–fatty acid cycle* or *Randle cycle*. Upon exposure to physical stress (i.e., exercise), fatty acid regulation of glucose metabolism is overridden, which is discussed later (31).

The endocrine influence on metabolism is for regulation of plasma glucose. During feeding, insulin levels increase, which inhibits fatty acid oxidation and shifts metabolism toward glucose uptake into the muscle and oxidation. The hormone glucagon is released during nutrient deprivation and prolonged exercise and preserves plasma glucose by stimulating liver glycogenolysis and adipose tissue lipolysis, leading to fatty acid oxidation. Thyroid hormone increases metabolism from both nutrient sources by activating glycolysis/glycogenolysis and mobilizing fatty acids. Both growth hormone and cortisol increase fatty acid oxidation and reduce glucose uptake into the skeletal muscle.

Short Bouts of Intense Exercise

ATP demand increases dramatically at the onset of intense exercise, and the resting metabolic rate is not prepared for this shift. The immediate source of ATP is provided through the ATP-PCr system, which is activated due to the increase in ADP levels from ATP hydrolysis. The capacity of this system is small, and it will support ATP demand for roughly 5 to 15 s (figure 1.7). In addition, activation of the adenylate kinase reaction causes an increase in cytosolic AMP concentrations. The increase in AMP and ADP (continued ATP breakdown) activates PFK and PK and accelerates the glycolytic rate. In 6 s of maximal exercise ATP turnover rates may exceed 15 mmol of ATP/s, with almost 50% supplied by the PCr system and the remaining supplied by nonaerobic glycolytic sources (lactate production) (62). Nonaerobic contribution to energy production is close to 90% in events lasting 20 s or less (25). Roughly 60% to 70% of energy is supplied by nonaerobic metabolism during 30 s of maximal cycling, and approximately 30% to 40% is supplied by aerobic energy sources (6,25,42). In maximal events lasting greater than 1 min, the majority of energy is supplied through aerobic metabolism (25).

Submaximal Exercise

Several regulatory mechanisms must take place to achieve a metabolic steady state of exercise. **Steady state** is considered a level of exercise at which ATP supply is being met through the aerobic energy system. The initial activation of the ATP-PCr system and nonaerobic glycolysis fills the ATP gap until oxidative phosphorylation can meet the energy demand. This commonly is called the ***oxygen deficit***. At submaximal intensities of 50% to 80% $\dot{V}O_2$max, oxidative phosphorylation becomes the major source of ATP supply.

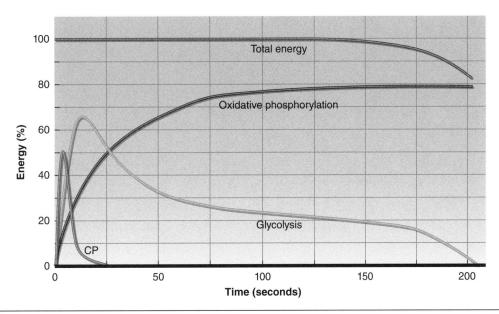

Figure 1.7 The contribution of energy systems during a short, high intensity bout of exercise lasting roughly 200 seconds.

Reprinted, by permission, from V. Mougios, 2006, *Exercise biochemistry* (Champaign, IL: Human Kinetics), 243.

TCA enzyme activity accelerates due to a slight decrease in the NADH/NAD and ATP/ADP ratios. In addition, plasma epinephrine levels increase, which activates hormone-sensitive lipase and lipolysis. At approximately 55% to 65% $\dot{V}O_2$max, fatty acid oxidation accounts for roughly 50% of energy production (figure 1.8). Evidence has shown markedly higher levels of plasma free fatty acids and glycerol during prolonged exercise at 50% $\dot{V}O_2$max (75), which was higher in fasted subjects than in fed subjects (75). The remaining energy supply comes from plasma glucose and stored glycogen (69). However, only small amounts of lactate (an end product of nonaerobic glycolysis) are released from the contracting muscle at this lower intensity (52). The lactate release may be small because pyruvate is not being reduced to lactate and entering the mitochondrial PDH reaction or because the lactate produced is being shuttled into the mitochondria and oxidized back to pyruvate for entry into the PDH reaction. In both scenarios, glucose/glycogen is supplying ATP through the aerobic pathway. The PDH reaction is highly regulated by changing levels of acetyl-CoA, NADH, and ATP. If the contracting muscle is supplied with adequate ATP from fatty acid acetyl-CoA and NADH levels are maintained, then PDH will not be highly active. However, if a slight decline occurs in acetyl-CoA/CoA, ATP/ADP, and NADH/NAD due to increased contractility, then PDH activity will increase. These enzymes are constantly active, but the level of activity varies based on changing exercise intensity.

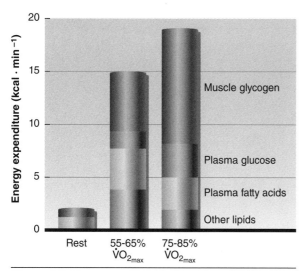

Figure 1.8 Energy sources at various exercise intensity ranges. As intensity increases a greater percentage of energy is derived from muscle glycogen. While the percentage of fat oxidation decreases at higher exercise intensities, the overall amount of fat being used is greater because more overall energy is being expended.

Reprinted, by permission, from L.J.C. van Loon et al., 2001, "The effects of increasing exercise intensity on muscle fuel utilization in humans," *The Journal of Physiology*, 536.1: 295-304. © Wiley Company.

Maximal Exercise Intensity

As exercise intensity increases, a profound shift occurs toward carbohydrate metabolism; this has been termed the *crossover concept* (10). Several regulatory mechanisms are responsible for this shift. The rate of ATP turnover is greatly enhanced, causing changes in the ratios of NADH/NAD and ATP/ADP and further increases in AMP. Shifts in these ratios activate the PFK and PK reactions. The increase in NAD and ADP, along with increasing pyruvate and intracellular Ca^{++}, activates the PDH reaction. Further, glycogen phosphorylase is activated by the increase in plasma epinephrine and inorganic phosphate along with the increase in Ca^{++}. This releases stored glycogen for entry into glycogenolysis and is the major reason why glycogen is the dominant carbohydrate during prolong and intense exercise (figure 1.9).

At high-intensity exercise, the RER value approaches 1.00, which we learned is 100% contribution of aerobic

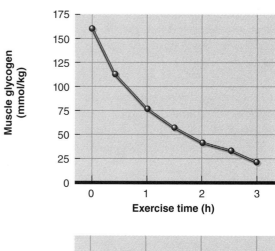

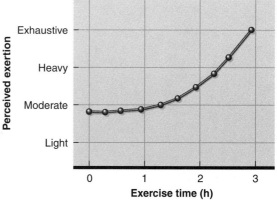

Figure 1.9 Skeletal muscle glycogen levels decrease after three hours of treadmill running at 70% of $\dot{V}O_2$max.

Reprinted, by permission, from L.W. Kenney, J.H. Wilmore, and D.L. Costill, 2015, *Physiology of sport and exercise,* 6th ed. (Champaign, IL: Human Kinetics), 135. Adapted from D.L. Costill, 1986, *Inside running: Basics of sports physiology* (Indianapolis: Benchmark Press). By permission of D.L. Costill.

metabolism from carbohydrate oxidation. The additional CO_2 produced from carbohydrate oxidation is from the PDH reaction, so we can conclude that the RER of 1.00 is representative of full activation of the PDH.

A point of controversy remains about the rate of lipolysis during intense exercise. It is well documented that fatty acid oxidation rates decline substantially at 85% $\dot{V}O_2$max; however, the rate of lipolysis does not change between intensities of 65% and 85% (53). Although the lipolytic rate remains high, adipose tissue blood flow may be reduced, indicating less fatty acid movement into the bloodstream. Further interpretations include reduced fatty acid delivery to the contracting muscle or a decrease in fatty acid transport into the muscle. In summary, lipolysis continues during high-intensity exercise, but delivery to the contracting muscle for beta-oxidation is reduced.

At maximal exercise intensity or at the end of a $\dot{V}O_2$max test, the large amount of ATP turnover can be attributed to nonaerobic glycogenolysis (63). As the glycolytic rate speeds up, cytosolic NADH will accumulate, and for glycogenolysis to continue pyruvate will accept the protons and be reduced to lactate (63). Again, this allows NAD to move back up into the G3P dehydrogenase reaction (previously discussed). Several metabolites begin to accumulate, including ADP, Pi, H^+, AMP, and lactate (3); all may play a role in muscle fatigue and termination of exercise. Skeletal muscle ATP levels remain stable even during maximal exercise,

which lends to the idea that exercise termination may not be due to a lack of ATP but rather to a decrease in ATP hydrolysis. If the skeletal muscle did not limit ATP hydrolysis, it would essentially continue to work until it was destroyed.

Is this shift toward nonaerobic metabolism attributable to lack of oxygen or to the fact that oxidative phosphorylation cannot meet the ATP demand of the intensely contracting muscle? Lactate formation is considered an anaerobic process, which can be interpreted as lack of oxygen. It has been shown that the PO_2 of skeletal muscle does indeed decline during exercise but remains well above the threshold for ATP synthesis (52). In addition, lactate levels increase in the muscle before the decline in skeletal muscle PO_2 (52). In support of this, performing maximal contractions in hyperoxia (higher oxygen content) did not increase peak power output (46). It may be that the rate of ATP supplied by oxidative metabolism cannot meet the demand of the contracting muscle, and it may not be due to lack of oxygen. In addition, at high-intensity exercise there is a greater recruitment of Type II fibers, which have lower oxidative capacity. The maximal rate of ATP synthesis from aerobic metabolism (glycogenolysis) is roughly 0.5 to 2.9 ATP/s (mmol $\cdot$ kg^{-1} $\cdot$ s^{-1}) (51,71) compared with approximately 6.0 to 9.0 and 6.0 to 9.3 ATP/s (mmol $\cdot$ kg^{-1} $\cdot$ s^{-1}) for nonaerobic glycogenolysis and ATP-PCr, respectively (60). We previously stated that skeletal muscle ATP turnover rates may reach 15 mmol $\cdot$ kg^{-1} $\cdot$ s^{-1} (62).

Aging and Metabolism

Age-related reductions in exercise capacity and muscle metabolism have been demonstrated among human subjects. It has been reported that elderly humans have reduced TCA enzyme activity (17), resulting in reduced skeletal muscle oxidative capacity (12). This may be due to muscle atrophy from disuse and to reduced capillary density and blood flow to the skeletal muscle. In addition, exercise glycogenolysis is enhanced in elderly untrained subjects, which indicates a shift toward greater carbohydrate breakdown. Beginning aerobic exercise training in middle to late ages increases mitochondrial density (11) and enzyme activity of both the TCA cycle and ETC (4). Evidence suggests that these levels may be equal to those in younger counterparts. In addition, aerobic training enhances glycogen sparing and further improves glucose uptake and glycogen synthesis in previously sedentary older men (21). These adaptations improve the oxidative capacity of elderly muscle and shift metabolism toward greater fat oxidation.

Interestingly, elderly men who have maintained an aerobic training routine throughout life have greater expression of Type I muscle fibers compared with sedentary men (~71% vs. 49%, respectively) and a lower expression of Type II fibers (~29% vs. 51%). This demonstrates that maintaining exercise throughout life preserves oxidative metabolism (7).

FACTORS THAT INFLUENCE EXERCISE METABOLISM

Several important aspects that affect metabolism, including variations in skeletal muscle fibers as well as metabolic adaptations to aerobic and resistance exercise training, are discussed in the following sections. This will help us understand and develop training programs for athletes participating in all types of sporting events.

Skeletal Muscle Fiber Type

The intensity of exercise dictates the type of skeletal muscle recruited during exercise and regulates the rate of metabolism. Muscle fiber types are characterized based on their metabolic profiles (table 1.3).

Type I (or slow oxidative) skeletal muscle fibers are designed for prolonged, repetitive contractions and support endurance-related activities (e.g., running, cycling). Type I fibers are metabolically designed for large amounts of ATP synthesis via oxidative metabolism. This fiber type has the largest volume of mitochondria, capillary density, and myoglobin levels. In addition, Type I fibers have greater expression of TCA and ETC enzymes such as citrate synthase, succinate dehydrogenase, and cytochrome c (30) and expression of TCA enzymes. Further, oxidative fibers have a greater expression of FABP and carnitine transferase, indicating a greater capacity to transport and utilize fatty acids. Intermediary fibers (Type IIA) have moderate characteristics of oxidative metabolism, whereas Type IIX fibers have low expression of oxidative markers.

During high-intensity exercise, larger muscle fibers are recruited such as Types IIA and IIX, which have greater expression of glycolytic enzymes to support faster rates of ATP synthesis. However, Type I fibers do express moderate levels of glycolytic enzymes. Interestingly, during prolonged low-intensity exercise (30% $\dot{V}O_2$max, 3 h), Type I fibers were more depleted of glycogen, whereas Type IIX/IIA fibers had minimal depletion. In addition, during short bouts of high-intensity exercise (1-min intervals at 120% $\dot{V}O_2$max), Type IIA/IIX fibers were depleted and Type I fibers had only modest depletion (13). Glycogenolysis occurs at high rates in Type I fibers between intensities of 60% and 75% $\dot{V}O_2$max. At higher intensities, glycogenolysis is more active in Type IIA and Type IIX fibers (27).

Markers of the ATP-PCr system, such as stored PCr and enzyme expression (creatine kinase, adenylate kinase), are markedly higher in Type IIX fibers and subsequently decrease across Type IIA and Type I fibers, respectively. Some studies have shown that predominately Type II muscles (i.e., digitorum longus) have twice the amount of stored PCr compared with Type I muscles (i.e., soleus) (39,45). After maximal contractions, Type IIA/X fibers demonstrate lower levels of PCr, and levels remain lower into recovery (35). Additionally, in single fiber work, after 10 s of maximal contractions PCr levels decrease approximately 46%, 53%, and 63% in Type I, IIA, and IIX fibers, respectively (35).

It is important to understand that whole skeletal muscles are heterogeneous, meaning that they contain a mixture of the three fiber types. In other words, each energy system is expressed in all skeletal muscle.

Table 1.3 Skeletal Muscle Characteristics

Characteristic	Type I	Type IIa	Type IIx
ATP-PCr	Low	Medium	High
Glycolytic enzymes	Low	Medium-high	High
PDH	Medium	Medium	High
TCA enzyme	High	Medium	Low
ETC enzyme	High	Medium	Low
FABP	High	Medium	Low
Carnitine shuttle	High	Medium	Low
Mitochondrial volume	High	Medium	Low
Glycogen storage	Low	Medium-high	High
Lipid storage	High	Medium	Low

Summary

- Humans must be able to support the ATP requirement of the contracting muscle from the immediate onset through maximal exercise intensity. Learning how each of the systems is designed, regulated, and integrated and how they adapt is key to advancing one's knowledge in the field of exercise physiology.

- The ATP-PCr system and nonaerobic glycolysis/glycogenolysis generate ATP at rapid rates and in the absence of oxygen. Combined, they provide most of the energy during high-intensity exercise.

- Oxidative phosphorylation is the dominant source of energy during low- to moderate-intensity exercise and utilizes the stored energy from both lipids and carbohydrate for ATP synthesis.

- These systems are highly regulated by small changes in substrates (i.e., ADP, AMP, NADH, acetyl-CoA), which are constantly changing based on the intensity of the bout of exercise.

- Chronic exercise training causes tremendous adaptations to the metabolic systems. Aerobic exercise results in greater mitochondrial volume that includes greater expression of TCA and ETC enzymes, whereas adaptation to the glycolytic system requires more intense training of a longer duration (i.e., high-intensity interval training).

- Resistance training has also been shown to improve metabolic function by enhancing the rate of glucose uptake and breakdown.

Definitions

endergonic reaction—The reaction requires energy input to move toward product formation and is fueled by exergonic reactions.

enthalpy—Measure of total energy transfer; separated into energy released as heat and energy used to perform cellular work.

entropy—Measure of disorder or randomness; governs the directionality of all reactions.

equilibrium—A reaction in equilibrium is in balance, and free energy change does not occur.

equilibrium constant—The point during a reaction at which the concentration of reactants and products is no longer changing. The higher the equilibrium constant, the greater the formation of products from substrates.

exergonic reaction—The reaction occurs spontaneously toward product formation without the need for energy input.

free energy—Energy used to perform cellular work (symbolized by ΔG); quantified by the difference between enthalpy and entropy.

oxygen deficit—The nonaerobic ATP-generating reactions that must occur to support the muscle at the onset of exercise, which bridges the gap until the aerobic system catches up. These include ATP-PCr and nonaerobic glycolysis/glycogenolysis.

steady state—A level of exercise at which the ATP demands of the contracting muscle are met through aerobic metabolism; usually takes 3 to 5 min.

Chapter 2

Dynamics of Skeletal-Neuromuscular and Gastrointestinal Physiology

Knowledge of skeletal muscular control and function is essential for the exercise physiologist. Better understanding of the various components that make up the skeletal-neuromuscular system can allow for real-world application that may lead to improvement in athletic performance, rehabilitation outcomes, and disease prevention. Beginning within a single **muscle fiber**, the neuromuscular system is analogous to an automobile with a motor (i.e., the contractile fibers), a fuel source (i.e., adenosine triphosphate; ATP), and a source of ignition (i.e., the **alpha motoneurons**). The interactions of this integrated system seamlessly allow an individual to go from fine motor movements (e.g., texting on a phone) to gross motor movements (e.g., pushing a broken car down the street). Furthermore, the neuromuscular system has tremendous plasticity. Depending on the type of stimulus (or lack of stimulus), changes may occur at the cellular, neural, or endocrine levels.

The skeletal-neuromuscular system of course does not operate in a vacuum but rather is dependent on the other body systems, including the gastrointestinal (GI) system. Proper functioning of the GI system is dependent on the coordinated actions of smooth muscles, which, like the skeletal muscular system, are controlled by the nervous system, but as we will discuss, the neuromuscular and functional properties of smooth muscle have unique differences when compared to the skeletal-neuromuscular system. Better understanding of those differences and the other properties of the GI system has led to informed dietary practices which has been shown to improve athletic performance. However, the ability to absorb fluids and macronutrients has implications not only for the optimization of athletic performance, but it also can have an effect when exercising individuals have various chronic health conditions.

STRUCTURE OF SKELETAL MUSCLE AND MECHANISMS OF CONTRACTION

The adult human has 660 skeletal muscles, which comprise roughly 80% water, 12% myofibrillar proteins, and 8% additional proteins, enzymes, inorganic salts, and other. In the following sections we describe these various components that make up muscle tissue, beginning with a whole muscle, which comprises many thousand individual contractile fibers. We then discuss the functional unit of muscle cells, known as the *sarcomere*, as well as the contractile proteins actin and myosin. Additionally, we discuss the intricate chemical mediators and other components that are essential for contraction and relaxation.

Anatomy of Muscle

If you remove a muscle and section it (as in figure 2.1), you would notice tightly packed bundles that look like fiberoptic cables; these are known as *fascicles*. Inside the fascicle are **muscle fibers** (or muscle cells) that, like myocardial muscle, contain light and dark striations. The muscle fiber length can vary from as little as 1 mm up to several centimeters, depending on the type of muscle. In the sartorius muscle (the longest human skeletal muscle), a single muscle fiber can be 6 cm long (32).

Unlike cardiac myocytes, muscle fibers are multinucleated and contain approximately 200 to 300 nuclei per millimeter of fiber length. The ability to be multinucleated is a major factor that allows the muscle to have myoplasticity. The cell nuclei of the muscle fibers are located inside the **sarcolemma**, which is made up of an outer membrane known as the *basement membrane* and an innermost membrane known as the *plasmalemma*. Within the sarcolemma are hundreds to thousands of **myofibrils**. It is thought that increasing the number of myofibrils—not muscle fibers—is what allows individu-

als such as Denis Cyplenkov, a Ukrainian strongman and a World Armwrestling Champion, to have 64-cm biceps! Although experts still debate the topic of muscle hypertrophy, so far studies have not found direct evidence for **muscle hyperplasia** (i.e., the growth of new muscle fibers) in humans. Interestingly, hyperplasia has been shown across different animal populations such as birds, rats, and cats (3,4,54).

Satellite Cells

Similar to many other adult cells (e.g., myocytes and neurons), adult muscle fibers are postmitotic cells that are unable to reproduce. However, muscle fibers show tremendous plasticity in response to mechanical load, stretch, and injury. Because of this ability, researchers have long believed that the adult stem cells of the muscle, known as **satellite cells**, are responsible for muscle adaptation and regeneration (47). Satellite cells are named because of their location in the outermost layer of the muscle fiber (i.e., the basement membrane). They contain the same genetic material as the nuclei located on

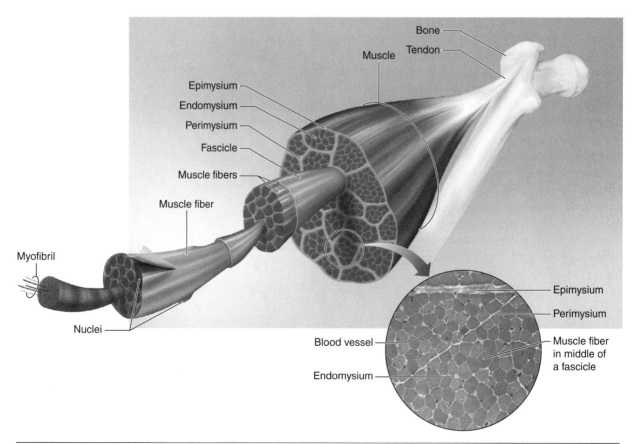

Figure 2.1 The basic muscle structure. Muscle is contained in a tight sheath of connective tissue and organized in bundles known as *fascicles*.

the other side of the sarcolemma, and their "addition" to existing myofibrils is why muscle cells are multinucleated.

Proliferation of satellite cells is greatest early in life, remains relatively stable throughout adulthood, and starts to decline toward the end of life (19). Perhaps one reason why satellite cell activity is greater earlier in life is to help support the growth of the myofibrils in series as a response to increasing bone length (47). Other factors that activate satellite cells from a quiescent (i.e., dormant) stage to an active stage are cell damage and exercise. In fact, satellite cell production has been shown to increase 4 days after a single exercise bout, with higher levels seen following a few weeks of exercise training (19). Once activated, a satellite cell begins to express three important myogenic regulatory factors: myogenic factor 5, MyoD, and myogenin (19). The myogenic regulatory factors essentially commit the satellite cell down the myogenic pathway leading to muscle growth. The steps of this pathway include the following:

1. Mitosis into additional satellite cells
2. Formation of a myotube, which contains the materials for contraction (i.e., contractile proteins)
3. Translocation of the myotube across the sarcolemma
4. Fusion of the myotube with an existing myofibril

It has long been believed that satellite cells are responsible for hypertrophy and regeneration after injury (36). However, work from Jackson et al. (36) has shown evidence of hypertrophy independent of satellite cells. Specifically, it seems that muscle regrowth after atrophy can occur in the absence of satellite cell involvement. However, satellite cells still seem to have an indirect role because the amount of regrowth found was attenuated without satellite cell involvement.

Contractile Properties of Muscle

In 1674, using a light microscope, Anton van Leeuwenhoek first described the repeating muscle striations on the myofibril. The dark striations are denoted as *A bands* and *Z discs* (also known as *Z lines*), and the light striations are denoted as *I bands* and *H zones* (figure 2.2). The sarcomere (German for "small boxes") runs from Z disc to Z disc and is the contractile unit of the muscle (figure 2.2). When viewing the cross-striations along a two-dimensional plane, each sarcomere contains a single set of thick filaments made up of the protein **myosin** and two sets of thin filaments consisting of **actin**. However, if you were to look at a sarcomere in a three-dimensional plane, each myosin wound be surrounded with six actins. The interaction of these two contractile proteins creates the molecular motor allowing contraction of the muscle.

A closer look at the myosin molecule reveals a six-subunit protein consisting of two myosin heavy chains (MHCs) and two myosin light chains (MLCs; figure 2.3). They are denoted as *heavy* and *light* because of

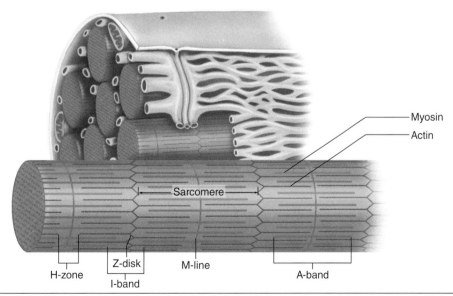

Figure 2.2 The contractile properties in a muscle fiber. Here is a microscopic view of a muscle fiber, a myofibril, and the myofilaments actin and myosin. Note the striations in both the fiber and myofibril; these alternating light and dark bands are created by the geometric arrangement of the protein filaments actin and myosin.

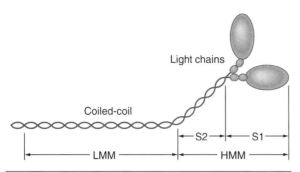

Figure 2.3 The myosin molecule containing light mero-myosin and heavy meromyosin components.

their respective molecular weights. There are three isoforms of MHC that exist in humans and vary depending on the sequence of amino acids. These isoforms are known as slow-twitch (Type I) and fast-twitch (Types IIa and IIx) fibers. (We discuss more about these fiber types later.) The MHC can be further dissected into the light meromyosin, which is the "coiled" tail of the protein, and the heavy meromyosin, which contains the S1 fragment (i.e., the binding site to actin) and the S2 fragment (i.e., the lever arm that produces torque during muscular contraction). Similar to the MHC, the MLC contains four isoforms, two of which are always the same. In vitro studies show that removal of the MLC does not prevent contraction, but it does slow down the contraction speed. Therefore, it is believed that although MHC is still important for gross velocity of contraction, the role of the MLC is to help regulate the fine-tuning velocity (43).

In addition to myosin and actin, other proteins make up muscle. Two other major subgroups are the *structural proteins* (table 2.1), which act as anchors to support and stabilize the sarcomeres, and the *regulatory proteins* (i.e., **troponin** and tropomyosin), which regulate the binding of myosin to actin.

Cross-Bridge Cycle

Using an electron microscope, Dr. Hugh Huxley in 1953 first identified the thick and thin filaments that make up the dark A zones and light I zones of the myofibrils (31). As Huxley was observing the "disappearance" of the I zone under the microscope, he noted that the thin and thick filaments must somehow slide past each other during contraction (35). This is now known as the *sliding filament theory* of muscle contraction. Parenthetically, another Dr. Huxley (A.E. Huxley), who was doing similar work in 1954, is also credited with the sliding filament theory.

At rest, tropomyosin proteins provide a molecular, or steric, block on the actin binding sites where the cross-bridge of the actin–myosin bond occurs. A single tropomyosin strand covers up seven actin molecules. Also bound to each tropomyosin is a troponin complex (figure 2.4), which consists of three subunits: troponin C (which binds to calcium), troponin I (which inhibits actin–myosin binding), and troponin T (which is bound to tropomyosin). To initiate contraction, calcium molecules (i.e., Ca^{+2}) are released from the sarcoplasmic reticulum and bind to four calcium sites located on troponin C. Interestingly, cardiac troponin C has only three calcium binding sites, thus allowing for faster contraction. The binding of Ca^{+2} to troponin C leads to a conformational change in the troponin complex, which moves the adjacent tropomyosin away from the actin binding site, thus allowing the actin–myosin cross-bridge to occur.

However, before binding to actin, myosin must first be activated through the hydrolysis of ATP—that is,

Table 2.1 Selected Structural Myofibrillar Proteins

Protein	Characteristics
Nebulin	A large, nonelastic molecular protein coinciding with the thin filament actin. Extends the length of the thin filaments
Titin	The largest known protein. Connects the Z disc to the M disc.
Myosin binding protein C	Links the end of myosin to the Z disc
M line protein	Holds thick and thin filaments in correct spatial arrangement
α-actin	Cross-links adjacent actin filaments
Desmin	Links Z discs of myofibrils
Dystrophin	Connects contractile proteins to the sarcolemma. Mutation in the dystrophin gene results in the loss of muscle integrity for individuals with Duchenne muscular dystrophy.

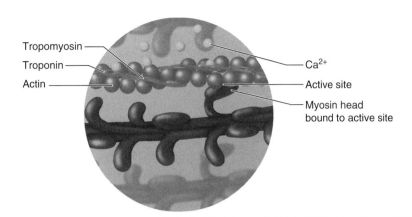

Figure 2.4 Troponin and tropomyosin regulate contraction by providing a steric block on active actin binding sites. When Ca²⁺ is present it binds to troponin C, leading to a conformational shift that moves tropomyosin off actin, thus allowing myosin to bind to actin.

the splitting of ATP by water to adenosine diphosphate (ADP) and inorganic phosphate (Pi). This reaction is catalyzed by the enzyme ATPase and occurs on the S1 fragment of the myosin head. The hydrolysis of ATP to ADP and Pi creates a conformational change in the myosin head, causing it to move from a 45° angle to 90° (figure 2.5). Once myosin binds to actin, the freed phosphate molecule is released from the myosin ATP site along with energy that causes the myosin head to rotate and ratchet back to 45° while pulling on actin. This process is called the *power stroke*. At the end of the power stroke, the remaining ADP is released from the S1 myosin site, and myosin stays bound to actin until it is released by the attachment of another ATP molecule. This is also how rigor mortis occurs a few hours after death: ATP is no longer present to detach myosin from actin, thus causing the muscles to stay in a state of contraction.

Calcium Handling for Contraction and Relaxation

All muscle requires Ca^{+2} as a second messenger to commence contraction (9). The storage, release, and resequestering of Ca^{+2} is dependent on a network of membranes collectively referred to as the *sarcoplasmic reticulum* (SR), which runs parallel to the myofibrils (figure 2.6). To initiate the release of calcium, the SR depends on a system of adjacent tubules known as the *transverse tubules* (or T *tubules*), which run perpendicular to the myofibrils. The function of the T tubules is to propagate the electrical depolarization that occurs as a result of an action potential sent from a motor neuron.

Transforming the electrical message from the T tubules into a chemical message are the dihydropyridine receptor and the ryanodine receptor (also known as the *foot proteins*; figure 2.6). These receptors form a bridge to the terminal cisternae (i.e., the outer edge of the SR). This allows the SR to release calcium and, in doing so, increases cytosolic levels inside the myocyte by 100 times compared with rest (9). Interestingly, it has been shown with aging that the mechanical connection between the dihydropyridine receptor and the ryanodine receptor to the SR becomes "uncoupled" over time, resulting in less calcium release. This explanation is one likely determinant of why muscle contraction speed slows over time as we age (22).

Similar to how increases in cytosolic calcium lead to muscle contraction, a decrease in calcium levels precipitates muscle relaxation. However, the speed of relaxation (known as **lusitropy**) and the mechanisms involved with the translocation of calcium can vary slightly depending on the muscle tissue (e.g., cardiac vs. skeletal) as well as the muscle fiber type. Similar to the cross-bridge cycle, the process of calcium resequestering is dependent on ATP. The **SR calcium-ATPase pump** (SERCA; figure 2.6) is responsible for the reuptake of calcium into the SR. In adult skeletal muscle there are two isoforms of SERCA: SERCA1a, found in fast-twitch Type II muscles, and SERCA2a, found in slow-twitch Type I, heart, and smooth muscle (9).

Additionally there are a number of modifying proteins which can speed up or even attenuate muscle lusitropy. Sarcolipin is a protein attached to SERCA that causes Ca^{+2} to slide off the molecule, thus preventing some Ca^{+2} from entering into the SR. The reason for this inefficiency is not entirely understood, but it may have implications with regard to body fat regulation because the uncoupling

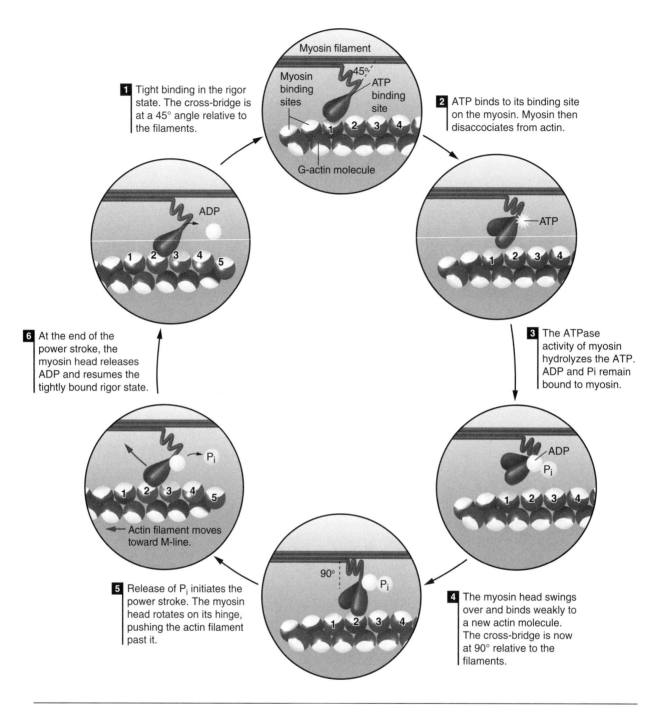

Figure 2.5 The cross-bridge cycle.

of Ca^{+2} from SERCA uses more energy (i.e., ATP), thus leading to higher metabolic rates (27). Other proteins associated with the SR that do improve the speed of Ca^{+2} reuptake include parvalbumin, which is not found in humans except for the intrafusal fibers (i.e., fibers associated with the muscle spindles), and **calsequestrin**, which Ca^{+2} binds to in the SR (9).

The ability to release and resequester Ca^{+2} may be one of the factors leading to fatigue with exercise. Supporting this notion is a study by Ortenblad et al. (46), who showed that a 5-wk program of all-out sprint bouts on a cycle resulted in a 12% improvement in peak power and was associated with an increased amount of SERCA1 (41%) and SERCA2 (55%) in the SR.

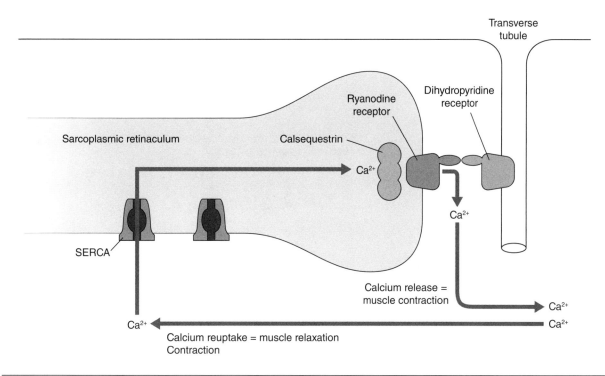

Figure 2.6 The sarcoplasmic reticulum. Depolarization of the muscle membrane activates the release of calcium stored in the SR, which leads to muscle contraction. Relaxation takes place when calcium is returned to the SR via the SR calcium–ATPase pump.

NEUROMUSCULAR CONTROL OF MOVEMENT

In this section we discuss how conductions travel across neurons and innervate muscle. However, beyond simply transmitting signals like electoral wires in a house, we also discuss how neural influence can affect the functional makeup of the fiber types themselves. This point is illustrated by the fact that all the muscle fibers innervated by a single motoneuron are homogenous.

Organization of the Nervous System

The nervous system can be organized broadly into the central nervous system (CNS), which includes the brain and spinal cord, and the peripheral nervous system (PNS), which includes everything outside of the brain and spinal cord (figure 2.7). The PNS can be further broken down into the efferent division, which consists of nerves transmitting information from the CNS to effector organs, and the afferent division (i.e., sensory nerves), which sends impulses in the opposite direction back to the CNS.

Some peripheral nerves (e.g., the olfactory nerves, which are responsible for sending smells to the brain) contain only afferent fibers. However, like all nerves emanating from the spinal cord, muscle contains both efferent and afferent nerves. The efferent nerves of skeletal muscles are also a part of what is known as the *somatic nervous system* (or the *voluntary nervous system*). Later in this chapter we discuss the roles of the parasympathetic and sympathetic nervous systems, which together make up the autonomic (or involuntary) nervous system.

Neuron

The transmission of afferent and efferent signals depends on the basic functional and anatomical unit of the nerve known as the *neuron* (i.e., nerve cell). The neuron contains three parts: the body (or soma), the dendrites, and the axon (figure 2.8). In large nerve fibers, such as the motor neurons that innervate skeletal muscle, the axon is surrounded by a lipid and protein substance known as

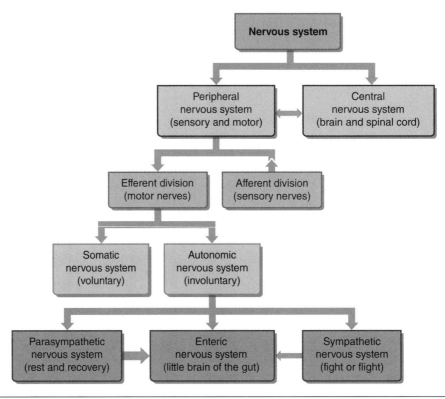

Figure 2.7 Organization scheme of the nervous system.

a *myelin sheath*. Myelin sheaths are organized in 1-mm segments separated by very small gaps (approximately 1 μm long) known as *nodes of Ranvier* (figure 2.8).

Both the myelin sheath and a larger axon diameter help speed up the transmission of signals in the form of electric impulses called *action potentials* (56). Some neurons (typically shorter ones) do not contain myelin and are known as *unmyelinated nerve fibers*. However, as mentioned above, most motor neurons contain myelin, which is especially important because these fibers—like those connecting the spinal cord to the lower limb muscles—can exceed 1 m in length (56). In fact, myelinated fibers conduct electronic impulses 10 times faster (at speeds between 217 and 362 km/h) than unmyelinated fibers!

Not all the cells that make up the nervous system are nerve cells. The largest group of cells in the nervous system is the **glial cells**. *Glia* comes from the Greek word meaning "glue," which is exactly what glial cells do: structurally support and protect the surrounding neurons (59). Glial cells also provide nutrition for the adjacent neurons. One type of glial cell specific to the PNS is *Schwann cells,* which provide structural and nutritional support and help maintain the myelin sheaths on the axon. Researchers are currently investigating Schwann cells in individuals with multiple sclerosis, an

autoimmune disease characterized by the loss of myelin (i.e., demyelination) that can lead to muscle weakness, cognitive defects, and pain (56).

Propagation of an Action Potential

At rest, there is an electric gradient between the inside and outside of a nerve fiber, known as the *resting membrane potential*. Is called a *resting* potential because no signal is being propagated; however, the neuron itself must expend a tremendous amount of energy to maintain this. At rest, a differential charge between the inside and outside of a nerve is maintained through an energy-dependent ATPase ion pump. The inside of a resting neuron is electronically negative (around −70 mV); there is a high concentration of negatively charged organic molecules, which are trapped, and positively charged K^+ ions, which slowly leak out through K^+ channels. Conversely, the outside of the neuron is positive (hence the gradient); there is a high concentration of Na^+ ions, which slowly leak into the neuron through passive Na^+ channels. The ATPase pump works to maintain this differential charge by exchanging three Na^+ molecules for two K^+ ions. The

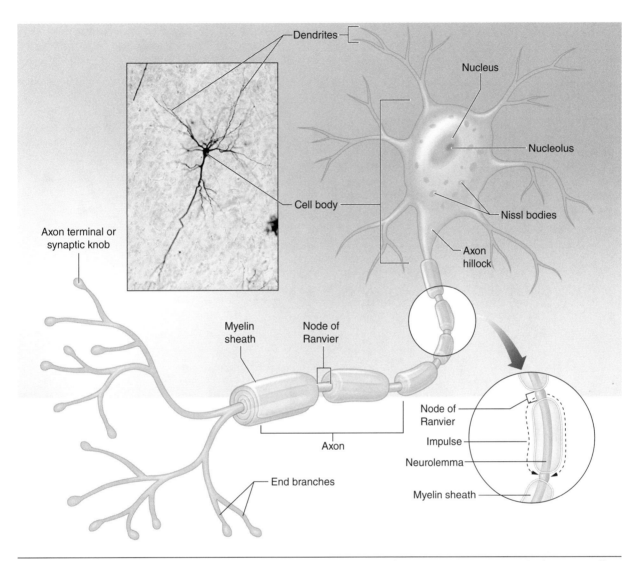

Figure 2.8 The structure of a motoneuron. Note the cell body (or soma) containing the nucleus, the long axon fiber that conducts impulses away from the cell body toward the axon terminal.

3:2 ratio of positive ions (three Na+ out and two K^+ in) keeps the outside of the neuron positive and the inside negative. Much of our knowledge regarding the electrical chemical properties of nerves and how they function was discovered through early studies in the 1930s on giant nerves found inside squids (38).

The change from resting potential to an action potential occurs when the cell membrane permeability changes, allowing an influx of Na^+ into the cell. This temporarily makes it so the inside of the neuron has a positive charge and the outside has a negative charge (figure 2.9). As the cell changes its polarity, it begins to propagate this signal along the entire length of the nerve fiber. This happens in both myelinated and unmyelinated cells. However, in myelinated cells, because an impulse cannot pass through the thick, insulated myelin, the electric current "jumps"

the axon, traveling around the myelin via the adjacent cytoplasm from one node of Ranvier to the next. This is known as *salutatory conduction* and is why impulses travel much faster along myelinated nerves.

Neurotransmitters and the Neuromuscular Junction

The transmission of a nerve impulse is initiated by chemical messengers known as *neurotransmitters*. Neurotransmitters are contained inside synaptic vesicles located at the terminal branches of axons (also known as the *synaptic knobs*). As the wave of depolarization propagates to the synaptic knob, Ca^{2+} channels open, signaling the synaptic vesicles to undergo exocytosis and release the

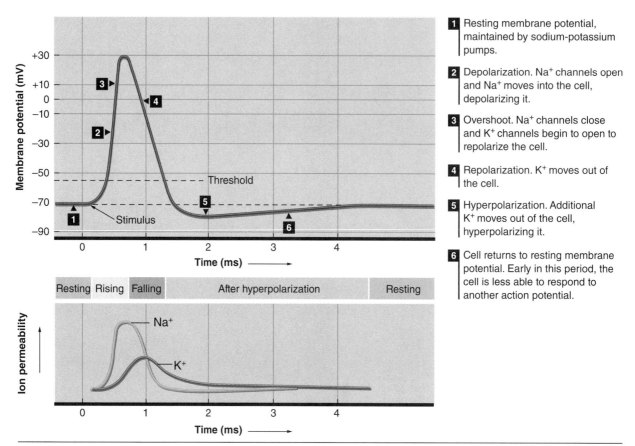

Figure 2.9 Development of an action potential.

Reprinted, by permission, from L.W. Kenney, J.H. Wilmore, and D.L. Costill, 2015, *Physiology of sport and exercise,* 6th ed. (Champaign, IL: Human Kinetics), 77.

neurotransmitters across the synaptic cleft (figure 2.10) (55). The neurotransmitters then cross over the synaptic cleft and bind to receptors located on the adjacent cell or postsynaptic neuron. Depending on the neurotransmitter and the effector cell, this will trigger a specific biological response.

Acetylcholine is a ubiquitous neurotransmitter responsible for many functions in the body, including muscular contraction. When an action potential is propagated to muscle along an alpha motoneuron (i.e., the type of efferent motor unit that innervates nearly all skeletal muscles), acetylcholine is released from the terminal axon, crosses the synaptic cleft, and binds to nicotinic receptors located on an area of the plasma membrane known as the *motor end plate.* The binding of acetylcholine opens voltage-gated Na^+ channels at the motor end plate, which depolarizes the plasma membrane, propagating the action potential across the muscle fiber and eventually leading to contraction. The remaining

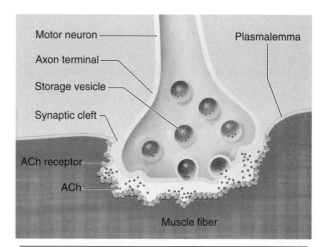

Figure 2.10 Chemical signaling between cells is mediated through neurotransmitters, which are released from synaptic vesicles, translocate across the synaptic cleft, and bind to receptors on effector cells. For muscle cells the neurotransmitter used is acetylcholine, which binds to nicotinic receptors on the ends of motor end plates.

acetylcholine molecules not bound to nicotinic receptors are degraded by the enzyme *acetylcholinesterase* (see the sidebar).

Acetylcholine is but one of more than 50 known neurotransmitters, most of which are found in the CNS. Neurotransmitters in the PNS are only excitatory; however, in the CNS neurotransmitters can be either excitatory (allowing the neuron to depolarize and an action potential to occur) or inhibitory (causing hyperpolarization and the cell to remain at rest; figure 2.11). These excitatory and inhibitory signals are known as *excitatory postsynaptic potentials* and *inhibitory postsynaptic potentials*. The sum of these potentials determines whether a nerve cell reaches the "all or nothing" threshold leading to an action potential.

Motor Unit

An alpha motoneuron (i.e., motor neuron) and the muscles it innervates are known as a **motor unit**. A single alpha motoneuron can innervate anywhere from 1 to 1,000 muscle fibers. Muscles requiring very fine movements (e.g., the muscles of the eye) may have only one muscle fiber per motoneuron, whereas muscles used for heavy lifting (e.g., the quadriceps) can have hundreds of muscle fibers per α-motoneuron. An interesting and

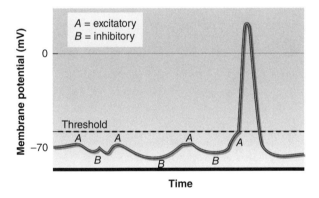

Figure 2.11 General example of an action spike in a nerve cell. Excitatory postsynaptic potentials (*A*) and inhibitory postsynaptic potentials (*B*) determine whether a neuron will reach its action potential threshold and hyperpolarize. The depolarization spike (action potential) is caused by an alternation in membrane semipermeability, allowing an influx of sodium.

Reprinted, by permission, from L.W. Kenney, J.H. Wilmore, and D.L. Costill, 2015, *Physiology of sport and exercise,* 6th ed. (Champaign, IL: Human Kinetics), 46.

important phenomenon is the homogeneity of the muscle fibers in a motor unit.

Previously we introduced slow- and fast-twitch muscle fiber types when discussing the MHC isoforms (Ia, IIb, IIx). These fiber types are uniform in a single motor unit, meaning that some alpha motoneurons innervate only Type I fibers, whereas others innervate exclusively Type IIa or Type IIx fibers. This homogenous relationship between nerve and muscle suggests that neural input has a strong influence on the muscle fiber type makeup. In fact, studies show that when a denervated muscle fiber is reinnervated by a different motor unit, that reinnervated muscle fiber type will adopt the same kinetic (e.g., speed of contraction) and molecular (e.g., myosin ATPase) properties as the other muscle fibers of that same motor unit, even if the reinnervated muscle fiber originally was a different fiber type (6,23).

Although a single motor unit may innervate only muscle fibers of one type (e.g., Type IIa), the distribution of different motor units within a muscle group is mixed. This mixture of innervated fibers, known as the *mosaic distribution* (figure 2.12), is what allows for a more fluid contraction as motor units are recruited during activity.

Henneman's Size Principle

Earlier we discussed the "all or nothing" principle of an action potential and how it eventually leads to a muscle contraction. However, not all motor neurons have the same excitability threshold. Depending on the stimuli (e.g., lifting a 2-kg weight vs. a 20-kg weight), some motor neurons will activate, whereas others stay dormant. The size principle is an important exercise physiology concept that provides a rule for how muscles (or, more specifically, motor units) are activated during activity (33). Stimulating stretch receptors in cats, Henneman et al. (33) found that smaller, less fatigable motor units (i.e., Type I) are recruited first followed by larger, more fatigable motor units (i.e., Type II) as the frequency and intensity of the stimulus increases. Thus, when designing an exercise program for strength and power, it is key to understand that larger fast-twitch muscles may not be recruited when training at moderate intensities.

The frequency and order of motor unit recruitment may also explain individual and cross-species differences in fiber type sizes. For example, in cats the order of cross-sectional fiber type size from smallest to largest is as follows: Type I < Type IIa < Type IIx (12). Because of this, it previously has been assumed that all species follow

---

Pharmaceuticals, Poisons, and Disorders Involving the Neuromuscular Junction

The neuromuscular junction is a common mechanistic site for a number of pharmaceuticals, particularly neuromuscular blockers, which are used widely to provide muscle relaxation during surgery (10). Unlike some commonly used medications such as aspirin, which was widely used for centuries before knowing the underlying mechanism, the knowledge of how neuromuscular blockers work preceded their clinical use (10).

A well-known example of this is curare, a plant-derived poison that South American natives use on the tips of hunting arrows or blow darts to paralyze prey. In the early 1800s, British surgeon Sir Benjamin Collins Brodie observed how the poison led to death by asphyxia while the heart continued to beat (2). Through later experiments by Claude Bernard and others, it was discovered that the curare molecule is an antagonist that competes with acetylcholine at nicotinic receptors on the postsynaptic neuromuscular junction. When this occurs, curare blocks acetylcholine and prevents the opening of Na^+ channels, which inhibits the action potential across the sarcolemma. This leads to paralysis of skeletal muscle, including those muscles responsible for respiration. However, because the neurotransmitter receptors responsible for the contraction of cardiac muscle are different, the heart is unaffected by the poison. The discovery of this mechanism eventually led to tubocurare, the first neuromuscular blocker used for clinical purposes in the 1940s.

Myasthenia gravis (MG) is an autoimmune disorder that produces symptoms similar to those of curare toxicity. MG has a prevalence of 15 per 100,000 individuals and occurs when immunoglobulin G antibodies bind and destroy nicotinic receptors on the postsynaptic neuromuscular junction (55). Clinical symptoms of MG include fatigue and muscle weakness, which tend to worsen toward the end of the day and with exercise. The drugs most commonly used to improve symptoms of MG are anticholinesterases, which work by deactivating the enzyme acetylcholinesterase. The result is a prolonged life of acetylcholine molecules in the synaptic cleft, which allows acetylcholine more time to bind to unaffected nicotine receptors. For this reason, anticholinesterases can also be used as an antidote for curare.

---

the same order with respect to fiber type size. However, other studies have found this not to be the case. Studies conducted in human females have found Type I fibers to be larger in cross-sectional area compared with both Type IIa and IIx fiber types (12). This is important because the type of training or stimulus (e.g., long-distance running vs. sprinting) can determine the frequency of fiber type activation, which in turn can affect the diameter of individual fiber types.

Wave Summation

Muscle fiber diameter and number of motor units recruited certainly are important determinants for generating strength in an individual muscle. Another important component is wave summation. A single nerve impulse (i.e., action potential), known as a *twitch*, causes a very brief period of contraction followed by relaxation (figure 2.13). When a second twitch occurs consecutively before the muscle completely relaxes, the force production summates, leading to increased force or strength of contraction. As more force is needed for a given task, summation continues. Eventually, if the consecutive twitches continue, the individual twitches become completely fused, which is known as *tetanus* (figure 2.13). The fusion of a motor unit can produce forces three times greater than a single twitch. However, prolonged contraction can lead to fatigue, especially in Type II fibers.

To combat muscle fatigue when performing a prolonged isokinetic task (e.g., carrying a heavy bag of groceries), another strategy the muscles use is **motor unit rotation** (28). During a prolonged muscular contraction, some motor units deactivate and others become active, maintaining the strength of contraction. The synchronization between muscle groups was once thought to occur primarily in postural muscles. However, Bawa and Murnaghan (8) found that this phenomenon can be found in other muscle groups as well.

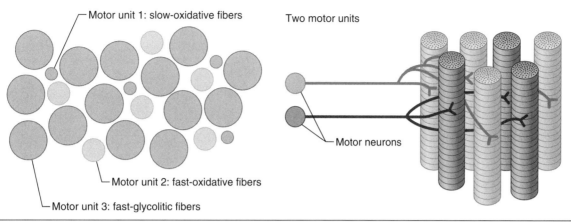

Figure 2.12   An example of a cross-section of muscle showing the mosaic distribution of the fiber types. According to Henneman's size principle, smaller, less fatigable motor units are recruited first, followed by large, more easily fatigable motor units. All muscle fibers of a given motor unit contain the same properties (e.g., Type I).

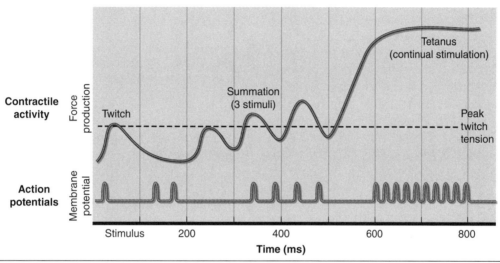

Figure 2.13   Increased force and tension through wave summation. Summation occurs as a result of consecutive twitches (i.e., action potentials), which increase the force of contraction. Note how the muscle does not return to a resting state during summation. When the twitches become completely fused, peak force occurs; this is known as *tetanus.*

FUNCTIONAL PROPERTIES OF MUSCLE

Muscular strength may be defined as the force or tension a muscle or, more correctly, a muscle group can exert against a resistance in one maximal effort. There are four basic types of muscular contraction: isotonic, isometric, eccentric, and isokinetic. Muscle fiber type often is focused on when describing factors related to force production. However, other properties of muscle, such as elasticity, length, and speed of contraction, can also affect force. Knowledge of these basic muscle principles can help when designing specific training programs.

Types of Muscular Contractions and the Force–Velocity Relationship

Imagine trying to lift a suitcase without knowing its contents and realizing that it is much heavier than you assumed. This is an example of an *isometric contraction.* During an isometric contraction, the myosin–actin cross-bridges are activated and the muscle is generating force, but the object (in this example, the suitcase) is not displaced and the muscle length stays constant. The two other types of contractions—eccentric and concentric—both involve the displacement of an object. An eccentric contraction occurs during lengthening of the muscle when the force of an object is greater than the force generated by the muscle (e.g., the biceps brachii when lowering an object to the ground). Concentric contractions, on the other hand, involve the shortening of a muscle when the force applied to the object is greater (e.g., lifting the same object off the ground). A contraction (eccentric or concentric) that is maintained at a constant speed is known as an *isokinetic contraction*. Because the force and speed of contraction are not uniform throughout the entire range of motion of a muscle, isokinetic contractions typically require the use of an apparatus that provides accommodating resistance, meaning that the speed of contraction stays constant regardless of the force applied.

Based on the kinetics of how myosin and actin attach and detach, Huxley was able to predict in 1957 the forces produced during a concentric contraction at various speeds (34). This led to the idea of the force–velocity relationship. When an object is very heavy (e.g., a suitcase), it takes additional time for cross-bridge attachment and detachment to occur, which slows the velocity of shortening. Conversely, the faster the muscle shortens, the less force is produced. This phenomenon occurs regardless of the fiber type (figure 2.14). Of course, there is variation when comparing fiber types. Thus, at similar speeds of contraction, a Type II fiber will have greater force than a Type I fiber.

The force–velocity relationship is helpful for predicting and explaining the forces and speeds of concentric contractions but is less helpful when examining eccentric contractions (34) because of the inclusion of passive forces (i.e., titin) along with the actin–myosin cross-bridge detachment and reattachment. In fact, studies have shown that eccentric contractions are capable of producing the highest forces compared with isometric and concentric contractions (24). Because of this, several training strategies involve the use of eccentric or a combination of eccentric and concentric (known as *stretch–shortening cycle*) exercises to potentiate athletic performance (26). However, the higher forces during muscle lengthening result in damage to both contractile and structural myofibril proteins, which can result in delayed-onset muscle soreness and greater recovery time.

Muscle Fiber Types

The term *fast-twitch muscle* has become part of sport lexicon for describing athletes such as Usain Bolt, who broke the world record in 2009 for the 100-m sprint with a blazing time of 9.58 s (1). For centuries, even before muscle fiber types were known, there was an understanding that some athletes have innate properties that lead to athletic success. Costill et al. (16) conducted one of the

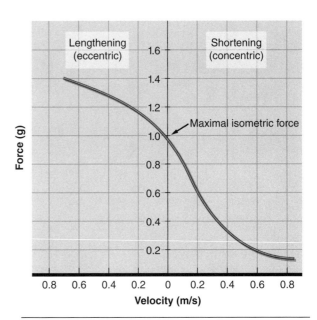

Figure 2.14 Force–velocity relationship. During concentric contractions, the force of contraction increases as the speed of velocity slows down. This is based on the kinetics of the cross-bridge cycle and the time it takes for actin and myosin to attach and detach. Notice how this relationship does not apply to eccentric contractions.

Reprinted, by permission, from L.W. Kenney, J.H. Wilmore, and D.L. Costill, 2015, *Physiology of sport and exercise*, 6th ed. (Champaign, IL: Human Kinetics), 47.

first of many studies that described how athletes have a different fiber type makeup compared with the average population. They showed that endurance athletes had 60% to 70% Type I fibers compared with untrained controls, who displayed a 50/50 ratio of Type I to Type II fibers. In the same study, they also reported that sprinters had approximately 80% fast-twitch fibers.

Numerous studies since then have found moderate to strong associations between fiber type makeup and performance measures (60). One example is a study by Kaczkowski et al. (37), who demonstrated a strong relationship between anaerobic capacity on the Wingate cycle test ($r = .81$) and the percentage of Type II fibers. The degree to which muscle fiber type can be altered with training is explored further in chapter 7.

The innate properties that differentiate slow-twitch Type I fibers from fast-twitch Type IIa and IIx fibers are both biochemical and physical. In general, the enzymes and physical properties of Type I fibers support endurance-type activities, whereas the enzymes and physical properties of Type II fibers support high-velocity, high-power activities (table 2.2). An example of this is myosin ATPase, which is an enzyme found mainly in fast-twitch white Type 2 fibers. As discussed earlier, myosin ATPase

Table 2.2 Characteristics of Fiber Types Found in Humans

	Characteristic	Slow-twitch red (Type I)	Fast-twitch red (Type IIa)	Fast-twitch white (Type IIx)
Favors aerobic capacity	Capillary density	↑↑↑	↑↑	↑
	Mitochondria density	↑↑↑	↑↑	↑
	Myoglobin content	↑↑↑	↑↑	↑
	Oxidative enzymes	↑↑↑	↑↑	↑
	Fatigue resistance	↑↑↑	↑↑	↑
Favors muscular speed and power	Contractile velocity	↑	↑↑↑	↑↑↑
	Force production	↑	↑↑	↑↑↑
	Myosin ATPase	↑	↑↑	↑↑↑
	Motor neuron size	↑	↑↑↑	↑↑↑
	Calsequestrin	↑	↑↑	↑↑↑

catalyzes the hydrolysis of ATP on the myosin head, which allows it to bind to actin. The greater the amount of myosin ATPase, the faster the contraction velocity. Consequently, Type II fibers have two to three times greater ATP hydrolysis rates compared with Type I fibers (60).

The molecule **myoglobin** also illustrates the biochemical differences between fiber types. Similar to hemoglobin, myoglobin can store and help diffuse oxygen into skeletal muscle, thus supporting aerobic metabolism. In fact, the red appearance seen in dark meat (which comprises mainly Type I fibers) is attributable to the molecule myoglobin. This is also why Type I muscles sometimes are referred to as "slow-twitch red" and Type II muscles are called "fast-twitch white."

Laboratory Techniques for Muscle Fiber Typing

Fiber typing can be done through several methods, including morphological, physiological, histochemical, and immunocytochemical methods. We have already discussed how fiber types can be differentiated through the molecular weights of the MHC isoforms. This is done using the analysis known as *sodium dodecyl sulfate–polyacrylamide gel* (SDS-PAGE) (53). Briefly, the SDS denatures the MHC protein, creating a negatively charged molecule. The MHC protein is then placed in wells on the top of the gel with a positive electrode at the base. The lighter the molecular weight of the MHC protein, the farther it travels toward the positive electrode, thus differentiating the specific MHC isoforms (i.e., Type I, Type IIa, and Type IIx).

Because myosin ATPase is found in greater amounts in Type II fibers, another test commonly used to determine fiber types involves the histochemical analysis of this enzyme. To determine the amount of myosin ATPase, a muscle sample is placed in either an acidic solution or a basic solution. Under basic conditions fiber types with high amounts of ATPase appear dark or stained, whereas under acidic conditions those same fiber types appear white. Figure 2.15 shows the mosaic pattern of this muscle sample after incubation in a basic solution. Based on the number of stained versus light fibers, one can estimate the percentages of Type I versus Type II fibers.

Both SDS-PAGE and myosin ATPase staining (as well as many other types of muscle analyses) obtain tissue samples through the muscle biopsy procedure. In this procedure, a local anesthetic (e.g., lidocaine) is first applied. Then, a pencil-size needle, known as a *bioptome*, is inserted through a 1.0-cm incision of the skin and fascia. The bioptome contains a plunger-like tool that slices off 20 to 40 mg of muscle roughly 1.5 cm above the tip of the needle. Once a sample is collected, it typically is frozen using liquid nitrogen and stored for later analysis.

Although quantitative histochemical lab techniques have provided greater understanding of the underlying molecular properties of muscle, much of our understanding of skeletal muscle function also comes from the mechanical or physiological testing of either whole motor units or individual muscle fibers. Studies of individual muscle fibers often involve the use of force transducers and an electric stimulus to the fiber (figure 2.16). Historically, these studies were done on frog or rabbit muscle fibers (7). Table 2.2 shows some of the differences

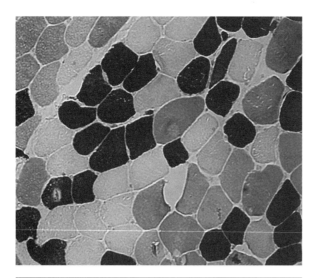

Figure 2.15 ATPase staining after incubation in a basic solution. The dark fibers are Type IIa, the light fibers are Type I, and the gray fibers are Type IIx.

Reprinted from L.W. Kenney, J.H. Wilmore, and D.L. Costill, 2015, *Physiology of sport and exercise,* 6th ed. (Champaign, IL: Human Kinetics), 135. ©David Costill.

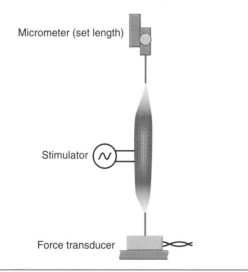

Figure 2.16 Example of a muscle fiber in a force transducer.

observed between fiber types with respect to speed of contraction, muscle fatigability, and force production (7).

Force–Length Relationship

Physiological testing has led to greater understanding of many other important skeletal muscle concepts, including the force–length relationship. In a classic study, using a force transducer and a single muscle fiber extracted from a frog, Gordon et al. (30) recorded the force production of isometric muscle contractions at various sarcomere lengths. They determined that maximal isokinetic force occurred at sarcomere lengths between 2.0 and 2.25 µm (figure 2.17). Knowledge of the actin–myosin cross-bridge theory is important for understanding why sarcomere length is related to the force of contraction. When the sarcomere is stretched too long, actin and myosin cannot attach; when the sarcomere's length is too close, there is overlap. The sarcomere length at which the greatest number of actin and myosin molecules align to produce maximal force is denoted as the *optimal length*.

Passive Length–Force Relationship

Another physical component of muscle that contributes to force production is the elastic components of the muscle itself. Earlier we described the largest muscle protein, **titin**, which is nearly 1 µm in length. In addition to its role of providing structural support, the tensile force produced by titin (along with some other structural proteins) is responsible for returning an overstretched muscle back to its resting length (45). Overall, muscle force is a byproduct of both passive force (produced from structural proteins) and active force (generated from the cross-bridging of actin and myosin). However, peak passive force occurs after the descending curve of the force–length relationship.

REGULATORY CONTROL OF THE GI SYSTEM

The GI tract functions as a conduit to process and deliver life-sustaining nutrients while simultaneously shielding the body from pathogens and toxins. From the moment a bite of food enters the mouth, the process of both mechanical breakdown (i.e., chewing) and chemical breakdown (i.e., amylase) begins. Shortly after a bolus of food moves to the rear of the pharynx (i.e., throat), the swallow reflex is activated and the regulation of digestion shifts from voluntary (i.e., the somatic nervous system) to involuntary (i.e., the enteric and autonomic nervous systems).

Enteric Nervous System

Below the surface of the gut lining, in the submucosal and myenteric plexus (figure 2.18), is the "little brain" of the GI system known as the *enteric nervous system* (ENS). The

Aging and Muscle Changes

It is clear that, in general, activity levels decline as we age (39). This is important in the context of skeletal muscle changes over a lifetime because some observed physiological changes attributed to aging are attenuated in older individuals who remain more active. In some studies, masters athletes showed muscle type makeup and enzymes similar to those of younger controls (15). However, even in the presence of exercise training, older adults succumb to the phenotypical hallmarks of age, which include reduced muscle strength, impaired speed of contraction, declines in endurance capacity, and prolonged muscle relaxation time (39).

Cross-sectional studies reveal that older individuals experience a loss in muscle mass, known as *sarcopenia,* mainly due to the loss of both the total number and the size of Type II (fast-twitch) fibers (42). Although Type I (slow-twitch) fiber size can also atrophy with age, this seems to happen to a lesser degree (42). This may be a result of muscle recruitment patterns and the size principle. Specifically, because everyday activities (e.g., walking, housecleaning, grooming) require only the activation of smaller Type I fibers, Type II fibers are more likely to atrophy from disuse. Interestingly, the opposite—a greater loss of Type I fibers and more reliance on Type II fibers—occurs during prolonged space flight (25). The latter, which is also seen in bed rest studies, appears to be a condition of unloading of the muscle (44).

This age-induced reduction in both the number and size of muscle fibers seems to be related, in part, to the loss of motor neurons (29). Whether caused by physical trauma or impaired signaling from the CNS, when a motor unit dies, the muscle fibers previously innervated by it are no longer capable of voluntary contraction (29). When this occurs, muscle fibers release autocrine signals, which stimulate the growth of new dendrites (in a process known as *sprouting*) from adjacent motor units. The new dendrites reinnervate to the abandoned muscle fiber, which takes on the properties of the other fibers in that motor unit. However, as we age, the process of sprouting and reinnervation begins to fail, leading to the death or apoptosis of muscle fibers that are abandoned (29). In fact, individuals over age 65 show decreased activity in areas of the brain associated with motor control as a result of fewer motor neurons (29). It should be noted again, that some of these changes may be the result of inactivity, as masters athletes have been shown to have a higher number of functional motor units compared with age-matched controls. In other words, use it or lose it!

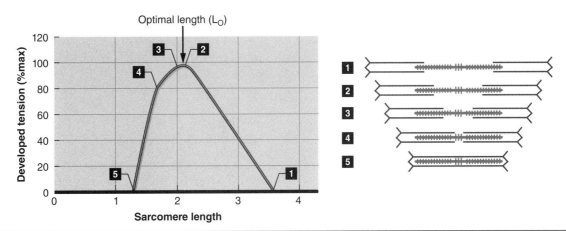

Figure 2.17 Force–length relationship. (1) No overlap of contractile proteins; thus, no force production. (2,3) Optimal overlap (2.0-2.25 µm); thus, optimal force. (4,5) Excessive overlap means that force production decreases.

Reprinted from L.W. Kenney, J.H. Wilmore, and D.L. Costill, 2015, *Physiology of sport and exercise,* 6th ed. (Champaign, IL: Human Kinetics), 47. Adapted, by permission, from B.R. MacIntosh, P.F. Gardiner, and A.J. McComas, 2006, *Skeletal muscle: Form and function,* 2nd ed. (Champaign, IL: Human Kinetics), 156.

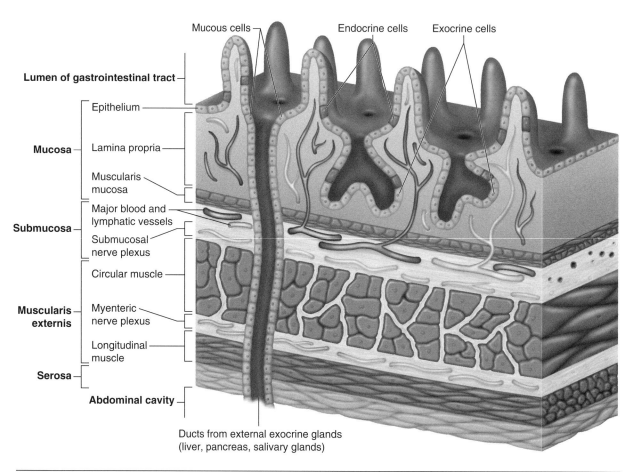

Mucous cells
Endocrine cells
Exocrine cells

Lumen of gastrointestinal tract

Mucosa
- Epithelium
- Lamina propria
- Muscularis mucosa

Submucosa
- Major blood and lymphatic vessels
- Submucosal nerve plexus

Muscularis externis
- Circular muscle
- Myenteric nerve plexus
- Longitudinal muscle

Serosa

Abdominal cavity

Ducts from external exocrine glands (liver, pancreas, salivary glands)

Figure 2.18 A longitudinal view of the GI wall. Note the internal muscular layers and their associated nerve plexus, which together are responsible for peristalsis.

ENS contains 100 million neurons and connects nerves along the GI path, allowing for synchronized interactions throughout the gut (41). Innervated by the ENS are specialized cells known as the *interstitial cells of Cajal*, which, similar to the sinoatrial node in the heart, acts as the pacemaker. In fact, just like an electrocardiogram measures the electrical activity of the heart, the electrical rhythm of the gut can be measured, as was first done by Alvarez et al. in 1914 (41).

Interestingly, if you were to measure the electrical activity of the gut, you would notice various waveform frequencies throughout the components of the GI tract. For example, within the stomach these waveforms typically are measured at 3 intervals · min^{-1}, whereas in the small intestines they are measured at closer to 13 intervals · min^{-1} (figure 2.19). These electrical waves, called **migrating motor complexes** (also known as *GI slow waves*) are what you feel during hunger when your stomach grumbles. However, unlike the pacemaker cells of the heart, the migrating motor complexes generated by the interstitial cells of Cajal do not propagate action potentials by themselves. For smooth muscle contractions to occur, efferent signals in the form of stretch and chemoreceptors are needed to push the action potential above its threshold (figure 2.19). Once stimulated above the action potential threshold, the migrating motor complexes coordinate **peristalsis**—rhythmic waves of smooth muscle contractions that help grind, mix, and transport the meals we ingest (13).

Autonomic Nervous System

The autonomic nervous system also greatly influences GI motility. Although the ENS can function without input from the CNS (hence why it is called the "little brain"), neurons from both the sympathetic nervous system and

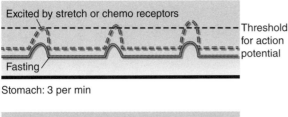

Stomach: 3 per min

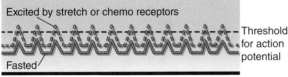

Small intestines: 12 per min

Figure 2.19 The migrating motor complexes are the underlying electrical rhythm of the GI system. For an action potential to occur, the cells need to be depolarized through afferent signals from stretch or chemoreceptors.

parasympathetic nervous system (PNS) also innervate the ENS. By, far the PNS has the larger influence on digestion and absorption because acetylcholine released from PNS neurons can trigger stronger peristaltic waves as well as greater release of digestive enzymes.

Norepinephrine, on the other hand, is released by the sympathetic nervous system and is known as the "fight or flight" neurotransmitter. It is responsible for functions such as increasing heart rate during exercise. Although it is true that increased sympathetic tone can slow down peristaltic waves through the release of norepinephrine, typically this occurs only during high-intensity exercise. Exercise releases norepinephrine and epinephrine in proportion to exercise intensity; therefore, one proposed mechanism to how exercise slows down digestion is the release of these neurotransmitters.

SMOOTH MUSCLE

Three distinct muscle types—skeletal, cardiac, and smooth—are found in humans. Table 2.3 illustrates some important similarities and differences between the two muscle types discussed in this chapter (i.e., skeletal and smooth). With a few exceptions (e.g., external anal sphincter), the smooth muscles are the major muscle type associated with the GI system.

Smooth Muscle Structure

Smooth muscle was named because it lacks the striations created by overlapping sarcomeres seen in skeletal

muscle. Instead of a linear arrangement, the contractile proteins of smooth muscles are arranged diagonally, creating a unique pattern of contraction when compared with skeletal muscle (figure 2.20). Specifically, smooth muscle contraction is nonlinear as the muscle cell both shortens and widens. This is advantageous for changing the diameter of tubular structures, as seen with smooth muscle found in blood vessels, intestinal walls, and bronchi.

Similarities to skeletal muscle include the use of the sliding filament mechanism for contraction and the contractile proteins myosin and actin (59). Both also operate along specific length–tension curves; however, smooth muscle has the ability to contract when stretched at much greater lengths than skeletal muscle (59). Another important similarity between all muscle cells is the use of calcium signaling for muscle contraction. All three muscle cells rely to some degree on the SR; however, smooth muscle contraction (as well as cardiac muscle) also relies on calcium from outside the cell.

Contraction of Smooth Muscle

In skeletal muscle, when Ca^{+2} enters the cytosol (via the SR), it binds to troponin, leading to a conformational shift of tropomyosin off the active actin site and allowing the cross-bridge of actin and myosin to occur. Smooth muscle, however, does not contain troponin. Instead, it relies on **calmodulin**, an important molecule responsible for the handling of calcium across many different cell types. In smooth muscle, Ca^{+2} enters the cytosol from the SR as well as from outside the cell. Once inside the cell, the Ca^{+2} binds to calmodulin. The Ca^{+2}–calmodulin complex binds to MLC kinase and, with the assistance of an ATP molecule, phosphorylates the myosin thick filament, which allows it to bind to actin (figure 2.21).

OVERVIEW OF THE GI ORGANS

In this section we discuss the organs of the gastrointestinal track.

Stomach

A typical empty stomach has a volume of approximately 50 mL and can expand 20 to 30 times that amount, (especially following an extra grande latte). The stomach

Table 2.3 Properties of Smooth Versus Skeletal Muscle

	Smooth muscle	Skeletal muscle
Striation pattern	No	Yes
Nervous system regulation	Automatic and enteric	Somatic
Multinucleated	No	Yes
Automaticity	Yes, the cells of Cajal	No
T tubules	No	Yes
Gap junctions	Yes	No
SR	Yes	Yes
Source of Ca^{+2} for contraction	SR and extracellular	SR
Activation of troponin for contraction	No	Yes

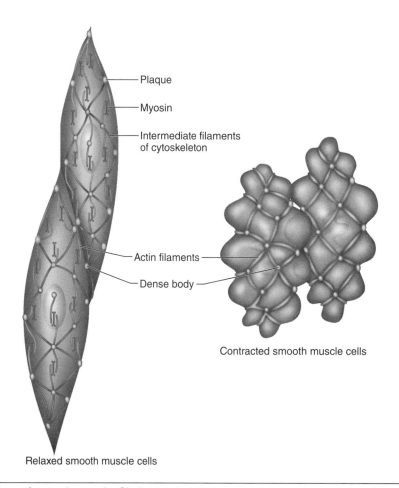

Plaque

Myosin

Intermediate filaments of cytoskeleton

Actin filaments

Dense body

Contracted smooth muscle cells

Relaxed smooth muscle cells

Figure 2.20 Structure of smooth muscle. Similar to skeletal muscle, actin and myosin proteins slide past each other during contraction. Based on the diagonal chains of smooth muscle, contraction causes the muscle to both shorten and widen.

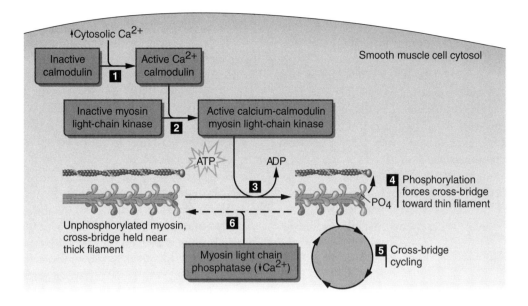

Figure 2.21 The role of calmodulin and MLC kinase in calcium handling and smooth muscle contraction.

has four main sections: the cardia, fundus, body, and antrum. Although some absorption (mainly water) can occur through the stomach, its main roles are to break down food (both chemically and physically), kill bacteria (through hydrochloric acid), and store food while allowing small amounts to pass though the pyloric sphincter and into the small intestines.

The lumen of the stomach is covered with a protective layer known as the *mucosa* (figure 2.18). The outermost layer of the mucosa consists of an epithelium lining, which contains an abundance of foveolar cells. The mucous cells are very important in the acidic environment of the stomach because they secrete mucus, which covers and protects the stomach lining from gastric acid. In addition to mucous cells, there are two other main gastric cells: parietal cells and chief cells. Parietal cells pump hydrogen cations (H^+) through an ATP H^+/K^+ transport along with chloride anions (Cl^-) to produce hydrochloric acid. Chief cells secrete pepsinogen, which is converted to the protein digestive enzyme pepsin.

When food passes through the lower esophageal sphincter into the stomach, the body of the stomach expands in response to stretch receptors. This is known as *accommodation*. While the upper portion of the stomach expands, the antrum provides forceful peristaltic contractions that churn the food in a mixture of gastric acid and enzymes known as *chyme*. A collection of videos titled "The Moving Gut" (http://humanbiology.wzw.

tum.de/index.php?id=22&L=1) shows peristalsis in the stomach and how it differs depending on the contents in the stomach. For example, the emptying of liquids from the stomach through the pyloric sphincter occurs at an exponential rate (i.e., greater volume = faster emptying).

Conversely, the stomach needs more time to adequately grind solid food into tiny (<2 mm) particles; a large, complex-nutrient meal may take hours to pass through the stomach (13). In cases where the stomach is surgically altered, as in weight loss surgery, individuals may experience dumping syndrome, which occurs when a meal concentrated with simple carbohydrate (e.g., glucose) is released too quickly into the duodenum, leading to symptoms such as weakness, nausea, and profuse sweating. This occurs because the rapid release of carbohydrate creates a hyperosmolar bolus that affects the flow of fluids across the intestinal mucosa.

The rate of stomach emptying determines how quickly fluids and glucose can enter the body. This is vital because providing sufficient water and glucose during prolonged exercise (i.e., >60 min) has been shown to improve performance and delay fatigue for both continuous (e.g., running) and intermittent (e.g., soccer) sport activities (5). Thus, when a beverage is consumed during exercise, the volume, osmolarity, and glucose concentration of the solution affect the rate of emptying. In general, greater volume, lower osmolarity, and lower glucose concentration help promote faster emptying

(17). Most studies report that consuming greater than an 8% carbohydrate solution can delay gastric emptying during exercise (5). With respect to exercise intensity, Costill and Saltin (17) reported that exercise intensities above 70% of $\dot{V}O_2$ peak slowed gastric emptying. Interestingly, a study by Rehrer et al. (51) found that gastric emptying was attenuated further during exercise if a person was also dehydrated. Translated, this means that waiting until one is dehydrated to drink fluids while exercising may cause gastric cramping because the fluid remains in the stomach longer (20).

Small Intestines

Most absorption of food and water (including reabsorption of saliva and gastric secretions) occurs through the small intestines. The small intestines are approximately 9 ft (2.7 m) long and have the surface area of a tennis court (59). Contributing to this large surface area is a thin outer layer of fingerlike projections along the lumen known as *villi* and even smaller hairlike projections along the villi known as *microvilli*. The outer surface of the intestines regenerates rapidly: A new cell layer replaces the old layer every 5 days!

The entire gut wall, including the intestines, has a layer known as the *muscularis externa* (figure 2.18), which contains two distinct layers of smooth muscle known as *circular muscle* and *longitudinal muscle*. As their names suggest, the two smooth muscle layers are arranged perpendicular to each other and thus have different actions on the GI walls. The circular muscle layer constricts the lumen, causing it to become narrower, whereas the longitudinal layer shortens segments of the GI wall. Both layers work together during peristalsis to propel contents through the gut.

In addition to peristalsis, which involves moving food in one direction, the small intestines have a pattern of contraction known as *segmentation*. Segmentation involves back-and-forth mixing of chyme, which breaks down food more and allows more transit time for absorption across the intestinal border.

However, before nutrients are absorbed, they must be further digested in the small intestines, largely with the assistance of the pancreas and liver. The pancreas releases digestive enzymes as well as the stomach acid–neutralizing compound bicarbonate, and the liver produces an essential secretion known as *bile*. Bile contains bile salts, which are important for fat digestion because they emulsify fat, allowing it to be absorbed.

With the exception of fat, which is absorbed by passive diffusion through the epithelial wall, proteins and carbohydrate have specific active transporters in the small intestines. The main transporters for carbohydrate are those for the monosaccharides glucose (i.e., **sodium-dependent glucose transporter-1**; SGLT1) and fructose (i.e., glucose transporter-5). Consuming a sport beverage that contains only glucose during exercise is thought to oversaturate the SGLT1 and thus limit total carbohydrate absorption (61). However, incorporating a combination of saccharides (e.g., fructose, glucose, and galactose) has been shown to improve the amount of carbohydrate absorbed (>50-60 g/h). For this reason, many sport drinks incorporate **multiple transportable carbohydrate** to help maintain glycogen stores and blood glucose levels during exercise.

Large Intestines

Undigested food moves from the small intestines into the cecum. While most digestion occurs in the stomach and small intestines, some digestion occurs in the colon with the help of bacteria. Segmentation, similar to contractions in the intestines, also occurs in the colon which allows the opportunity for absorption of mainly fluids and electrolytes back into the body. In fact, only about 1% (or approximately 100 mL) of fluid is lost in the feces each day (59). Before elimination happens, the undigested food must travel through the main three segments of the colon: the ascending, transverse, and descending colon.

Approximately three times per day, a unique type of peristaltic movement called *mass movements* generates the propulsion of intestinal contents. This contraction, which occurs over long distances, is stimulated by food in the stomach and is mediated by the ENS. When the intestinal contents reach the rectum, stretch receptors stimulate afferent signals to the brain, leading to the urge to defecate, which is back under voluntary muscle control through the external anal sphincters.

Increased physical activity accelerates movement of stool through the colon and has been shown to improve constipation (11,40). This has been reported in both

GI Problems in Athletes

GI complaints often are cited as a hindrance to performance in athletic events, particularly those involving long-distance endurance (21). Depending on the study, the prevalence of GI symptoms among endurance athletes ranges between 30% and 70%; more elite athletes report a higher prevalence (21). Commonly reported GI symptoms with exercise include nausea, abdominal cramps, diarrhea, and vomiting. The etiology of these GI issues is multifactorial. Nonexercise factors such as use of nonsteroidal anti-inflammatory drugs, poor timing and composition of preworkout meals, and existing GI disorders (e.g., Crohn's disease) can increase the risk of these symptoms. With respect to exercise, the main factors often cited as potential causes of GI distress are hypoperfusion, hypomotility, and mechanical stress.

Splanchnic circulation refers to circulation to and from the stomach, intestinal tract, pancreas, liver, and spleen. During exercise, norepinephrine released from sympathetic nerve endings increases vascular resistance in the splanchnic circulation, reducing blood flow proportionally to the intensity of exercise (49). During vigorous exercise this reduction has been reported to be between 60% and 80% of resting blood flow (49). Reduction in blood flow during prolonged exercise is thought to lead to GI ischemia, which in turn leads to injury of the mucosal layer of the gut (58). Damage to the mucosa breaks down the intestinal barrier and can create an inflammatory response (58). This can result in ischemic colitis, a serious GI condition resulting in blood loss in the stool (52).

Another possible contributing factor to GI symptoms with exercise is mechanical stress and strain that occurs with repetitive motions (e.g., running) or compression activities (e.g., weightlifting). It is believed that mechanical stress and strain, along with poor perfusion, is a contributing factor to why GI bleeding can occur with exercise (21). Supporting this argument are studies that have reported fewer GI complaints in cyclists compared with runners (21). However, there is still uncertainty regarding this, as a study by Pfeiffer et al. (50) found no difference in GI complaints between cyclists and marathoners.

Another GI condition found to differ in runners and cyclists is gastroesophageal reflux disease (GERD) (14). A study by Clark et al. (14) found that exercise (compared with rest) induced lower intraesophageal pH (<4.0) and was associated with greater symptoms of GERD. Further, they found that running and resistance training induced more episodes of acid reflux than did cycling. Pandolfino et al. (48) found similar results and suggested that the mechanism leading to these greater episodes of GERD was mechanical stress.

The effect of exercise on GI motility often is cited as a source of GI problems (57). Exercising at high intensities can affect the rate of gastric emptying, but other factors, such as the caloric density of a meal and osmolarity, also have an effect—probably to a greater extent. The effect of exercise on the motility of the small intestines needs further investigation. Although evidence shows that absorption of glucose is impaired during high-intensity exercise, it is unclear whether this is attributable to a reduced transit time due to altered GI motility or other factors (57). As stated earlier, evidence shows that exercise (both aerobic and resistance training) can reduce colonic transit time. However, whether this leads to GI problems such as diarrhea is yet to be elucidated.

The cause of many GI symptoms during exercise probably is multifactorial. More than likely, mild exercise has very little effect on adverse GI symptoms. High-intensity exercise, especially in the context of poor pre-event meal planning and predisposition to GI disorders, can increase the risk of problems ranging from mild diarrhea to ischemic colitis.

aerobic and resistance training studies. While the mechanism for this improvement is unknown, it is thought to be, in part, due to increased vagal tone, which enhances peristalsis. Regardless of how exercise helps with motility, several large epidemiological studies have found that exercise may also reduce the risk of colon cancer by approximately 27% (11). For individuals diagnosed with colon cancer, evidence suggests that exercise may reduce mortality (18).

Summary

- Skeletal muscle fibers (cells) are organized inside bundles of fascicles, which are contained in a tight sheath of connective tissue. Each muscle fiber is multinucleated and within it contains hundreds of threadlike proteins called *myofibrils*.
- Adult stem cells of skeletal muscle, known as *satellite cells*, contribute to muscle hypertrophy and regeneration, especially earlier in life.
- Located between Z discs on the muscle fiber, the sarcomere is the functional contractile unit of the muscle. Within the sarcomere are the two contractile proteins, actin and myosin.
- The myosin protein comprises MHCs and MLCs. Different MHC and MLC isoforms influence force and velocity of contraction to various degrees.
- The troponin complex (which is made up of the subunits C, I, and T) regulates muscle contraction along with the protein tropomyosin.
- All muscular contraction requires calcium as a second messenger. When calcium is released in the muscle cytoplasm, it binds to troponin C, creating a conformational change that allows the cross-bridge cycle to occur. This involves ATP and results when the actin filaments are pulled over the myosin filaments (i.e., muscular contraction).
- Cytosolic calcium is released by the SR and resequestered by the SERCA pump.
- The efferent nerves of skeletal muscle (motoneurons) are part of the somatic nervous system (also known as the *voluntary nervous system*). To assist in increasing the speed of transmission, motoneuron axons have a wide diameter and are covered with a myelin sheath.
- The resting membrane potential inside a motoneuron is negative. When it reaches its "all or nothing" threshold, the polarity inside (and outside) the motoneuron changes, allowing an action potential to be propagated along the axon.
- The neurotransmitter acetylcholine translocates across the synaptic cleft and binds to nicotinic receptors on the motor end plate located on the plasma membrane.
- The motor unit, which is the motoneuron and muscles it innervates, contains homogeneous muscle fibers and can be large or small depending on the muscle.
- Henneman's size principle states that smaller, less fatigable motor units are recruited first, followed by larger ones.
- In addition to muscle fiber type, other functional properties of muscle (e.g., contraction type, passive forces, and speed of contraction) determine the ability of muscle to generate force.
- Regulatory control of the GI system mainly involves the ENS and the parasympathetic nervous system. Exercise, which triggers the sympathetic nervous system, can interfere with GI functions at intensities greater than 70% peak $\dot{V}O_2$.
- Electrical waves known as *migrating motor complexes* stimulate smooth muscle contractions along the GI tract (known as *peristalsis*).
- Unlike in skeletal muscle, calcium handling in smooth muscle is regulated by calmodulin.
- Consuming glucose during prolonged exercise can prevent the onset of fatigue. However, in general, consuming beverages with lower glucose concentrations (i.e., <8%), greater volume, and lower osmolarity helps promote faster gastric emptying, thus supporting hydration and carbohydrate stores while exercising.
- Poor splanchnic blood flow, hypomotility, and mechanical stress have all been cited as potential causes of GI distress during exercise.

Definitions

actin—One of two principal contractile protein found in skeletal muscle.

alpha motoneuron—A large myelinated neuron of the somatic nervous system.

calmodulin—A calcium-binding protein that regulates muscle contraction in smooth muscles.

calsequestrin—A calcium-binding protein that increases the reuptake and storage of calcium.

glial cells—Specialized cells of the of the nervous system that structurally support and provide nutrients to the surrounding neurons.

lusitropy—The speed of myocyte relaxation.

migrating motor complexes—Slow electrical waves in the gut that are generated by the cells of Cajal and trigger peristalsis.

motor unit—An alpha motoneuron (i.e., motor neuron) and the muscles it innervates.

motor unit rotation—During a prolonged muscular contraction, some motor units will deactivate and others become active, maintaining the strength of contraction.

multiple transportable carbohydrate—A strategy for increasing carbohydrate concentration by consuming a combination of saccharides (e.g., fructose, glucose, and galactose).

muscle fiber—The multinucleated myocyte of skeletal muscle.

muscle hyperplasia—The growth of new muscle fibers.

myelin sheath—A lipid and protein substance that surrounds the axon of a neuron and speeds up transmission.

myofibrils—Long, threadlike proteins in a muscle fiber that run the length of the fiber and contain the contractile components.

myoglobin—An iron-containing protein that binds to oxygen in muscle fibers.

myosin—One of two principal contractile proteins found in skeletal muscle.

peristalsis—Rhythmic waves of smooth muscle contractions that help grind, mix, and transport food.

sarcolemma—The functional contractile unit of the muscle.

sarcoplasmic reticulum—A network of membranes within myocytes that is responsible for releasing and re-sequestering calcium.

sarcoplasmic reticulum calcium–ATPase pump—Responsible for the reuptake of calcium into the SR.

satellite cells—Skeletal muscle adult stem cells that are responsible for muscle adaptation and regeneration.

sodium-dependent glucose transporter-1—The main carbohydrate transporter specific for monosaccharides.

titin—The largest known protein; connects the Z line to the M line and provides passive elastic force.

transverse tubules—Specialized tubules that run perpendicular to the myofibril and propagate the electrical depolarization, leading to the release of calcium and subsequent muscular contraction.

troponin—A regulatory protein that, along with tropomyosin, plays an important role in the actin–myosin cross-bridge cycle

Chapter 3

Cardiovascular System: Function and Control

You're driving north on Interstate 75 through southeast Michigan and are taking Exit 59—8 Mile Road. While exiting, your cell phone rings. As you check the phone to see who is calling, you take your eyes off the road. When you look up, you see a car stalled in the middle of the exit ramp. You slam on the brakes and just miss hitting the stalled car. What might you feel in your chest and why? How are the palpitations (i.e., sensation that your heart is pounding or racing) you feel also related to the increase in heart rate and force of contraction that occur when you move from sitting and watching television to walking outside to get into your car? Which of the body's systems regulates these changes in heart rate to two totally different situations—one associated with exertion and one associated with anxiety or fear? And how would the cardiac responses differ if you were a well-trained college athlete or a patient who received a heart transplant just 8 wk ago?

Consider also the cardiovascular responses and regulatory mechanisms involved when one blushes (dilatation of cutaneous precapillary vessels), becomes lightheaded and feels like fainting upon standing (orthostatic hypotension), or notices that their resting heart rate upon waking in the morning is lower after 6 wk of regular exercise training. Advanced knowledge of cardiovascular physiology and its acute responses to exercise will benefit the person entering a career in clinical exercise physiology, human performance, coaching, or athletic training. Such knowledge is needed regardless of whether that person is prescribing exercise for a friend or conducting research aimed at better understanding the adaptations that occur when athletes push themselves to the upper limits of human performance

GENERAL ANATOMY OF THE CARDIOVASCULAR SYSTEM

The cardiovascular system serves both nutritive and nonnutritive functions. The nutritive function involves the transport of proteins, fats, carbohydrate, vitamins, and minerals. Nonnutritive functions include maintenance of a pressure in a closed system; thermoregulation; and the transport of gases, blood cell elements, and gland secretions. To accomplish this, the cardiovascular system comprises a number of components: great vessels, arterioles, venules, veins, and, of course, the heart. The latter is made up of cardiac muscle, four distinct valves, a specialized electrical conducting system, arteries, and veins; is innervated by the autonomic nervous system; and is located in a fibroserous sac called the *pericardium*. Figure 3.1 shows an overview of the cardiovascular system. The heart is really two pumps separated by a septum that prevents blood from mixing.

The adult heart weighs, on average, 325 g in men and 275 g in women. Most of this weight is derived from the cardiac muscle (myocardium) responsible for contracting and generating a pressure for blood to flow to a level that is sufficient to meet the demands of even maximal

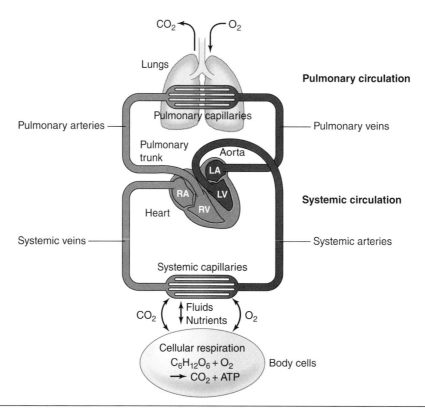

Figure 3.1 Overview of pulmonary and systemic circulation. The right side of the heart (pulmonary circulation) delivers blood to the lungs in order to eliminate carbon dioxide (CO_2) and oxygenate hemoglobin. The left side of the heart accommodates systemic circulation.

exercise. The innermost portions of the myocardium that form the four chambers of the heart are covered by a relatively thin layer of tissue, called the **endocardium**, that is in direct contact with blood. The cells that make up the endocardium share the same embryology and many of the functions ascribed to vascular endothelium, including the release of agents that can prolong or shorten myocardial contractions. The endocardium also provides protection to the thick myocardium and acts as a blood–heart barrier that regulates the nutrients and ion composition of the extracellular fluid that surrounds cardiac muscle cells, or myocytes.

The outermost layer of heart tissue, called the **epicardium** (figure 3.2), comprises mostly connective tissue and serves as a protective layer. The epicardium fuses with and can be thought of as being part of the innermost layer of the **pericardium**, a double-layered membrane or sac that surrounds and helps protect the heart from external forces. The pericardium is filled with approximately 10 to 50 mL of a clear lubricating fluid, called *pericardial fluid*, that moistens the various surfaces.

The right and left atria fill with blood at the same time to accommodate the blood returning from the systemic and pulmonic circulations, respectively. The atria initially allow approximately 80% of the blood they receive to pas-

sively pass through to their respective ventricles, then contract at the same time to fill the ventricles by about another 20%. The ventricles then contract, with the right ventricle contracting milliseconds before the left ventricle. The four heart valves are one-way valves designed to prevent retrograde blood flow and ensure that blood flows forward along the paths just described. If a valve is damaged due to disease and does not close properly (i.e., regurgitant or insufficient), retrograde blood flow can occur. Likewise, if a valve does not open properly (i.e., stenotic), the atria or ventricles must generate more tension, which leads to other abnormalities. In both instances, a valve that is not functioning properly causes noise, called a *heart murmur*, that can be heard through a stethoscope. The nature and severity of a heart murmur is evaluated based on its location, pitch, configuration, and intensity.

The right side of the heart receives blood in the thin-walled right atrium from the body via the inferior and superior vena cavas and then moves it across the open tricuspid valve into the right ventricle. From the right ventricle, blood is pumped across the pulmonic valve and through the pulmonary arteries to the lungs. Once in the lungs, CO_2 moves down its pressure gradient and across the alveoli–capillary interface and enters the air that is soon to be exhaled. At the same time, oxygen (O_2) moves down its pressure gradient from ambient or inhaled air into the blood, where it binds to hemoglobin (Hb) in the blood before it flows to the left side of the heart.

The left side of the heart receives blood returning from the lungs via the pulmonary veins into the left atrium, which pumps the blood across the open mitral valve into the left ventricle. When the left ventricle contracts, the blood is pumped under higher pressure across the aortic valve and into the aorta. This pressure is sufficient to ensure that all of the body's tissues are adequately perfused with nutrient-rich blood both at rest and during maximal exercise.

MECHANISMS OF CONTRACTION

Lying between the endocardium and epicardium layers is the myocardium, a thick layer of cardiac muscle that carries out the functions that are most often associated with the heart—contraction (or systole) and relaxation

(or diastole). Rather than being attached to bone like skeletal muscle, cardiac muscle is attached to connective tissue rings located around the four heart valves. Cardiac muscle is laid out not in a linear arrangement but rather in a circular or semispherical manner. Figure 3.2 shows a gross view of the heart as well as a microscopic view of cardiac cells, or myocytes. Note that, like skeletal muscle, cardiac myocytes are also striated in appearance due to the actin and myosin protein filaments involved with contraction. Table 3.1 compares and contrasts key anatomical and functional characteristics of cardiac and skeletal muscle. Individual cardiac myocytes are shorter than skeletal muscle cells and are anatomically connected in series (i.e., end to end) by **intercalated discs**.

These discs represent the fusion of cell membranes and provide some structural strength to hold cells together. They also allow for cell-to-cell communication, which in turn allows for the heart to work as a functional (rather than anatomical) syncytium ("together cell"). The connection between two adjacent cardiac myocytes via an intercalated disc contains within it two junctions:

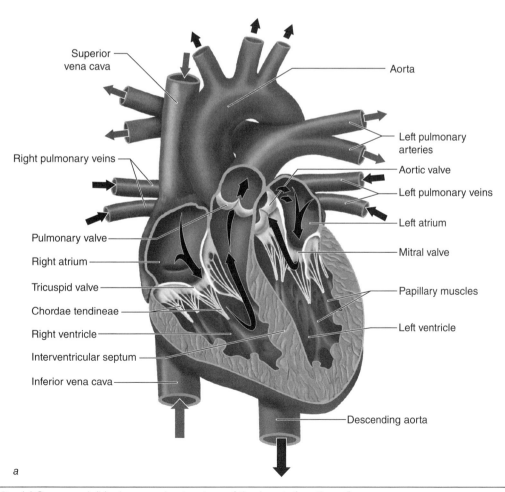

Figure 3.2 *(a)* Gross and *(b)* microscopic structure of the heart. *(continued)*

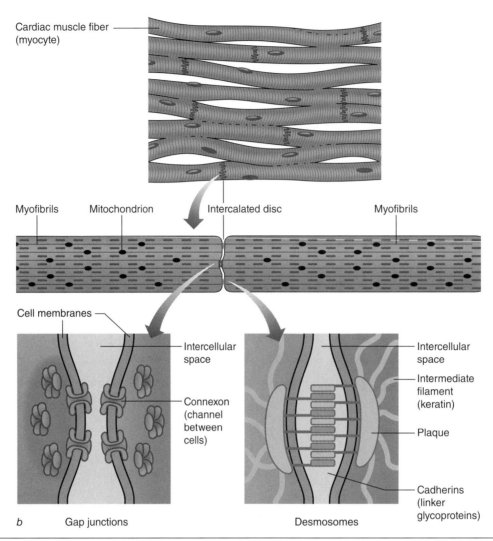

Figure 3.2 *(continued)*

the gap junction and the desmosomes (figure 3.2). The former contains protein-lined tunnels or connexons—low-resistance physical connections between two adjacent cells that allow for the transmission of ions, amino acids, sugars, and nucleotides from one cardiac myocyte to another. These connexons allow for the electrical impulse (i.e., depolarizing action potential) to flow from myocyte to myocyte so that the entire myocardium contracts in unison. Desmosomes include protein filaments arising from each respective cell to help anchor or hold adjacent cells together—a much needed characteristic when the cells are contracting.

Excitation–Contraction Coupling

Broadly, there are two kinds of cardiac cells in the myocardium. The first kind possesses contractile capabilities and is responsible for generating force. The vast majority of cardiac myocytes (atrial myocytes, ventricular myocytes) are of this type. The second type of cardiac myocyte, discussed in the next section, is a slightly modified cell that does not generate force. Instead, it initiates (pacemaker cells) or rapidly transmits the action potential (AP) or electrical impulse across the entire myocardium. The cardiac cells that do contract to generate force do so in a manner that is quite similar to skeletal muscle—that is, through a process known as *excitation–contraction coupling*. These cardiac myocytes contain cylindrical myofibrils, within which are found sarcomeres (figure 3.3). Sarcomeres represent the basic contractile or functional units in cardiac muscle. Anatomically, they are lined up end to end (in series) in the myofibril, defined as extending from one Z disc to the next Z disc. Close inspection reveals that sarcomeres contain the protein-rich myofila-

Table 3.1 Comparison of Cardiac Versus Skeletal Muscle Characteristics and Function

	Cardiac	Skeletal
Mechanics of excitation–contraction coupling and sliding filament theory; myofibrils and sarcomeres	Yes	Yes
Muscles connected to	Connective tissue around valves	Bone
Array or pattern	Circular, spherical	Linear
Length of fiber	Shorter (up to 100 cm), with branching	Larger, longer
Nuclei	Mostly single	Multiple
Automaticity	Yes (sinoatrial and atrioventricular nodes)	No
Ca^{2+} concentration	Higher extracellular and in the sarcoplasmic reticulum	Higher intracellular and in the sarcoplasmic reticulum
Sarcoplasmic reticulum and T tubules	Less dense	More dense
Intercalated discs	Yes	No
Transmission of action potential	Cell to cell via gap junctions in intercalated discs	From innervating alpha motor neuron; not transmitted from cell to cell
Action potential/contraction time	Slow/long (200-400 ms)	Fast/short (1-2 ms)
Refractory period	Long; no tetani possible	Short; tetani possible
Fiber recruitment pattern	Action potential from sinoatrial node leads to all cells contracting together.	Motor units are recruited and muscles contract based on force requirements needed.
Metabolism	Aerobic metabolic pathways (more mitochondria, density of which does not change appreciably with exercise training); high myoglobin content	Aerobic and anaerobic pathways based on energy demand; mitochondrial density may increase with exercise training
Predominant myosin heavy chain isoform	β	I, IIa, and IIx
Fatigable	No	Yes

ments myosin (thick) and actin (thin). Actin also contains two other proteins, troponin and tropomyosin. All of these proteins work in concert, along with calcium, to help regulate cardiac muscle contraction.

The AP originates from cardiac pacemaker cells and is transmitted to an adjacent cell via the gap junctions located in the intercalated discs. Once an AP reaches the sarcolemma of a cell, it is carried into the interior of the cell via invaginations of the cell membrane that are located at the Z discs, called T tubules (figure 3.4). Depolarization of the sarcolemma allows calcium ions (Ca^{2+}) to enter the cell through specialized L-type voltage-dependent channels (sometimes known as *dihydropyridine receptors,*

named for the drug that led to their discovery). The net effect is that Ca^{2+}, once in higher concentration on the outside of a cell (i.e., extracellular), moves to the interior of the cell, where it triggers or induces the release of Ca^{2+} stored in the sarcoplasmic reticulum (SR) network of the cell. Within a cardiac myocyte, the T tubules and portions of the SR are extremely close to one another. When activated by the Ca^{2+} entering the cell from the T tubules, SR releases its Ca^{2+} stores through the SR calcium release channels or ryanodine channels. The AP in cardiac myocytes is prolonged or slow (200-400 ms; see table 3.1) in order to allow Ca^{2+} to cross the T tubule membrane. Other portions of the SR are responsible for the storage

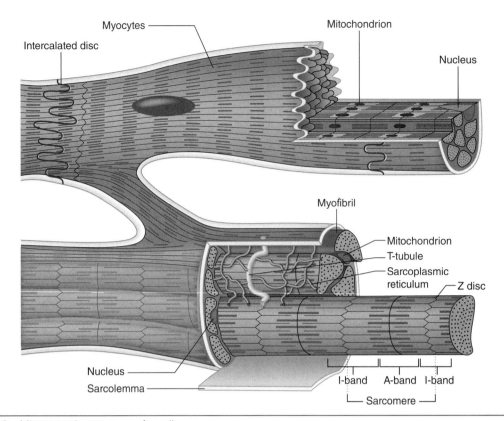

Figure 3.3 Microscopic structure of cardiac sarcomeres.

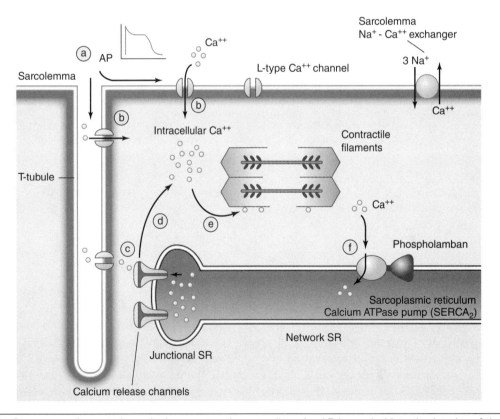

Figure 3.4 Sequence of events in excitation–contraction coupling. An AP is carried into the interior of the cell via T tubules *(a)*. Calcium ions enter the cell through L-type receptors *(b)*. Calcium binds to calcium release channels on the junctional sarcoplasmic reticulum (SR) *(c)*, causing the release of Ca^{2+} from the junctional SR *(d)*. Intracellular calcium binds to the regulatory protein troponin, leading to cross-bridge formation *(e)*. At the end of a cardiac AP, SR calcium–adenosine triphosphatase (ATPase) pumps located in the network of SR actively pump Ca^{2+} back into the SR, and the sarcoplasmic sodium–calcium exchanger ejects Ca^{2+} from the cell *(f)*.

and reuptake of Ca^{2+}, the latter of which is accommodated by the SR calcium–ATPase pumps. Note that in skeletal muscle, the mechanism of Ca^{2+} release is similar to cardiac muscle, except the L-type Ca^{2+} channels in T tubules do not actually contain or carry Ca^{2+}. However, they still trigger Ca^{2+} release from the SR—that is, skeletal muscle does not involve a Ca^{2+}-induced release of Ca^{2+}.

The Ca^{2+} released from the SR and, to a lesser extent, the Ca^{2+} that moves into the cell help initiate and regulate contraction and relaxation of the cardiac myocyte. Once in the sarcoplasm of the cell, calcium binds with the troponin C portion of the troponin complex. Other troponin subunits are troponin T, which is attached to tropomyosin, and troponin I, which in the resting state inhibits the formation of an actin–myosin cross-bridge. Once troponin C is activated by Ca^{2+}, the configuration between troponin I and troponin T is altered such that the inhibitory effect of troponin I is lost and a binding site on actin, once covered by tropomyosin, is exposed and available to bind with the head of the thicker myofilament myosin.

Force occurs each time myosin binds to actin and uses one molecule of adenosine triphosphate (ATP) before splitting. This is known as *cross-bridge cycling*. The actual mechanical action that occurs as a result of the chemical interaction between actin, myosin, and ATP leads to the thin myofilament being pulled over or sliding across myosin and toward the center of the sarcomere (sliding filament theory), a shortening of the sarcomere (Z discs are pulled toward each other and the H zone shrinks), and a generation of force. Once myosin binds to actin and a cross-bridge is formed and ATP energy is used (inorganic phosphate is released from myosin), the myosin head shifts or reconfigures. In doing so, the thin filament is pulled across the myosin. When completed, however, if Ca^{2+} is still present inside the cell, actin and myosin re-engage to again generate force. When the AP ceases and Ca^{2+} is resequestered into the SR or is expelled from the cell, tropomyosin again inhibits the binding site on actin and tension ceases, and the sarcomere returns to its original resting length.

It is important to review the role that calcium plays in regulating cardiac contraction and relaxation. First, at rest and in the absence of calcium in the sarcoplasm of a cardiac myocyte, tropomyosin blocks the actin–myosin binding site. Once calcium becomes available, the inhibitory effect of tropomyosin is lost, an actin–myosin cross-bridge is formed, and shortening and contraction of the sarcomere ensues. However, if the cell is no longer depolarized, then Ca^{2+} is either sequestered by the SR or expelled from the cell. Second, calcium also influences

the magnitude or force of contraction. Specifically, the tension developed by a sarcomere is influenced by both the length (degree of stretch on the myocyte) and the concentration of Ca^{2+} in the cell. Concerning the former, up to a certain length, the longer the length of the sarcomere (i.e., filling of the left ventricle during diastole), the greater the tension that can develop.

With respect to intracellular Ca^{2+} concentration, if Ca^{2+} remains present inside the cell, the binding of myosin and actin continues to repeat itself (i.e., cross-bridge recycling). Also, the higher the concentration of Ca^{2+} in the cell, the greater the amount of cross-bridge cycling during the time that Ca^{2+} is elevated inside the sarcoplasm. Essentially, the greater the delivery of Ca^{2+} during a single heartbeat, the more cross-bridges that cycle and the greater the force of contraction. As one might expect, different chemical agents can influence the amount of Ca^{2+} made available during Ca^{2+} release from the SR inside the cell. For example, norepinephrine—a common catecholamine that is both released during exercise at increased levels from the postganglionic sympathetic fibers innervating the heart and found at increased levels in the plasma during exercise—increases the entry of Ca^{2+} into the cell. The net effect of increased contractile force is referred to as a positive **inotropic state** for the heart. Conversely, agents that diminish Ca^{2+} available for contraction—for example, by blocking the entry of Ca^{2+} into the cell (e.g., the slow-channel calcium entry blockers diltiazem and verapamil)—can decrease the force of myocardial contraction. Such drugs are known as *negative inotropes*.

CARDIAC ELECTROPHYSIOLOGY

The previously described contraction of heart cells begins with an AP from the heart's intrinsic pacemaker tissue called the **sinoatrial (SA) node**. The bulk dispersal of the depolarization across the heart is facilitated by specialized conduction cells and pathways (e.g., His-Purkinje system), then propagated as a depolarization wave or electrical impulse from one cell to the next via the gap junctions. The net result is a uniform contraction of the myocardium. In nonpacemaker cardiac myocytes, the AP can be divided into five phases that correspond to the opening of a series of protein channels in the cell membrane (figure 3.5). At rest (Phase 4), a typical atrial or ventricular myocyte has a membrane potential such that the inside of the cell is −90 mV relative to the outside of the cell. This potential is due to the concentration of

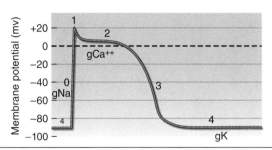

Figure 3.5 Transmembrane conductance (g) of sodium (Na⁺), potassium (K⁺), and calcium (Ca²⁺) during various phases of an AP in a ventricular myocyte.

Adapted, by permission, from P. Brubaker, L. Kaminsky and M. Whaley, 2001, *Coronary artery disease: Essentials of prevention and rehabilitation programs* (Champaign, IL: Human Kinetics), 115.

three key ions on the inside versus the outside of the cell (chemical or concentration gradients), the relative permeability of the cell membrane to these ions (i.e., electrical conductance), and the activity of electrogenic pumps (e.g., Na⁺/K⁺ ATPase). Inside the cell, K⁺ concentration is higher and Ca²⁺ and Na⁺ concentrations are much lower. In contrast, outside the cell, Na⁺ and Ca²⁺ concentrations are higher and K⁺ concentration is lower. At rest, the cell is highly permeable to K⁺, which is flowing down its chemical gradient from inside to outside the cell. However, several membrane ion pumps bring the K⁺ back into the cell and move Ca²⁺ and Na⁺ out to maintain the −90 mV membrane potential.

In Phase 0, when the AP *threshold* of approximately −70 mV is reached, the AP is rapidly initiated as Na⁺ rushes inward through fast channels (and potassium channels close) and the polarity across the cell membrane is reversed. Partial repolarization of the AP takes place during Phase 1, with loss of sodium conductance and a brief, slight outward movement of K⁺. Calcium conductance predominates during Phase 2 of the AP, as Ca²⁺ moves into the cell through the slow, L-type voltage-dependent channels, which maintain depolarization and result in a plateau. During Phase 3, rapid repolarization occurs as Ca²⁺ conductance is decreased and K⁺ conductance is increased, all of which re-establish resting membrane polarity at −90 mV (Phase 4).

In addition to cardiac myocytes that contract when stimulated by the AP of an adjacent cell, other slightly modified cardiac myocytes (pacemaker cells) can initiate an AP on their own in the SA node. The SA node is ellipsoid in shape (3 mm wide, 15 mm long, 1 mm thick) and located in the superior right atrium. If left free of neural and hormonal influences, the SA node would depolarize at a rate of approximately 90 to 100 beats · min⁻¹. If the

SA node falters for some reason, its characteristic rate of automaticity (i.e., ability to generate an AP on its own) is replaced by a slower rate from the **atrioventricular (AV) node** or Aschoff–Tawara node of the heart.

Cells of the SA node differ from myocyte contractile cells of the atria or ventricles in several ways. First, depolarization of the SA node contains only three (not five) phases: Phase 4, or the slow change in resting membrane potential to a level that triggers the AP; Phase 0, or the rapid depolarization portion of the AP; and Phase 3, or repolarization. Second, the resting membrane potential of cells in the SA node is less negative than that of other cardiac myocytes (−60 mV vs. −90 mV) and does not remain stable due to movement of Na⁺ into the cell during Phase 4. This causes the membrane potential to slowly progress to the threshold level of −35 mV, at which point an AP is generated and the cell depolarizes. The electrophysiology of the AV node is similar to that of the SA node and, as a result, capable of automaticity (rate of 35-60 beats · min⁻¹). (Note that the pacemaker cells that depolarize the most frequently are the ones that lead, so the SA node sets the heart rate.)

Several factors can modify the resting membrane potential of SA and AV nodes during Phase 4 to move them either closer to or farther away from the threshold potential of -35mV. These factors include chemicals and hormones normally found in the body (e.g., norepinephrine, acetylcholine), certain drugs (e.g., calcium channel entry blockers, digoxin), an insufficient supply of O₂ to the cells (i.e., ischemia), and even age. The factor that influences SA node function the most is the autonomic nervous system, which has two branches: the parasympathetic nervous system and the sympathetic nervous system. The vagus nerve (cranial nerve X) is part of the parasympathetic system, a branch of which influences the SA node through the release of acetylcholine. This neurotransmitter hyperpolarizes or moves resting membrane potential farther away from the threshold needed for generating an AP during Phase 4, the net effect of which is a decrease in heart rate. Conversely, the SA node (and many other portions of the heart) is also influenced by the slightly slower acting sympathetic nervous system. When activated, the cardiac postganglionic fibers release norepinephrine, an agent that improves permeability of the cell membrane to both Ca²⁺ and Na⁺. This causes the resting potential to move closer to the threshold potential needed to depolarize the cell during Phase 4; thus, heart rate is increased. At rest, the SA node is under the predominant influence of the parasympathetic nervous system (rate of 60-70 beats · min⁻¹).

CONDUCTION SYSTEM AND ELECTROCARDIOGRAM

All of the previously described electrical activity involving the heart, which triggers the mechanical contraction and relaxation phases, can be recorded on the surface of the skin using an electrocardiogram (ECG). Measuring heart rate and rhythm in athletes, apparently healthy people, and patients with known disease during different types of laboratory and field tests is important. Therefore, an understanding of the basic information associated with ECG measurement and interpretation is needed.

Before discussing in detail how changes in electrical activity can be graphically recorded as a function of time using an ECG, let's review the intrinsic electrical conduction system of the heart (figure 3.6). The electrical impulse is initiated in the SA node and then spreads throughout the atria (right and left) at a velocity of approximately 0.05 m·s⁻¹. Some of the impulse also travels along internodal tracts, which spread the impulse more quickly to multiple cells throughout the atria than if the signal had to travel from myocyte to myocyte. Once the depolarization impulse reaches the AV node, it is delayed for approximately 0.10 s before being allowed to move into the ventricles. The delay of the electrical signal at the AV node provides sufficient time for the atria to finish filling the ventricles before atrial contraction or systole is initiated (the "atrial kick"). If the AV node is not first activated by a depolarization wave generated by an SA node, cells in the AV node will depolarize on their own (a nodal rhythm).

From the AV node, the depolarization wave progresses to the proximal portion of the interventricular septum via more specialized conducting cells called the bundle of His, which transmits the signal at a velocity of approximately 2 m·s⁻¹. The bundle of His then divides into the right and left bundle branches, which conduct the electrical impulse to the right and left ventricles, respectively. The right bundle branch and left bundle branch give off many subbranches called Purkinje fibers, collectively known as the Purkinje system. These fibers rapidly (approximately 4 m·s⁻¹) carry the electrical signal throughout the entire ventricular myocardium and depolarize surrounding contraction-capable cardiac myocytes such that a synchronized contraction ensues. The Purkinje fibers also have automaticity and can depolarize at a very slow rate of 15 to 40 beats·min⁻¹.

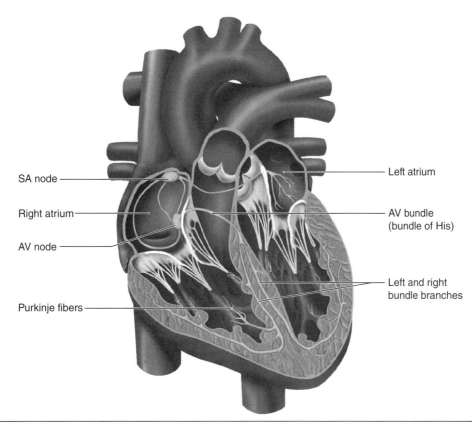

SA node —
Right atrium —
AV node —
Purkinje fibers —
Left atrium —
AV bundle (bundle of His) —
Left and right bundle branches —

Figure 3.6 The intrinsic electrical conduction system of the heart.

ECG: Recording Electrical Events in the Heart

The standard ECG is created by placing on the body surface 10 electrodes that are connected to an amplifier and a recorder called an *electrocardiograph*. The electrical activity is recorded on specially marked paper such that the horizontal direction or *x*-axis of the paper represents paper speed, which usually is set at 25 mm · s⁻¹. The paper has a grid pattern consisting of both very small boxes (1 mm × 1 mm) and larger boxes (5 mm × 5 mm). Horizontally, this arrangement allows for the conversion of distance in millimeters to time in seconds. At a paper speed of 25 mm · s⁻¹, each small box equals 0.04 s (40 ms) and each large box equals 0.2 s (200 ms). At a paper speed of 25 mm · s⁻¹, how many seconds is represented by 50 mm and 75 mm?

On the vertical direction or *y*-axis, the squares represent voltage. Specifically, 1 mV is represented by 10 small boxes (10 mm or 1 cm) of vertical deflection. An upward deflection is termed *positive,* and a downward deflection is termed *negative.* A deflection that is partly positive and partly negative is called *biphasic.* The portion of the ECG tracing that is flat (no deflection) signifies no electrical activity (i.e., no voltage current).

Using electronic switches built into the ECG machine, the electrodes can be made to act as positive electrodes or as ground electrodes. When the SA node depolarizes and initiates an AP or depolarization wave, it moves across the atria and then the ventricles as described in the previous section. If the depolarization wave is predominantly flowing toward an electrode set by the machine as positive, then a positive deflection or upward deflection is noted on the ECG. Similarly, if the electrical wave of depolarization is moving away from a positive electrode, the deflection on the ECG will be predominantly downward or negative.

Figure 3.7 provides an example of a normal ECG for two consecutive cardiac cycles or heartbeats. The following explains the key features of the ECG waveform.

- The P wave captures the time (approximately 0.06 s, or 60 ms) and voltage associated with depolarization of the atrial myocytes, which preceeds atrial contraction.

- The PR interval normally is between 0.14 and less than 0.20 s in duration. It represents both the depolarization of the atrial myocytes and the delay in the transmission of the depolarization wave as it moves from the atria and is held up for about 0.10 s before being allowed to move on to the ventricles.

The QRS complex represents the time (between 0.08 and 0.12 s) and voltage associated with depolarization of the ventricular myocytes. The Q wave (when observable) is the first negative deflection after the P wave and reflects initial activation of the ventricular septum.

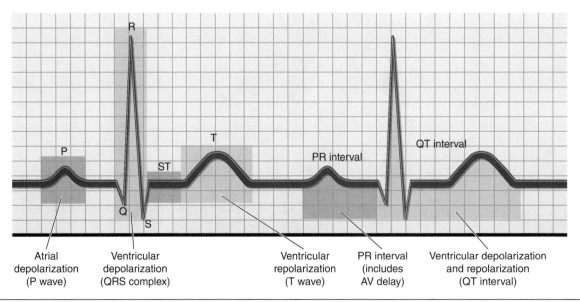

Figure 3.7 Normal resting ECG for two complete cardiac cycles.

Reprinted, by permission, W.L. Kenney, J.H. Wilmore, and D.L. Costill, 2015, *Physiology of sport and exercise,* 6th ed. (Champaign, IL: Human Kinetics), 159.

This is followed by the R wave, which is the first positive deflection and represents depolarization of the majority of the ventricle myocardium—thus, it generates the greatest voltage and is the tallest wave. The QRS complex concludes with the S wave, which reflects the last portions of the ventricles to be activated.

Other important features of the ECG waveform are the T wave (ventricular repolarization); the ST segment (measured from the end of the QRS complex to the beginning of the T wave and representing a period of time free of electrical activity); and the RR interval, which is the time it takes to complete one cardiac cycle.

The standard 12-lead ECG is collected while the subject is resting and lying supine. Although the ECG machine typically uses 10 electrodes, it provides the reader with a picture of the same electrical event viewed from 12 different angles. Imagine you are standing in front of a rectangular house that has a circumference of more than 130 ft (40 m), which is too big for you to easily observe what the house looks like on the left or right sides. The front of the home has white wood trim that appears to be freshly painted; can the same be said for the sides of the house that you can't see? Unlike what you might experience standing in front of this house, the 12-lead ECG provides different views of the heart's electrical activity simultaneously.

Exercise ECG

Consistent with the increase in heart rate that occurs during exercise, RR intervals are shortened on an ECG recorded during exercise. A rate above 100 beats · min^{-1} is called a *tachycardia*. If the source of the depolarization wave is the SA node (as it normally should be), it is referred to as **sinus tachycardia**, even if the rate approaches 200 beats · min^{-1}. (A rate at rest below 60 beats · min^{-1} that originates from the SA node is called **sinus bradycardia**.) Other normal changes observed during exercise are a shortened PR interval, a shortened QRS duration, a slightly taller P wave, a decrease in the height of the R wave, and initially an increase in the height of the T wave that gives way to a decrease in amplitude near maximum rate. The exercise-induced shortening of the PR interval and QRS duration provides a good example of a key characteristic of heart function called **dromotropicity**, which refers to the velocity of the electrical activity. An agent or factor (e.g., norepinephrine) that increases conduction velocity (i.e., slows conduction time) has a positive dromotropic effect, and an agent or intervention that slows conduction velocity has a negative dromotropic effect.

Abnormalities in the ECG waveforms at rest and during exercise can also provide invaluable information about the following:

1. Possible abnormal enlargement or hypertrophy of the muscle mass of the atria or ventricles

2. The function of the SA or AV nodes and other aspects of electrical conduction

3. Whether the rate increases in a normal fashion or suddenly speeds up or becomes irregular (called an *arrhythmia*) during exercise

4. Whether the heart is sufficiently supplied with all the O_2 it is demanding during exercise (if not, this is referred to as **myocardial ischemia**, or a temporary lack of blood flow to the heart)

5. Whether there is a portion of the heart in which the cells have died and are no longer functional (i.e., **myocardial infarction**)

CORONARY BLOOD SUPPLY

Like all human tissue and organs, the heart requires a blood supply of its own to provide it with nutrients and O_2 and remove waste products. The blood supply to the heart is referred to as the *coronary circulation* (figure 3.8). The heart is supplied by two arteries, the left main coronary artery and the right coronary artery. Each artery originates from an **ostium**, or opening in the wall of the aorta, at a point that is just superior to the aortic valve. The left main coronary artery branches into the left anterior descending artery and the circumflex artery. The former predominantly supplies the anterior portion of the left ventricle, and the latter serves the lateral wall of the left ventricle and left atrium. The right coronary artery supplies the right atrium, the inferior areas of both ventricles, and a portion of the ventricular septum. In general, coronary veins run alongside the arteries, and most of the venous blood (approximately 75%) drains into a very large vein called the *coronary sinus,* which eventually flows into the right atrium of the heart.

Because the heart is contracting constantly at rest and even more so during exercise, there is an equally constant and high demand for ATP—the energy needed to operate the actin–myosin cross-bridges. In addition, energy is needed during both the contraction and relaxation phases to maintain the energy-dependent pumps involved with ion transport across cell membranes. The ATP needed for cardiac metabolism is supplied almost entirely through aerobic metabolic pathways (because anaerobic

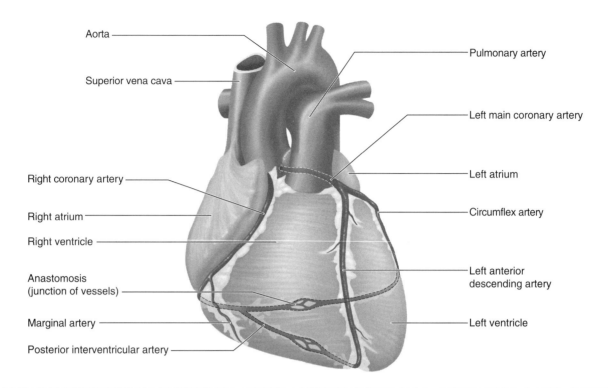

Figure 3.8 The coronary circulation. The heart is supplied by two major arteries, the left main coronary artery and the right coronary artery.

capabilities are very limited), which is why the density of mitochondria in cardiac myocytes is high. The substrates used for the heart's production of ATP are many and vary based on availability. For example, for several hours after eating a mixed-food meal the primary substrate for energy production can be as high 70% fatty acids, with the balance coming from carbohydrate in the form of glucose. Conversely, if predominantly carbohydrate is eaten, the energy produced in the cardiac myocytes predominantly comes from glucose. Other sources that the cardiac cells can use to generate ATP include lactate (especially during exercise when blood lactate levels are increased), ketones, and amino acids (16,26).

Consistent with the high demand for energy and, therefore, the high production of energy using aerobic pathways in the cardiac myocyte, myocardial O_2 consumption (even at rest) is relatively high. At rest, cardiac cells extract approximately 75% of the O_2 available in the arterial blood that is supplied by the coronary arteries. In contrast, at rest, only approximately 25% to 30% of the available O_2 is extracted from the blood that perfuses the peripheral organs and skeletal muscles. Therefore, as a result, when cardiac work increases, such as during exercise, the demand for increased O_2 (and ATP) is predominantly met by increasing the volume of blood that flows to the cells per minute (vs. extracting more O_2 from the blood). For example, at rest flow through a typical coronary artery is approximately 250 mL · min⁻¹, but during exercise coronary flow is increased threefold or more to 750 mL · min⁻¹ (24,39).

The main mechanisms responsible for the improved flow are locally mediated responses that override the vasoconstrictive effect of increased sympathetic stimulation. Specifically, the diameter of the coronary arterioles is dynamic and as a result, these vessels play an important role in determining blood flow when acted upon by increased sympathetic stimulation (i.e., norepinephrine causes vasoconstriction). However, during exercise this process is overridden by the influence of reduced O_2 content in venous blood and locally released chemicals (e.g., adenosine) on vascular function (9,24,39). The latter are released at greater levels during exercise due to the increase in metabolism that supports faster and more forceful contracting cardiac cells. These local agents cause the smooth muscles in the coronary arterioles to relax (i.e., less constriction), thus improving blood flow. In addition, the increased shear stress on the vascular endothelium of the arterioles that occurs due to increased blood flow results in the release of agents called *autacoids* (e.g., nitric oxide), which also leads to less vasoconstriction of the surrounding vascular smooth muscles (8).

When the heart contracts (systole), it produces pressure on the blood in the arterial system that is maintained even between beats when the heart is relaxing (diastole). Therefore, during diastole, pressure does not fall to zero but instead is maintained at lesser levels because elastic recoil in the blood vessels maintains blood flow. In coronary arteries, blood flow is actually halted (and possibly reversed) during systole due to the high pressures generated inside the heart and the associated effect of compressing or collapsing surrounding vessels. Therefore, blood flow through coronary arteries predominantly occurs during diastole or between contractions, when compression or pressure on the coronary arteries is much lower. Increases in heart rate (e.g., increases that occur with exercise) shorten the period of time devoted to diastole during a normal cardiac cycle and therefore reduce the period of time for blood to flow in the coronary arteries. In the healthy heart, this reduction in coronary flow poses no problem for the delivery of O_2 and nutrients. However, in patients whose arteries are already narrowed due to coronary heart disease, the decrease in coronary filling that accompanies the increase in heart rate can cause myocardial ischemia, which typically is associated with the symptom of angina pectoris or chest discomfort during exertion. With regular exercise training, heart rate at rest and during submaximal exercise is reduced—a nice training effect that also helps with the maintenance of adequate coronary blood flow.

CARDIAC CYCLE AND MECHANICS

In a rhythmic and alternating manner, the heart contracts and relaxes hundreds of time each hour and thousands of time each day—all due to the changes in electrical activity described previously. During each cycle of systole and diastole, corresponding changes occur in the pressures and volumes of blood throughout the heart. Table 3.2 and figure 3.9 describe and depict these changes for the left ventricle during one cardiac cycle. Each cardiac cycle has six distinct phases. The isovolumic phase (Phase 1) is sometimes called the *isometric phase*; however, technically this is not correct because although some cardiac sarcomeres do develop tension without shortening (isometric), others are lengthening and still others are shortening (isotonic). Also, the ejection phase (Phase 2) is actually split into two parts, the rapid (early) and slow (late) portions, which correspond to more rapid and slower changes in ventricular volume.

At the bottom of figure 3.9, an ECG conveys the relationship between electrical activity and mechanical event. The P wave is associated with atrial depolarization and preceeds the contraction of the atria (Phase 6), which

Table 3.2   Summary of Events During the Cardiac Cycle

Line number	Phase of cardiac cycle	Pressure/volume changes
1	Beginning of ventricular systole, isovolumic ventricular systole	Ventricular pressure greater than atrial pressure; ventricular pressure increases rapidly; little change in aortic and atrial pressures
2	Ventricular ejection (rapid and slow phases)	Ventricular pressure greater than aortic pressure; ventricular and aortic pressures reach peak, then begin to decrease; rapid and then slower decrease in ventricular volume; aortic valve closes
3	Beginning of diastole and isovolumic relaxation	Ventricular pressure less than aortic pressure; ventricular pressure decreases rapidly; little change in aortic and atrial pressures
4	Diastole	Ventricular pressure lower than atrial pressure; little change in ventricular and atrial pressures; aortic pressure decreases gradually; rapid and then slower increases in ventricular volume
5	Diastasis	Rest
6	Atrial contraction	Small increases in ventricular pressure with no change in aortic pressure; small increase in ventricular volume; mitral valve closes

Changes in contractile force cause changes in intracardiac pressures that result in the opening and closing of the cardiac valves.

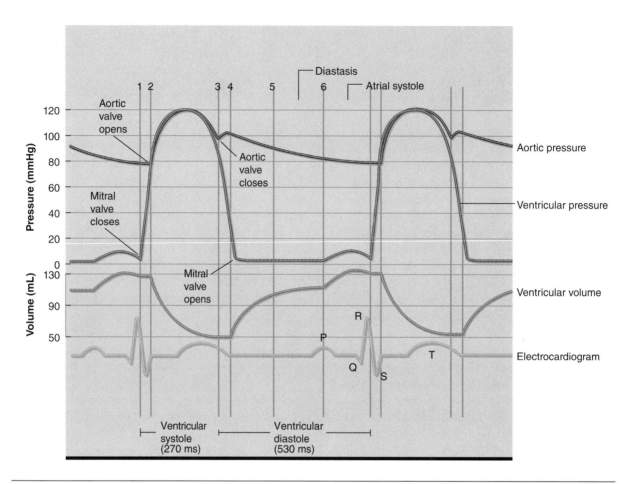

Figure 3.9 Electrical and mechanical changes in the left heart during the cardiac cycle. The upper two curves depict pressure changes in the aorta and the left ventricle, respectively. The middle curve shows changes in the volume of blood in the left ventricle. The bottom curve represents the ECG recording. The numbered vertical lines correspond to important events of the cycle, which are detailed in table 3.2.

Adapted, by permission, W.L. Kenney, J.H. Wilmore, and D.L. Costill, 2015, *Physiology of sport and exercise,* 6th ed. (Champaign, IL: Human Kinetics), 161.

tops off ventricular volume by providing the last 15% to 20% of its final volume. This atrial kick that occurs during late ventricular diastole is of lesser importance at rest but plays a bigger role when heart rate is increased during exercise. Similarly, the QRS (depolarization of the ventricular cardiac myocytes) complex preceeds and coincides with Phase 1, ventricular systole. At a rate of 75 beats · min[-1], systole (Phases 1 and 2) composes about one third (270 ms) of the total cardiac cycle, whereas the majority of time during the cardiac cycle is spent in diastole (530 ms). During exercise, at a rate of 150 beats · min[-1], systole is modestly shortened to approximately 210 ms and diastolic time is reduced by more than 50% to approximately 190 ms.

Keep in mind that the cardiac cycle for the left heart (ventricle) is very similar to that of the right heart (ventricle) but is not identical. One big difference pertains to

ventricular systolic and diastolic pressures, which are two to four times higher in the left ventricle (110/8 mmHg) than in the right ventricle (25/4 mmHg).

CARDIAC PERFORMANCE

The two major determinants of cardiac performance are **stroke volume** and heart rate. Given the dynamic state of the functioning heart, both must act in concert to optimize cardiac performance. Stroke volume is the amount (milliliters) of blood pumped per beat of the heart, usually by the left ventricle and in the upright position. As shown in table 3.3 and figure 3.10, left ventricular stroke volume represents the difference between how much blood is in the left ventricle at the end of filling (end-diastolic volume, Phases 4 and 5) and how much blood remains in the left ventricle at the end of systole (end-systolic

Table 3.3 Typical Left Ventricular Characteristics in the Normal, Increased Preload, and Increased Inotropicity States

State	End-diastolic volume (mL)	End-systolic volume (mL)	Stroke volume (mL)	Ejection fraction (%)
Normal	120	55	65	55
Increased preload	140	55	85	61
Increased inotropicity	120	35	85	71

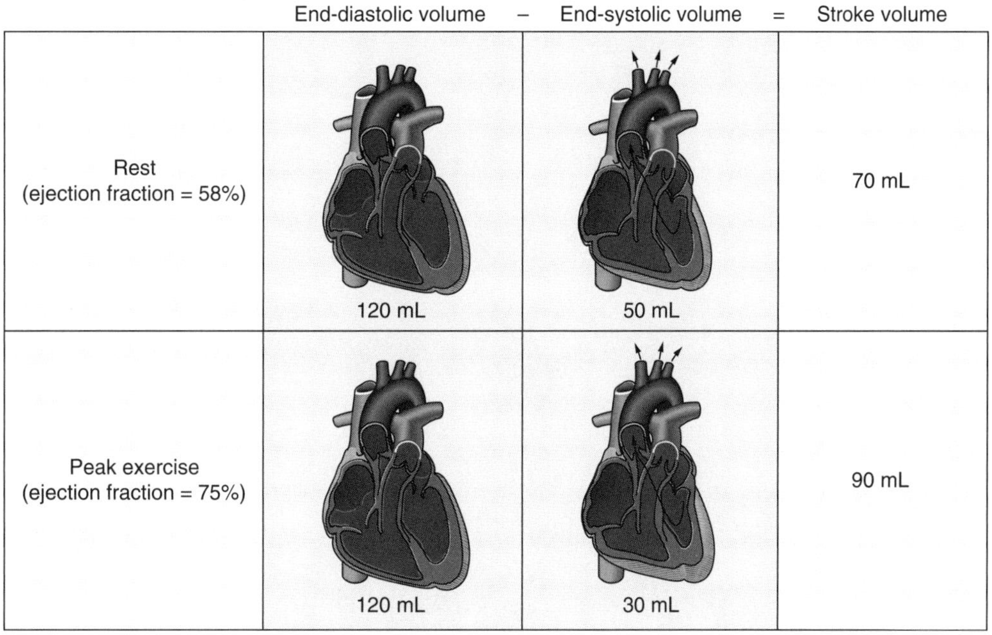

End-diastolic volume — End-systolic volume = Stroke volume

Rest (ejection fraction = 58%) 120 mL 50 mL 70 mL

Peak exercise (ejection fraction = 75%) 120 mL 30 mL 90 mL

Figure 3.10 Left ventricular end-diastolic volume, end-systolic volume, stroke volume, and ejection fraction at rest and during exercise. Note that stroke volume increases in this example due to better emptying of the left ventricle at the end of systole (end-diastolic volume unchanged). This is attributable to increased myocardial contractility (positive inotropic effect) with exercise.

Reprinted from M.L. Foss, S.J. Keteyian, and E.L. Fox, 1998, *Fox's physiological basis for exercise and sport,* 6th ed. (Pittsburgh, PA: William C. Brown). By permission of the authors.

volume). At rest, stroke volume in untrained men and women ranges between 70 and 90 mL · beat⁻¹ and 50 and 70 mL · beat⁻¹, respectively (3,10,18,43).

Figure 3.11 depicts the combination of stroke volume and what we previously discussed about the cardiac cycle. Ventricle volume (*x*-axis) is plotted against ventricular pressure (*y*-axis). Once the left ventricle is filled, isovolumic contraction follows and pressure increases to a level that is sufficient to exceed pressure in the aorta. The aortic valve opens, ejecting blood into the aorta.

Given end-diastolic and end-systolic volumes, we can compute **ejection fraction**—the percentage of end-diastolic blood pumped with each contraction or systole (figure 3.10; table 3.3)—as follows:

$$\text{Ejection fraction } (\%) = \frac{\text{end-diastolic volume} - \text{end-systolic volume}}{\text{end-diastolic volume}} \times 100$$

$$= \frac{\text{stroke volume}}{\text{end-diastolic volume}} \times 100$$

(Equation 3.1)

The instrument commonly used to measure end-diastolic and end-systolic volumes, stroke volume, and ejection fraction is called an *echocardiograph*. It is important to note that the echocardiogram ordinarily is recorded with the subject in the supine position, which induces a higher end-diastolic volume and, therefore,

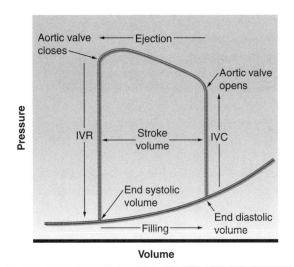

Figure 3.11 Normal pressure–volume loop during different phases of the cardiac cycle. IVC = isovolumic contraction; IVR = isovolumic relaxation.

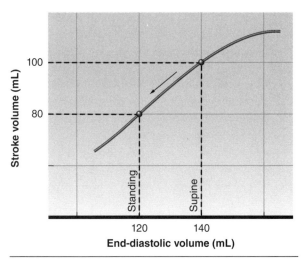

Figure 3.12 The Frank-Starling mechanism showing that a change in left ventricular end-diastolic volume results in a change in stroke volume, such as that which occurs when moving from the supine position to the standing position.

Reprinted from M.L. Foss, S.J. Keteyian, and E.L. Fox, 1998, *Fox's physiological basis for exercise and sport,* 6th ed. (Pittsburgh, PA: William C. Brown). By permission of the authors.

a greater stroke volume. The greater stroke volume is attributable to blood that was once pooled or stored in the veins of the lower extremities and that entered circulation when the person moved from an upright position to a supine position. This increase in total blood volume leads to an increase in end-diastolic volume and a subsequent increase in stroke volume. Echocardiograms can be performed with the subject in an upright or semirecumbent position or even during exercise on a stationary cycle, but technically the procedure is more difficult to undertake.

The four factors that affect stroke volume are preload, inotropic state, hypertrophy, and afterload. The first two are briefly discussed here. Hypertrophy of the cardiac muscle, which can result from months to years of exercise training, is discussed in chapter 6. **Afterload** is the load that opposes or is applied against the ventricle after the muscle begins to contract. For the left ventricle, afterload is a function of aortic blood pressure, compliance and resistance of the aorta, and size of the left ventricle. It is not discussed here because its influence on stroke volume during exercise is minimal (20,36).

Otto Frank (1895) and Frank Starling (1915) showed that a change in the length (stretch or preload) of the cardiac muscle fiber influences cardiac performance (stroke work, or the product of mean ejection pressure and stroke volume). This is known as the *Frank-Starling law of the heart.* Note that stroke work is used in this law; however, we take a more simplified approach and use stroke volume. Figure 3.12 shows that as end-diastolic volume is changed by moving up or down the ventricular function curve, the stretch or **preload** (length–tension relationship) on the cardiac muscle fibers changes in a manner that leads to a commensurate change in stroke volume. Something as

simple as changing from the supine position to the standing position influences preload. In this example, end-diastolic volume decreases (less stretch on the sarcomeres) because several hundred milliliters of blood pools in the venules (the capacitance vessels of the venous system) of the legs, which means that less total blood is in circulation to preload the heart. Table 3.3 provides an example of increased preload, evidenced by greater end-diastolic volume when compared to normal (140 mL · beat^{-1} vs. 120 mL · beat^{-1}, respectively), with no change in end-systolic volume due to increased preload. In this instance, the cardiac sarcomeres are stretched farther and yield a greater stroke volume with no meaningful change in ejection fraction. Other factors can influence preload, such as a change in total blood volume (e.g., increased with training and pregnancy) and changes in intrathoracic pressure (e.g., those that occur during respiration). Because the heart can pump only what is returned to it, venous return (and its maintenance during exercise) is an important factor when discussing preload. Issues pertinent to preload during exercise are discussed later in this chapter.

The second factor that influences stroke volume is inotropic or contractile state. By definition, *inotropic* indicates a shift in the Frank-Starling curve upward and to the left (positive inotropicity) or down and to the right (negative inotropicity). A change in inotropic state means a greater (positive) or lesser (negative) force of contraction and therefore a greater or lesser stroke volume for any given end-diastolic volume.

Ejection fraction is used as a measure of inotropic state and normally exceeds 55%. Table 3.3 provides an example of increased inotropic state (ejection fraction increased from 55% to 71%), evidenced by no change in end-diastolic volume and a greater emptying (lowering) of end-systolic volume. Such an example typifies what might occur with the administration of a drug such as norepinephrine —an agent known to have a positive inotropic effect.

The second major determinant of cardiac performance is heart rate or **chronotropicity**. At rest, heart rate is predominantly regulated by the parasympathetic branch of the autonomic nervous system—specifically, the vagus nerve and its neurotransmitter, acetylcholine. Withdrawal or inhibition of vagal nerve input to the heart results in an increase in heart rate at rest. Any agent (e.g., norepinephrine) that increases heart rate is said to have a positive chronotropic effect, whereas any factor that lowers heart rate, such as a regular endurance-type exercise training regimen, is said to have a negative chronotropic effect.

Heart rate is almost always the same value as pulse rate, which is the frequency of pulses (due to heart contractions) that we are able to palpate (feel) in any artery (i.e., radial pulse in the wrist, carotid pulse in the neck, and apical pulse on the chest wall). However, in some people, heart rate on the ECG and pulse rate palpated at an artery are not always exactly the same value. For some people with a very fast heart rate or an irregular heart rhythm, not all contractions of the heart are forceful enough that an equal number of pulse beats are palpated in a peripheral artery.

CHANGES IN HEART RATE, STROKE VOLUME, AND CARDIAC OUTPUT DURING AN ACUTE BOUT OF EXERCISE

Cardiac output is the volume of blood pumped by the ventricles (left and right ventricles should be just about equal) and represents the product of heart rate and stroke volume.

$$\text{Cardiac output } (L \cdot min^{-1}) = \text{heart rate } (beats \cdot min^{-1}) \times$$
$$\text{stroke volume } (mL \cdot beat^{-1})$$
$$\text{(Equation 3.2)}$$

At rest and independent of level of exercise training, cardiac output is approximately $5 \ L \cdot min^{-1}$ (heart rate of 70 beats $\cdot min^{-1} \times$ stroke volume of 70 mL $\cdot beat^{-1}$). Twenty percent of this forward flow is diverted to the skeletal muscle, and the balance nourishes the visceral organs

(liver, spleen, kidney, and intestines; approximately 45%), heart (5%), brain (13%), and other organs. During exercise, cardiac output can increase up to 20 to 25 L $\cdot min^{-1}$ in healthy, untrained college-age adults and can exceed 40 L $\cdot$ min^{-1} in elite male endurance-trained athletes (2,3,11,42,46).

The increase in cardiac output during exercise closely parallels the increase in external work rate or effort intensity up to maximum (i.e., peak). Peak cardiac output $(\dot{Q})$ for a 20-yr-old untrained person with a peak heart rate (HR) of 190 beats $\cdot min^{-1}$ and a peak stroke volume (SV) of 150 mL $\cdot beat^{-1}$ would be computed as follows.

$$Q = HR \times SV$$
$$= 190 \times 150$$
$$= 28.5 \ L \cdot min^{-1} \qquad \text{(Equation 3.3)}$$

The change in heart rate during exercise is also linear and increases as work rate or intensity of effort increase. This differs slightly from stroke volume, where an increase is observed in the transition from rest to moderate work, after which it may stay the same (plateau) or increase slightly. The increase in cardiac output from the onset of exercise through moderate work rate is attributable to increases in both heart rate and stroke volume. Then, the further increase in cardiac output that occurs from moderate work up to peak is predominantly due to the increase in heart rate alone. Figure 3.13 summarizes the changes in cardiac output, heart rate, and stroke volume that occur during exercise.

Broadly, the changes in cardiac output are similar in men and women throughout an acute bout of exercise up to maximum. If exercising at the same oxygen uptake $(\dot{V}O_2)$, cardiac output is similar or slightly higher in women compared with in men. Any difference may be attributable to the lower O_2 carrying capacity of blood in women, which is secondary to the lower level of Hb in women (2,13). Also, peak cardiac output at maximal exercise is uniformly lower in women than in men due to the smaller volume of their left ventricle.

Stroke Volume

After an initial increase, stroke volume (SV = Q/HR) typically plateaus during exercise. This usually occurs at a level approximating 40% to 60% of peak $\dot{V}O_2$. To some extent this plateau is observed during both arm and leg exercise and in both men and women regardless of whether they are trained or untrained (3,5,23,38,43). Peak stroke volume can exceed 200 mL $\cdot beat^{-1}$ in endurance-trained males (11); this is substantially more than the peak volumes of 120 to 140 mL $\cdot beat^{-1}$ observed in elite trained women. Similarly, when exercising at the same

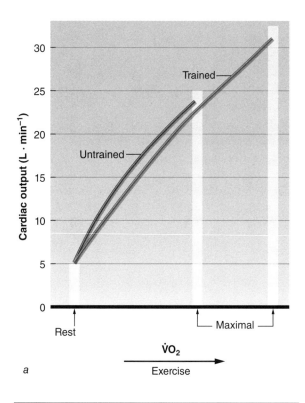

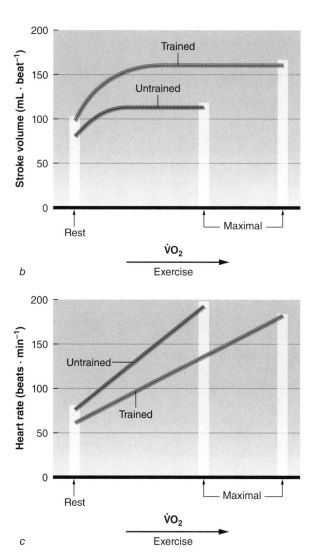

Figure 3.13 (a) Cardiac output, (b) stroke volume, and (c) heart rate during exercise in trained and untrained subjects. Cardiac output and heart rate are closely related to $\dot{V}O_2$ over the entire range from rest to maximal exercise. Maximal stroke volume is usually reached at a submaximal exercise $\dot{V}O_2$.

Reprinted from M.L. Foss, S.J. Keteyian, and E.L. Fox, 1998, *Fox's physiological basis for exercise and sport,* 6th ed. (Pittsburgh, PA: William C. Brown). By permission of the authors.

submaximal work rate, stroke volume is lower in women than in men (46). In both instances, the smaller stroke volume in women is attributable to their smaller left ventricular volume. The two main reasons for the increase in stroke volume during exercise can be tied back to two of the four determinants of stroke volume described earlier: preload and inotropic state.

Preload

The response of left ventricular volume during exercise appears to vary based on the training state of the individual. Several studies suggest that left ventricular end-diastolic volume is essentially unchanged during exercise (17,41). Thus, the Frank-Starling mechanism (i.e., preload) would play a limited role relative to increasing stroke volume. However, prior work involving well-trained cyclists and swimmers showed that left ventricular end-diastolic volume increased from rest to submaximal

exercise in athletes compared with age-matched controls (38) (figure 3.14). These findings are consistent with the work of others (47) who investigated older individuals, suggesting that among untrained younger persons the Frank-Starling mechanism acts, at minimum, to maintain end-diastolic volume at pre-exercise levels.

The left side of the heart can pump per beat (stroke volume) or per minute (cardiac output) only an amount that is equal to the amount of blood that is returned to the right heart via the venous system, called *venous return*. Therefore, a cardiac output of 20, 25, or 30 L · min⁻¹ during exercise means that venous return must be increased to the same amount. The mechanisms responsible for making this happen are threefold and include the skeletal muscle pump, the respiratory and abdominal pumps, and venoconstriction.

Often called the "second heart," the skeletal muscle pump results from the mechanical action produced by the

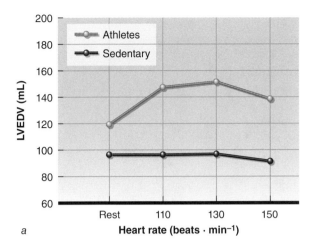

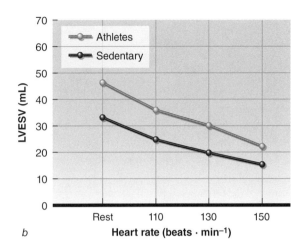

Figure 3.14 *(a)* Changes in left ventricular (LV) end-diastolic volume (EDV), *(b)* end-systolic volume (ESV), and *(c)* stroke volume (SV) during submaximal exercise in endurance-trained swimmers and cyclists versus sedentary controls.

Reprinted from M.L. Foss, S.J. Keteyian, and E.L. Fox, 1998, *Fox's physiological basis for exercise and sport,* 6th ed. (Pittsburgh, PA: William C. Brown). By permission of the authors.

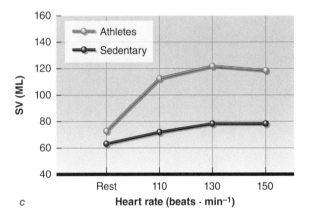

rhythmical contraction of skeletal muscles to compress surrounding veins and drive and propel blood within the veins toward the heart. Blood is prevented from flowing backward (i.e., retrograde flow) in the veins because they contain valves. When the muscles relax, blood again fills the surrounding veins until the next muscle contraction, which then again compresses the veins and moves blood toward the heart. We cannot emphasize enough that the muscle pump is a major factor contributing to venous return and, therefore, the maintenance of preload during exercise. Venous return must equal cardiac output because if it does not, cardiac output would not sufficiently increase during exercise. This description of the skeletal muscle pump pertains only to dynamic activities such as walking, running, and other similar exercise that is rhythmic in nature. During weightlifting or other activities associated with static or sustained muscular contraction that compress and occlude the surrounding veins, venous return can be hampered due to the absence of a true muscle pump.

The respiratory and abdominal pumps are two other mechanical pumps that facilitate venous return during exercise and at rest. With the respiratory pump, the veins of the thorax are emptied during inhalation because intrathoracic pressure becomes more subatmospheric during that phase of respiration. This serves to widen the pressure gradient between the right atrium and the

location where the inferior vena cava enters the thorax through the diaphragm. The net effect is an acceleration of blood in the vena cava during inspiration and movement of blood toward the right atrium.

The flattening of the diaphragm during inspiration also compresses the abdomen, increasing intra-abdominal pressure. As a result, blood in the veins in the abdomen empties toward the heart (i.e., abdominal pump). The reverse occurs during exhalation, which allows veins in the thorax and abdomen to fill. Thus, the simple act of breathing enhances venous return, which operates at an even greater level during exercise when the depth and frequency of breathing are increased.

Like arterioles in the arterial system, the walls of the venules in the venous system contain smooth muscles that are capable of contracting and causing constriction (i.e., venoconstriction). Venules are capacitance vessels capable of storing blood. However, when constricted, the volume of blood in the venules is reduced and more blood is entered into circulation and available to preload the heart. Like other aspects of the circulatory system, the venules are influenced by the sympathetic nervous system and its neurotransmitter norepinephrine, which leads to more venoconstriction when released during exercise.

Inotropic State

Normally, ejection fraction is 55% to 60% at rest. During exercise, the percentage of blood pumped from the left ventricle per beat can increase to 75% or more. This increase in ejection fraction during exercise reflects a positive inotropic effect, consistent with an almost doubling of stroke volume and a decrease in end-systolic volume. All of this is mediated through the increase in activity of the sympathetic nervous system, which leads to a greater amount of calcium entering the myocyte and a more forceful contraction. During exercise, the net effect associated with improved inotropicity and the maintenance of preload due to increased venous return is an increase in stroke volume.

Heart Rate

A host of factors and agents influence chronotropicity, including respiration (inspiration, exhalation), hormones, body temperature and fever, excitement or fear, breath holding or diving underwater, blood pressure, and—yes—exercise. In both men and women and trained and untrained people, heart rate increases in a generally linear manner as one progresses from rest to maximum effort. Like stroke volume, the changes in heart rate during exercise are greatly influenced or regulated by sympathetic and parasympathetic activity. From rest up to a heart rate of approximately 100 beats · min⁻¹ (which is the intrinsic rate of the SA node), the increase is predominantly attributable to parasympathetic withdrawal (i.e., less stimulation of the vagus nerve).

During exercise at a fixed submaximal work rate, men tend to have lower heart rates than women, which is also the case for trained versus untrained individuals (46). In fact, a lowering of one's heart rate during exercise is a classic sign of improved physical fitness or conditioning. Peak heart rate can approach or exceed 200 beats · min⁻¹ in teenagers and young adults and can be broadly estimated using various formulae. A common formula for estimating peak rate is 220 − age, but this approach may over- or underestimate measured maximal heart rate by up to 11 beats · min⁻¹ (4). Another formula for estimating maximum heart rate is 207 − 0.7 × age (15).

After exercise ceases, heart rate should decrease in a two-phase manner. Within the first 2 min a marked decrease in rate should occur due to a marked increase in parasympathetic or vagal stimulation. Heart rate should decrease by 12 or more beats within 1 min of recovery or by 22 beats or more by 2 min of recovery. Smaller decreases during early recovery may suggest an imbalance between the parasympathetic and sympathetic branches of the autonomic nervous system and are associated with increased risk for mortality in the future (6).

Stroke volume is difficult to measure during exercise without sophisticated equipment such as an echocardiograph machine. Conversely, heart rate during exercise can be easily recorded in the laboratory, field, or clinical settings using a heart rate monitor or ECG machine, and pulse rate can be measured via palpation of the radial or carotid arteries. Given the direct and close relationship between heart rate and work rate or $\dot{V}O_2$, heart rate is an excellent, simple-to-measure indicator of exercise intensity. In fact, the correct training intensity can be easily guided in athletes, patients, and apparently healthy people using a heart rate–based method (covered in chapters 6 and 7).

CIRCULATORY HEMODYNAMICS AND DISTRIBUTION OF BLOOD FLOW

The study of the factors that govern the flow of fluids is called *hemodynamics*. Two factors that influence the flow of blood in the circulatory system are blood pressure and resistance. The relationship between blood flow (cardiac output), blood pressure, and resistance is captured in the following two equations, the latter of which rearranges the first in order for you to better appreciate the factors that contribute to the maintenance of blood pressure both at rest and during exercise. In the second equation, mean arterial pressure (MAP) is influenced by an increase or decrease in cardiac output or an increase or decrease in resistance.

$$\text{Cardiac output} \left(\text{L} \cdot \text{min}^{-1} \right) = \frac{\text{MAP - right atrial pressure}}{T_S P_R}$$

(Equation 3.4)

MAP (mmHg) = $\dot{Q} \times T_S P_R$,

where $\dot{Q}$ = cardiac output, MAP = the difference between pressure in the systemic circulatory system during a complete cardiac cycle and pressure in the right atrium (because right atrial pressure usually is very close to zero [2-8 mmHg], MAP alone can be used in the equations), and

$T_S P_R$ = total systemic peripheral resistance.

Blood pressure is the pressure that the blood exerts against the inside of the arterial walls. It is also the force that moves blood through the cardiovascular system. Blood pressure is higher in the feet than in the head. To standardize any discussion about blood pressure, typically it is measured with the arm at heart level when seated.

Key to the understanding of blood flow is an appreciation for the fact that fluid flows along pressure gradients from an area of higher pressure to one of lower pressure. As shown in figure 3.15, systemic pressure is highest in the left ventricle of the heart during systole (e.g., 100-140 mmHg) and lowest in the right atrium. Therefore, blood flow follows the same path, exiting the left ventricle into the aorta during systole and then arriving in the right atrium. Although pressures are lower in pulmonary circulation, the flow of blood in the pulmonary system keeps pace with that of the systemic circulation because resistance is lower in the pulmonary circuit.

Systolic, Diastolic, and Mean Arterial Pressures

Pressure in the arterial system is highest during systole. As shown in figure 3.15, as blood is distributed during diastole, pressure decreases to its lowest levels (approximately 80 mmHg). During the cardiac cycle, blood enters the aorta in a pulsatile manner—all of stroke volume

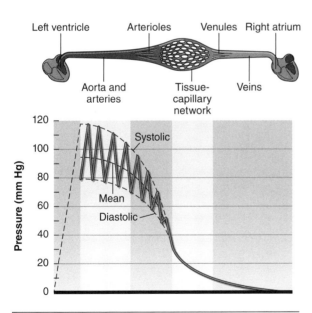

Figure 3.15 Blood pressure differential along the systemic vascular tree. Blood always flows from an area of higher pressure to one of lower pressure. Note that the pressure (and thus the flow of blood) fluctuates in the arteries and arterioles but is steady in the capillaries.

Reprinted from M.L. Foss, S.J. Keteyian, and E.L. Fox, 1998, *Fox's physiological basis for exercise and sport,* 6th ed. (Pittsburgh, PA: William C. Brown). By permission of the authors.

enters during systole and none enters during diastole. During systole, although some of the forward flow into the aorta supports forward flow into the capillaries, the vast majority of the ejected blood distends the walls of the larger arteries, which have an elastic property due to the protein elastin (29). During ventricular diastole, the elastic recoil of the arterial walls maintains forward flow into the capillaries throughout the remainder of the cardiac cycle. This distribution of blood through the capillaries throughout the cardiac cycle is important because the diffusion of gases and transfer of nutrients occurs in the capillaries.

MAP is used in our prior equations, and it is the pressure that best represents the driving pressure that influences blood flow perfusing the tissues at any given moment. At rest, MAP is usually between 70 and 110 mmHg, and a value above 60 mmHg is considered sufficient to perfuse organs and tissues in a manner that maintains normal function. During the cardiac cycle diastole lasts longer than systole, so estimated MAP is not simply the average of systolic and diastolic blood pressures. Instead, resting MAP can be fairly closely approximated using the following equation, illustrated here using a person with a resting blood pressure of 120/80 mmHg. In the equation, note that diastolic pressure is weighted twice that of systolic pressure because about two thirds of the cardiac cycle is spent in diastole.

$$
\begin{aligned}
\text{MAP} &= \frac{(2 \times \text{diastolic blood pressure}) + \text{systolic blood pressure}}{3} \\
&= \frac{160 + 120}{3} \\
&= 93.3 \, \text{mmHg} \quad\quad\quad \text{(Equation 3.5)}
\end{aligned}
$$

The second factor that greatly influences the flow of fluids is resistance, which in humans is referred to as *total systemic peripheral resistance* (T_sP_R). As the name implies, T_sP_R reflects the sum (total) of resistance imposed by the peripheral vasculature. It is directly proportional to both the viscosity of the blood and vessel length and is inversely proportional to the radius of the vessel to the fourth power.

Blood viscosity or thickness, such as an increase in the number of red blood cells (polycythemia) that might occur in patients with emphysema, causes greater friction in the fluid layers within the vessel and between the fluid and the vessel walls, causing greater resistance to flow. Note the statement that the fluid (i.e., blood) in the vascular system flows in layers. Briefly, there are two types of blood flow: laminar and turbulent. One can think of laminar flow (also called *streamlined flow*) as different layers flowing in the same direction, one on top of the others. The layers toward the center of the vessel flow faster, whereas the layers closer to and in contact with the arterial wall flow slower. Turbulent flow is a swirling of the blood and a mixing of the aforementioned layers. It is influenced by the viscosity and density of the blood, rate of blood flow (more likely to occur during high blood flow states, such as exercise), size of the blood vessel (more likely to occur in larger vessels such as the aorta), and length of the blood vessel (the longer the blood vessel, the greater the surface area that is in contact with the blood and therefore the greater the resistance to flow).

The radius of the lumen of a blood vessel is the most important factor that influences T_SP_R. A decrease in luminal size means that a greater portion of the blood is in contact with the wall, which means slower flow. Conversely, dilation of the vessel leads to less of the blood being in contact with the vessel wall and, therefore, less friction and less resistance to blood flow. This relationship between vessel lumen size and blood flow cannot be overstated. For example, increasing the radius of the vessel by twofold decreases resistance by 16-fold. Conversely, if the radius is decreased by half, then resistance is increased 16 times. Changes in vessel diameter (decrease = vasoconstriction; increase = vasodilation or less vasoconstriction) occur in the arterioles of the arterial system and therefore are the primary gate keepers of blood flow throughout the systemic circuit. In fact, the majority of T_SP_R and changes in T_SP_R are due to arteriolar tone (vs. blood viscosity or vessel length). At rest, the dominant controller of arteriolar tone and vessel radius is the sympathetic nervous system.

Changes in Pressure, Resistance, and Distribution of Blood Flow During an Acute Bout of Exercise

During exercise there is an intricate interplay between pressure and resistance and their influence on total and regional blood flow. As mentioned before, at rest most of the cardiac output is diverted to tissue other than the skeletal muscles. At peak exercise, this changes greatly such that more than 80% of cardiac output, which may be upward of 25 to 30 L · min^{-1}, is diverted to the more metabolically active skeletal muscles (figure 3.16) (1). Like the skeletal muscles, the heart is metabolically more active during exercise, so blood flow increases to match metabolic demand. Also as the body's core temperature increases, the sympathetic nervous system causes arterioles in the skin to vasodilate, allowing cutaneous blood flow to increase and facilitate heat dissipation.

This redistribution of blood flow to the metabolically more active skeletal muscles results from both vasoconstriction of the arterioles supplying the less metabolically active tissues (e.g., visceral organs) and vasodilation of the arterioles nourishing the metabolically more active tissues. The former primarily is attributable to the increase in sympathetic tone that accompanies modest to vigorous exercise—specifically, an increase in plasma norepinephrine and the release of norepinephrine from postganglionic fibers innervating the arterioles. Vasodilation of the arterioles in the working muscles mostly is attributable to the overriding or autoregulatory effects of factors and agents that develop during exercise and act on the arterioles (e.g., an increase in adenosine, temperature, CO_2, and H^+ and a decrease in O_2) and the release of vasodilatory agents (e.g., endothelial-derived factors such as nitrous oxide) from the endothelium of the arterioles. The latter is stimulated by the presence of some of the previously mentioned factors but also by the physical or mechanical effect of increased blood flow on the endothelium (shear stress).

Figure 3.17 reviews the changes in systolic, diastolic, and mean pressures during exercise. The increase in systolic blood pressure may exceed 200 mmHg, whereas diastolic blood pressure is essentially unchanged or may decrease slightly and MAP increases modestly. Overall, T_SP_R represents the sum of both regional resistance in metabolically more active tissue (e.g., the legs while jogging, where resistance is low due to vasodilation of the arterioles) and metabolically less active tissues (e.g., the viscera, where the vasoconstrictive effect of norepinephrine is not overridden by local autoregulation). The net result in the body during exercise is a more than fourfold decrease in T_SP_R, which accounts for the absence of any change in diastolic blood pressure and the modest increase in MAP.

You might be wondering whether cardiac output and T_SP_R, which change in opposite directions, would simply negate one another and lead to no change in MAP during exercise. However, MAP does increase modestly because the increase in cardiac output ($\dot{Q}$) during exercise is greater than the decrease in T_SP_R.

$$\uparrow MAP \, (mmHg) = \uparrow\uparrow\uparrow \dot{Q} \times \downarrow\downarrow T_SP_R$$

(Equation 3.6)

This means that even during vigorous exercise, when resistance is low and one might expect the arterioles to be as vasodilated as possible, the arterioles still experience some constriction in order to maintain MAP at a level that is adequate to perfuse the tissues.

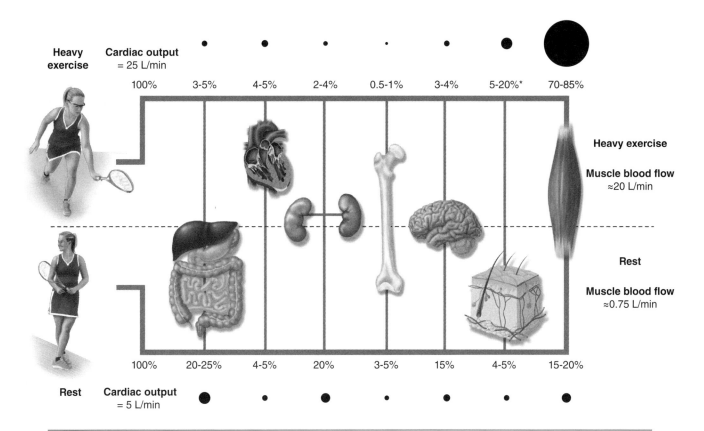

Figure 3.16 Relative distribution of cardiac output through the major body organs at rest (lower scale) and during exercise (upper scale). Does not include estimated blood flow to fatty tissues at rest (5%-10%) and during exercise (1%-5%).

Reprinted, by permission, from P.O. Åstrand et al., 2003, *Textbook of work physiology: Physiological bases of exercise,* 4th ed. (Champaign, IL: Human Kinetics), 143.

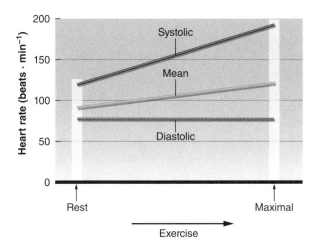

Figure 3.17 Changes in systolic, mean, and diastolic blood pressures during exercise.

Reprinted from M.L. Foss, S.J. Keteyian, and E.L. Fox, 1998, *Fox's physiological basis for exercise and sport,* 6th ed. (Pittsburgh, PA: William C. Brown). By permission of the authors.

Recall from earlier in the chapter that it is difficult to measure MAP. We provided a simple formula for estimating MAP: $[(2 \times \text{diastolic blood pressure}) + \text{systolic blood pressure}]/3$. Note that diastolic pressure is weighted twice that of systolic pressure because, at rest, about two thirds of the cardiac cycle is spent in diastole. During exercise, both systolic and diastolic times shorten as heart rate increases, but the decrease is proportionately much greater during diastole than during systole. To adjust for this, when we estimate MAP during exercise the equation is $1/2$(pulse pressure) + diastolic blood pressure, where pulse pressure is systolic blood pressure minus diastolic blood pressure.

Within minutes after exercise, all pressures return to the values observed before exercise and may even fall below pre-exercise values (i.e., postexercise hypotension). Compared with values measured before exercise, systolic and diastolic blood pressures after exercise may be up to 12 and 8 mmHg lower, respectively, and these lowered values may persist for several hours after exercise was stopped (25,44).

TRANSPORT OF O_2 AND CO_2

A primary nonnutritive function of the cardiovascular system is the transport of gases. O_2 diffuses across alveoli in the lungs to pulmonary capillary blood and then is transported to the tissues. CO_2 diffuses from peripheral tissues to the tissue capillary blood and then is transported to the lungs, where most of it diffuses from the pulmonary capillaries into the alveoli and is eventually exhaled. Before explaining more about the transport of these gases, we need to first discuss some important points about the constituents of blood.

All of the body's fluids can be categorized into one of three compartments: intracellular (within the approximately 75 million cells that make up the body), extracellular (the vast majority of which is the fluid found between cells, with a lesser amount being the plasma within the blood), and transcellular (e.g., fluid within the joints and other spaces such as the abdomen and pericardium). The typical 72-kg human has about 42 L of water, more than 95% of which is found intra- and extracellularly.

Blood contains both extracellular fluid (plasma) and intracellular fluid (inside the blood cells). The total blood volume for the average adult is approximately 5 L. The portion of the blood comprising blood cells and formed elements is called *hematocrit*; it makes up about 45% and 40% of total blood volume in men and women, respectively. Blood volume can change during exercise in that fluid moves from the plasma into the active muscle cells and surrounding interstitial spaces. This process, referred to as **hemoconcentration**, can decrease plasma volume by up to approximately 10%. Blood volume is directly linked to the amount of O_2 transported to the tissues and, therefore, peak aerobic power. Blood volume is about 25% lower in untrained women than in untrained men but only about 15% lower in trained women than in trained men.

Transport of O_2

O_2 is carried in the plasma both chemically bound to Hb (which by far is the main mechanism) and in physical solution as a dissolved gas (table 3.4). Concerning the latter, the ability of O_2 to dissolve in plasma is quite low—less than $0.3 \text{ mL} \cdot 100 \text{ mL}^{-1}$ of blood. Translated to resting values, only about 3% of the O_2 we consume each minute comes from O_2 shuttled to the tissues in plasma. During maximal exercise, dissolved O_2 is even less at about 2% of consumed O_2. We are not trying to convey that dissolved O_2 is of little importance; it does determine the partial pressure of O_2 (PO_2) in the blood, which is critically important in gas exchange and diffusion.

The O_2 that diffuses into the red blood cells combines chemically with Hb to form **oxyhemoglobin** (HbO_2), which increases the O_2 carrying capacity of the blood some 65-fold more than simply dissolving the O_2 into solution (table 3.4). Hb is a complex molecule containing iron (heme) and protein (globulin); the ability to combine with O_2 is related to the heme component. One molecule of Hb combines with four molecules of O_2, as expressed in the following equation:

$$Hb_4 + 4O_2 \leftrightarrow Hb_4\left(O_2\right)_4 \text{, simplified as } Hb + O_2 \leftrightarrow HbO_2$$

(Equation 3.7)

When all of the heme units in a molecule of Hb are fully saturated, each gram of Hb can carry 1.34 mL of O_2. For the typical man and woman with an Hb concentration in their blood of 15 and $13.5 \text{ g} \cdot \text{dL}^{-1}$, respectively, we can calculate total O_2 carried in the blood as approximately $20.1 \text{ mL} \cdot \text{dL}^{-1}$ for men (1.34×15) and approximately $18.1 \text{ mL} \cdot \text{dL}^{-1}$ for women (1.34×13.5). As mentioned, during exercise hemoconcentration occurs such that some of the fluid in plasma moves from the blood to extracellular space outside of the vascular system. This increases Hb concentration by about 10%. For a man with an Hb concentration of $15 \text{ g} \cdot \text{dL}^{-1}$ at rest, during exercise it increases to $16.5 \text{ g} \cdot \text{dL}^{-1}$: ($15 \times 0.1$) + 15 = 16.5. In this scenario, the O_2 carrying capacity is increased as well, from 20.1 in the previous example to $22.1 \text{ mL} \cdot \text{dL}^{-1}$ (1.34×16.5).

In addition to understanding Hb carrying capacity, it is important to appreciate the concept of percentage saturation of Hb ($\%SO_2$), sometimes simply called O_2sat. It is the ratio of O_2 that is actually combined with Hb to the maximal capacity of Hb to bind to O_2. At sea level at rest, all Hb is almost fully saturated, so $\%SO_2$ normally is approximately 98% (19.7/20.1). If an individual's O_2 capacity is $20.1 \text{mL} \cdot \text{dL}^{-1}$ and the amount of O_2 actually combined with Hb is $17 \text{ mL} \cdot \text{dL}^{-1}$, which is typical for someone standing on top of Pikes Peak in Colorado (elevation = 14,100 ft or 4,300 m), their $\%SO_2$ would be 85% (17/20.1).

At least four main factors affect the saturation of Hb with O_2, especially during a bout of exercise:

- PO_2 in the blood
- The temperature of the blood
- The pH of the blood (acidity, H^+ concentration, or the Bohr effect)
- The amount of CO_2 in the blood

Table 3.4 O_2 and CO_2 Content of Blood at Rest and During Exercise

O₂ OR CO₂ CONTENT (ML · 100 ML⁻¹ OF WHOLE BLOOD)			
Transport mechanism	Arterial blood	Mixed venous blood	Arterial mixed venous O₂ difference
REST ($\dot{V}O_2$ = 0.246 L · MIN⁻¹, $\dot{V}CO_2$ = 0.202 L · MIN⁻¹)			
Total O_2	19.8	15.18	4.62
In solution	0.3	0.18	0.12
As oxyhemoglobin	19.5	15.0	4.5
Total CO_2	48.0	51.8	3.8
In solution	2.3	2.7	0.4
As HCO_3^-	43.5	45.9	2.4
As carbamino compounds	2.2	3.2	1.0
MODERATE TO VIGOROUS EXERCISE ($\dot{V}O_2$ = 3.2 L · MIN⁻¹, $\dot{V}CO_2$ = 3.03 L · MIN⁻¹)			
Total O_2	21.2	5.34	15.86
In solution	0.3	0.06	0.24
As oxyhemoglobin	20.9	5.28	15.62
Total CO_2	45.0	60.0	15.0
In solution	2.1	3.1	1.0
As HCO_3^-	40.8	53.2	12.4
As carbamino compounds	2.1	3.7	1.6

The relationship between %SO_2 and the first three factors is nicely depicted using the oxyhemoglobin dissociation curve (figure 3.18), which in this instance assumes an Hb concentration of 15 g·dL^{-1}. The curve at the top of figure 3.18 also assumes normal resting values, which means a pH of 7.4, a body core temperature of 37 °C, an arterial blood PO_2 of 100 mmHg, and a mixed venous blood PO_2 of 40 mmHg. Note that the higher the PO_2, the greater the loading or association of O_2 with Hb; conversely, the lower the PO_2, the greater the unloading or dissociation of O_2 from Hb. For example, at an arterial PO_2 of 100 mmHg (x-axis), %SO_2 is 97%. In mixed venous blood, at a PO_2 of 40 mmHg, %SO_2 decreases to 75%. The difference between the two, called the ***arterial mixed venous O_2 difference*** (a-v̄O_2 diff), represents how much O_2 was extracted—in this example, at rest, approximately 4.5 mL of O_2·dL^{-1} of blood flow (19.5 mL·dL^{-1} – 15 mL·dL^{-1}).

There are two other important features of the oxyhemoglobin dissociation curve to point out. First, the upper part of the curve is almost flat and, as such, is somewhat protective in nature. This means that a fairly large change in PO_2 (from 100 mmHg to 70 mmHg) actually has little effect on the release of O_2 from Hb. Second, the middle and lower portions of the curve are

quite steep, such that a PO_2 below 50 mmHg or so is associated with a large change in %SO_2, able to provide the tissues with a large amount of much-needed O_2. At a mixed venous PO_2 of 40 mmHg at rest, note the large decrease in %SO_2 (from 75% to 12%) when PO_2 falls to 10 mmHg, which one might find in the peripheral muscle tissues during exercise. In fact, PO_2 in metabolically active muscles may fall as low as 5 mmHg during vigorous exercise!

Speaking of exercise, changes in blood chemistry (decrease in pH, increase in temperature) during exertion shift the oxyhemoglobin dissociation curve down and to the right (figure 3.18b). Such a shift is very favorable for the peripheral muscle tissues in that even more O_2 disassociates from Hb and is made available. For example, if no shift in the curve were to occur, the a-v̄O_2 diff during exercise would be as follows:

$$19.5 - 11.6 = 7.9 \text{ mL of } O_2 \cdot dL^{-1} \text{ of blood}$$
$$\text{(Equation 3.8)}$$

With the shift to the right, the a-vO_2 diff is increased by about 30%:

$$19.0 - 8.8 = 10.2 \text{ mL of } O_2 \cdot dL^{-1} \text{ of blood}$$
$$\text{(Equation 3.9)}$$

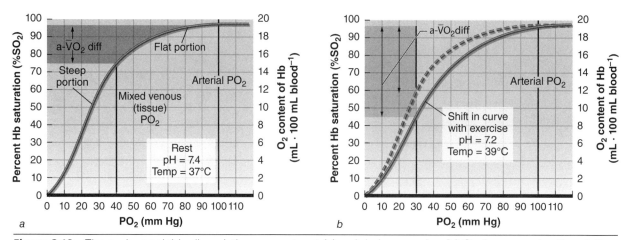

Figure 3.18 The oxyhemoglobin dissociation curve at rest *(a)* and during exercise *(b)*. Such curves give the relationship between PO_2 and how much O_2 associates and dissociates with Hb. During exercise, the curve shifts downward and to the right, which facilitates the diffusion of additional O_2 into muscle.

Reprinted from M.L. Foss, S.J. Keteyian, and E.L. Fox, 1998, *Fox's physiological basis for exercise and sport,* 6th ed. (Pittsburgh, PA: William C. Brown). By permission of the authors.

In the pulmonary capillary blood of the lungs, where PO_2 equals 100 mmHg, the rightward shift in the curve that occurs during exercise has little effect on the loading of Hb with O_2.

Transport of CO_2

Like O_2, CO_2 is carried by the blood both in chemical combination and in dissolved or physical solution. And like O_2, the amount of CO_2 dissolved in the plasma is quite small (table 3.4)—about 5% of total CO_2 transported by the blood, the amount of which also regulates or determines blood and tissue PCO_2.

The bulk of CO_2 is transported in chemical combination; however, the reactions that CO_2 is involved with are very different from those involving O_2 and include carbonic acid, bicarbonate ions (HCO_3^-), and carbamino compounds (table 3.4). When CO_2 diffuses from muscle tissue into plasma, it reacts with water to form carbonic acid (H_2CO_3). It does so quickly when in the presence of an enzyme called *carbonic anhydrase,* which is predominantly found inside red blood cells:

$$CO_2 + H_2O \leftrightarrow H_2CO_3 \qquad \text{(Equation 3.10)}$$

However, H_2CO_3 ionizes as quickly as it is formed, indicating that CO_2 is carried predominantly in the plasma in the form of bicarbonate ions (HCO_3^-), where it diffuses out of the red blood cells. (Note in figure 3.19 that as HCO_3^- diffuses out into the plasma, CL^- moves into the cells to maintain ion concentration across the cell membrane.)

$$CO_2 + H_2O \leftrightarrow H_2CO_3 \leftrightarrow H^+ + HCO_3^-$$
$$\text{(Equation 3.11)}$$

The reaction is driven to the right as CO_2 is added to the capillary blood from the surrounding metabolically active peripheral tissues. Conversely, in the lungs the reaction is driven to the left when CO_2 diffuses down its pressure gradient from the blood to the alveoli. In the previous equation, note that the breakdown of carbonic acid also yields H^+, which decreases the pH of venous blood. The H^+ are buffered by available globin or the protein portion of Hb—the same protein that is more available after dissociating O_2 from the Hb molecule at the tissues.

In addition to buffering H^+, the globin portion of Hb (and other proteins in plasma) can chemically react with CO_2 to form carbamino compounds (e.g., carbaminohemoglobin). When formed, more H^+ are generated, which in turn are buffered as described previously. Thus, the Hb molecule is capable of combining with and transporting both O_2 (to heme) and CO_2 (to globin). However, an Hb molecule alone is able to carry more CO_2 than is a molecule of Hb already combined with O_2 (HbO_2), thus facilitating the loading of CO_2 where it is produced (and O_2 is unloaded) in the peripheral tissues.

Regarding both O_2 and CO_2, remember that they are expressed as volume per volume of blood (mL · dL^{-1} of blood). Table 3.4 draws attention to the large increase in total $\dot{V}O_2$ that occurs from rest to exercise—in this case, from 0.246 L · min^{-1} to 3.2 L · min^{-1} (a 13-fold increase). One way to accommodate such an increase is to increase O_2 extraction by the skeletal muscle, evidenced by the increase in a-vO_2 diff from 4.62 mL · dL^{-1} to 15.86 mL · dL^{-1}— an increase that is important but by itself insufficient to account for the 13-fold increase in overall $\dot{V}O_2$ uptake. To accomplish this, total blood flow (i.e., cardiac output) must increase.

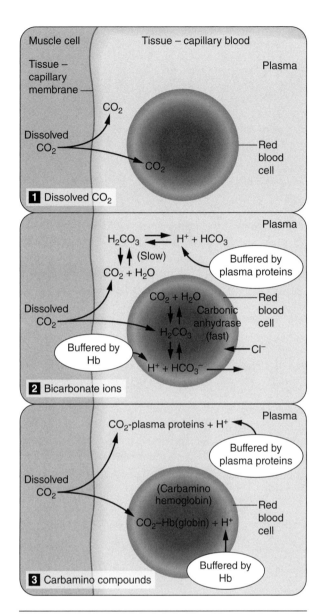

Figure 3.19 CO_2 is transported as (1) dissolved CO_2 or in chemical combination as (2) bicarbonate and (3) carbamino compounds.

Reprinted from M.L. Foss, S.J. Keteyian, and E.L. Fox, 1998, *Fox's physiological basis for exercise and sport,* 6th ed. (Pittsburgh, PA: William C. Brown). By permission of the authors.

CARDIORESPIRATORY FITNESS: THE WHOLE PICTURE

The ability of the body to transport O_2 to the peripheral tissues and the ability of those tissues to utilize the O_2 that is made available to them represents the best overall measure of cardiorespiratory fitness and function. Measuring $\dot{V}O_2$ can require sophisticated invasive equipment; therefore, a simplified, indirect method—measuring indirect spirometry at the mouth—is commonly used.

Peak or maximal $\dot{V}O_2$ has been given many names, such as cardiorespiratory fitness, peak $\dot{V}O_2$, $\dot{V}O_2$max, aerobic capacity or power, and exercise tolerance. To best illustrate the concept of O_2 transport and utilization, we draw upon the Fick equation (Adolph Fick, circa 1870):

$$\text{Cardiac output } (\dot{Q}; L \cdot min^{-1}) = \frac{O_2 \text{uptake } (\dot{V}O_2; L \cdot min^{-1})}{\text{a-}\overline{v}O_2 \text{diff} (mL \cdot L^{-1})}$$

$$\text{(Equation 3.12)}$$

Rearranging the equation to solve for $\dot{V}O_2$, we get the following:

$$\text{Cardiorespiratory fitness} = \text{transport} \times \text{utilization}$$

$$\text{(Equation 3.13)}$$

or

$$\dot{V}O_2 = \dot{Q} \times \text{a-}\overline{v}O_2 \text{diff}$$
$$= (\text{heart rate} \times \text{stroke volume}) \times \text{a-}\overline{v}O_2 \text{diff}$$

$$\text{(Equation 3.14)}$$

Table 3.5 provides examples of O_2 transport (heart rate × stroke volume) and O_2 utilization (a-$\overline{v}O_2$ diff), with each variable making a contribution to overall O_2 uptake. Note, for example, that $\dot{V}O_2$ increases more than 10-fold from 252 to 3,100 mL · min⁻¹; this is accomplished by the 1.6-fold increase in stroke volume, a greater than 2.5-fold increase in heart rate, and the nearly 3-fold increase in O_2 extraction. Also notice that the increase in heart rate and stroke volume combine to increase cardiac output from 5 L · min⁻¹ at rest to more than 22 L · min⁻¹ at peak, clearly indicating the increased transport of O_2 with exertion.

CARDIOVASCULAR REGULATION

Regulation of MAP at rest and during exercise and the distribution of blood to the more metabolically active tissues during exercise are the net results of a complex series of afferent stimuli or input and efferent actions. Most of these actions are regulated by the autonomic nervous system, the essential constituents of which are located in the cardiovascular control center of the medulla oblongata of the brain stem. Afferent input to this area of the brain is quite varied, and the stimuli can be classified as central, humoral, physical, and peripheral (table 3.6). The medulla integrates changes in these various stimuli, such as what occurs during exercise, and makes changes in blood pressure and blood flow via adjustments in chronotropicity, inotropicity, and vascular tone.

Table 3.5 Components of the O_2 Transport System at Rest and During Maximal Exercise

Condition	$\dot{V}O_2$ (mL·min⁻¹)	=	Stroke Volume* (L·beat⁻¹)	×	Heart rate (beats·min⁻¹)	×	a-vO₂ diff (mL·L⁻¹)
Rest	252	=	0.070	×	72	×	49.9
Maximal exercise	3,100	=	0.112	×	200	×	138.4

*Usually expressed in milliliters per beat (e.g., 0.070 L · beat⁻¹ = 70 mL · beat⁻¹).

Table 3.6 Summary of Stimuli on the Cardiovascular Control Area of the Brain Stem

Stimuli	Description
Central motor command or central command	Originate in the motor cortex and irradiate or spill over as they pass through the cardiovascular control center in the medulla to activate skeletal muscle.
Humoral	Chemical (PO_2, PCO_2, pH) changes in the blood influence chemoreceptors in the medulla, aortic bodies of the aortic arch, and carotid bodies of the carotid arteries, all of which provide afferent input to the cardiovascular control center.
Physical	Volume- and pressure-sensitive mechanoreceptors (baroreceptors) located in the aortic arch and carotid sinus of arterial system. Short-term decrease in MAP (such as what occurs with moving from supine to standing) at the aortic and carotid baroreceptors results in afferent input to the medulla with corresponding parasympathetic withdrawal (to increase rate) and sympathetic activity (to peripheral arterioles to vasoconstrict). Net effect is maintenance of MAP at normal levels.
Peripheral	Mechanoreceptors (group III afferent) and metaboreceptors (group IV afferent) respond to stimuli in and around various peripheral tissues and provide afferent input to the medulla. Mechanoreceptors respond mostly to movement and changes in tension in skeletal muscles and metaboreceptors are activated by changes in local chemistry (H^+, lactate, phosphate), especially during exercise of moderate intensity and greater. The responses of these two receptors are not isolated in that mechanoreceptors respond somewhat to chemical changes and metaboreceptors respond somewhat to mechanical stimuli.

These adjustments are mediated through the two branches of the autonomic nervous system—the sympathetic and parasympathetic systems. The former sometimes is referred to as the *thoracolumbar lumbar system* because its preganglionic fibers exit the spinal cord at the level of the thoracic and upper lumbar vertebrae, whereas the latter often is called the *craniosacral system*.

When stimulated, the sympathetic cardioaccelerator nerves that innervate the heart cause a positive chronotropic and inotropic response due to the release of norepinephrine, an agonist agent that stimulates the cardiac β-1 receptors on the membrane of cardiac myocytes. Norepinephrine causes more Ca^{2+} to enter the cells of the SA node during Phase 4, hypopolarizing them and increasing rate. This increased influx of Ca^{2+} also increases the force of contraction.

Conversely, the cardiac vagus nerve of the parasympathetic nervous system, when stimulated, releases acetylcholine and decreases rate by hyperpolarizing the cells of the SA node. At rest, the SA node is under the predominant influence of the parasympathetic nervous system. As an individual starts to exert himself (e.g., getting up from a chair and simply walking to another room) or finds himself in a stressful situation (e.g., almost getting into a car accident), the accompanying increase in rate (up to about 100 beats · min⁻¹) is predominantly due to inhibition or withdrawal of parasympathetic input (i.e., inhibition of the vagus nerve). During exercise, further increases in heart rate above about 100 beats · min⁻¹ and up to maximum are predominantly due to stimulation of the cardiac sympathetic efferent fibers that innervate the

Fitness Across the Lifespan

Aging is a normal biological process that often is associated with disease and disability. The ability of older persons to function independently depends on many factors, including maintenance of sufficient aerobic capacity, as measured by peak O_2 uptake ($\dot{V}O_2$). After age 30, typical aging is associated with a decline in $\dot{V}O_2$ of approximately 8% to 10% per decade. To compute this 10-yr rate of change in peak $\dot{V}O_2$, the Baltimore Longitudinal Study on Aging followed a cohort of people aged 21 to 87 yr who were free of known heart disease (12). Results showed a continued decline in fitness across six age decades; however, the rate of decline was greatest among those over age 70 (20% per decade) compared with those under age 30 (3%-6% per decade). After age 40, the rate of decline was greater in men than in women. Whether changes in peak heart, peak stroke volume, or a-$\bar{v}O_2$ are chiefly responsible for the loss of fitness with age has been investigated; it is likely that, to a varying level, the loss is due to changes in all three factors over time (12,21,27). It is important to point out that regular exercise training blunts the loss in cardiorespiratory fitness with age to the extent that it reduces by about one half the loss experienced by nonexercising aging men (21). Additionally, aerobic capacity can be improved (i.e., exercise intolerance reversed), usually between 10% and 20%, at almost any age (40,45).

heart and release norepinephrine, which stimulates the cardiac β-1 receptors as described previously (30).

Another mechanism that influences cardiac function during exercise is the level of catecholamines (norepinephrine and epinephrine) in the plasma. Both agents are released by the adrenal medulla, with the preponderance being epinephrine. In addition, a portion (about 20%) of the norepinephrine released elsewhere in the body from postganglionic sympathetic terminals is not taken back up in the terminal and instead spills over into circulation (19). Thus, plasma norepinephrine levels can go quite high during exercise. Regardless, these catecholamines have both cardiac and peripheral effects, such as increasing chronotropic and inotropic states and inducing more vasoconstriction in the peripheral arterioles. After exercise is stopped, parasympathetic or vagal activity is promptly restored, resulting in a marked decrease in heart rate early on during recovery.

It is important to note that in addition to autonomic input to the heart, the medulla oblongata regulates efferent sympathetic nerve traffic to the peripheral arterioles that, when stimulated, lead to vasoconstriction. During exercise, sympathetic traffic is increased, resulting in a decrease in blood flow in the metabolically less active tissue (e.g., kidney and splanchnic circulations). In the metabolically more active tissues, such as the working skeletal muscles, the attempt at sympathetically mediated vasoconstriction is overridden by local autoregulation. The net effect of this is much less vasoconstriction (even

below resting levels) and an increase in much-needed blood flow.

Cardiovascular Control During Exercise: Putting It All Together

A primary objective of the cardiovascular system during exercise is to maintain MAP and blood flow in a manner that ensures adequate perfusion pressure and adequate blood flow to all tissues at levels that are commensurate with metabolic demand. Accomplishing this task, especially during strenuous dynamic exercise, is challenging and complex, and the regulatory stimuli responsible for maintaining MAP and blood flow at rest (i.e., arterial pressure, PO_2, PCO_2, pH) may not all be primary factors for controlling MAP during exercise.

Even before exercise begins, an increase in heart rate can be observed when a person simply anticipates exercise. This anticipatory response is due to parasympathetic withdrawal and allows heart rate to increase. Additionally, the increases in heart rate and stroke volume (and therefore cardiac output) that are known to occur with exercise contribute to an increase in arterial pressure. If the increase in MAP (which is normal and expected during exercise) were to evoke the arterial baroreceptors (aortic arch and carotid sinus) in the same manner that it does at rest, heart rate and stroke volume would, through the autonomic nervous system, not increase. Fortunately, baroreceptor function as it normally operates at rest is temporarily reset to a higher pressure or operating point once exercise is started; this allows heart rate and MAP to increase without opposition (28,31,32,34,35). Overall, these increases are allowed and managed by the medulla through the complex interaction of mainly central command, physical afferent input (the arterial baroreflex and resetting thereof), and peripheral afferent input (skeletal muscle mechano- and metaboreceptors).

OTHER IMPORTANT TOPICS IN ADVANCED CARDIOVASCULAR EXERCISE PHYSIOLOGY

Several important topics in the advanced study of exercise physiology are worth mentioning briefly. Each deserves more attention than can be provided in this textbook and is worthy of further exploration based on your level of interest.

Double Product or Rate Pressure Product

The term *peak* $\dot{V}O_2$ should now bring to mind whole-body O_2 uptake as the product of cardiac output and a-vO_2 difference. However, measuring whole-body O_2 uptake does not isolate or identify the contributions of individual organs, such as O_2 uptake of the brain, the kidneys, or—relevant for this chapter—the heart. During exercise the heart clearly requires more energy and therefore O_2, but, just like whole-body $\dot{V}O_2$, directly measuring myocardial O_2 consumption is impractical because it requires instruments and invasive methods that are not routinely available in most exercise physiology laboratories.

That said, invasive studies performed years ago using catheters placed in the coronary arteries and coronary sinus (vein) showed that the heart consumes about 70% of the O_2 it is provided and, when it demands more during exercise, it is met by increasing blood flow. Factors that influence myocardial O_2 consumption are heart rate, left ventricular size or stretch, inotropic state, and wall thickness. Fortunately, the studies involving invasive methods also showed that noninvasive measures such as heart rate and systolic blood pressure (i.e., **rate pressure product** or **double product**, which is the product of heart rate and systolic blood pressure) correlate well with measured myocardial O_2 consumption (37). During exercise, as both heart rate and systolic blood pressure increase, the estimated myocardial O_2 consumption increases as well, all of which indicates the heart's greater demand for the O_2 needed to produce energy.

Comparative Responses During Arm and Leg Exercise

The response of the cardiovascular system during submaximal and peak exercise differs when one is performing arm versus leg exercise. Learning about such

---

Cardiovascular Physiology: A Clinical Perspective

Consider the 64-yr-old man who regularly experiences angina pectoris (due to myocardial ischemia, or temporary lack of O_2 to the heart) while carrying a trash can out to the curb in front of his house or hurrying into work in the morning—activities that for most people are fairly light but sufficient enough to increase heart rate and systolic blood pressure and, therefore, estimated myocardial O_2 consumption (heart rate × systolic blood pressure = rate pressure product). If you were measuring this man's vital signs when angina began during one of these activities, you might observe that his heart rate increased to 109 beats · min^{-1} and systolic blood pressure increased to 160 mmHg; the product of this is 17,400 mmHg · beat^{-1} · min^{-1}. For this man, any combination of heart rate and systolic blood values that produces a rate pressure product of approximately 17,400 or greater drives myocardial O_2 consumption beyond what O_2 can be supplied and will likely produce his symptom of angina. Note that we can use this information not only to quantify his angina threshold but also to guide exercise intensity. He should exercise at a level that is below 17,000 to ensure that he is symptom free and training at a level where the O_2 supplied to the heart is equal to or greater than the O_2 demanded.

differences might not seem important at first glance, but consider all the tasks one does with the upper limbs, such as shoveling snow, raking leaves, and even propelling oneself in a wheelchair. As shown in table 3.7, exercise completed with the arms at the exact same work rate as the legs yields a higher heart rate, systolic blood pressure, rate pressure product, $\dot{V}O_2$, and minute ventilation. Conversely, stroke volume is lower during arm exercise performed at the same work rate; therefore, cardiac output is the same. The factors potentially responsible for these differences include higher mechanical efficiency during leg exercise, smaller muscle mass involved with the arms, and increased central command or a greater increase in plasma catecholamines with arm exercise.

The same cannot be said if one compares arm versus leg response during peak or maximal exercise. In fact, across all variables, the magnitude of the increase is greater with leg versus arm exercise, which is a consistent finding in healthy people and patients with a cardiovascular disease (i.e., heart failure) (table 3.8).

Cardiac Output During Prolonged Exercise

The changes in heart rate, stroke volume, and cardiac output that occur during a single bout of submaximal steady-state exercise are described earlier in the chapter and are, for the most part, what one would observe if the exercise bout is less than 30 to 40 min in duration. Heart rate increases and stroke volume increases and then plateaus; this contributes to an increase in cardiac output that is able to match the demand for blood flow to the working muscles. However, when steady-state exercise continues over several hours, and especially if the exercise occurs in a warm or hot environment, the response of heart rate and stroke volume drift over time.

Table 3.7 Cardiorespiratory Responses of 10 Moderately Active Males* After 4 Min of Submaximal Arm or Leg Exercise at 30 Watts

	Heart rate (beats · min⁻¹)	Stroke volume (mL· beat⁻¹)	Cardiac output (L·min⁻¹)	Systolic blood pressure (mmHg)	Rate pressure product × 10³ (mmHg · beat⁻¹ · min⁻¹)	$\dot{V}O_2$ (L· min⁻¹)	Ventilation (L · min⁻¹)	Respiratory exchange ratio	Blood lactate (mmol·L⁻¹)
Arm	97	107	10.4	169	16.4	0.84	28.7	1.01	4.3
Leg	92	115	10.6	148	13.6	0.73	21.0	0.85	2.8

*Average age = 51 yr.

Based on data from Keteyian et al. (22)

Table 3.8 Cardiorespiratory Responses to Peak Arm and Leg Exercise in Healthy Persons and Patients With Heart Failure

	Power output (watts)	Heart rate (beats · min⁻¹)	Rate pressure product × 10³ (mmHg · beat⁻¹ · min⁻¹)	$\dot{V}O_2$ (L· min⁻¹)	Ventilation (L· min⁻¹)	Respiratory exchange ratio
HEALTHY						
Arm	62	140	28.5	1.5	63.6	1.14
Leg	162	154	32.5	2.28	89.8	1.14
HEART FAILURE						
Arm	43	128	20.5	1.08	51.2	1.15
Leg	101	144	24.1	1.48	66.5	1.15

Based on data from Keteyian et al. (22)

Known as *cardiovascular drift,* stroke volume decreases and heart rate increases gradually over many minutes to hours despite no change in work rate. Because these changes are in opposite directions of proportionately equal magnitude, cardiac output is unchanged, which is what one would expect if no change occurs in pace or external work rate.

Although the exact mechanisms responsible for the changes in heart rate and stroke volume are not completely understood, prior work suggests that the increase in heart rate may be attributable to several factors, including the effect of an exercise-induced increase in body temperature on SA node function or cardiac sympathetic nerve activity and increased cutaneous blood flow (7). Combined with a decrease in plasma due to sweating, the increase in blood flow to the skin eventually results in a decrease in venous return (decreased end-diastolic volume or preload) and a lower stroke volume, thus resulting in a compensatory increase in heart rate to maintain cardiac output. Regardless of the exact mechanism, the drift or increase in heart rate during prolonged submaximal exercise can be quite dramatic such that over the course of a 3-h marathon event, the last hour or so may be run at or near maximal heart rate even though the race pace is submaximal during that time period.

THE KIDNEYS AND EXERCISE

We would be remiss if we did not at least briefly mention another important organ that can influence cardiovascular function and fluid status: the kidneys. Despite the fivefold or greater increase in cardiac output that occurs with exercise, blood flow to the kidneys is decreased from approximately 20% (or 1 L · min⁻¹) of total cardiac output at rest to approximately 3% (0.75 L · min⁻¹) or less during exercise. The factors responsible for this decrease likely are a combination of increased renal sympathetic nervous activity and the increase in plasma norepinephrine, both of which can lead to vasoconstriction of the arterioles in the kidneys.

This decrease in blood flow also leads, to a slightly lesser extent, to a decrease in glomerular filtration rate, or the amount of blood that passes through the tiny glomeruli per minute. Glomeruli are the filters that are freely permeable to water, ions, and glucose. The glomerular filtration rate influences excretory function and is itself influenced by blood pressure and an individual's hydration status. In general, and depending somewhat on exer-

cise intensity (14,33), urine volume decreases with exercise from approximately 1.0 mL · min⁻¹ to 0.5 mL · min⁻¹ or less. Likewise, compared with resting values, sodium excretion during exercise is decreased by 50% or more.

Summary

- Cardiac muscle and skeletal muscle fibers have several similarities (e.g., proteins, mechanics of excitation–contraction coupling) and many differences (e.g., fiber length, density of the SR, AP or contraction time, presence of intercalated discs).
- Calcium ions play an important role in regulating cardiac muscle contraction and relaxation; intracellular concentration influences strength of contraction.
- Cardiac AP originates in the SA node through a mechanism referred to as *automaticity*—a property also present in the cells of the AV node. The autonomic nervous system greatly influences the function of the SA node.
- An ECG records the electrical activity (voltage and time intervals) of the heart and is used to identify normal function and the influence of disease.
- The cardiac cells (myocytes) receive their nutritive blood supply through coronary circulation, which comprises the right coronary artery and the left main coronary artery. The coronary veins generally run alongside the arteries, and most of the blood empties into the right atrium.
- Each cardiac cycle of systole and diastole results in changes in contractile force, which then cause changes in intracardiac pressures that lead to the opening and closing of the cardiac valves.
- Two of the main determinants of cardiac performance are heart rate and stroke volume. At rest, the former is predominantly regulated by the parasympathetic nervous system. During exercise, such influence is mostly withdrawn, and further increases in heart rate are greatly attributable to direct sympathetic tone. Stroke volume is influenced by preload, afterload, ionotropic state, and hypertrophy, the first three of which are altered to some extent during a bout of exercise.
- Pressure in the arterial system is at its highest during systole, and pressure decreases to its lowest levels as blood is distributed during diastole. MAP, which represents the driving pressure that influences blood

flow perfusing metabolic tissues, is estimated at rest as $[(2 \times \text{diastolic blood pressure}) + \text{systolic blood pressure}] / 3$. Other factors that influence the flow of blood at rest and during exercise are viscosity of the blood, vessel length, and radius of the blood vessel.

- A nonnutritive function of blood is to transport gases such as O_2 and CO_2. O_2 is carried bound to Hb (oxyhemoglobin) and, to a much lesser extent, dissolved in plasma. With exercise, changes in blood chemistry (i.e., decrease in pH and increase in temperature) facilitate a greater unloading of O_2 from Hb at the more metabolically active peripheral muscle tissues.

- Peak cardiorespiratory fitness, as measured by peak O_2 uptake, represents the ability of the body to transport (peak cardiac output) and utilize (peak a-$\bar{v}O_2$ diff) O_2.

- A major objective of the cardiovascular system during exercise is to maintain MAP. This is accomplished through an intricate interplay between the autonomic nervous system, baroreceptors, and skeletal muscle mechano- and metaboreceptors, all orchestrated by the medulla oblongata located in the lower half of the brain stem.

Definitions

afterload—The load that opposes or is applied against the ventricle after the muscle begins to contract. For the left ventricle, this is a function of aortic blood pressure, compliance and resistance of the aorta, and size of the left ventricle.

arterial mixed venous O_2 difference—Difference (in milliliters per liter) between arterial concentration of O_2 and mixed venous blood. The latter often is sampled from the pulmonary artery to ensure a sample representative (mixed) from both the superior and inferior vena cavas.

atrioventricular node—Group of cardiac muscle cells located below the right atrium capable of automaticity. If the SA node falters, the AV node becomes the predominant pacemaker (nodal rhythm) at a slower rate than that of the SA node.

chronotropicity—Heart rate, measured in beats per minute. Agents or actions that increase heart rate are referred to as inducing a positive chronotropic effect.

dromotropicity—The conduction velocity (measured in milliseconds) of electrical activity of the heart.

ejection fraction—The percentage of end-diastolic blood pumped with each contraction or systole.

endocardium—A thin layer of tissue that is in direct contact with blood and that covers the innermost portions of the myocardium that form the four chambers of the heart. Also provides protection to the thick myocardium and acts as a blood–heart barrier that regulates the nutrients and ion composition of the extracellular fluid that surrounds cardiac muscle cells, or myocytes.

epicardium—The outermost layer of heart tissue; comprises mostly connective tissue and serves as a protective layer. The epicardium fuses with (and can be thought of as being part of) the innermost layer of the pericardium.

hemoconcentration—The concentration of blood elements due to a decrease in plasma volume—for example, the movement of blood from plasma into interstitial space that occurs during exercise.

inotropic state—One of the four main determinants of stroke volume. By definition, inotropic is a shift in the Frank-Starling curve upward and to the left (positive inotropicity) or down and to the right (negative inotropicity). A change in inotropic state means a greater (positive) or lesser (negative) force of contraction and, therefore, a greater or lesser stroke volume for any given end-diastolic volume.

intercalated discs—The fusion of the cell membranes between two cardiac cells that provides some structural strength to hold cells together via desmosomes and allows two adjacent cardiac myocytes to exchange ions, amino acids, sugars, and nucleotides.

myocardial infarction—Portion of the heart in which the cells have died and are no longer functional.

myocardial ischemia—Temporary lack of blood flow to the heart.

oxyhemoglobin—Chemical combination of O_2 and Hb. Fully saturated, one molecule of Hb can combine with four molecules of O_2.

ostium—An opening in the walls of the aorta, located just above the aortic valve, that originates the two main coronary arteries: right coronary artery and left main coronary artery.

pericardium—A double-layered membrane or sac that surrounds the heart and helps protect it from external forces. The pericardium is filled with approximately 10 to 50 mL of a clear lubricating fluid called *pericardial fluid* that moistens the various surfaces.

preload—The stretch or tension (length–tension relationship) on the cardiac muscle fibers at the end of ventricular filling. Changes in preload are associated with commensurate changes in stroke volume.

rate pressure product or double product—Product of heart rate and systolic blood pressure (mmHg · beat^{-1} · min^{-1}); often serves as a noninvasive estimate of myocardial O_2 consumption.

sinoatrial node—Group of specialized cardiac muscle fibers that serves as the heart's intrinsic pacemaker. It initiates the electrical AP on its own through the property of automaticity. If left free of neural and hormonal influence, the SA node depolarizes at a rate of approximately 90 to 100 beats · min^{-1}.

sinus bradycardia—Heart rhythm originating from the SA node at a rate less than 60 beats · min^{-1}.

sinus tachycardia—Heart rhythm originating from the SA node at a rate of 100 beats · min^{-1} or more.

stroke volume—The amount of blood ejected from the right or left ventricle per beat of the heart.

Chapter 4

Pulmonary Exercise Physiology

The respiratory system is an incredible arrangement: conducting airways starting at the mouth and nose and millions of air sacs surrounded by a network of capillaries, all working in synch with muscles to facilitate inspiration and expiration. It is capable of providing the required amount of gas exchange both during rest and during increasing intensity of exercise to provide the body with needed oxygen (O_2) delivery and removal of carbon dioxide (CO_2). This system, although limited, is capable of adapting to exercise training to accommodate increasing exercise intensities as an individual becomes better trained. The amazing ability of this system to accommodate a person both at rest and under extreme exercise conditions such as a marathon (26.2 mi, or 42 km) at a pace of less than 5 min per mile exemplifies its extraordinary range. Ventilating a volume of 10 to 15 L · min^{-1} at rest, a world-class endurance athlete can increase this volume by 10 to 15 times to a level of often more than 200 L · min^{-1}. That is the equivalent of breathing in and out about two hundred 1-L soda bottles in 1 min! This occurs while the lungs are working in synchronization with the cardiovascular system to achieve the needed gas exchange. The pulmonary system can also help to stabilize the body during activity that requires a rigid torso (e.g., heavy lifting) and regulate blood pH and is vital in both speech and smelling. It is a wonderfully designed system. This chapter describes its structure as well as its function at rest and during exercise and presents factors that can enhance or negatively affect its function.

LUNG STRUCTURE AND FUNCTION

The pulmonary system is made up of the mouth and nasal cavity, the pharynx and larynx, the trachea, the right (three lobes) and left (two lobes) lungs, the bronchi and bronchioles, and the alveoli (figure 4.1). Each breath moves air from the beginning to the end of this anatomical structure. Gas exchange takes place at the alveoli level. The intake of air can begin at the nose, the mouth, or both. Resting breathing typically uses the nose, which allows for warming and humidifying the air. The nasal passage also allows cleansing of the air; the nasal mucosa that lines the nasal cavity traps dust and other debris. However, the nasal passages are limited by their size. The mouth is required to breathe when larger volumes of air exchange are required, as with increasing intensities of physical activity.

Structure

The lungs are not directly attached to the inner walls of the thoracic cavity. They actually ride on a very thin layer of serous fluid. This fluid is produced and secreted by two membranes known together as the pleurae (singular = *pleura*). The space between these membranes is known as the *pleural cavity*. These pleural membranes surround each lung; the inner layer of the internal pleura is attached

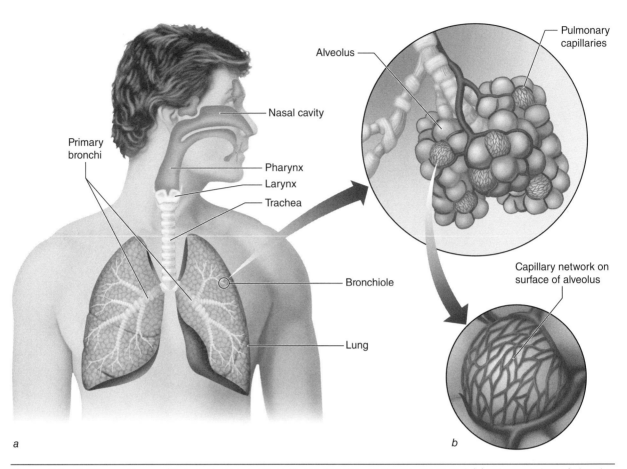

Figure 4.1   *(a)* The anatomy of the respiratory system, illustrating the respiratory tract and *(b)* enlarged view of alveolus where gas exchange occurs between the alveolus and blood in the pulmonary capillaries.

to the lung, and the outer layer of the external pleura is attached to the ribcage and diaphragm. The purpose of the pleura is to prevent lung collapse during the time the pressure in the pleural cavity (i.e., the intrapleural pressure) is less than the atmospheric pressure.

As air is inhaled it travels down the trachea, a tube structure that runs from the larynx to the fork to the right and left bronchi. The trachea maintains its shape with the support of cartilage. The trachea bifurcates to begin the bronchial tree and within the bronchi and subsequent bronchioles (terminal and respiratory) is supported by cartilage. There is also a layer of smooth muscle surrounding each bronchiole that allows them to contract and dilate as needed. From the nose and mouth to the terminal bronchioles there is no gas exchange, only air flow. This section of the pulmonary system is known as the **conducting zone**. The conducting zone filters (via mucus entrapment), warms, and humidifies the inhaled air to protect against lung damage.

Distal to the terminal bronchioles lie (in order) the respiratory bronchioles, the alveolar ducts, and the alveoli, where gas exchange takes place. The area from the terminal bronchioles to the alveoli is termed the **respiratory zone**.

A network of capillaries surrounds each alveolus, which brings deoxygenated blood containing high CO_2 to the area for gas exchange via diffusion. The alveolus is an air sac with almost direct contact with the capillary, allowing for gas diffusion to take place. Upwards of 300 million alveoli provide a huge area (60-80 m^2, or about the size of a tennis court) for gas exchange to occur. Each alveolus is only a single cell layer thick, which, in addition to the partial pressure gradient, promotes effective gas exchange. Between neighboring alveoli there are small holes known as the *pores of Kohn* that allow an even distribution of **surfactant** throughout the respiratory membranes. This is important in the control of alveolar expansion and pressure and in lung inflation.

Respiratory Muscles and Breathing

Breathing at rest and at low levels of physical activity primarily is initiated by contraction of the diaphragm, a large, dome-shaped muscle that lies just below the lungs. It separates the chest region from the abdomen and is innervated by the phrenic nerve. When stimulated, it results in the contraction and movement of the diaphragm downward, which expands the chest cavity by causing reduced intrapleural pressure (by about 2-3 mmHg) and expansion of the lungs. This allows air to flow in from the higher pressure outside the body. The external intercostal muscles, which lie between the ribs, also contract to cause the ribs and sternum to swing forward and outward. During increasing levels of exercise, the rate and intensity of contractions increase. This expands the thoracic region to a greater degree and allows an increase in both breathing frequency (breaths·min^{-1}) and depth of the tidal volume (volume of air displaced in a normal exhalation; L·breath^{-1}), thus allowing more total air per minute (L·min^{-1}) to be moved into and out of the lungs. Once adequate gas exchange occurs within the alveoli, expiration occurs. At rest, expiration is passive and occurs when the diaphragm relaxes and the elastic properties of the chest wall and the ribs return to their resting state. This results in an intrapulmonary pressure of 2 to 3 mmHg above the atmospheric pressure, which results in air moving out of the lungs. Figure 4.2 depicts the breathing process with changes in intrathoracic pressures.

To better understand the process of air flow into and out of the lungs, it is useful to consider the mathematical relationship between pressure differences, resistance, and flow.

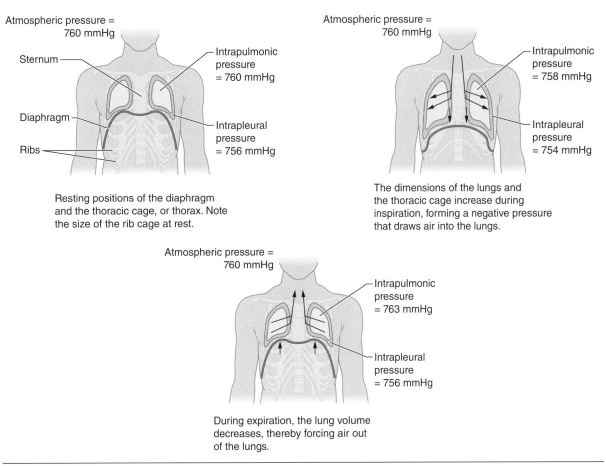

Figure 4.2 Process of inspiration and exhalation.

$$\text{Flow} = (\text{pressure inside lungs - pressure outside lungs})/\text{airway resistance} \qquad \text{(Equation 4.1)}$$

The only variable in this equation not yet discussed is airway resistance. This refers to the resistance offered by the anatomic structures of the airway, including the diameter of the nasal passages, mouth, trachea, bronchiole tree, and alveoli. Airflow occurs when the pressure difference is positive and greater than the airway resistance. The flow can be increased by a greater pressure difference (i.e., increased diaphragm contraction) or a reduction in resistance. Recall that some of the air passages have smooth muscle that allows them to contract or dilate, which can increase and decrease, respectively, the resistance to airflow. Chronic obstructive pulmonary disease (COPD), asthma, and other obstructive lung diseases result in narrowed airway passages and limited ability to dilate and thus result in an increased airway flow resistance (see section titled "Illness" for more information).

Lung Volumes and Capacities

Understanding lung volumes is important for assessment of functional limitations and exercise capacity.

A pulmonary function test (PFT) is an assessment performed using a spirometer to measure lung volumes and functional capabilities. Spirometers can cost as little as a few hundred dollars for a limited-capability device to as much as tens of thousands of dollars for whole-body analyzers (e.g., plethysmograph) with multiple capabilities. Additionally, measures can be taken at rest or during submaximal and maximal exercise. The technique of using a spirometer for assessment is known as **spirometry**. The spirometer can assess both inspired and expired airflow and report it in terms of both volume and time. Figure 4.3 depicts the spirogram and the various lung volumes and capacities. Many types of spirometers are available. Original devices were known as *bell spirometers* and had two metal containers, one inverted over the other. The inverted container was made airtight by sealing it in a column of water. Exhaled air entered the container and caused the bell to move up on exhalation and down on inhalation. This movement corresponded to the volume of air movement and was recorded on a chart by a stylus marker attached to the outside of the bell container.

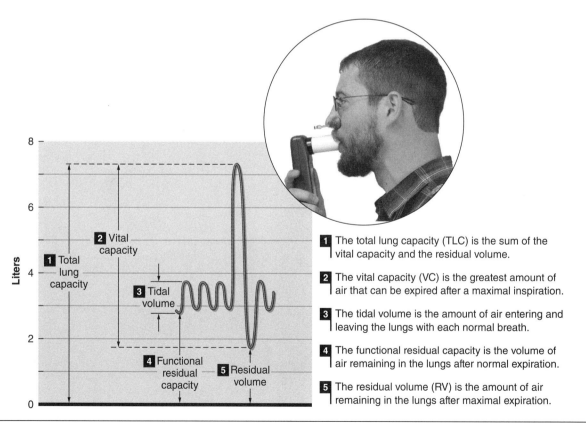

1 The total lung capacity (TLC) is the sum of the vital capacity and the residual volume.

2 The vital capacity (VC) is the greatest amount of air that can be expired after a maximal inspiration.

3 The tidal volume is the amount of air entering and leaving the lungs with each normal breath.

4 The functional residual capacity is the volume of air remaining in the lungs after normal expiration.

5 The residual volume (RV) is the amount of air remaining in the lungs after maximal expiration.

Figure 4.3 Lung volumes and capacities as determined using a spirometer to develop this spirogram.

Reprinted, by permission, W.L. Kenney, J.H. Wilmore, and D.L. Costill, 2015, *Physiology of sport and exercise*, 6th ed. (Champaign, IL: Human Kinetics), 179.

Spirometry Terminology

$\dot{V}$ = volume per unit of time (typically 1 min)

T = tidal

V = volume

A = alveolar

D = dead space

I = inspired

E = expired

P = pressure

Q = perfusion or flow

These types of devices rarely are used today. Today's spirometers are stand-alone, fully automated computerized devices that are quite portable; other spirometers can be part of a gas exchange unit known as an *indirect calorimeter*. When well maintained and calibrated, these devices are very easy to use and provide accurate information.

The reader should become very familiar with a few of the volumes and capacities determined by spirometry. The tidal volume (V_T) is the volume of air inspired or expired in each breath. When the V_T is multiplied by the respiratory frequency (f_R) in breaths per minute, the value is termed the *minute ventilation* $(\dot{V}_E)$. Thus, the $\dot{V}_E$ is the volume of air inhaled or exhaled per minute and typically is expressed in liters per minute. At rest, breathing f_R typically is in the range of 8 to 16 breaths $\cdot$ min^{-1}. The V_T varies by individual sex, age, and size. Thus, the $\dot{V}_E$ can be calculated as follows:

$$V_T \times f_R = \dot{V}_E$$

Example: 1.2 L $\cdot$ breath^{-1} $\times$ 12 breaths $\cdot$ min^{-1} = 14.4 L $\cdot$ min^{-1}

(Equation 4.2)

Relevant resting spirometry assessments include the forced vital capacity (FVC), which is the total amount of air that can be exhaled in a maximally forceful manner, and the forced expiratory volume in the initial second of an FVC (FEV$_1$). The FVC and FEV$_1$ can be used to assess for abnormal lung function and to predict other lung values. For instance, the maximal ventilation volume (MVV), the theoretical maximal total ventilation possible for an individual over 1 min, can be estimated by the FEV$_1$ × 40.

Table 4.1 presents normal values for several common static and active measures. The ranges provided for each measure are from healthy, nonsmoking individuals and take into account sex and age, both of which can influence these values. Normal values can be used to assess whether an individual has lung function values that are expected or outside of the normal range. Hankinson et al. (15) reported normal values specific to race (as well as sex and age) for FVC and FEV$_1$ in a group of white, black, and Hispanic individuals. This is important because differences in values may exist based on race. For instance, they report that white subjects had higher FVC and FEV$_1$ values than either black or Hispanic individuals. This type of specific information can be important when determining whether someone has reduced lung function secondary to disease or other factors.

Ventilation

Pulmonary ventilation refers to the movement of air between the external environment and the lungs via inhalation and exhalation. At rest the $\dot{V}_E$ of a reference male (i.e., 70 kg, 175 cm) is approximately 7 to 8 L $\cdot$ min^{-1}. It will be lower for someone smaller in stature based on a smaller lung size and smaller V_T. During inhalation, all of the inhaled air does not result in gas exchange within the alveoli (i.e., **alveolar ventilation**; V_A). Because the entire airway from the mouth to the alveoli contains the inhaled air, some air will remain in the nonconducting structures above the alveoli. Space in which air is conducted but gas exchange does not occur is termed *dead space*. In this example, this space is known as **anatomical dead space**. Some of the alveoli into which air is conducted have very little or no blood flow, and therefore no gas exchange takes place; this is termed **alveolar dead space**. When this amount is added to the anatomic dead space, the sum is termed the **physiologic dead space**. During resting breathing, the pulmonary ventilation is distributed primarily to the alveoli at the bottom region of the lungs.

Blood Flow

Circulation is a vital part of O_2 and CO_2 gas exchange. The pulmonary circulation moves from the right side of the heart, and the mixed-venous blood then moves through the lungs and back to the left side of the heart and into the systemic circulation. Figure 4.4 shows the circulatory pressures of the pulmonary and system circulations. At rest the cardiac output to both the lungs is in the range of 4 to 6 L $\cdot$ min^{-1}; it is about the same for systemic circulation from the left ventricle to the visceral organs and skeletal muscles. It is important to note that even though blood flow is similar, a large pressure difference exists between the pulmonary and systemic vascular systems. The pulmonary vasculature is a low-pressure system because the vascular resistance

Table 4.1 Important Lung Function Tests and Volumes

Test name	Purpose	Volume measured	Range	Measured at rest or during exercise?
Vital capacity (VC)	To assess the maximum amount of air that can be exhaled	Liters	4-5 L	Rest
Residual volume (RV)	May be used in body composition analysis during underwater weighing	Liters	1.3-2.4 L	Rest
Tidal volume (V_T) or milliliters	To assess the volume of air inhaled or exhaled with each breath as well as changes with increasing exercise	Liters per breath	400-600 mL · breath^{-1} Body size dependent	Both
Forced vital capacity (FVC)	To assess the total volume that can be forcefully expired	Liters	2.6-5.1 L	Rest
Forced expiratory volume in 1 s (FEV_1)	To measure the amount of air forcefully exhaled in 1 s to assess for potential airway collapse	Liters	1.9-4.1 L	Rest
Respiratory frequency (f_R)	To assess the rate of breaths per minute	Not applicable	8-12 breaths · min^{-1} at rest and 20-50 breaths · min^{-1} during exercise	Both
Maximal ventilatory volume (MVV)	To compare with exercise maximum ventilation for pulmonary limitation	Liters per minute Assesses the maximal depth and rate of breathing for 10-15 s and extrapolates the final value to 60 s	90-170 L · min^{-1} Age and sex dependent	Rest

Based on Neder (26); Neder (27).

is very low. Even during periods of higher flow, such as during exercise, the pulmonary pressures stay relatively low due to the vasculature's ability to dilate and recruit unused capillaries from the upper (apex) regions of the lung that are not perfused at rest in the upright position. The major deterrent to perfusion of the lung apex at rest is gravity. When an individual assumes a supine (lying) position, the effect of gravity is nullified and these lung regions are perfused uniformly with the rest of the lung.

Blood flow to the capillaries surrounding an alveolus can be described in relation to the ventilation to the same alveolus. An ideal alveolar ventilation-to-perfusion ratio ($\dot{V}_A/\dot{Q}$) is about 1.0. This means that the ventilation matches the perfusion or blood flow equally and likely would result in the best conditions for maximal gas exchange. At rest, the upper (apex) region of the lungs has a high $\dot{V}_A/\dot{Q}$ due to lower amounts of blood perfusion compared with the base of the lung. A high $\dot{V}_A/\dot{Q}$ ratio means that there is poor gas exchange because much of the ventilation does not come in close contact with blood. To put these $\dot{V}_A/\dot{Q}$ ratios into perspective, consider the following:

Lung apex: $\dot{Q} = 0.1\ L \cdot min^{-1}$ and $\dot{V}_A = 0.25\ L \cdot min^{-1}$; therefore, $\dot{V}_A/\dot{Q} = 2.5$ (inadequate gas exchange)

Lung base: $\dot{Q} = 1.25\ L \cdot min^{-1}$ and $\dot{V}_A = 0.75\ L \cdot min^{-1}$; therefore, $\dot{V}_A/\dot{Q} = 0.6$ (adequate gas exchange)

(Equation 4.3)

Because a portion of the right and left lungs lies above the heart, a hydrostatic force (up to 25 mmHg) opposing blood flow must be overcome to perfuse the upper portions of the lungs. The $\dot{V}_A/\dot{Q}$ increases (i.e., ventilation is greater than perfusion) in the middle and upper regions of the lungs.

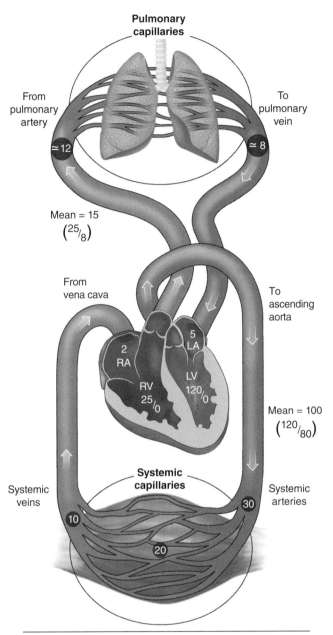

Figure 4.4 A comparison of mean pressures (mmHg) in the pulmonary and systemic circulation.

Gas Diffusion

Inhaled air contains approximately 20.93% O_2 and 0.03% CO_2; the majority of the remainder is made up of nitrogen. Nitrogen and the other gases that make up a very small amount of the air we breathe are considered inert. This means that they do not exchange with the blood in the lungs and therefore are not a part of metabolism. Diffusion of O_2 and CO_2 in the lungs is dependent, in part, on the **partial pressure (P)** difference between these gases and the blood passing by an alveolus. The partial pressure of O_2, CO_2, and nitrogen (N_2) at sea level (760 mmHg) is as follows:

$$PO_2 : 0.2093 \times 760 \, \text{mmHg} = 159 \, \text{mmHg}$$
$$PCO_2 : 0.0003 \times 760 = 0.23 \, \text{mmHg}$$
$$PN_2 : 0.7904 \times 760 = 600.7 \, \text{mmHg}$$
$$\text{(Equation 4.4)}$$

Fick's law of diffusion states that the rate of gas diffusion is proportional to the diffusion coefficient of the specific gas, the tissue area and its thickness that is available for diffusion, and the partial pressure difference of the gas:

$$\text{Rate of gas diffusion} = A/T \times D \times \left(P_1 - P_2\right)$$
$$\text{(Equation 4.5)}$$

where A is the tissue area, T is the tissue thickness, D is the gas diffusion coefficient, and $P_1 - P_2$ is the partial pressure difference between two tissues. Figure 4.5 depicts the anatomy of the alveolar and capillary membranes—the location of gas diffusion in the lungs.

This equation shows us that the greater the difference in pressure across the tissue (i.e., the driving pressure), the larger the surface available for diffusion, and the thinner the tissue, the higher the rate of diffusion of a gas. Diffusion in a normal, healthy lung is effective because the alveoli have a very thin membrane and the surface area available for diffusion is very large. Figure 4.6 provides an illustration of the pulmonary blood flow and gas exchange that take place in the lungs and the cells of the body. Note the difference between PO_2 and PCO_2 at the lungs, driving O_2 across the capillary membrane and CO_2 across the alveolar membrane. A similar situation exists at the cell level, driving O_2 into the cell and CO_2 out of the cell and into the blood.

The term *ventilation,* or breathing, refers to the movement of air into and out of the lungs. This, along with the diffusion of O_2 into and CO_2 out of the blood, is referred to as *external respiration.* **Internal respiration** (also called *cellular respiration*) is the transport of the O_2 and CO_2 in the blood and the vascular system and the diffusion of these gases with tissue at the capillary level. Although both O_2 and CO_2 primarily are carried in the blood, very little of either is dissolved directly in the blood. O_2 has a high affinity for hemoglobin (Hb), which is a protein in the red blood cell (i.e., erythrocyte). Four molecules of O_2 can attach or bind to a single Hb, thus forming a fully saturated molecule of **oxyhemoglobin**. About 70 times more O_2 is carried by Hb than is dissolved in the blood. The term *deoxyhemoglobin* describes Hb without bound O_2. The amount of Hb can vary within an individual, but the Hb molecule is located in each of the 30 trillion red blood cells in the body. Low levels of Hb are associated with **anemia**. The normal amount of Hb is lower in females than in males (approximately 12-15 vs.

Life Span

As with all other systems of the body, the pulmonary system undergoes changes secondary to aging. These begin to occur after full lung maturity is achieved by the age of 20 to 25 yr (33). Structural changes occur in the bones and muscles supporting both the pulmonary structure and its function. Bones of the ribcage thin and change shape with aging, which can affect their ability to expand and contract with breathing. The diaphragm and accessory breathing skeletal muscles lose mass and weaken, which negatively affects inspiratory and expiratory ability. Smooth muscle of the airways and its ability to maintain airway **patency** are also affected. Changes in lung tissue can result in deformities of the alveoli and may lead to air trapping and reduced O_2 and CO_2 exchange between the lungs and blood. Changes to the nervous system may affect the control of breathing and the ability to remove particles from the airways by coughing. An example of functional change with aging is the noted decline in the PFT value "forced expiratory volume in 1 s" (FEV_1), which declines at a rate of 3% to 5%/yr beginning at about age 25 and decreases more rapidly with older age (39). All other common PFT measures also decline similarly (30). Finally, immune system changes that reduce the body's ability to fight infection can lead to an increase in lung infections and the development of pneumonia.

During maximal exercise, most respiratory measures are reduced in the elderly compared with younger individuals. These include total ventilation volume ($\dot{V}_E$), V_T (L · breath^{-1}), breathing rate, and inspiratory capacity (19). The respiratory system typically is considered overbuilt for exercise because in normal conditions, even maximal exercise does not approach the system's full capability. For instance, most individuals have 40% to 60% of their respiratory reserve (i.e., breathing reserve) remaining when they fatigue from high-intensity exercise. However, with aging, the ability to breathe comfortably can be affected. Up to 30% of healthy individuals over age 65 yr report excessive breathlessness during activities of daily living and light exercise (19). With aging there is an accelerated ventilatory response to exercise. $\dot{V}_E$, $\dot{V}_E/\dot{V}O_2$, and $\dot{V}_E/\dot{V}CO_2$ values are consistently higher at similar relative intensities, which likely is the combined result of increased physiological dead space, earlier lactic acidosis reflected by an earlier ventilatory threshold, and reduced efficiency of the locomotor skeletal muscles (19). Additionally, a relationship exists between the level of breathlessness experienced and the inspiratory muscle effort performed during exercise in fit elderly individuals (20).

The good news is that performing regular exercise can influence the age-related decline in pulmonary function (13). In a **prospective** study examining the influence of regular physical activity on the decline of pulmonary function with aging, it was noted that a slower rate of decline of FEV_1 occurred in those who reported regular physical activity (28). This was most evident in the highest caloric expenditure **tertile** of daily physical activity and was not affected by smoking. Although no mechanisms were sought, it has been postulated that physical activity might reduce the aging effect on bone and skeletal muscle changes that occur, reduce the aging-related elastic recoil loss, and improve inspiratory muscle endurance (6).

14-17 g · dL^{-1} of blood, respectively); thus, the O_2 carrying capacity of a healthy male is greater than that of a female. Assuming 100% oxygenation of Hb, then the following are true:

Male : 15.0 g of Hb · dL^{-1} of blood × 1.34 ml of O_2 · g^{-1} of Hb = 20.1 mL of O_2 · dL^{-1} of blood

Female : 13.0 g of Hb · dL^{-1} of blood × 1.34 ml of O_2 · g^{-1} of Hb = 17.4 mL of O_2 · dL^{-1} of blood

(Equation 4.6)

At sea level, the normal alveolar O_2 pressure (P_AO_2) is approximately 100 mmHg. When the PO_2 of the air decreases (e.g., altitude, pollution), the pressure gradient between the air and the alveoli is reduced. This possibly will result in a reduction of both PO_2 and P_AO_2 and thus arterial pressure of O_2 (PaO_2).

Figure 4.7 illustrates the oxyhemoglobin dissociation curve. This shows the effect of changes in P_AO_2 on Hb saturation and PaO_2 concentration. Both Hb and PaO_2 concentration remain nearly unchanged even with a decrease in PO_2 to 80 mmHg. There is still little change with a drop to 60 mmHg. When the PO_2 decreases further, the Hb saturation and PaO_2 concentration decrease rapidly and significantly affect exercise capacity. The steep portion of the curve reflects the unloading of O_2 in the metabolically

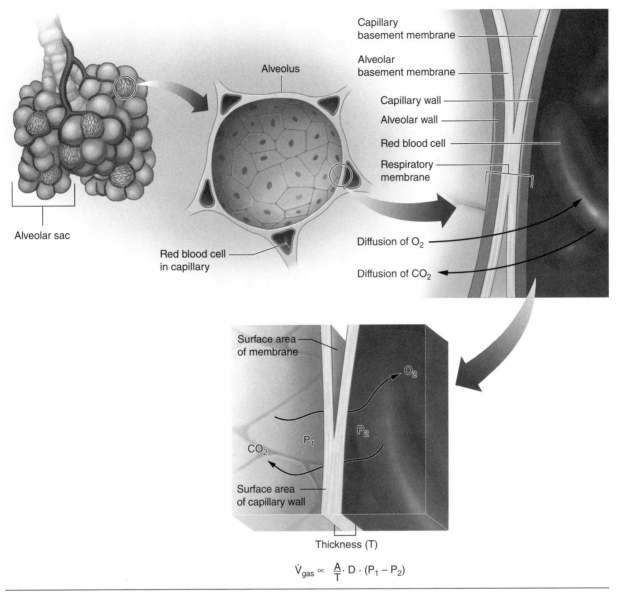

Figure 4.5 The diffusion anatomy and process depicting gas exchange (oxygen [O_2] and carbon dioxide [CO_2] diffusion). The rate of diffusion ($\dot{V}$gas) of these gasses is proportional to the area (A), the gas specific diffusion constant (D), and the difference in partial pressure of the specific gas (P1-P2) and inversely proportional to the thickness (T) of the tissue through which diffusion occurs.

active tissues, which have a low O_2 concentration. Other factors affecting O_2 and Hb dissociation are temperature, PCO_2 levels, 2,3-bisphosphoglycerate concentration, and a reduced blood pH. Each of these contributes to the Bohr effect, which results in a shift of the oxyhemoglobin dissociation curve down and to the right (see figure 4.7). The net result is a greater dissociation of O_2 from Hb at the tissue level. Other factors contributing to the shift of the curve are listed in figure 4.7.

CO_2 is also dissolved in the plasma and in the red blood cells, but the majority of CO_2 is carried in the blood in the form of carbonic acid (H_2CO_3). Carbonic acid dissociates to bicarbonate (HCO_3^-) and a free

proton (H^+), both of which are stored in the blood. The Haldane effect also affects ability to carry CO_2 in the blood. The Haldane effect states that when the partial pressure of O_2 decreases (i.e., Hb unloading), affinity of Hb for CO_2 increases. This results in an upward and leftward shift of the CO_2–Hb dissociation curve and a greater loading of CO_2 onto Hb at any given PCO_2 level.

At a PO_2 of less than 60 mmHg, the affinity of O_2 for Hb is less than that for myoglobin. This results in the transfer of O_2 from Hb to myoglobin, allowing for continuous oxygenation of the cell's mitochondria. At the mitochondria (m) the PmO_2 is 10 mmHg or less. Figure 4.8

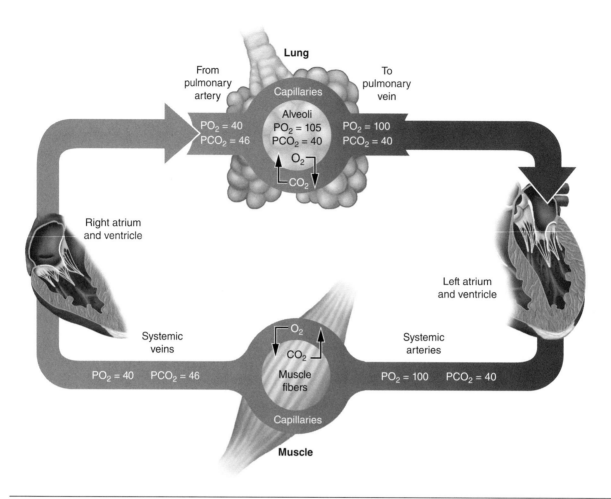

Figure 4.6 Partial pressure of oxygen (PO_2) and carbon dioxide (PCO_2) in the blood throughout the circulatory system.

shows a comparison of the dissociation of O_2 from Hb and myoglobin. The entire processes of external and internal respiration are carried out in a single, continuous manner in order to maintain a constant transport of O_2 to the mitochondrion for the production of adenosine triphosphate (ATP) via oxidative phosphorylation.

EXERCISE AND LUNG FUNCTION

Although much research has been performed to understand the control of the respiratory response to physical activities and exercise, much more remains to be determined. Debate exists over the primary mechanisms for controlling the respiratory system during exercise and the exercise **hyperpnea** at the onset of an increase in work rate. The goal of regulation is to appropriately supply O_2 to the working muscles and remove CO_2 at a rate that maintains physiologically appropriate arterial and tissue

gas levels. Thus, these responses must adjust as exercise intensity increases (or decreases) to reduce the risk of O_2 debt and CO_2 surplus. For this to occur, there must be an adequate increase of ventilation volume, which is matched by the circulation via an increased cardiac output and an increase in the arterial–venous O_2 difference ($A\text{-}\dot{V}O_2$ difference) (38). This needs to occur at a rate that is sufficiently matched with the rate of change of exercise intensity. As stated, the control producing tightly regulated changes is controversial. The three primary competing hypotheses are as follows (38):

1. Feedback mechanisms involving factors such as sensing chemical factors that are closely related to changes in the metabolic rate

2. Feedback from nonrespiratory sources, including the heart, changes in body temperature, and chemical or mechanical receptors in active skeletal muscle

3. A feed-forward mechanism in the brain that drives both movement and respiratory adaptations

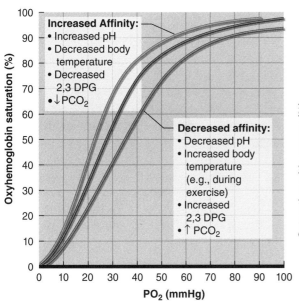

Figure 4.7 The oxyhemoglobin dissociation curve showing the effects on oxygen loading and unloading of oxygen to and from hemoglobin due to a variety of dynamic factors (e.g., temperature, H+ ion concentration, and 2-3 DPG concentration).

Reprinted, by permission, W.L. Kenney, J.H. Wilmore, and D.L. Costill, 2015, *Physiology of sport and exercise,* 6th ed. (Champaign, IL: Human Kinetics), 186.

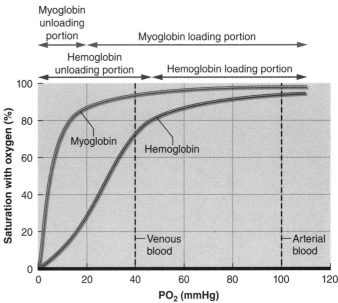

Figure 4.8 Comparison of the affinity of oxygen for myoglobin and hemoglobin.

Reprinted, by permission, W.L. Kenney, J.H. Wilmore, and D.L. Costill, 2015, *Physiology of sport and exercise,* 6th ed. (Champaign, IL: Human Kinetics), 189.

It has been proposed that all three mechanisms may play a role in respiratory control during exercise (11).

Ventilation changes from rest to exercise have a pattern that reflects both mechanical (e.g., airway adaptations, flow rate changes) and muscular control changes. The pattern of these changes depends on, among other factors, the mode of exercise (e.g., cycling, running, swimming), positioning of the body (upright, seated, supine), exercise intensity, and underlying pathologies (e.g., lung or heart disease). This process is continually under the control of a complex neural system with both negative and positive feedback.

Both V_T and f_R increase as exercise begins and progresses in intensity. As a result, the product of V_T and f_R (total $\dot{V}_E$) also increases. This increase occurs immediately and proportionately to a change in exercise intensity. As with O_2 consumption $(\dot{V}O_2)$, during steady-state submaximal exercise, $\dot{V}_E$ levels off after an abrupt increase. As exercise intensity reaches vigorous levels, the increase in $\dot{V}_E$ is almost entirely attributable to an increased f_R. The f_R may become very rapid (two to three times the resting rate and up to 60 breaths · min^{-1}) in some individuals, particularly athletes and very fit individuals who require very high $\dot{V}_E$ values at peak exercise. High

f_R may also be seen in individuals with diseases such as COPD and congestive heart failure because their ability to increase V_T is markedly low. In these cases, $\dot{V}_E$ at peak exercise is considerably below normal. The increase in V_T typically levels off at about 50% to 60% of the vital capacity. In some cases, the V_T may decrease due to the reduced time for airway filling during an inhalation at a very high f_R. It is important to note that the increase in f_R is attributable to a reduction in both inspiratory and expiratory time, with expiratory time decreasing more than inspiratory time.

During quiet rest and lower levels of exercise intensity, the required respiratory muscle contraction needed for breathing occurs only for inspiration and involves only the diaphragm. Resting breathing is regulated by intrinsic respiratory neurons located in the medulla oblongata. In anticipation of exercise, there is a small increase in $\dot{V}_E$ as a result of voluntary stimulation from the higher brain cerebral cortex (i.e., central nervous system command acting on the medulla oblongata). Expiration is passive at rest; the use of expiratory abdominal muscles, as observed using electromyography, begins at very low levels of exercise intensity. This can be observed as the end-expiratory lung volume begins to decrease. This is beneficial during exercise because it allows for a greater V_T without an increase in inspiratory pressure. Thus, the accessory inspiratory muscles are not invoked until much higher intensity levels. At high levels of $\dot{V}_E$ during high-intensity exercise, the ability to inspire at high pressures (and thus volume) is limited by the ability of the inspiratory muscles to contract sufficiently and in a coordinated manner. Expiration, however, is not limited by the ability to generate pressure but rather by airway mechanics. For instance, those with airway obstructive disease (e.g., COPD, asthma) experience airway obstruction in the form of inflammation or airway collapse during exhalation.

In summary, the following occur during exercise to result in needed ventilation changes:

- $\dot{V}_E$ increases rapidly in the initial several seconds of exercise via the effects of central command and possibly joint or muscle receptor activation.

- A slower $\dot{V}_E$ increase occurs after the initial rapid increase, which can level off during submaximal exercise. These responses are likely controlled by the central command as well as by chemical stimuli such as changes in the partial pressure of CO_2 (PCO_2) and H^+ (i.e., reduction in pH) in the cerebral spinal fluid and arterial blood that stimulate chemoreceptors in the medulla. Additionally, peripheral chemoreceptors in the aorta and carotid arteries also provide feedback to the medulla. Evidence for this may be seen in the very tightly coupled relationship between CO_2 output ($\dot{V}CO_2$) and $\dot{V}_E$ during incremental exercise (32).

The $\dot{V}_E$ can increase to values exceeding 150 and 200 L/min (Body Temperature Pressure Saturated [BTPS]) in female and male athletes, respectively. Highly trained athletes often can use nearly 100% of their predicted peak exercise ventilatory capacity, whereas moderately trained or untrained individuals will fatigue with 30% to 50% of their predicted ventilatory capacity remaining. During recovery, as soon as exercise intensity is reduced or exercise is stopped, a sudden decrease occurs in $\dot{V}_E$ via central command. This is followed by a gradual reduction in $\dot{V}_E$ toward resting values that is proportional to the decrease in chemoreceptor stimulation as PCO_2 and pH return to resting values. Table 4.2 provides a summary of the ventilatory changes that occur during and after exercise.

Hyperventilation

Although many believe that the term *hyperventilation* refers to excessive breathing (i.e., hyperpnoea) during exercise, it actually refers to the increased ventilation associated with an increase in arterial PCO_2 and the need to maintain PCO_2 levels at normal (40 mmHg). Hyperventilation also can occur at rest when ventilation rate increases and excessive CO_2 is exhaled. This may occur when someone is experiencing anxiety or pain. Resting hyperventilation can reduce PCO_2 levels down to 20 mmHg. Individuals who hyperventilate at rest may feel lightheaded and breathless. A common treatment is the rebreathing of air (often using a paper bag) to quickly increase PCO_2 levels.

Blood Gases

During exercise, the body strives to maintain the partial pressure of both CO_2 ($PvCO_2$ = approximately 50 mmHg in venous blood) and O_2 (PaO_2 = approximately 80-100 mmHg in arterial blood). Hypoxemia can occur during normal exercise in those with gas diffusion abnormalities (e.g., COPD). It also has been noted to occur at sea level in well-trained endurance athletes (9). A well-trained individual may reduce their PaO_2 to below 80 mmHg. This hypoxemia may occur in up to 50% of well-trained endurance runners when their exercise intensity exceeds 80% of $\dot{V}O_2$max (29). One theory is that this may be related to very high cardiac outputs, which increase the transit rate of each red blood cell in the pulmonary capillaries and leads to a reduced time for gas exchange. Other possible mechanisms exist, including inadequate ventilation (i.e., $\dot{V}_E/\dot{Q}$ mismatch), impaired gas diffusion, and a reduction of the diffusion capacity of the lungs for O_2. It is possible that no one of these proposed mechanisms is solely responsible but rather that multiple changes contribute to the hypoxemia observed in a portion of the well-trained athletic population.

Regulation of Blood pH: Acid–Base Balance

The respiratory system serves as a mechanism to control blood pH. Molecules that contribute protons (i.e., H^+) to a solution are called *acids*. Alternately, a base contributes hydroxyl ions (OH^-). Each is capable of combining with the other (i.e., H^+ and OH^-) to offset a change in pH. More H^+ leads to acidity and a reduction of pH, whereas OH^- increases pH. At rest blood is slightly alkalinic, with a pH of approximately 7.4 (7.0 is a neutral pH). During vigorous-intensity exercise, H^+ accumulates in the muscle and the blood, which can cause the pH to decrease below 7.0. Traditional thought is that lactic acid accumulation is the primary source of the H^+ and the cause of acidosis (21). It has been suggested that the H^+ are produced from two major sources during exercise—glycolytic reactions

Table 4.2 Ventilatory Changes Before, During, and After Exercise

Phase	Change	Controlling mechanism
Rest	---	Central and peripheral chemoreceptors influencing intrinsic pattern established by the medulla
Before exercise	Moderate increase	↑ Central command (cerebral cortex)
During exercise Immediate	Rapid increase	↑ Central command and possibly
Mid	Steady-state or slower rise	↑ neural stimuli to medulla caused by activation of muscle and joint receptors
End	Continued or rapid (hyperventilation) increase	Central or peripheral chemoreceptors reacting to ↑ PCO_2 and ↓ pH in blood or cerebral spinal fluid ↓ Central command ↓ Input from central and peripheral chemoreceptors as PCO_2 and pH normalize
Recovery Immediate	Rapid decrease	↓ Central command
Later	Slower decrease toward rest	↓ Chemoreceptor stimulation

Reprinted from M.L. Foss, S.J. Keteyian, and E.L. Fox, 1998, *Fox's physiological basis for exercise and sport*, 6th ed. (Pittsburgh, PA: William C. Brown). By permission of the authors.

and adenosine diphosphate (ATP) breakdown (ATP + $H_2O \rightarrow ADP + Pi + H^+$ [Pi = inorganic phosphate])—and that the production of H^+ from pyruvate actually serves as a cellular proton buffer that attempts to maintain a normal pH (31). Additionally, carbonic acid (H_2CO_3) is formed from the CO_2 produced via aerobic metabolism. Reductions in muscle and blood pH impair exercise performance and can quickly result in fatigue. The chemical formulas depicting carbonic acid production and H^+ buffering are as follows:

Aerobic: $CO_2 + H_2O \leftrightarrow H_2CO_3 \leftrightarrow H^+ + HCO_3^-$

Anaerobic glycolysis: $H^+ + HCO_3^- \rightarrow H_2O + CO_2$

(Equation 4.7)

Increased H^+ leads to a reduction in pH in both the arterial blood and the cerebral spinal fluid. This increased H^+ concentration stimulates chemoreceptors, resulting in increased pulmonary ventilation (i.e., both f_R and V_T increase). This excessive respiration facilitates the removal of CO_2, resulting in the reduction of H^+ concentration and an increase in pH. For example, alterations in the rate and depth of respiration can have immediate effects on body fluid pH, as noted in figure 4.9. When

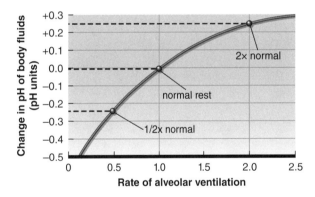

Figure 4.9 Hyperventilation, which results in a reduced PCO_2 in the blood, will cause the blood and body fluid pH to increase. A reduced alveolar ventilation will cause pH to decrease.

Reprinted from M.L. Foss, S.J. Keteyian, and E.L. Fox, 1998, *Fox's physiological basis for exercise and sport*, 6th ed. (Pittsburgh, PA: William C. Brown). By permission of the authors.

hyperventilating at twice the rate of normal, the result is an increase in blood and body fluid pH by as much as 0.25 pH units. A similar opposite change in pH occurs when respiration is reduced to about 50% of normal. During exercise the production of CO_2 from the bicarbonate buffering of lactic acid results in the increase of exhaled CO_2 and drives the respiratory exchange ratio, which is determined by dividing the $\dot{V}CO_2$ by the $\dot{V}O_2$. A value of at least 1.0—and preferably 1.10—often is used as a marker of maximal effort because it reflects the product of a high rate of anaerobic metabolism.

The degree to which the pH of the body fluids is affected by the buildup of CO_2 and the subsequent formation of carbonic acid (H_2CO_3) depends on the amount of bicarbonate (HCO_3^-) available for the buffering process. The amount of HCO_3^- available is known as the *alkali* or *buffer reserve*. The pH of the body fluids is related to the ratio of the concentration of HCO_3^- to the amount of dissolved CO_2, as depicted in this formula:

$$pH = pK + \log\left(HCO_3^- / CO_2\right) \qquad \text{(Equation 4.8)}$$

where pK refers to a constant of the buffer. This is 6.1 for HCO_3^-. This means that when the concentration of HCO_3^- is equal to the amount of dissolved CO_2 (i.e., a 1:1 ratio), the pH of the solution (i.e., body fluids) will be 6.1. Normally, at rest the ratio of HCO_3^- to CO_2 is 20:1 with a pH of 7.4 ($pH = 6.1 + \log^{10} 20 = 6.1 + 1.3 = 7.4$). As shown in figure 4.10, an increase in the concentration of HCO_3^- causes an increase in pH, whereas an increase in dissolved CO_2 (acid added) decreases the pH.

Ventilatory Derived Anaerobic Threshold

During incremental exercise to exhaustion, the body responds metabolically by producing ATP at an ever-increasing rate to meet the exercise demand. In general, blood lactate (see chapter 1) remains relatively unchanged until an individual's $\dot{V}O_2$ exceeds about 50% to 60% of maximum. At these higher exercise intensities, the rate of ATP production requires a greater use of glycolysis, a product of which is pyruvate. Once the rate of pyruvate production is greater than the ability to convert to acetyl-CoA and move into the Krebs cycle, the pyruvate is converted to lactate and H^+ is produced, likely by the accelerated rate of glycolysis and ATP breakdown. Lactate and H^+ move from the skeletal muscle cells and into the blood. The point where the nonlinear increase in blood lactate occurs during exercise is called the *lactate threshold* (also known as the *anaerobic threshold*) or the *onset of blood lactate accumulation*. Assessing blood lactate (see chapter 7) requires the collection and analysis of blood during exercise. Noninvasive methods for identifying the anaerobic threshold are easier than having to take a blood sample. Because ventilation, O_2, and CO_2 are linked to metabolism and are relatively easy to measure, they represent an easier tool for assessing changes in metabolism during exercise.

Ventilation during incremental submaximal exercise increases linearly, to a point, with both O_2 uptake and CO_2 output. As pyruvate production increases during glycolysis, there is an exercise intensity at which it begins to accumulate because the Krebs cycle cannot use it fast enough. This occurs because cytosolic nicotinamide adenine dinucleotide (NADH $+H^+$, which is reduced from NAD^+ during glycolysis) cannot be oxidized quickly enough by the mitochondrial membrane proton shuttle; this is possibly because O_2 required by the exercising muscles cannot be supplied at a sufficient rate. The result is an increased NADH $+H^+/NAD^+$ ratio in the cytosol. When NADH $+H^+$ accumulates, it can be used to oxidize pyruvate to form lactate: pyruvate + NADH $+H^+ \rightarrow$ lactate $+ NAD^+$. Lactate then begins to accumulate in the blood along with H^+; this can lead to a reduction in blood pH as exercise intensity increases. The H^+ are buffered in the blood by bicarbonate (HCO_3^-), which reduces the H^+ and produces the end products H_2O and CO_2. (Details of this process are discussed in the previous section.) Ventilatory drive increases as a result of the reduced blood pH and acts to remove excess CO_2 produced from H^+ buffering. Because of this process, increases in $\dot{V}CO_2$ relative to $\dot{V}O_2$ during exercise can be used to detect the development of lactic acidosis. This detection often is termed the *ventilatory derived anaerobic threshold* (*VDAT*; also known as *anaerobic threshold, gas exchange threshold*, or *gas exchange*

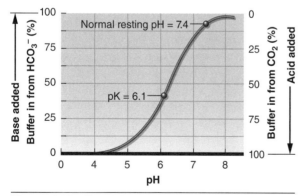

Figure 4.10 The relationship of bicarbonate (HCO_3^-) buffer change on blood pH.

Reprinted from M.L. Foss, S.J. Keteyian, and E.L. Fox, 1998, *Fox's physiological basis for exercise and sport*, 6th ed. (Pittsburgh, PA: William C. Brown). By permission of the authors.

lactate threshold), which is considered by some to be synonymous with the lactate or anaerobic threshold (i.e., the point at which blood lactate accumulates at a rate faster than can be buffered) (5, 40). However, this proposed mechanism is controversial (3, 8).

Several gas exchange methods have been developed to detect the VDAT. The ventilatory equivalent method assesses the $\dot{V}_E/\dot{V}CO_2$ and $\dot{V}_E/\dot{V}O_2$ ratio responses and determines the anaerobic threshold to be the point at which the $\dot{V}_E/\dot{V}CO_2$ ratio remains stable as the $\dot{V}_E/\dot{V}O_2$

ratio increases. The stable $\dot{V}_E/\dot{V}CO_2$ ratio depicts what is termed the *isocapnic buffering period*, where bicarbonate buffering of lactate results in a similar rate of increase of $\dot{V}_E$ and $\dot{V}CO_2$ and thus a flattened $\dot{V}_E/\dot{V}CO_2$ ratio line (figure 4.11*a*). As the bicarbonate stores deplete and blood pH decreases, the ventilatory drive increases and the $\dot{V}_E/\dot{V}CO_2$ ratio begins to increase. This is known as the *respiratory compensation point* (figure 4.11*a*).

As a gas exchange technique for determining anaerobic threshold, the V-slope method has risen in popular-

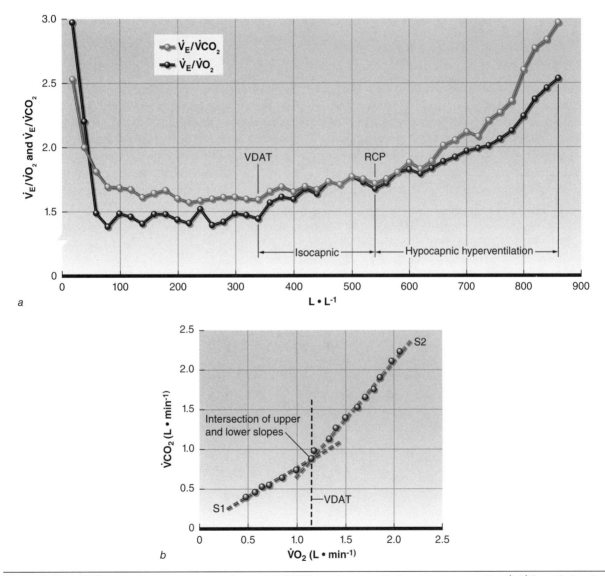

a

b

Figure 4.11 *(a)* The respiratory compensation point (RCP) is where ventilation increases as the $\dot{V}_E/\dot{V}O_2$ ratio begins to rise. *(b)* The modified V-slope method of determining the ventilatory derived anaerobic threshold (VDAT) identifies the point at which $\dot{V}CO_2$ begins to increase at a faster rate than $\dot{V}O2$. The change in slope of the line depicts the VDAT.

ity over the past 30 yr and generally is considered the method of choice for determining the VDAT. It was initially developed by Beaver et al. (1) using computerized regression analyses of CO_2 output ($\dot{V}CO_2$) versus O_2 uptake ($\dot{V}O_2$) slope. Sue et al. developed a visual method of the V-slope (i.e., the modified V-slope) that is strongly correlated with the original method (figure 4.11b) (35). This method, using data from each breath, detects a change in the slope of the line that signifies the point at which CO_2 output (produced from buffering by bicarbonate) increases at a greater rate than O_2 uptake (1,35). The initial slope is slightly below 1 and is referred to as *S1*. The change in slope increases it to greater than 1 and is termed *S2*.

Dyspnea

Dyspnea refers to the sensation of difficulty breathing or feeling out of breath. In healthy individuals, heavy physical activity (e.g., walking stairs) or exercise (e.g., running, swimming) typically result in dyspnea. Several common diseases and conditions can also lead to dyspnea with light activity or even at rest. These include poor fitness levels, congestive heart failure, chronic bronchitis, COPD or emphysema, and asthma. Cardiopulmonary exercise testing (CPET) can help determine which of these possibilities is the cause of an individual's dyspnea.

Exercise Stitch

A common occurrence during exercise is a sharp pain typically located on the right or left sides at the level of the lower ribcage. The pain is sometimes described as being under the ribs. It may occur during early exercise or as exercise intensity increases. It often subsides as exercise is continued. Specific breathing techniques such as pursed-lip breathing (which lowers the airway pressures, thus reducing effort on the next inhalation) and breathing control (e.g., inhaling and exhaling with the rhythm of the legs in running or cycling) may help alleviate the pain. The specific origin of the pain is not well understood, but one suggestion includes a lack of O_2 to the respiratory muscles, including the diaphragm. This may cause the diaphragm to spasm, resulting in pain.

Smoking

Smoking, particularly one or more packs per day for many years, can greatly affect the O_2 cost of ventilation.

At rest, the work required of the ventilatory muscles to overcome the elastic recoil of the lungs and resistance to air flow of the pulmonary passages is minimal—only 1% to 2% of total body $\dot{V}O_2$. This can increase to 8% to 10% during high-intensity or vigorous exercise (34). Exercise training enhances ventilatory efficiency (figure 4.12) so that the ventilation volume is lower at any submaximal workload. Smoking, particularly cigarette smoking, can affect the ventilation–O_2 consumption relationship both acutely (i.e., one cigarette) and chronically (i.e., packs of cigarettes per day for many years). Acutely, the O_2 cost to breathe is higher in those who smoke directly before an exercise bout than in those who abstain or who never smoke (figure 4.13). Chronic smoking can result in obstructive lung disease (i.e., COPD), which leads to air becoming trapped in the lungs, a reduced area for gas exchange, and a necessary increase in ventilation at any exercise workload. For those who smoke, the added O_2 cost of excessive use of the ventilatory muscles to breathe reduces the O_2 availability to the skeletal muscles during exercise, leading to a reduced exercise capacity. This is compounded by the carbon monoxide (CO) inhaled, which has a 240 to 300 times greater affinity for Hb than does O_2, thus reducing the O_2 carrying capacity of Hb.

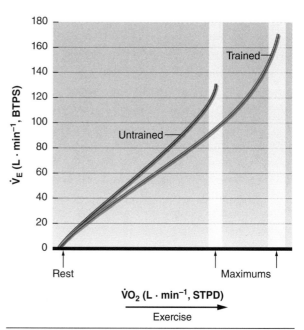

Figure 4.12 Effects of acute exercise on minute ventilation ($\dot{V}_E$) in relation to $\dot{V}O_2$ in trained and untrained subjects.

Reprinted from M.L. Foss, S.J. Keteyian, and E.L. Fox, 1998, *Fox's physiological basis for exercise and sport,* 6th ed. (Pittsburgh, PA: William C. Brown). By permission of the authors.

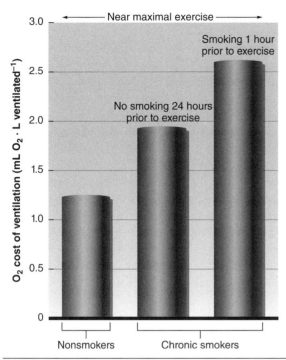

Figure 4.13 The oxygen cost of ventilation in chronic cigarette smokers is greatly increased during near-maximal exercise. This is increased further if smoking occurs within one hour of an exercise bout. Abstinence for 24 hours will improve the oxygen cost of ventilation, but not to nonsmoking levels.

Reprinted from M.L. Foss, S.J. Keteyian, and E.L. Fox, 1998, *Fox's physiological basis for exercise and sport,* 6th ed. (Pittsburgh, PA: William C. Brown). By permission of the authors.

These acute effects of smoking on exercise ability can be reduced by abstaining from smoking for 24 h or more. This is a standard practice request in clinical programs such as cardiac and pulmonary rehabilitation.

Exercise Training Responses

Detraining, disease, and aging are known to lead to a reduced coordination of the lung–heart muscle axis (4). This means that the movement of air into and out of the lungs, the exchange of O_2 and CO_2 with the circulation, and the delivery of blood to the lungs and the skeletal muscle by the heart, which can occur with remarkable synchronization, uncouple in a coordinated fashion. Once the initial precipitating factor begins this deadaptation process, a vicious cycle often occurs in which the individual begins to use their locomotor muscles less often, leading to a decline in functional capacity, including a loss of adequate respiratory function. In addition, changes that occur with aging include

an increased resistance of the chest wall to expansion during an inhalation, declines in respiratory muscle strength and endurance, reduced elastic recoil, reduced gas exchange surface area, and a reduction of capillaries in the lungs (36). Although one might assume that this loss of respiratory function is a limiting factor in exercise, this likely is not the case in healthy aging. For instance, the breathing reserve ($\dot{V}_E$max:MVV) is not influenced by age and is approximately 40% to 50% at peak exercise in healthy younger and elderly individuals. This means that plenty of $\dot{V}_E$ reserve is left at peak exercise in most healthy individuals (12). The MVV also declines at a slower rate (approximately 6% per decade) than peak $\dot{V}O_2$ (approximately 10% per decade) (36). Additionally, pulmonary gas exchange typically is not a factor in fitness decline; the lung diffusion capacity declines at a rate of approximately 5% per decade. This lack of decline in respiratory function, which is different from overall fitness decline, suggests that the deadaptation previously described does not affect the respiratory system to a significant degree during healthy aging.

Unlike the cardiovascular and neuromuscular systems, the respiratory system has much less of a response to long-term exercise training. No structural changes occur in the lung due to training. Age-related declines in respiratory function appear to be attenuated in those who remain active and exercise throughout their life span. Avoidance of risk factors (e.g., smoking, particle inhalation) leading to disease (i.e., COPD, restrictive disease) is an important factor in maintaining adequate respiratory function with age. Evidence shows that the respiratory muscles used during vigorous-intensity exercise for both inhalation and exhalation are affected by exercise training. However, the degree to which a healthy respiratory muscle system can be affected likely is minimal, particularly in those performing physical activities of daily living and moderate-intensity exercise. Despite this, a common training effect is a reduced $\dot{V}_E$ at a given submaximal work rate after an exercise training regimen (figure 4.12). The specific mechanism of this change is not well understood but may be related to improved O_2 delivery to the skeletal muscles so that the relative intensity of exercise is reduced at a given absolute submaximal work rate. (See "Performance" for more information on respiratory muscle training.)

ILLNESS

Respiratory-related illness can be very debilitating for the affected individual. Treatment, including medical

Performance

High-performance training for team sports (e.g., volleyball, basketball, soccer) and individual sports (e.g., running, cycling, swimming, and triathlon) typically focus on skill development, skeletal muscle strength and power, and cardiorespiratory endurance specific to the particular sport. In today's ultracompetitive athletics environment, the possibility of a unique method of enhancing fitness and performance can be very intriguing. Recently, the notion of respiratory muscle training has received attention. This type of training involves breathing against resistance in an attempt to enhance both inspiratory and expiratory endurance. But why might this be a target of training?

Traditionally used as a rehabilitative technique for those recovering from thoracic surgery and those with various lung diseases, there is evidence (i.e., systematic reviews and meta-analyses) that respiratory muscle training provides a training and performance benefit in both athletic and healthy nonathletic individuals (14, 17). Studies regularly demonstrate the benefit of inspiratory and general respiratory muscle training in different athletes competing in a variety of anaerobic and aerobic sports. However, this benefit is not consistently noted in swimmers, who traditionally have larger lung volumes and more total alveoli compared with similar sedentary individuals and runners. This is true even when accounting for age and body size. Understanding the unique training environment of a swimmer provides some perspective on this finding. Most of the breathing during swim training occurs while the torso is submerged, meaning that each inhalation must overcome resistance to expand the thoracic cage. Additionally, proper breathing mechanics in competitive swimming require the exhalation to occur while the mouth and nose are underwater. Thus, the breath out occurs against the resistance of the water. Might it be that this resistance during inhalation and exhalation mimics what is found when other types of athletes perform resistive breath training? Research suggests that swimmers can develop inspiratory muscle fatigue when exercising at very vigorous intensities at or above the critical velocity (23). This fatigue likely occurs due to a high reliance on anaerobic metabolism and the subsequent accumulation of metabolic byproducts (e.g., H^+, CO_2).

Breath trainers have been developed and marketed to aid in both inspiratory and expiratory muscle fatigue training. Kilding et al. (22) used a commercial inspiratory device in swimmers who performed 30 dynamic inspiratory efforts against a pressure threshold of 50% of their measured maximal inspiratory pressure twice per day for 6 wk. This was compared against a **sham** group, who also performed this training but against only 15% of their maximal inspiratory pressure capacity. After 6 wk, the trained group showed a significantly improved swimming velocity over sprint, middle-distance, and distance events and a lower perceived respiratory exertion. If an individual is healthy and performing a considerable amount of aerobic and resistance training, a question arises about the possible mechanisms allowing for an increase in respiratory performance in very fit individuals. Although the mechanisms are not clearly understood, this study does suggest that highly trained athletes can improve performance secondary to inspiratory training. Several hypotheses have been postulated, including structural adaptations in the inspiratory muscles (e.g., increased proportion of Type I fibers and size of Type II fibers), the prevention or delay of diaphragmatic muscle fatigue, the modification of sensory input to the central nervous system (which may reduce limb and peripheral effort sensations), and reduced chemosensitivity of Type III/IV receptors that lead to vasoconstriction in the limbs. Although uncertainty exists about these mechanisms, the body of research into respiratory muscle training in highly trained athletes supports potential performance benefits.

therapy and exercise, can help to relieve symptoms and improve functionality. This section briefly reviews the diseases and treatments.

Asthma

Asthma affects 4% to 5% of the U.S. population. Worldwide, there are regions in which the incidence is low (<2.5%) and others where it exceeds 7.5%. Conservatively, asthma may affect 235 million around the globe (37). Those with asthma and subsequent "attacks" have a hyperirritability of the tracheobronchial tree, which can reduce airflow. This airway reactivity may be the result of allergens (e.g., pollens), air pollutants, occupational

exposures, breathing cold air, and even exercise. When an attack occurs due to exercise, the term *exercise-induced bronchospasm* (EIB) is used.

In the case of exercise-induced bronchospasm, exercise typically results in dyspnea after exercise has ceased. This often occurs within the first 5 min of rest after a bout of exercise. A common example is a basketball player who comes out of a game and sits on the bench and feels short of breath after a few minutes. Up to 30 min may be required for full recovery. Precipitating factors during exercise include high levels of $\dot{V}_E$; cool, dry air; and environmental stimuli such as trichloramines, which are produced in pools when chlorine reacts with ammonia compounds from urine or an unshowered body. Like

---

Clinical Pulmonary Rehabilitation

Individuals with pulmonary disease or limitations will benefit from a structured exercise program. In clinical settings (e.g., hospitals, physician clinics), supervised pulmonary rehabilitation programs provide a multitude of beneficial services for those with a chronic pulmonary disease. Commonly treated diseases and conditions include COPD, asthma, bronchitis, cystic fibrosis, and pulmonary fibrosis. Additionally, individuals with conditions of nonpulmonary origin that can affect pulmonary function (e.g., ankylosing spondylitis, Guillain-Barré syndrome, and paralysis of the diaphragm) are prime candidates for referral to pulmonary rehabilitation. These conditions must moderately to severely reduce a person's ability to perform activities of daily living. Clinically, this is assessed by an exercise test with a peak work capacity less than or equal to 5 metabolic equivalents of task (METs) and general PFT values of less than 60% of predicted values for FVC and FEV_1.

It is well established that regular exercise can reduce the feeling of shortness of breath during activity. In addition to exercise, pulmonary rehabilitation provides a comprehensive intervention program that includes smoking cessation assistance and support, general education about the disease process, breathing retraining, and psychosocial support. Therapeutic modalities are emphasized and include medication and O_2 use. Patients report a reduction in associated symptoms (e.g., dyspnea and fatigue), an improved general quality of life, and an improved exercise tolerance (2,25). Although optimal prospective randomized studies have not yet been performed, observational research suggests that there may be an improvement in mortality in those who participate in pulmonary rehabilitation as well as a reduction in health care utilization such as hospitalizations and length of a hospitalization (7,16).

Both leg and arm aerobic exercise training are performed in pulmonary rehabilitation. Improved endurance is consistently demonstrated in studies assessing the effects of this type of training. The mechanisms of exercise improvement are not fully clarified, but investigations point toward improvements in skeletal muscle oxidative enzymes leading to reductions in anaerobic acidosis (i.e., blood lactic acid levels) and ventilation (multiple possible mechanisms, including desensitization to dyspnea and altered chemoreceptor sensitivity) (24). Arm exercise is important for these patients because of the number of activities of daily living that require use of the arms and hands. The limited research performed suggests that arm training results in task-specific (e.g., carry an object) benefits of reduced fatigue and dyspnea. Resistance training with the arms may result in improved aerobic exercise capacity as well as improved strength and general function (18).

traditional asthma, exercise-induced bronchospasm can be treated with medications that result in bronchodilation as well as those that stabilize the cells (mast cells) that produce chemicals that result in bronchospasm. Avoidance of precipitating factors also is an effective treatment strategy.

Chronic Obstructive Pulmonary Disease

Two diseases—emphysema and chronic obstructive bronchitis—are considered in the spectrum of COPD. Emphysema is most often the result of long-term cigarette smoking; more than 50% who smoke develop emphysema. It results in the destruction of the bronchioles, alveolar ducts, and alveoli. This results in overdistension of these airways and the trapping of air in the lungs because the airways have a reduced, or absent, elastic recoil. Pulmonary function testing of these patients demonstrates reductions in expiratory variables, particularly FEV_1.

Chronic obstructive bronchitis is caused by excessive production of mucus for more than 3 mo. Typical symptoms include coughing and **expectoration** of the mucus. The mucus secretion causes the bronchial walls to thicken and negatively affects the ability to exhale. In some situations an individual can have both emphysema and chronic bronchitis. Ultimately, COPD reduces the physical capacity of an individual, often because their arterial O_2 saturation decreases from a normal range of 93% to 99% to under 90% and as low as under 80%. This results in significant dyspnea. Often, a vicious cycle occurs in which a reduced level of physical activity due to COPD results in increasingly more sedentary activity, which also reduces physical fitness and exercise ability. Individuals with COPD often can improve their exercise capacity by participating in pulmonary rehabilitation (see "Clinical Pulmonary Rehabilitation" for more information).

Restrictive Diseases

Restrictive lung diseases are those that reduce the ability of the lungs to fully expand upon inhalation. The cause of this type of disease can be extrapulmonary (e.g., nonmuscular disease of the thorax, including kyphosis and pectus excavatum, and obesity) or intrinsic to the parenchyma of the lung (e.g., various causes of pulmonary fibrosis, including radiation, long-term exposure to dust or asbestos, or sarcoidosis). When assessing a PFT, both the FEV_1 and the FVC are reduced. The reduction of the FVC is greater than that of the FEV_1, reflecting the inability to fully inflate the lungs. This reduction is defined as a measured FVC of more than 20% below the predicted FVC value. These individuals have a reduced exercise capacity as measured by the 6-min walk test, a standard functional evaluation (10). Proposed mechanisms for exercise intolerance include elevated pulmonary vascular resistance leading to an inadequate O_2 delivery, a reduced ability for cardiac output to increase, and pulmonary diffusion impairment. These individuals also can benefit from participating in pulmonary rehabilitation.

Summary

- The respiratory zone of the lungs has a very large area for gas exchange between the alveoli and the blood.
- Through a variety of feedback information, ventilation is responsive to changes from rest to maximal exercise effort with a goal of adequate tissue oxygenation and CO_2 removal.
- The pulmonary system works in synchrony with the skeletal muscles and cardiovascular system to achieve its functional goals, particularly during exercise.
- Exercise training has minimal effect on the pulmonary system, which has less ability to adapt compared with the neuromuscular and cardiovascular systems.
- Several conditions and diseases, including asthma and COPD, can affect pulmonary function and the ability to perform exercise.
- Participation in a pulmonary rehabilitation program can help individuals become more functional and less symptomatic when they are physically active.

Definitions

alveolar dead space—A poorly perfused alveoli in which air reaches the alveoli but does not exchange with the blood.

alveolar ventilation—The total amount of gas (air) entering the lungs per minute.

anatomical dead space—The total volume (approximately 150 mL in average-size humans) of the conducting airways from the nose or mouth to the level of the terminal bronchioles. This air does not exchange with the blood.

anemia—Condition often attributable to a deficiency of red blood cells or Hb.

conducting zone—Made up of the nose, pharynx, larynx, trachea, bronchi, bronchioles, and terminal bronchioles; filters, warms, and moistens inhaled air as it conducts to the lungs.

deoxyhemoglobin—A deoxygenated Hb.

expectoration—Ejecting matter, such as phlegm, from the throat or lungs by coughing and spitting.

external respiration—Occurs at the lung level where O_2 (in) and CO_2 (out) are exchanged with the blood.

Fick's law of diffusion—States that the rate of diffusion across a membrane is directly proportional to the concentration gradient of a substance on the two sides of the membrane and inversely related to the thickness of the membrane.

hyperpnea—Increased depth and rate of breathing.

hyperventilation—The increased ventilation associated with an increase in arterial PCO_2 and the need to maintain PCO_2 levels at normal (40 mmHg).

internal respiration—Occurs at the cellular level where O_2 diffuses in and CO_2 diffuses out of the metabolizing cells.

oxyhemoglobin—An oxygenated Hb. When loaded with O_2, it is red in color.

partial pressure (P)—The pressure a gas exerts by the fraction of a specific gas in a mixture if it occupied the same volume on its own.

patency—Condition of being open, expanded, or unobstructed.

physiologic dead space—The sum of anatomic and alveolar dead space.

prospective—Concerned with or applying to the future.

respiratory zone—Made up of the bronchioles, alveolar ducts, and alveoli; carries air into the lungs for gas exchange.

sham—Something meant to deceive a person.

spirometry—Use of a calibrated instrument to measure the air capacity of the lungs.

surfactant—In the lungs, a substance that lowers surface tension, allowing for lung compliance and expansion; keeps the alveolar spaces dry.

tertile—The division of an ordered distribution into three statistically similar portions.

Chapter 5

Immune and Endocrine System

We would like to thank Micah Zuhl, PhD, and Rachael Nelson, PhD, for their contributions to this chapter.

The study of exercise and immune function recently has come to the forefront of the field of exercise physiology. Intense exercise has been shown to promote a proinflammatory effect, whereas moderate-intensity exercise may have a dominant anti-inflammatory effect and serve as a treatment for inflammatory diseases (e.g., coronary heart disease, diabetes, irritable bowel disease). Improving one's aerobic fitness level has profound positive effects on immune function. This chapter discusses the effect of acute and chronic exercise on the immune system.

IMMUNE SYSTEM COMPONENTS AND FUNCTION

The immune system is complex, and many masters-level students in exercise science have not studied the basics of immunology. Learning the components of proper immune function is essential for appreciating the influence of exercise on this system.

Every day humans are exposed to numerous **pathogens**, which are infectious agents in the form of bacteria, viruses, fungi, parasites, or toxins. Every time a person touches an object, consumes food, or comes into contact with biological substances, they put themselves at risk for infection. Fortunately, humans have a highly developed immune system that is able to identify, target, and remove harmful agents. After exposure to a harmful pathogen, immunity—resistance against future exposures—can be established. For example, a vaccine is an active or inactive form of a virus or toxin that is injected into the body for the purpose of building immunity. Many inflammatory diseases are caused by abnormal or insufficient immune responses.

The main components of the immune system include barriers against the environment (e.g., skin, mucosal membranes), leukocytes (e.g., immune cells), and plasma proteins (e.g., cytokines, chemokines). A pathogen, or infectious agent, must initially break through the physical barrier to gain access to the internal environment. Once this occurs, the leukocytes and plasma proteins work to identify and remove the agent. Several types of leukocytes exist and have various biological functions, which are described in table 5.1. Granulocytes account for the majority of circulating leukocytes (60%-70%); monocytes (10%-15%) and lymphocytes (20%-25%) make up the remainder. Plasma proteins are secreted from tissue epithelial cells, blood monocytes, tissue macrophages (developed monocytes), lymphocytes, and the liver. These molecules produce a variety of immune responses.

The two major arms of the immune system include the innate and adaptive systems. The **innate immune system** is the initial defense system against pathogens and is activated immediately when surface barriers are compromised. Innate immunity is capable of handling most biological intruders by recognizing cellular components of microorganisms and recruiting immune cells and plasma proteins to remove the invader. If the innate system is incapable of removing the foreign substance, then the adaptive system responds through activation of lymphocytes and the production of antibodies. These two systems do not function independently but rather

Table 5.1 Leukocyte Descriptions

Leukocyte	Function	Blood count Total count = 4-11 million · mL^{-1} of blood
Neutrophil (granulocyte)	Phagocytosis	60%-70% leukocyte count 2-8 million · mL^{-1}
Eosinophil (granulocyte)	Phagocytosis of parasites	2%-5% leukocyte count 0.1-0.4 million · mL^{-1}
Basophil (granulocyte)	Involved in inflammation	0%-2% leukocyte count 0.1 million · mL^{-1}
Monocyte (mononuclear cell)	Develops into tissue macrophage, phagocytosis, secretes cytokines	10%-15% leukocyte count 0.2-0.8 million · mL^{-1}
Lymphocyte	Immune response produces cytokines, recognizes antigens, produces antibodies; regulatory role	20%-25% leukocyte count 1.5-3.0 million · mL^{-1}

are in constant communication; for example, the innate system presents an antigen to the adaptive arm, which may respond by synthesizing an appropriate antibody. An antigen is any substance that causes the immune system to produce antibodies. Further details on innate and adaptive immunity are discussed in the following paragraphs.

The innate system provides early defense against infection, provides information to the adaptive system, and responds to instructions from the adaptive system. The components of the innate immune system include the epithelial barrier, phagocytes (neutrophils, macrophages, dendritic cells), specialized lymphocytes (natural killer cells), and the complement system. These components are capable of identifying general features that are shared by common microbes. For example, lipopolysaccharide (LPS) is a component of the cell wall of many bacteria. Cells of the innate system have LPS receptors that allow them to identify a large number of bacterial species. The innate system is capable of identifying general patterns of various bacteria, fungi, and certain antigens (termed *pathogen-associated molecular patterns*). Each time the body is exposed to a certain bacterium, the innate system responds in the same fashion.

Cells of innate immunity contain receptors called *toll-like receptors* (TLRs) that are capable of identifying nonself invading microbes. Several types of TLRs exist and are specific to different types of bacteria. TLR-4, which is expressed mainly on macrophages, neutrophils, and dendritic cells, recognizes LPS. Recognition by the TLRs results in immune cell activation of transcription factors that lead to synthesis of inflammatory cytokines and chemokines. Once produced, cytokines serve several roles, including neutrophil activation, immune cell prolif-

eration, endothelial cell coagulation, and hypothalamus-induced fever. Chemokines mainly serve to stimulate migration of neutrophils and other leukocytes through the endothelium wall to sites of infection. After the pathogen is identified and transcription of inflammatory proteins (cytokines, chemokines) has occurred, how does the innate system remove the invader? If the microbe is in the blood or moves into the tissue, neutrophils and macrophages ingest and destroy it through **phagocytosis**. In this process, the phagocyte engulfs the microbe and destroys it by releasing reactive oxygen species. This event is a major cause of tissue swelling and tenderness. Another method for removal is destroying the host cell if it has become infected. This event is performed by natural killer cells, which recognize the stressed cell through release of cytokines from tissue macrophages. Last, release of **cytokines** begins communication with the adaptive system, enhancing the immune response. The cytokines and chemokines serve important roles during these events, which is why they are important markers for exercise immunologists (see table 5.2 for detailed roles of plasma proteins).

The **adaptive immune system**, which comprises the B and T lymphocytes, rids the body of foreign substances through humoral and cell-mediated responses. Humoral immunity protects the extracellular compartments against microbes by secretion of antibodies from B cells. For example, if a virus is identified in the bloodstream, the B cells respond by producing antibodies that neutralize the virus and by binding to surrounding cells to protect them from infection. Cell-mediated immunity is the process of removing infection by activation of T cells, which remove infected cells or activate phagocytes (macrophages, neutrophils) to remove the invading microbes.

Table 5.2 Important Exercise Plasma Proteins

Plasma protein	Site of release	Function	Exercise response
TNF-α	Monocytes, macrophages, T cells	Activates neutrophils, induces fever, coagulation, acute phase protein release	↑ Intense and prolonged exercise ← → Low-to moderate-intensity exercise
IL-6	Monocytes, macrophages, T lymphocytes, skeletal muscle	Proinflammatory: proliferation of B cell and antibody production Anti-inflammatory: increases synthesis of IL-10 and IL-1ra from monocytes and macrophages	↑ Anti-inflammatory IL-6 Low- to high-intensity exercise
IL-8	Endothelial cells, neutrophils, macrophages	Recruitment of neutrophils	↑ Intense exercise ← → Low- to moderate-intensity exercise
IL-1β	T cells, monocytes, macrophages	Causes fever, pain, T cell proliferation	← → No exercise response
IL-1ra	Monocytes, macrophages, neutrophils, endothelial cells	Inhibits IL-1β release	↑ Moderate- to high-intensity and prolonged exercise
MCP-1	Endothelial cells	Recruitment of monocytes	↑ Moderate- to high-intensity exercise
IL-10	Monocytes, T cells	Inhibits activation of T cells, monocytes, and macrophages	↑ Intense exercise
IL-12	Natural killer cells	Proliferation of T cells	↑ Intense exercise ← → Low, moderate exercise
Acute phase proteins (CRP)	Liver	Activates complement system and phagocytes	↑ Intense exercise

TNF = tumor necrosis factor; IL = interleukin 1; MCP = monocyte chemoattractant protein; CRP = C-reactive protein.

Data from Pedersen (52); Rowbottom (54).

For example, if a virus penetrates and infects a host cell, specific T cells (cytotoxic T cells) destroy and remove the entire cell. The helper T cells and cytotoxic T cells are the main controllers of cell-mediated immunity and are differentiated by a component of their cell membrane called the *cluster of differentiation* (CD). Helper T cells are labeled CD4[+], and cytotoxic T cells are labeled CD8[+]. The CD4[+] helper T cells identify an invading microbe and act by recruiting phagocytic macrophages or activating B cells to produce antibodies. The CD8[+] cytotoxic T cells identify cells that have been infected by a microbe and act by destroying the cell (and thus the invader). Human immunodeficiency virus works by infecting the CD4[+] helper T cells, which triggers destruction by CD8[+] cytotoxic T cells and ultimately leads to low levels of CD4[+] T cells. Acquired immunodeficiency syndrome typically is diagnosed when CD4[+] T cells fall below 200 cells/μl of blood.

An extremely important function of the adaptive system is the ability to identify a pathogen. The lymphocytes (T and B cells) must be presented with invaders to initiate the appropriate response. This is accomplished by **antigen presenting cells** (APCs) that contain **major histocompatibility complex** (MHC) molecules that provide specific instruction to the T cell lymphocytes. The MHCs are separated into class I MHCs, which present intracellular antigens (infected cells), and class II MHCs, which present extracellular antigens (microbes in circulation).

Dendritic cells (DCs) and macrophages are considered the main APCs and work by allowing the lymphocytes access to the antigen. The DCs mainly are located in the epithelial wall and contain receptors that "sample" the environment. Once the DCs identify an invader, they detach from their location and travel to the lymph nodes to present the antigen to helper T cells. Cytokines produced from the innate system by macrophages and DCs help govern the mobilization of the DCs. The DCs begin to express class II MHC on the cell surface and are concentrated in the lymph nodes (the reason for swollen lymph nodes). The CD4$^+$ helper T cells recognize the microbe presented by the DC and produce cytokines that serve two major roles: activation of phagocytic macrophages to engulf and destroy the microbe, and stimulation of antibody production from B cells. B cells are also capable of identifying a specific microbe and can coordinate antibody production in the absence of helper T cell stimulation. If a virus has infected a cell, then that specific cell will present the class I MHC, which activates the CD8$^+$ cytotoxic T cells and results in the destruction of the cell.

The adaptive immune system is specific to the microbe that has invaded the body. Upon an initial exposure, an individual may experience symptoms of illness (i.e., fever, body aches) while the immune system begins producing an appropriate antibody. Once the attack has been fought off, memory B cells, and to a lesser extent memory T cells, remain in circulation to combat any successive attacks.

This introduction has provided a basic understanding of the immune system. The remainder of this chapter discusses how acute and chronic exercise influence the immune system. Much of our discussion centers on the production of various plasma cytokines in response to exercise. In addition, we highlight the possible benefits and detrimental effects of exercise on the immune system.

EXERCISE AND IMMUNE FUNCTION

Exercise training is an established treatment for cardiovascular and metabolic disease states that is used to improve hemodynamics and blood glucose regulation,

respectively. We are now beginning to learn that various types of exercise may play a role in modulating the immune system. Thus, exercise is becoming a treatment for chronic inflammation, which underlies many disease states. It has been shown that both acute and chronic aerobic exercise reduce symptoms and improve recovery of mammals with the influenza virus (58). Those who are physically fit demonstrate lower levels of inflammation (6) and respond better to infection. However, chronic high-intensity exercise without adequate recovery may put one at greater risk for infection.

Several researchers have studied how aspects of the immune system react to exercise of various intensities, durations, and modes, which makes it difficult to summarize the exercise responses (22,52,54). We discuss the acute and chronic responses separately in this section and differentiate between low- and high-intensity exercise.

ACUTE EXERCISE AND THE IMMUNE RESPONSE

The total number of circulating immune cells increases after an acute bout of aerobic exercise, and levels remain elevated up to 5 h postexercise. This response, termed *leukocytosis*, has been shown to be proportional to the intensity of the exercise bout (figure 5.1). For example, a 30-min bout of cycling at 85% peak $\dot{V}O_2$ increases circulating leukocyte levels more than a 30-min bout at 50% peak $\dot{V}O_2$ (41). Due to the diverse leukocyte subsets, it is important to understand which immune cells drive the leukocytosis. At lower intensities, it generally has been shown that T lymphocyte (CD4$^+$ and natural killer T cells) levels decrease slightly after exercise but return to normal several hours afterward (2,41). Neutrophil levels increase slightly after these light workloads and similarly return to near-baseline levels in recovery (2,41). From these results, you can conclude that the leukocytosis after low-intensity exercise is attributable to the large increase in neutrophil levels as T cells decrease.

A robust leukocytosis response occurs immediately after intense exercise. Shortly into recovery there is a brief decline, followed by rebound leukocytosis that remains for several hours after exercise. This initial response is caused by increased lymphocytes (CD4$^+$ and NK T cells) and neutrophils. The biphasic response is a result of a decrease in T lymphocytes below pre-exercise levels during early recovery (1 hour postexercise), whereas the rebound leukocytosis is driven by continued increasing neutrophil levels that last up to 5 h into recovery (2,41). The T lymphocytes remain depressed for up to 24 h postexercise, which may put an individual at greater risk of attack from a pathogen (72). This, termed the *open window,* may be one reason why endurance athletes suffer high rates of respiratory infection after competition (59).

The leukocytosis responses are caused by exercise-induced activation of the sympathetic nervous system, leading to increases in plasma catecholamine and cortisol levels (69). This has two main effects that promote the increase in circulating white blood cells:

1. Increased cardiac output and sheer stress, causing cells to migrate from the vascular wall, pulmonary circulation, bone marrow, and spleen (17,41)

2. Downregulation of adhesion molecules on the immune cells, which promotes deployment from the vascular wall into circulation

In addition, it generally has been thought that the early increase in immune cell circulation is a result of the catecholamine activation, whereas the late phase increase is a result of cortisol (46). These mechanisms have been supported by giving human subjects an adrenergic-blocking drug, which inhibited the increase in circulating leukocytes during exercise (1). Other cells, such as monocytes, DCs, and B lymphocytes, have not been consistently shown to migrate profoundly after an acute bout of exercise.

Mobilization of leukocytes into circulation during and after acute exercise changes the functionality of these cells. Moderate-intensity exercise enhances the phagocytic and oxidative burst capacity of neutrophils (49), which may support the ability of these cells to destroy pathogens (49). Researchers have reported that neutrophils respond more aggressively to a bacterial exposure after exercise (47); however some evidence also suggests that this effect is diminished several hours into recovery from intense exercise but not moderate-intensity exercise (55). Acute heavy exercise (90% and 120% peak $\dot{V}O_2$) has also been shown to decrease the responsiveness of T cells to pathogens, whereas low-intensity exercise does not have this effect (21,45). In fact, T cell response to a viral infection is enhanced after 30 min of moderate-intensity aerobic exercise (60). The research on B cell function is less clear but points toward similar findings that low-intensity exercise may enhance B cell antibody production (43), whereas a decline occurs after high-intensity exercise (69). In summary, it appears that acute low-intensity aerobic exercise improves immune cell responsiveness and function and, conversely, that immune cell function may be depressed after an acute bout of intense exercise (see figure 5.1).

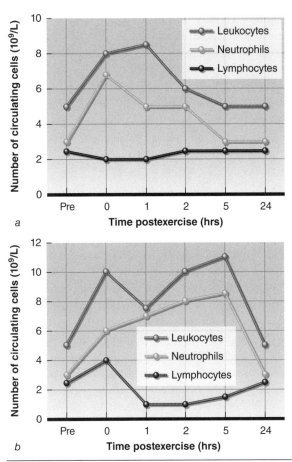

Figure 5.1 The levels of circulating leukocytes in response to *(a)* acute moderate-intensity aerobic exercise and *(b)* high-intensity aerobic exercise.

The cytokine and intracellular inflammatory signaling response to acute exercise recently has come to the forefront in exercise physiology. Cytokines play complex roles in immune function by coordinating the recruitment and activation of both the innate and adaptive cells. They are released from a variety of cells during exercise, including immune, endothelial, and skeletal myocytes. The most studied cytokine in exercise research is IL-6, which has been shown to increase proportionally to the duration of the exercise bout. In normal immune function, IL-6 is released primarily from macrophages, T cells, and endothelial cells and has a proinflammatory response by instructing B cells to proliferate and stimulate antibody production. However, during exercise, IL-6 is released from the skeletal muscle and has an anti-inflammatory response by increasing the production of IL-1ra and IL-10 from T cells and macrophages. These cytokines are considered anti-inflammatory because they reduce T cell production of the proinflammatory IL-12 (63). In addition, high-intensity exercise can dominate a proinflammatory cytokine response, which has been demonstrated by increased plasma levels of tumor necrosis factor alpha (TNF-α), monocyte chemoattractant protein-1 (MCP-1), interleukin-8 (IL-8), and interleukin-1 beta (IL-1β) (48,75). These cytokines may be coming from monocytes, T cells, epithelial cells, and possibly skeletal muscle. Each one plays a role in activation of immune cells (TNF-α), recruitment of neutrophils (TNF-α, IL-8) and macrophages (MCP-1), and secretion of acute phase proteins (IL-1β). Also, C-reactive protein (CRP), which is an acute phase proinflammatory protein released from the liver, consistently has been shown to increase after strenuous exercise (30). The consequence of these elevated proinflammatory cytokines in response to exercise remains to be understood. However, the anti-inflammatory effect of IL-6 may override the proinflammatory effects of these plasma proteins. This has been demonstrated by improved recovery from TNF-α exposure after exercise-induced IL-6 release (61).

During prolonged exercise, immune cells may have a reduced ability to produce cytokines in response to bacterial stress. This has been demonstrated by extracting and culturing cells from a runner after a marathon-length race. The cells were then exposed to a bacterial stress, and lower levels of cytokines were measured in the cell media (62). This demonstrates that excessive exercise may inhibit the immune cells' responsiveness, thus putting one at risk for infection.

In summary, an acute bout of exercise has profound effects on the immune system. Low-intensity exercise appears to improve immune function by improving neutrophil and lymphocyte function, along with the release of IL-6. As exercise intensity and duration increase, immune function may become compromised, thus increasing the susceptibility to an invading pathogen.

Chronic Exercise and Immune Function

It generally has been accepted that aerobic training of moderate intensity and volume enhances immune function and reduces one's risk for infection (45); however, high-volume training can put one at greater risk for infection. Athletes commonly experience upper respiratory tract infections and flu-like symptoms (e.g., coughing, sore throat, fever) after periods of chronic long training sessions (>2 h) combined with intense competition. Conversely, exercise of moderate intensity and duration has been shown to improve immune profiles among those with low-grade inflammation, such as obese individuals and cardiac and diabetic patients (24,29).

It appears that minor changes occur in immune function after healthy sedentary humans engage in aerobic exercise training for several weeks to months. Aerobically fit individuals tend to have lower resting levels of overall circulating leukocytes, as demonstrated by decreased monocytes, neutrophils, and lymphocytes (28). However, the pattern of leukocytosis that occurs during acute exercise is not different after training. It also has been shown that exercise training elevates DC counts and improves this cell's ability to produce IL-12 cytokine. This may indicate improvement in the ability to identify pathogens along with recruiting additional immune cells upon exposure; however, this has been shown only in animal model research (11). Overall, this demonstrates that moderate levels of aerobic training improve overall immune function. In a large-scale study of roughly 4,000 athletes, it was demonstrated that those who participate in jogging and aerobic dancing have a lower likelihood of elevated inflammatory markers in comparison to nonaerobic athletes (i.e., weightlifters) (31).

Periods of excessive training and competition consistently have been shown to compromise immune function. When athletes increase their training loads, they have demonstrated depressed neutrophil and natural killer cell function and reduced ability of T and B cells to respond to infection (4). In addition, latent viruses have been shown to reactivate after long high-intensity interval sessions. The cause for these changes in response to chronic heavy training is unclear but may be due to a combination of increased proinflammatory cytokines in comparison to anti-inflammatory cytokines along with a reduced ability of survey cells (APCs) to detect antigens (23). The alterations in cytokine expression may be a result of chronically elevated cortisol and catecholamine levels causing alterations in T cell production of cytokines (35). In summary, periods of heavy intense training have a suppressive effect on the immune system, which increases the likelihood of illness. It is important for coaches and athletes to recognize symptoms during their long competitive seasons and adjust workload accordingly.

The greatest positive effect of exercise training on immune function has been demonstrated in those with underlying disease. It was stated that the major cause of chronic illness is lack of exercise, which was based on evidence of exercise as a preventative measure against 35 disease conditions (7). Moderate-intensity aerobic exercise has been shown to modulate plasma markers of inflammation among patients with coronary artery disease (24) and type 2 diabetes (29). This is demonstrated by reduced proinflammatory plasma levels of CRP, IL-6, TNF-α, and IL-1β and increased levels of

Aging Perspective: Exercise Modulates Immune Among Elderly

Aging is associated with higher levels of circulating pro-inflammatory cytokines such as IL-6, TNF-α, and CRP. This puts individuals at greater risk for developing a host of diseases later in life. The mechanism may take place through enhanced oxidative stress, which may indicate overactive phagocyte and endothelial activity combined with reduced levels of antioxidants. It has been shown that elderly individuals have an exaggerated proinflammatory response to bacterial stress and are incapable of downregulating the immune response in a timely fashion after the pathogen has been removed. This may be a reason why it takes longer for the elderly to recover from illness. In regards to an acute bout of exercise, older individuals demonstrate slightly reduced leukocytosis response and have shown reduced T cell responsiveness after exercise. It is not clear what this differing response compared with young people indicates. However, it has been established that chronic exercise training improves immune function among those aged 60 to 90 yr (74). Trained elderly individuals have shown improved lymphocyte responsiveness to bacterial stress along with enhanced phagocytic activity, which was comparable with that of younger adults. Further, a profound inverse relationship exists between the level of physical activity and inflammation among older adults. Again, this indicates the positive benefits of exercise on the immune system.

Clinical Perspective: Exercise as Treatment for Chronic Inflammatory Diseases

Due to the anti-inflammatory effects of low- to moderate-intensity aerobic exercise, researchers are beginning to explore exercise as a treatment for those suffering from chronic inflammatory diseases. It is not uncommon for those with inflammatory disease to have elevated levels of IL-6, TNF-α, CRP, and other cytokines and chemokines. The most profound benefits have been demonstrated among those with coronary artery disease and diabetes, where levels of plasma cytokines are substantially lower after consistent exercise ranging from 4 to 24 wk (51). Further, exercise training has been shown to reduce the size of atherosclerotic plaque among patients with coronary artery disease, indicating disease regression (39). Exercise training has also been shown to improve inflammatory profiles of those with rheumatoid arthritis, inflammatory bowel disease, and cancer. In fact, when serum from physically fit humans is given to pancreatic cancer cells, the cells shrink in size (5). Although this experiment may be considered unorthodox, it demonstrates that something in the serum from those who participate in chronic exercise has anti-inflammatory affects. Most likely it is not just one chemical but rather a milieu of events. This is why we cannot put exercise into a pill!

anti-inflammatory IL-10 (24). Aerobic exercise also has been shown to benefit patients with human immunodeficiency virus by increasing helper T cell (CD4$^+$) counts and to improve immune function among cancer patients by enhancing NK and T cell activity (34). Interestingly, when serum from physically fit humans is cultured with cancer cells, it reduces the growth of the tumor cell (5). The mechanisms behind these improvements are unclear, but the improvements may be attributable to the release of the anti-inflammatory IL-6 from the contracting muscle along with endocrine modulation (cortisol and catecholamines).

Overall, it is apparent that both acute and chronic exercise have profound effects on immune function. In general terms, aerobic exercise of low to moderate intensity and shorter duration (<60 min) provides benefit, whereas strenuous heavy exercise may cause immunosuppression. This field of exercise physiology is growing rapidly as exercise therapy is becoming a positive treatment option for those with chronic inflammatory disease.

Sport Performance Perspective: Inflammation in Muscle Hypertrophy Process

It is not uncommon for those who participate in resistance exercise to experience localized skeletal muscle inflammation and soreness. Cells of the immune system play an important role in assisting the muscle regeneration process through several mechanisms. Immediately after exercise, the muscle releases various chemokines that recruit neutrophils, macrophages, and T helper cells into the damaged muscle. These cells begin releasing various cytokines such as IL-6, TNF-α, and other proinflammatory mediators (e.g., prostaglandins) that begin to activate satellite cells. This leads to coordination of muscle repair, differentiation, and ultimately growth. It has been shown that taking nonsteroidal anti-inflammatory drugs, which cause a decrease in prostaglandin synthesis, inhibits protein synthesis after resistance exercise (68). Athletes commonly take these drugs to reduce the pain and inflammation from training, but these results demonstrate that the inflammatory process is required for proper muscle hypertrophy. The total time for regeneration to occur varies based on the degree of muscle damage, but it may be as long as 8 d (67).

EXERCISE ENDOCRINOLOGY

The endocrine system is a complex network of organs, tissues, and glands that secrete hormones involved in the maintenance of homeostasis. Environmental conditions, nutrition status, disease state, and exercise are all examples of stressors that disrupt homeostasis. The endocrine response to exercise is often referred to as the "fight or flight" response, resulting in the activation of two main hormonal systems: the hypothalamus-pituitary-adrenal axis (HPA) and the adrenal system (10,25,66). Because activation of the HPA and adrenal system is initiated in the autonomic nervous system, it is impossible to discuss the endocrine system without discussing the nervous system. In fact, the nervous and endocrine systems often are referred to collectively as the *neuroendocrine system*. Therefore, the nervous system may be mentioned at various points throughout this chapter because it relates to the activation of the endocrine system. However, here we focus primarily on the endocrine response to exercise; the nervous system is discussed in more detail in chapter 2. An important distinction between these two systems is that the nervous system produces an almost immediate physiologic response, whereas the endocrine system can take from minutes to hours to induce a physiologic change. The reason for this is that the nervous system directly innervates tissues it acts on, whereas the endocrine system produces hormones that (in most cases) must enter circulation before exerting their effect on various cell types. During exercise, there is a coordinated effort between the nervous system and the endocrine

system to release multiple hormones influencing substrate availability, blood pressure, blood volume, and, in the case of anaerobic exercise, both anabolic and catabolic pathways. Importantly, exercise mode (i.e., cycling vs. running), type (i.e., aerobic vs. anaerobic), and dose (i.e., intensity, duration, and volume) greatly influence the hormone profile during that particular exercise session. Further complicating the hormone profile during exercise are environmental conditions, nutritional status, disease state, age, sex, and training status. Consequently, understanding the endocrine response to exercise in its entirety takes years—if not a lifetime. Therefore, the major focus here is to begin to understand and appreciate the complexity of the endocrine system specifically during acute aerobic and anaerobic exercise in the untrained and trained states.

Endocrine System Structure

There are approximately 20 hormone-producing glands (e.g., adrenal gland), organs (e.g., pancreas), and tissues (e.g., adipose tissue) in the human body. Of particular importance during exercise is the **pituitary gland** (involved in the HPA) and the **adrenal gland** (involved in adrenal system activation as well as the pancreas, testes, and ovaries; see table 5.3). There are also more than 50 hormones in the human body that differ in both chemical structure and function. In general, hormones are classified as amine, peptide, or steroid based on their chemical structure. All **amine hormones** are derived from the specific amino acid tyrosine. Catecholamines (i.e., epinephrine and norepinephrine) are examples

Table 5.3 Important Exercise-Related Hormones

Hormone	Site of synthesis or secretion	Synthesis/secretion during exercise	Function during exercise
Catecholamines (epinephrine and norepinephrine)	Adrenal medulla	Increased	Cardiac contractility Peripheral vasoconstriction Bronchial vasodilation Lipolysis Skeletal muscle and hepatic glycogenolysis Skeletal muscle force production
Glucagon	Pancreas (alpha cells)	Increased	Hepatic glycogenolysis Gluconeogenesis Hepatic glycogen synthesis Hepatic amino acid uptake
Insulin	Pancreas (beta cells)	Decreased	Permissive of lipolysis
Angiotensin II	Circulation	Increased	Vasoconstriction Aldosterone secretion ADH secretion Sympathetic nervous system activity
Antidiuretic hormone	Posterior pituitary	Increased	Vasoconstriction Renal water reabsorption
Aldosterone	Adrenal cortex	Increased	Renal sodium reabsorption
Cortisol	Adrenal cortex	Increased	Proteolysis
IGF-1	Liver	Increased during heavy-load or high-intensity exercise	Cell growth and proliferation
Testosterone	Testes, ovaries, and adrenal gland	Increased during heavy-load or high-intensity exercise	Glucose metabolism Protein synthesis
Growth hormone	Pituitary gland	Increased during heavy-load or high-intensity exercise	Lipolysis Cell growth and proliferation IGF-1 production stimulation

of amine hormones that are secreted during exercise and involved in substrate availability, blood pressure control, and blood volume maintenance. All steroid hormones are synthesized from cholesterol. Three important steroid hormones secreted in response to exercise are aldosterone (involved in blood pressure regulation and blood volume), cortisol (involved in catabolic pathways), and testosterone (involved in anabolic pathways). **Peptide hormones**, often referred to as *protein hormones,* are the most abundant type of hormone in the human body. Examples of important exercise-related peptide hormones include glucagon and insulin (involved in substrate availability), renin and antidiuretic hormone (involved in blood pressure control during exercise), and insulin-like growth factor-1 (IGF-1; involved in anabolic pathways). A general distinction between peptide and protein hormones is size. Peptide hormones can be as small as a chain of three amino

acids, whereas protein hormones are several amino acids chained together.

Endocrine System Function

The chemical structure of a hormone influences how the hormone functions. Amine (e.g., epinephrine and norepinephrine) and peptide (e.g., glucagon, insulin, aldosterone, and IGF-1) hormones interact with receptors on the plasma membranes, trigging a cascade of intracellular events. Conversely, steroid hormones (e.g., aldosterone, cortisol, and testosterone) are lipid soluble and therefore can permeate plasma membranes and interact with intracellular receptors. Hormones are also classified by their primary function as intracrine, autocrine, and paracrine hormones. Intracrine (*intra* = "within") hormones influence the intracellular environment in which they are synthesized or internalized. This relatively new class of hormones was first described in 1984, and understanding of their physiologic action remains incomplete (53). Autocrine (*auto* = "self") hormones are released from, and exert their influence on, the same cell. Paracrine (*para* = "beside") hormones are released from one cell type and exert their effect on surrounding cells. Paracrine hormones are the most abundant type of hormone in the human body. Collectively, the endocrine system includes glands, organs, and tissues that release various types of hormones to help maintain homeostasis in the human body. In the context of exercise endocrinology, we are most interested in how the endocrine system helps influence blood pressure control, substrate availability, fluid balance, and catabolic and anabolic pathways during aerobic and anaerobic exercise in untrained and trained individuals.

HORMONE CONTROL OF THE CARDIORESPIRATORY SYSTEM DURING AEROBIC EXERCISE

The cardiorespiratory response to exercise is quite complex and is discussed in great detail in chapter 3. However, the overarching cardiorespiratory response to aerobic exercise is an increase in cardiac output, diversion of blood flow away from inactive tissue toward metabolically active tissue (i.e., skeletal muscle), and increased ventilation. These responses are necessary to meet the oxygen demands of actively contracting skeletal muscle for the production of adenosine triphosphate (ATP). Changes in cardiac output and blood flow during aerobic exercise primarily are mediated by the autonomic nervous system. However, circulating catecholamine (i.e., epinephrine and norepinephrine) hormones enhance this response. Catecholamines specifically interact with alpha-adrenergic (α-AR) and beta-adrenergic (β-AR) receptors. Two subtypes of α-AR (α1-AR and α2-AR) and three subtypes of β-AR (β1-AR, β2-AR, and β3-AR) have been identified. Within the scope of cardiorespiratory activity during exercise, β1-AR (located on the cardiac muscle), α1-AR (located on smooth muscle), and β2-AR (located on brachial smooth muscle) are of particular importance (66). During an acute session of aerobic exercise, circulating catecholamines influence cardiac contractility, blood pressure, and ventilation. Catecholamines work in concert with additional hormones, including angiotensin II, atrial natriuretic peptide, aldosterone, and antidiuretic hormone (ADH; also known as arginine vasopressin [AVP]) during acute exercise to further augment blood pressure and plasma volume. Importantly, exercise training augments the secretion of catecholamines and hormones involved in blood pressure and blood volume control.

Cardiac Output

The increase in cardiac output during aerobic exercise is a coordinated effort between the autonomic nervous system and the endocrine system. At the onset of aerobic exercise, the increase in heart rate toward 100 beats · min^{-1} stems from parasympathetic nervous system withdrawal. A further increase in heart rate is attributable to sympathetic nervous system activation, which also increases cardiac contractility. Concurrently, the sympathetic nervous system also stimulates the release of catecholamines from the adrenal medulla (the inner portion of the adrenal gland located around the kidneys). Catecholamines are not released proportionally; approximately 80% of adrenal secretion is epinephrine and 20% is norepinephrine (8). During aerobic exercise, the increase in circulating epinephrine levels further enhances heart rate and contractility by stimulating β1-AR on cardiac muscle tissue. When stimulated by catecholamines, the β-AR promotes calcium (Ca^+) flux into the cell, triggering greater Ca^+ release from the sarcoplasmic reticulum (SR) and ultimately activating cardiac muscle contractions. Importantly, systemic catecholamine concentration is proportional to the intensity (36) and duration of the aerobic activity (8). However, catecholamines tend to increase exponentially at exercise intensities corresponding to 60% to 70% peak $\dot{V}O_2$ (36). Therefore, catecholamines exert more influence over heart rate and contractility during more intense and longer duration aerobic exercise.

Blood Pressure

Systolic blood pressure and mean arterial pressure increase in proportion to the intensity of the exercise bout, whereas diastolic blood pressure remains constant or tends to decrease. The increase in systolic blood pressure and mean arterial pressure during exercise largely stems from vasoconstriction induced by the autonomic nervous system. However, α1-ARs on smooth muscle respond to circulating catecholamines. Similar to activation of the β-ARs, catecholamine-induced activation of α1-ARs also increases Ca^+ flux and promotes contraction of smooth muscle (i.e., vasoconstriction). Blood pressure is the product of cardiac output and peripheral vascular resistance. Therefore, catecholamines indirectly increase blood pressure by increasing peripheral vascular resistance. Local factors override both the autonomic nervous system and endocrine system in vascular beds of active tissue to promote vasodilation. However, the global vascular response to exercise is vasoconstriction, and the net effect is an increase in systolic and mean arterial pressures. The secretion of additional endocrine hormones (including angiotensin II, aldosterone, AVP, and ADH) further enhances vasoconstriction in vascular beds of inactive tissue. The sympathetic nervous system initiates this response by stimulating the secretion of the enzyme renin from the juxtaglomerular cells in the kidney and renin-angiotensin-aldosterone system.

Renin promotes the secretion of angiotensin II, a peptide hormone, which increases blood pressure in four ways. First, angiotensin II directly causes an increase in blood pressure as a potent vasoconstrictor. Second, angiotensin II indirectly increases blood pressure by stimulating the release of the steroid hormone aldosterone from the adrenal cortex. Aldosterone promotes the reabsorption of sodium and subsequently water in the kidneys, resulting in increased blood volume. This is also an indirect means of increasing blood pressure because aldosterone—rather than angiotensin II—results in the increase in blood volume. Furthermore, an increase in blood volume results in increased cardiac output and thus blood pressure. Third, angiotensin II also indirectly promotes an increase in blood pressure by enhancing sympathetic nervous system outflow. Last, angiotensin II further promotes an increase in blood pressure by stimulating the release of the peptide hormone ADH from the posterior pituitary. ADH also causes vasoconstriction as well as water reabsorption in the kidneys. In summary, the major hormones involved in increasing blood pressure during acute exercise are epinephrine, norepinephrine, angiotensin II, aldosterone, AVP, and ADH. These hormones work in concert to increase blood pressure both directly and indirectly by increasing peripheral vascular resistance and cardiac output.

Ventilation

Catecholamines also influence ventilation during exercise. Bronchial smooth muscle is lined by β2-ARs. Importantly, β2-ARs have a high affinity to epinephrine and, to a lesser extent, norepinephrine. Unlike β1-AR activation, β2-AR activation results in vasodilation (66). When epinephrine interacts with β2-AR it causes a cascade of intracellular events in bronchial smooth muscle that increases the release of nitric oxide, a very potent vasodilator. This response is directly proportional to circulating catecholamine concentrations.

HORMONE CONTROL OF SUBSTRATE AVAILABILITY DURING AEROBIC EXERCISE

ATP is synthesized by one of three metabolic pathways—the ATP–phosphocreatine system, glycolysis, and oxidative phosphorylation—and is required for muscular contraction. Importantly, glucose is required for the production of ATP via glycolysis, whereas glucose and nonesterified free fatty acids (NEFA) are major substrates for ATP production via oxidative phosphorylation. Consequently, substrate availability must be maintained during exercise in order to maintain ATP supply and prevent a decline in exercise performance. The endocrine system is vital in the mobilization of substrate during acute exercise. Hormones involved in substrate mobilization include insulin as well as epinephrine, glucagon, cortisol, and growth hormone, which collectively are referred to as *counterregulatory hormones* because they work in opposition to the action of insulin. For example, insulin promotes cellular uptake of substrates (e.g., glucose) and stimulates anabolic processes such as glycogen synthesis. Conversely, counterregulatory hormones promote the mobilization of stored substrate (i.e., glucose and NEFA) and catabolic processes such as glycogenolysis.

Factors that influence hormone-mediated substrate mobilization during aerobic exercise include hypoglycemia (i.e., low blood sugar) and exercise intensity and duration. Exercise inhibits insulin secretion and promotes the secretion of counterregulatory hormones. Epinephrine is secreted from the adrenal medulla as intensity and duration of aerobic exercise increase. Glucagon is secreted from the alpha cells in the pancreas in response to hypoglycemia and sympathetic nervous system stimulation. Growth hormone is secreted from the pituitary gland in response to increasing catecholamine levels during aerobic exercise. However, growth hormone secretion may also be influenced by metabolic byproducts (e.g., lactate and hydrogen ions) and core body temperature (64). Cortisol is secreted from the adrenal cortex (the inner portion of the adrenal gland located about the kidneys) in response to direct sympathetic stimulation and increasing catecholamine levels. These hormones work together through a series of complex interactions with the pancreas, liver, skeletal muscle adipose, and kidneys to increase systemic substrate availability during acute exercise.

Pancreas

The pancreas is responsible for secreting insulin in response to hyperglycemia and glucagon in response to hypoglycemia and epinephrine. Epinephrine also activates α-ARs in the pancreas, resulting in further secretion of glucagon and the inhibition of insulin secretion. The inhibition of insulin secretion during exercise is very important for preventing hypoglycemia (i.e., low blood sugar). Both insulin and muscle contraction promote the translocation of glucose transporter 4 (GLUT4) from the subcellular space to the cell membrane, allowing glucose to enter the cell from systemic circulation. Although both insulin and muscle contraction stimulate the translocation of GLUT4 to the cell membrane, each works through distinct pathways, resulting in an additive effect (38). Therefore, if insulin levels were to remain elevated during exercise, the combination of insulin and contraction-mediated glucose uptake would result in rapid and profound hypoglycemia.

Adipose Tissue

Insulin is a very potent antilipolytic agent. However, low insulin levels are permissive of lipolysis, thus facilitating a modest increase in systemic NEFA availability. Therefore, the inhibition of insulin secretion during exercise facilitates an increase in systemic NEFA availability. Further increases in systemic NEFA availability are attributable to increasing epinephrine levels. More specifically, epinephrine stimulates β3-ARs on adipose tissue. This results in the phosphorylation (and activation) of aptly named hormone-sensitive lipase (HSL). HSL catalyzes the breakdown of triglyceride to glycerol and NEFA. As previously discussed, epinephrine concentration increases in proportion to exercise intensity and duration. Consequently, more profound increases in NEFA are not observed until higher intensity (>50% $\dot{V}O_2$max) and longer duration activity is achieved. Therefore, the initial (and modest) increase in NEFA availability during aerobic exercise is permitted by low insulin levels, but increased NEFA availability during higher intensity and longer duration aerobic exercise is attributable to increasing epinephrine levels. In fact, β-adrenergic blockade results in a profound inhibition of NEFA availability during exercise, further suggesting that epinephrine is the major lipolytic stimulator during exercise (3). Growth hormone also stimulates lipolysis by increasing HSL activity, upregulating proteins involved

in lipolysis, and destabilizing lipid droplets, making them more vulnerable to HSL (8). Evidence also shows that cortisol stimulates lipolysis by increasing HSL activity in adipocytes (9). Overall, during an acute session of aerobic exercise, lipolysis is stimulated through an inhibition of insulin secretion and an increase in epinephrine and growth hormone secretion. Collectively, this increases systemic NEFA availability for ATP synthesis via oxidative metabolism and provides substrate (i.e., glycerol) for gluconeogenesis.

Skeletal Muscle

Within skeletal muscle, the endocrine response to exercise primarily stimulates glycogenolysis and indirectly increases substrate availability for glyconeogenesis. Epinephrine specifically stimulates glycogenolysis within skeletal muscle. When epinephrine binds to β-ARs on skeletal muscle it promotes the breakdown of ATP, resulting in elevated concentration of subcellular cyclic adenosine monophosphate (cAMP). The accumulation of cAMP initiates a cascade of intracellular reactions, including the phosphorylation of the enzyme glycogen phosphorylase from its inactive form (glycogen phosphorylase b) to its active form (glycogen phosphorylase a). Glycogen phosphorylase a catalyzes the breakdown of glycogen to glucose 6-phosphate (G6P). Ultimately, this increases G6P availability in skeletal muscle for glycolysis. Additionally, lactate and pyruvate are byproducts of glycolysis that can serve as substrate for gluconeogenesis during glycogen-depleting exercise. Another important counterregulatory hormone that affects skeletal muscle and influences substrate availability during exercise is cortisol. Cortisol promotes **proteolysis**—the breakdown of proteins to amino acids. Importantly, amino acids, and in particular alanine, also serve as substrate for **gluconeogenesis**. Therefore, within skeletal muscle, epinephrine increases glycogenolysis and cortisol increases proteolysis. This increases substrate availability for ATP production via skeletal muscle glycolysis and increases substrate for gluconeogenesis in other tissues.

Liver

The liver arguably is the most important organ involved in glucose maintenance during exercise. During exercise, glycogenolysis and gluconeogenesis increase in the liver to increase systemic glucose availability and glycogen synthesis is inhibited in the liver, thus conserving hepatic glucose utilization. Glucagon is the primary hormone responsible for stimulating glycogenolysis and inhibiting glycogen synthesis in the liver. However, epinephrine can also stimulate hepatic glycolysis. Both glucagon and epinephrine increase the activity of adenylate cyclase, promoting the breakdown of ATP and a subcellular increase in cAMP. Just as in skeletal muscle, the accumulation of cAMP in the liver leads to a cascade of intracellular reactions, resulting in the breakdown of glycogen to G6P. However, in the liver G6P is dephosphorylated, allowing glucose to exit the liver into systemic circulation. Importantly, dephosphorylation of G6P and the subsequent release of glucose into systemic circulation are exclusive to the liver because skeletal muscle lacks the enzyme involved in this process. Also, as previously discussed, GLUT4 facilitates skeletal muscle glucose uptake when stimulated by insulin or muscle contraction. However, GLUT2 is the primary glucose transporter in the liver. Unlike GLUT4, glucose transport via GLUT2 is facilitated by a glucose concentration gradient rather than insulin concentration. Therefore, as intracellular glucose concentrations increase in the liver via glycogenolysis and gluconeogenesis, this promotes the transport of glucose from the intracellular compartment to the extracellular compartment (systemic circulation). Activation of the gluconeogenic pathway in the liver occurs after hepatic glycogen stores are depleted. Key hormones involved in this process are glucagon, cortisol, epinephrine, and growth hormone. Glucagon stimulates both amino acid uptake and gluconeogenesis in the liver. Cortisol, epinephrine, and growth hormone indirectly promote gluconeogenesis by increasing systemic availability of substrate for the Cori cycle (i.e., pyruvate, lactate, glycerol, and amino acids).

HORMONE CONTROL OF FLUID AND ELECTROLYTE BALANCE DURING AEROBIC EXERCISE

Plasma is the liquid portion of the blood responsible for transporting nutrients, including glucose and NEFA, throughout the body. Conservation of plasma volume is essential for maintaining cardiac output and supplying working skeletal muscle with substrate for ATP production during exercise. However, during prolonged exercise plasma volume is reduced due to changes in osmotic pressure, hydrostatic pressure, and sweating. The endocrine system responds to sympathetic nervous system stimulation during exercise to conserve water and minimize the decline in plasma volume. Hormones involved in this process include ADH (secreted from the posterior pituitary gland) and aldosterone (secreted from the adrenal cortex).

Fluid Balance

An important advantage humans have over other mammals during exercise and physical activity is heat dissipation through sweating. However, this occurs at the expense of declining plasma volume. The endocrine system specifically responds to changes in plasma volume, plasma osmolality, and sympathetic nervous system stimulation to help preserve plasma volume during exercise. ADH secretion is stimulated first by increases in **plasma osmolality** and second by reductions in plasma volume. As sweating progresses during exercise, there is an increase in plasma osmolality sensed by osmotic receptors in the hypothalamus and a reduction in plasma volume is sensed by baroreceptors in the carotid sinuses and aortic arch. Activation of osmotic receptors and baroreceptors stimulates the hypothalamus to sends neural impulses to the posterior pituitary to release ADH. As previously discussed, ADH promotes water reabsorption in the kidneys.

Electrolyte Balance

Aldosterone is the primary hormone involved in maintaining electrolyte balance. As described previously, aldosterone is secreted from the adrenal cortex. Similar to ADH, aldosterone secretion is stimulated by a reduced plasma volume. However, aldosterone is also secreted in response to reductions in plasma sodium and, conversely, elevated plasma potassium concentrations. Aldosterone promotes sodium (and water) reabsorption in the kidneys as well as increased excretion of potassium from the kidneys. Aldosterone contributes to the maintenance of electrolyte balance as well as plasma volume.

HORMONE RESPONSE AFTER AEROBIC EXERCISE TRAINING

Aerobic exercise training results in important improvements in cardiorespiratory fitness and overall health. Improvements in cardiorespiratory fitness typically include a higher peak $\dot{V}O_2$ and time to exhaustion during aerobic exercise as a result of increased substrate storage (i.e., muscle glycogen and intramuscular triglycerides), greater reliance on fat oxidation (increased capillary and mitochondrial density), and, consequently, increased glycogen sparing after weeks to months of aerobic exercise training. Adaptations to the hormone profile during an acute session of aerobic exercise after exercise training also aid in glycogen sparing. For example, aerobic exercise training results in a reduction in sympathetic nervous system activation. This blunts the normal counterregulatory hormone response typically observed during acute aerobic exercise in the untrained state. Therefore catecholamines, cortisol, and growth hormone concentrations are reduced at the same relative and absolute exercise intensity after aerobic exercise training (15,73). Normally, epinephrine stimulates hepatic and skeletal muscle glycogenolysis during aerobic exercise. Therefore, reduced epinephrine secretion after aerobic exercise training favors glycogen sparing in the trained state. Similarly, a reduction in angiotensin II, aldosterone, AVP, and ADH secretion also occurs during submaximal exercise after aerobic exercise training. This is primarily due to training-induced increases in plasma volume (13,14). In fact, reductions in plasma volume associated with detraining appear to nearly restore angiotensin II, aldosterone, AVP, and ADH secretion during exercise to pretraining levels (57). Conversely, secretion of counterregulatory hormones, angiotensin II, aldosterone, AVP, and ADH appear to be similar or elevated at maximal exercise after aerobic exercise training, most likely because a higher absolute exercise intensity is achieved after aerobic exercise training. Overall, aerobic exercise training results in a reduction in hormone secretion during submaximal exercise but similar or elevated hormone secretion during maximal exercise in the trained state.

ENDOCRINE RESPONSE DURING ANAEROBIC OR RESISTANCE EXERCISE

Anaerobic exercise encompasses both sprint performance and resistance training exercise. Mode of exercise, intensity of exercise, number of sets (or sprinting bouts), volume, and quality and quantity of rest periods all dictate whether anaerobic exercise training will result in muscular strength, power, or hypertrophy. Furthermore, the training adaptations that result in increased muscular strength, power, or hypertrophy are partly attributable to the hormone profile elicited during each exercise session. However, the broad range of activities globally referred to as *anaerobic exercise* makes understanding the hormone response of a single session of anaerobic exercise more complicated. Moreover, age, sex, and training status also contribute to the hormone profile during an acute session of anaerobic exercise. Nevertheless, the general endocrine response to an acute session of anaerobic exercise contributes to the muscular phenotype we typically associ-

ate with anaerobic exercise training. This phenotype is largely due to the secretion and subsequent action of both catabolic (e.g., cortisol) and anabolic (including testosterone, growth hormone, and IGF-1) hormones during anaerobic exercise. Here we focus on the endocrine response during a single session of anaerobic exercise and in response to anaerobic exercise training.

HORMONE RESPONSE DURING ACUTE ANAEROBIC EXERCISE

Similar to aerobic exercise, anaerobic exercise results in adrenal gland secretion of catecholamines and cortisol proportional to the intensity and duration of the exercise session (19). In addition to increasing blood pressure and substrate availability, epinephrine appears to enhance force production and contraction rate during anaerobic exercise (20). Cortisol secretion is also workload dependent. It appears that heavy-load or high-intensity exercise accompanied by short rest periods may be necessary for stimulating cortisol secretion during anaerobic exercise (65). Cortisol mainly is involved in promoting catabolic activity, including proteolysis in skeletal muscle. Importantly, some but not all forms of anaerobic exercise produce an anabolic hormone response. Heavy-load or high-intensity exercise accompanied by short rest periods appears to be the most potent stimulus for the secretion of testosterone, growth hormone, and IGF-1 (33,70). In general, testosterone activates pathways in skeletal muscle that promote glucose metabolism and protein synthesis (56). Growth hormone also promotes cell growth and proliferation and stimulates substrate availability during exercise (e.g., lipolysis). Growth hormone also stimulates IGF-1 production in the liver; however, IGF-1 is also synthesized and secreted from adipose tissue and skeletal muscle. The exact mechanism leading to an increase in secretion of IGF-1 from adipose tissue and skeletal muscle during exercise remains unclear. However, it is thought that disruption of the structure of adipose tissue (e.g., degradation of lipid droplets) and skeletal muscle (e.g., glycogen breakdown, proteolysis) may facilitate IGF-1 secretion during anaerobic exercise. Regardless of the stimulus for IGF-1 secretion, it promotes cell growth and proliferation through activation of the protein kinase B pathway (also known as the *Akt pathway*). Overall, a sufficiently intense session of anaerobic exercise stimulates the secretion of catecholamines and cortisol as well as anabolic hormones, including testosterone, growth hormone, and IGF-1. Circulating catecholamines and cortisol influence blood flow and substrate availability similar to aerobic exercise, whereas testosterone, growth hormone, and IGF-1 stimulate anabolic pathways in response to sufficiently potent exercise stimulus.

ENDOCRINE ADAPTATIONS TO ANAEROBIC EXERCISE TRAINING

Sufficiently high anaerobic exercise stimulus results in the secretion of adrenal hormones (i.e., catecholamines and cortisol) as well as anabolic hormones (including testosterone, growth hormone, and IGF-1). However, the hormone profile during an acute session of anaerobic exercise changes in response to anaerobic exercise training. Similar to aerobic exercise training, anaerobic exercise training results in a blunted catecholamine response during submaximal exercise but higher concentrations of catecholamines during maximal (e.g., one-repetition maximum) exercise. The effect of anaerobic exercise training on cortisol levels, however, remains unclear. We observe transient increases in cortisol secretion during acute sessions of anaerobic exercise, which is advantageous for skeletal muscle remodeling. However, anaerobic exercise training associated with an exercise stimulus that continually elicits a cortisol response can lead to overtraining and thus overactive catabolic pathways. Therefore, the impact of anaerobic exercise training on cortisol levels is specific to the exercise stimulus. Furthermore, additional work needs to be done to fully understand the effect of anaerobic exercise training on cortisol to optimize its catabolic affects. A clear understanding of testosterone and growth hormone secretion after anaerobic exercise training is also complicated. Normally, secretion of testosterone and growth hormone fluctuates during a 24-h period, and these fluctuations differ between males and females. For example, females typically have lower (and steady) testosterone levels during the day, whereas testosterone levels in males are higher in the morning and decline throughout the day. In males, anaerobic exercise training appears to increase resting testosterone concentration (27). However, it should be noted that the potency of testosterone might wane over time as the physiologic capacity of testosterone-sensitive tissues are maximized. For example, skeletal muscle has a maximal capacity to hypertrophy. Therefore, despite elevated testosterone levels in the anaerobically trained state, further increases in muscle hypertrophy may be minimal (if any). Normal growth hormone secretion is pulsatile (i.e., bursts of secretion throughout the night and day). Therefore, resting growth hormone concentrations offer little, if any, insight on the effect of anaerobic exercise training. However, growth hormone secretion appears to be reduced after anaerobic exercise training at the submaximal workload (26). The effect of anaerobic exercise training on IGF-1 secretion also remains unclear. Currently available evidence suggests that IGF-1 secretion does not change in response to anaerobic exercise training (26). In general, the hormone profile during an acute session of anaerobic exercise is either reduced or unchanged after anaerobic exercise training.

Aging Perspective: Dehydration Among Elderly Populations

The ability to regulate fluid balance is compromised in older individuals due to the inability to respond to conditions of hypohydration (body water deficit) or dehydration. The normal response to declining plasma volume is reabsorption of water and concentration of urine by the kidneys through ADH release from the posterior pituitary. With aging, the regulatory failure to respond to conditions of decreasing plasma volume is caused by several mechanisms. Between the ages of 30 and 85 yr, an approximately 20% decrease in renal mass and number of nephrons occurs. This results in lower filtration and the kidneys' inability to reabsorb water. In addition, the kidneys become less responsive to ADH release, which heightens the failure to retain water and concentrate urine. Aging also has a negative effect on the thirst mechanism, indicating that older individuals will not feel thirsty in conditions of hydration. Given the inability to reabsorb fluids and the defect in the thirst mechanism, one can see that older individuals are at increased risk for dehydration. Exercise physiologists who work with elderly populations must be aware of this regulatory abnormality and encourage patients or clients to drink fluids during exercise (37).

Sport Performance Perspective: Nutrient Timing for Muscle Hypertrophy

The hormonal response to resistance training has a tremendous effect on protein synthesis and muscle hypertrophy (growth). Plasma levels of both growth hormone and testosterone increase after an intense bout of resistance exercise and promote repair and growth. Considerable research has been conducted to explore nutritional timing strategies that enhance the release of these anabolic hormones and increase in protein synthesis. It has been demonstrated that taking a protein and carbohydrate supplement before and immediately after a bout of resistance exercise increases both testosterone and growth hormone up to 30 min postexercise; however, this is followed by a sharp decline in testosterone levels. Another very important anabolic hormone is insulin, which promotes nutrient uptake into muscle. Insulin has been shown to increase substantially when protein and carbohydrate are consumed postexercise, optimally within 45 min to 1 h. In fact, it has been shown that protein synthesis can increase up to 400% when nutrients are ingested within 1 to 3 h after resistance training. In general, it is thought that a supplement should consist of 45 g of carbohydrate and 15 g of protein (3:1 ratio). Those participating in heavy training and competition must be aware of the importance of nutrient ingestion during the postexercise period to promote continued muscle recovery and growth (71).

Clinical Perspective: Exercise Improves Insulin Sensitivity

Insulin resistance is a condition in which tissues that normally are sensitive to insulin (e.g., adipose tissue, liver, and skeletal muscle) do not respond well to insulin. Normally, insulin promotes glucose uptake and prevents glucose output from the liver. Therefore, when someone is insulin resistant, their blood glucose levels can increase and contribute to the onset of type 2 diabetes mellitus (T2DM). Losing only 5% to 10% of initial body weight can improve insulin resistance (32); however, even this very modest weight loss can be very difficult to both achieve and sustain (16). Exercise can improve insulin resistance and impart important metabolic health benefits even without weight loss (18). A commonly held belief is that exercise has to be very intense in order to improve health and that people need to exercise train for weeks, months, or even years. However, this is not the case. A single session of moderate exercise (approximately 60% peak $\dot{V}O_2$) can reduce insulin resistance (44). Equally important, this improvement in insulin sensitivity occurs in the absence of weight loss or improved cardiorespiratory fitness. However, the insulin-sensitizing effects of exercise are short lived and last only 24 to 48 h postexercise (42). Consequently, in order to maintain these improvements, insulin-resistant and T2DM individuals need to exercise regularly. That is why the current exercise guidelines for preventing and treating T2DM suggest exercising most days of the week and exercising at least once every 48 h (12).

Summary

- Aerobic exercise of low to moderate intensity and shorter duration (<60 min) provides benefit, whereas strenuous heavy exercise may cause immunosuppression.
- Chronic aerobic training of moderate intensity and volume enhances immune function and reduces one's risk for infection.

- Athletes who engage in long training sessions and high-intensity competitions are more vulnerable to infection, as demonstrated by increased rates of upper respiratory tract infections.
- The greatest positive effect of exercise training on immune function has been demonstrated among those suffering from inflammatory diseases such as coronary artery disease or diabetes. Exercise training has consistently been shown

to decrease inflammatory markers among these populations.

- The endocrine response to exercise serves many functions, including cardiopulmonary response, blood pressure maintenance, metabolic regulation, fluid balance, and the release of anabolic agents.
- During an acute session of aerobic exercise, circulating catecholamines influence cardiac contractility, blood pressure, and ventilation.
- The secretion of additional endocrine hormones (including angiotensin II, aldosterone, AVP, and ADH) further enhances vasoconstriction in vascular beds of inactive tissue and helps regulate blood pressure and exercise hemodynamics.
- Hormones involved in substrate mobilization include insulin, epinephrine, glucagon, cortisol, and growth hormone.
- In general, insulin levels decrease and levels of the counterregulatory hormones increase during exercise to preserve blood glucose through mobilization of other substrates (e.g., lipids and liver glycogen).
- Hormones involved with fluid and electrolyte balance during exercise are ADH (secreted from the posterior pituitary gland) and aldosterone (secreted from the adrenal cortex). They help reabsorb water, retain sodium, and excrete potassium.
- Resistance exercise is a potent stimulus for the secretion of testosterone, growth hormone, and IGF-1, which are anabolic agents that promote protein synthesis and muscle hypertrophy.

Definitions

adaptive immune system—The arm of the immune system that rids the body of foreign substances through humoral and cell-mediated responses. These actions are carried out by B cells (humoral) and T cells (cell mediated).

adrenal gland—Endocrine gland located above the kidneys. Hormone secretion occurs from the outer cortex and inner medulla.

amine hormones—Hormones derived from the amino acid tyrosine. The catecholamines epinephrine and norepinephrine are examples.

antigen presenting cells—Immune cells that present an invading microbe to the lymphocytes.

counterregulatory hormones—Hormones that work in opposition to insulin. These include epinephrine, glucagon, cortisol, and growth hormone.

cytokines—Plasma proteins released by immune cells that communicate with surrounding cells and tissue. They have various functions that may be proinflammatory or anti-inflammatory.

gluconeogenesis—Metabolic pathway that generates glucose from noncarbohydrate sources. Common substrates for this pathway include lactate and amino acids.

innate immune system—The initial internal defense system against invading agents. The system recognizes and removes foreign substances and presents antigens to the adaptive arm to promote antibody production.

leukocytosis—An increase in the total number of circulating immune cells.

major histocompatibility complex—Molecules found on APCs that provide specific instruction to T cells. Separated into class I molecules, which present intracellular antigens (infected cells), and class II molecules, which present extracellular antigens (microbes in circulation).

neuroendocrine system—The collective and coordinated actions of the nervous and endocrine systems.

pathogen—Infectious agents in the form of bacteria, viruses, fungi, parasites, or toxins that produce an immune response.

peptide hormones—Hormones synthesized from amino acid chains; commonly called *protein hormones*. Common ones include insulin and glucagon.

phagocytosis—The process of engulfing a microbe and destroying it with reactive oxygen species. Performed by phagocytes such as neutrophils.

pituitary gland—A major endocrine gland that is located at the base of the brain and extends from the hypothalamus. It mainly comprises an anterior lobe and a posterior lobe.

plasma osmolality—A measurement of the amount of solutes in the plasma and detected by osmotic receptors. Higher values indicate more solutes and concentrated plasma.

proteolysis—The breakdown of proteins to amino acids.

toll-like receptors—Surface receptors expressed on innate immune cells (mainly macrophages and DCs) that are capable of identifying invading microbes and mobilizing an immune response.

Chapter 6

Principles for Testing and Training for Aerobic Power

The development and maintenance of aerobic power is important for both human performance and health-related physical fitness. Fitness for performance and health are easily observed and experienced throughout one's day. Walking briskly across campus to avoid an impending rain shower and running your first half-marathon both center on aerobic power or cardiorespiratory performance, which is the ability of the body to transport and utilize oxygen (O_2). Consider also the maladaptations or loss of aerobic function that develop if one has to stop training due to an injury, develops chronic kidney disease or heart failure, or trains too hard for too long and experiences overtraining syndrome. Appreciating this is important because it spans the breadth of human performance from health to disease and weekend warrior to world-class athlete.

This chapter first reviews the methods for accurately determining a person's aerobic power, as measured by the amount of energy he or she can expend. Such knowledge will serve you well regardless of whether you plan a career working with elite athletes, clients in fitness centers striving to improve their general health, or a patient recovering from colon cancer who is trying to reverse the fatigue associated with both the disease and the therapies often used to treat it. This chapter also reviews the changes or adaptations that develop in the circulatory, autonomic, and other organ systems as a result of training for improved aerobic power.

WORK AND POWER

Digging in a garden, vacuuming a carpet, folding clean clothes, running an 8-min mile, and changing a flat tire all have one thing in common: They all require the body to generate energy to perform the task. The capacity of the human body to convert the chemical energy in the food we eat to the adenosine triphosphate (ATP) needed to do muscular work can vary greatly from person to person, but it is a quantity that can be measured.

By definition, **work** is the application of a constant force through a distance (work = force × distance). For example, if a 70-kg person performed one pull-up on a bar that was 0.75 m above their chin, then the work performed would be 52.5 (70 kg × 0.75 m) kilogram-meters (kg-m). Converting this work to an energy unit that you might be more familiar with, such as kilocalories, requires you to know that 1 kcal = 426.85 kg-m. Therefore, this person expended 0.12 kcal (52.5/426.85) to complete that one pull-up.

Power is the rate at which the work is being performed, which equals work per unit of time:

$$\text{Work (force × distance)/time} \qquad \text{(Equation 6.1)}$$

For the previous example, if it took the person 0.5 s to perform one pull-up, their power would be as follows:

$$\text{Power (P)} = \frac{\text{work}}{\text{time}}$$

$$P = \frac{5.25 \text{ kg-m}}{0.5 \text{ s}}$$

$$P = 105 \text{ kg-m} \cdot \text{s}^{-1} \qquad \text{(Equation 6.2)}$$

Many scientists have demonstrated that the amount of O_2 consumed ($\dot{V}O_2$) at rest or when performing any type of task or exercise is equivalent to the heat (in kilocalories) produced. Therefore, the measurement of $\dot{V}O_2$ during the task is an indirect means for measuring heat produced or energy used. Although the precise amount of energy

Table 6.1 Energy Equivalents of the Energy Nutrients and Alcohol

Food	Energy (kcal $\cdot$ L^{-1} of O$_2$)	Energy (bomb calorimeter; kcal $\cdot$ g^{-1})	Net energy (physiological values; kcal $\cdot$ g^{-1})*
Carbohydrate	5.05	4.10	4.02
Protein	4.46	5.65	4.20
Fat	4.74	9.45	8.98
Alcohol	4.86	7.10	7.00
Mixed diet	4.83	—	—

*In the body a loss of kilocalories to digestion occurs as follows: carbohydrate, 2%; fat, 5%; and protein, 8% plus loss in urine. For alcohol, there is a small loss in urine and exhaled air.

(in kilocalories) used in a task depends on the nutrient being metabolized (i.e., protein, fat, or carbohydrate; see table 6.1), the energy yield is approximately 5 kcal $\cdot$ L^{-1} of O$_2$ consumed. For example, if a person walks 1 mi (1.6 km) and consumes approximately 22 L of O$_2$, then the energy yield would be approximately 110 kcal (22 × 5). If the person briskly walks that mile in 15 min, their power output (expressed as $\dot{V}O_2$) would be approximately 1.47 L $\cdot$ min^{-1} (22 L per 15 min). However, if tomorrow that person very slowly walks that same mile in 30 min, their power output would be approximately 0.733 L $\cdot$ min^{-1} (22 L per 30 min). For these two examples, power output expressed in kilocalories would be approximately 8 and 4 kcal $\cdot$ min^{-1}, respectively.

DIRECT AND INDIRECT MEASUREMENT OF ENERGY PRODUCTION

Using a variety of apparatuses and equipment, for well over a century scientists have been measuring humans' ability to produce energy and their capacity to perform work. Now, energy expenditure often is determined in athletes as O$_2$ uptake using indirect spirometry. It can also be measured directly using a calorimeter that assesses the change in the temperature of water that is surrounding a vessel or chamber. Regardless of the approach used, the ability of scientists to accurately measure energy expenditure provides useful information that has broad applications to public health, human performance, and the treatment of diseases.

Direct Measurement of Energy Production

When the body expends energy during work, heat is produced. The body does not differentiate whether the work performed is exercise for leisure, sport, or a household chore. Movement requires energy, and the transformation of chemical energy (ATP) to mechanical energy (movement) liberates heat. The source of the energy (and therefore the heat produced) is derived from the food we eat—protein, fat, and carbohydrate (see table 6.1).

The caloric values associated with each of these nutrients is equivalent to the heat liberated, as determined in the latter part of the 19th century by the German physiologist Max Rubner. He demonstrated this relationship using a closed chamber called a *calorimeter* (because it measured heat expressed in calories) with water circulating around the outside of the chamber. A dog was placed inside the chamber, and the heat produced by the dog was measured by noting the change (increase) in the temperature of the circulating water (figure 6.1). Each increase in water temperature of 1 °C per kilogram of water is equivalent to 1 kcal of energy. Based on this initial investigative work and much work that followed by Rubner, his colleagues, and others, we know that the energy expended by an individual performing any type of work is equal to the heat energy set free through metabolism of the energy nutrients. Along these same lines, to precisely determine how much energy is in an average-size potato or a medium pizza, that food is ignited and burned inside another type of calorimeter (called a *bomb calorimeter*), and the increase in circulating water is equivalent to the energy (in kilocalories) of the food. Although these approaches to measuring the production of heat energy are precise, they take much time, are somewhat insensitive to the rapid changes in energy expenditure that occur during exercise, and require expensive equipment. To gather the same information we can use a simpler indirect spirometric approach that measures the actual amount of O$_2$ consumed while performing an activity.

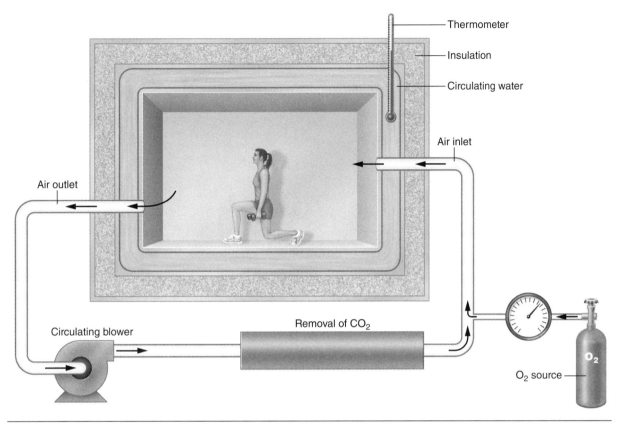

Figure 6.1 Calorimeter and closed-circuit spirometer.

Indirect Measurement of Energy Production Via O_2 Consumption

Earlier we mentioned that the energy associated with 1 L of O_2 consumed is approximately 5 kcal; according to table 6.1, this is a generally fair approximation for each of the three nutrients and alcohol. From table 6.1, we can also see that the exact kilocalories liberated per gram of the energy nutrient vary a bit depending on whether it is measured in a bomb calorimeter or expressed as net energy (physiologic value) after digestion. The reason for the difference between the calorimeter value and the net energy value is that some energy is needed to digest the nutrient (or, in the case of protein, to eliminate nitrogen in the form of urea), therefore yielding an even lower net energy value. Rounding the energy equivalents in table 6.1 to usable values, we can comfortably state that each gram of protein, fat, and carbohydrate is associated with 4, 9, and 4 $kcal \cdot g^{-1}$, respectively. Based on this information, a medium potato, which is almost all carbohydrate and dietary fiber, is about 180 kcal. That same potato sliced and then fried in oil (i.e., French fries) is 340 kcal. A dif-

ference of 160 kcal between the two methods of cooking doesn't seem like much, but that is an extra 16,000 kcal/yr for the person who eats one order of fries twice a week.

In chapter 3 we discussed the measurement of O_2 uptake during activity in the context of the Fick equation. In this chapter we've used O_2 uptake to describe energy production: Each liter of O_2 consumed is equivalent to approximately 5 kcal of energy liberated. This leads nicely into the question "How exactly does one measure O_2 uptake?" Recall from chapter 3 that O_2 uptake can be determined by measuring both cardiac output (heart rate × stroke volume) and the difference in the O_2 concentrations found in arterial blood and mixed venous blood (a-$\bar{v}O_2$ diff). Although highly accurate, this approach requires sophisticated equipment and highly trained personnel and is a complex undertaking if the measurements are collected during exertion.

A simpler approach to measuring O_2 uptake that is just as accurate is indirect open-circuit spirometry. Originally (and even in some labs today), one manual approach involved the collection of exhaled air in Douglas bags or meteorological balloons as a way to measure the volume of air and determine O_2 and carbon dioxide

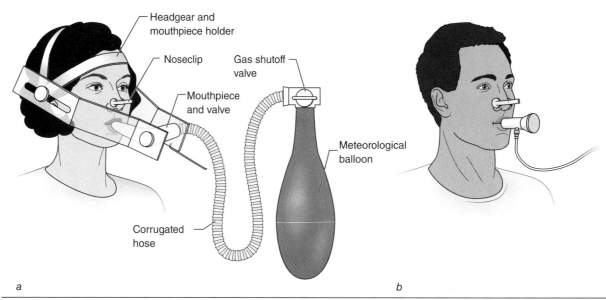

Figure 6.2 Gas collection. *(a)* A manual approach using a mouthpiece, valve, and gas collection balloon, and *(b)* a semiautomated method using a mouthpiece and a pneumotach for routing the air to a metabolic cart. (See also figure 6.3.)

(CO_2) concentrations. Figure 6.2 depicts two approaches for collecting exhaled air.

If gas collection bags are used, several may be needed if exhaled air is collected throughout exercise and recovery. The volume of air exhaled per minute is measured by emptying the collection bag through a gas or volume meter. Small samples are first extracted from the bags for CO_2 and O_2 analysis. Appendix A details the arithmetic used to calculate O_2 uptake ($\dot{V}O_2$) and volume of CO_2 ($\dot{V}CO_2$) exhaled. We already know the concentrations for O_2 and CO_2 in room air. After making some adjustments to the measured volume of exhaled air, we can compute the volume of these two gases in inhaled air. Similarly, if we measure the concentrations of O_2 and CO_2 in exhaled air, given our measured volume of exhaled air, we can also compute the volume for each of these two gases in exhaled air. Now we can compute the volume of O_2 consumed (volume of O_2 inhaled – volume of O_2 exhaled) and the volume of CO_2 produced (volume of CO_2 in exhaled air – volume of CO_2 in inhaled air).

Today, most university- and hospital-based exercise physiology or noninvasive laboratories avoid the manual method of collecting and analyzing expired gases and instead favor semiautomated electronic systems. Currently available systems are able to provide breath-by-breath analysis (versus drawing a sample from a bag filled with mixed air that was collected over a period of time); are compact, lightweight, and portable; provide advanced computing and graphic display capabilities; and are able to interface with existing commercial data management software. Figure 6.3 shows examples of measuring gas

exchange during exercise using a metabolic measuring system. Features common to almost all of the systems available today are semiautomated gas and volume calibration, continuous monitoring and reporting sorted by operator-preferred time intervals, and system capture of ambient temperature and barometric pressure data to be used in computing $\dot{V}O_2$ and $\dot{V}CO_2$.

NET O_2 COST OF EXERCISE

Figure 6.4 displays the time course of $\dot{V}O_2$ measured at rest (5 min), during a 15-min bout of steady-state submaximal exercise, and throughout a 30-min period of recovery. *Steady state* means a constant exercise pace that elicits heart rate (HR), $\dot{V}O_2$, and other physiologic responses that generally are unchanged over a moderate-duration (e.g., 10-30 min) bout of exercise. In figure 6.4, the gross $\dot{V}O_2$ during steady-state submaximal exercise is about 2 L · min^{-1}; however, subtracting resting $\dot{V}O_2$ (approximately 0.3 L · min^{-1}) from this amount provides a net O_2 uptake of 1.7 L · min^{-1}. Determining either the gross or net $\dot{V}O_2$ of an activity does require the individual to perform submaximal, constant-pace exercise, such as jogging on a treadmill at a fixed pace. Once physiologic steady state is reached, which may take between 2 and 4 min depending on an individual's fitness level and any underlying diseases, the $\dot{V}O_2$ and $\dot{V}CO_2$ values measured over a subsequent 1-min interval can be used for calculations.

There are a few more important points to discuss about the O_2 cost of an aerobic activity. First, note in figure 6.4

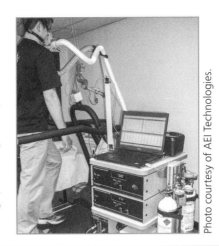

Photograph courtesy of Henry Ford Health System.

Photo courtesy of AEI Technologies.

Photo courtesy of MGC Diagnostics.

Figure 6.3 Examples of various gas collection systems and semiautomated analyzers used for assessing O₂ uptake and CO₂ production.

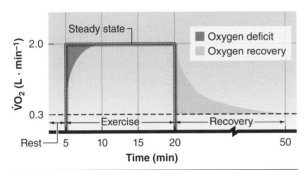

Figure 6.4 Time course of O₂ uptake (volume of O₂ used per unit of time, V̇O₂) at rest, during steady-state submaximal aerobic-type exercise, and in recovery. Subtracting resting V̇O₂ (L · min⁻¹) from exercise V̇O₂ permits us to report the net O₂ cost of exercise (L · min⁻¹).

that until physiologic steady state is reached, energy is produced in the absence of sufficient O₂ for a brief period until the aerobic system is revved up. The energy needed during this time comes from anaerobic pathways (mostly glycolysis) and is referred to as O₂ *deficit*. The energy needed to replenish the ATP and creatine phosphate (CP) stores and to clear the lactate that is produced during O₂ debt is why V̇O₂ does not immediately drop to resting values when exercise is stopped. The magnitude and duration of the O₂ recovery period essentially is equal to the energy required to replenish the energy stores and clear lactate from the blood.

Second, using the gross and net O₂ cost of an activity we can also compute the gross and net kilocalories expended. For example, the runner in figure 6.4 consumed approximately 2 L of O₂/min over 15 min for a gross O₂ cost of 30 L. Net O₂ cost during the same time period would be computed as follows:

Exercise O₂ of 2.0 L · min⁻¹ - resting O₂ of 0.3 L · min⁻¹

= 1.7 L · min⁻¹ × 15 min = 25.5 L · min⁻¹

(Equation 6.3)

Recall from our earlier discussion that each liter of O₂ consumed equals about 5 kcal. Therefore, for this person the gross energy associated with running for 15 min would be 150 kcal (30 L × 5 kcal · L⁻¹). The net energy expended above rest would be 128 kcal (25.5 × 5). Contemporary practice involves reporting gross O₂ cost or kilocalories expended instead of net values.

Third, performing at maximal (or peak) effort is certainly not a steady-state activity. When you read or hear about the maximal V̇O₂ of an athlete, the value measured usually refers to gross—not net—V̇O₂. Also, it is common for most university- and clinic-based laboratories to report V̇O₂ not only in absolute terms (L · min⁻¹) but also relative to body weight; the common unit of expression is mL · kg⁻¹ · min⁻¹. For example, a 70-kg cross country runner with an absolute maximal V̇O₂ of 5,040 L · min⁻¹ would have a relative V̇O₂ of 72 mL · kg⁻¹ · min⁻¹ (5,040/70). Because both are expressed per minute, either can be used to reflect this athlete's peak aerobic power.

CALORIC EQUIVALENT OF O₂

Previously we mentioned that the precise number of kilocalories liberated per liter of O₂ consumed varies based on the energy nutrient being metabolized; for simplicity, we use 5 kcal/L as an estimate. To more accurately determine the exact kilocalories associated with liters of O₂ consumed, we introduce the **respiratory exchange ratio** (RER), which is the ratio between the volume of CO_2 exhaled per unit of time and the volume of O_2 consumed during the same time interval (Note: a single cell respires, whereas the whole animal breathes. At the cellular level, $\dot{V}CO_2 / \dot{V}O_2$ is usually referred to as the respiratory quotient [RQ]. For our purposes, at the level of the lungs during steady state conditions, $\dot{V}CO_2 / \dot{V}O_2$ is termed respiratory exchange ratio [RER]).

$$RER = \frac{VCO_2}{VO_2} \qquad \text{(Equation 6.4)}$$

Carbohydrate, Fat, Protein, and Mixed Diet

Each molecule of carbohydrate contains carbon, hydrogen, and O_2; twice as much hydrogen than O_2 (i.e., the proper proportion to form water) is always present. Therefore, at rest and during steady-state exercise conditions, all of the O_2 consumed is used in the oxidation of carbon. Thus, for the typical molecule of carbohydrate (e.g., glucose):

$$C_6H_{12}O_6 + 6O_2 \rightarrow 6CO_2 + H_2O \qquad \text{(Equation 6.5)}$$

Because equal volumes of gases at the same temperature and pressure contain the same number of molecules (Avogadro's law), if an individual consumes a diet of only carbohydrate and thus metabolizes only carbohydrate, then:

$$1 \text{ mol of } C_6H_{12}O_6 + 6 \text{ mol of } O_2$$
$$= 6 \text{ mol of } CO_2 + 6 \text{ mol of } H_2O \qquad \text{(Equation 6.6)}$$

or

$$RER = \frac{\dot{V}CO_2\text{produced}}{\dot{V}O_2\text{consumed}} = \frac{6\,VCO_2}{6\,\dot{V}O_2} = 1 \qquad \text{(Equation 6.7)}$$

Simply, the oxidation of 1 mol of carbohydrate requires 6 mol of O_2 and produces 6 mol of CO_2. To identify the exact number of kilocalories liberated with the consumption of 1 L of O_2, we turn to table 6.2. At an RER of 1.0, only carbohydrate is metabolized, and the energy liberated is 5.047 (again, about 5). Restated, an RER of 1.0 indicates that only carbohydrate is being metabolized, and at that level 5.047 kcal is liberated for every liter of O_2 consumed.

The same approach is true for the oxidation of fat such that sufficient O_2 must again be available (from inhaled air or from within the nutrient) to combine with carbon to generate CO_2 and to combine with hydrogen to make water. Because fat molecules contain vastly more hydro-

gen than O_2, more O_2 has to be consumed. This is nicely exemplified in the following equation, which describes the complete combustion of palmitic acid:

$$1 \text{ mol of } C_{16}H_{32}O_2 + 23 \text{ mol of } O_2$$
$$= 16 \text{ mol of } O_2 + 16 \text{ mol of } H_2O \qquad \text{(Equation 6.8)}$$

or

$$RER = \frac{\dot{V}CO_2\text{produced}}{\dot{V}O_2\text{consumed}} = \frac{16\,\dot{V}CO_2}{23\,\dot{V}O_2} = 0.70 \qquad \text{(Equation 6.9)}$$

Again referencing table 6.2, we note that an RER of 0.70 indicates that only fats are being metabolized. In this instance, 1 L of O_2 consumed liberates 4.686 kcal (about $5 \text{ kcal} \cdot L^{-1}$).

People eat a mixed diet comprising carbohydrate and fat as well as protein. Because protein is not simply oxidized to just CO_2 and H_2O, another issue needs to be considered when determining the calories derived per liter of O_2 consumed. Less energy is available when protein is metabolized in the body because protein is deaminated, and nitrogen and a sulfur residue are excreted in the urine and feces. Therefore, energy production in the body from protein ($4.20 \text{ kcal} \cdot g^{-1}$; table 6.1) is less than what is liberated in a bomb calorimeter ($5.65 \text{ kcal} \cdot g^{-1}$). This difference must be taken into account when using RER to determine which energy nutrient is being metabolized.

Experiments were done years ago to determine the amount of CO_2 produced and the amount of O_2 consumed when protein is oxidized. Based on those experiments and given the amount of nitrogen excreted, the O_2 required and CO_2 produced when protein is oxidized could be subtracted from the total amount of O_2 required and CO_2 produced. This allowed scientists to generate a nonprotein RER for carbohydrate and fat only (which is what table 6.2 shows) (69). Therefore, given O_2 consumed and CO_2 produced and using nonprotein RER, we can compute kilocalories expended per liter of O_2 consumed and the relative contribution of carbohydrate and fats to energy liberation at rest or during steady-state exercise.

How many kilocalories are expended if someone exercises for 40 min and consumes 2 L of O_2/min (assume an RER of 0.84)? The answer is 388 kcal (80 L of O_2 consumed $\times$ 4.850 $\text{kcal} \cdot L^{-1}$ at a nonprotein RER of 0.84). Our prior estimate of $5 \text{ kcal} \cdot L^{-1}$ of O_2 consumed would have been a reasonable approximation in this case, at 400 kcal total.

Factors other than the oxidation of the nutrient can affect RER. Three common ones include hyperventilation, exercise condition (steady state vs. non–steady state), and exercise type (e.g., short-term exhaustive).

Table 6.2   Caloric Equivalent (kcal/L of O_2) and Percentage of Total Calories Provided by Carbohydrate and Fat at Each Nonprotein RER

Nonprotein RER	kcal of energy/L of O_2 consumed	Calories derived from carbohydrate (%)	Calories derived from fat (%)
0.70	4.686	0.00	100.00
0.71	4.690	1.10	98.90
0.72	4.702	4.76	95.20
0.73	4.714	8.40	91.60
0.74	4.727	12.00	88.00
0.75	4.739	15.60	84.40
0.76	4.751	19.20	80.80
0.77	4.764	22.30	77.20
0.78	4.776	26.30	73.70
0.79	4.788	29.90	70.10
0.80	4.801	33.40	66.60
0.81	4.813	36.90	63.10
0.82	4.825	40.30	59.70
0.83	4.838	43.80	56.20
0.84	4.850	47.20	52.80
0.85	4.862	50.70	49.30
0.86	4.875	54.10	45.90
0.87	4.887	57.50	42.50
0.88	4.899	60.80	39.20
0.89	4.911	64.20	35.80
0.90	4.924	67.50	32.50
0.91	4.936	70.80	29.20
0.92	4.948	74.10	25.90
0.93	4.961	77.40	22.60
0.94	4.973	80.70	19.30
0.95	4.985	84.00	16.00
0.96	4.998	87.20	12.80
0.97	5.010	90.40	9.58
0.98	5.022	93.60	6.37
0.99	5.035	96.80	3.18
1.00	5.047	100.00	0.00

Based on data from Zuntz and Schumberg in Lusk (69).

Concerning hyperventilation (voluntary or involuntary, the latter often being associated with anxiety or stress), higher than normal amounts of CO_2 are exhaled at the mouth during rapid and deep breathing, thus increasing the numerator ($\dot{V}CO_2$) in our calculation of RER. This often yields an RER value greater than 1.0. Likewise, before steady-state conditions are achieved during sub-maximal exercise (which may take 2-4 min), the body may be exhaling more CO_2 relative to the amount of O_2 consumed due to the increased ventilation associated with movement and exercise. This too could result in RER values that are temporarily higher until the amount of O_2 consumed increases to a value that meets the energy needs for the level of submaximal work being performed.

Finally, during short-term (20 s-2 min) exhaustive work, RER may quickly jump to 1.35 or higher as the body increases ventilation to buffer the large quantities of hydrogen ions being generated and CO_2 being released from the cell. Note that even though RER exceeds 1.0, only carbohydrate (glucose) is used, so each liter of O_2 consumed would be treated as an RER of 1.0 or liberating 5.05 kcal $\cdot$ L^{-1}.

PROTOCOLS FOR ASSESSING AEROBIC POWER

The measurement of aerobic power is much more widespread today than is the assessment of anaerobic power. Whereas the latter usually is limited to human performance settings (e.g., Olympic training centers, universities), measuring aerobic power applies not only to athletes but also to fitness enthusiasts, apparently healthy people in the general population, and patients with certain diseases. Aerobic power (maximal $\dot{V}O_2$) is determined using open-circuit gas exchange, performed while the individual exercises on a modality (e.g., motorized treadmill, cycle ergometer) that applies graded, increased work over time.

For most leisure-time athletes and apparently healthy people tested in the clinical setting, aerobic power typically is about 10% to 20% higher when measured using a weight-dependent modality such as a treadmill (vs. a weight-independent mode such as stationary cycling) (10). Conversely, elite athletes usually achieve their highest $\dot{V}O_2$ values on (and prefer to be tested on) a sport-specific modality, such as a treadmill for runners or swimming flume for swimmers (91). The higher values observed in athletes using sport-specific testing modes are partially attributable to highly developed task-specific neuromuscular patterns.

In most university, fitness, and clinical exercise laboratories, both a treadmill and a cycle ergometer are available for testing. The treadmill typically is the testing mode of choice in the United States, whereas the cycle ergometer typically is used in Europe and other countries. Table 6.3 summarizes typical max $\dot{V}O_2$ values measured in selected male and female athletes and provides values for generally healthy people and a few patients with selected health problems.

Regardless of the mode selected for testing, almost all testing protocols used today incorporate two important characteristics. First, although the progressive increments in workload imposed during a test vary based on starting workload and magnitude of increase per stage, selection of the correct protocol for athletes and nonathletes should aim to have peak or maximal effort achieved within 8 to 12 min (10,76). It is important to select a testing protocol that best aligns with an individual's current fitness level. To facilitate this, key questions can be asked such as current exercise training habits, type of occupation, health history, and types of routine activities engaged in throughout the day (e.g., flights of stairs climbed in a day, difficulty climbing up stairs). Table 6.4 provides examples of three commonly used testing protocols: one for an apparently healthy person (Bruce), one for a patient with a known, limiting cardiovascular disease (Naughton), and one for an athlete capable of running 6.2 mi (10 km) in 42 to 45 min.

Second, testing can be conducted using either a steady-state protocol or a ramp protocol. Both are continuous tests; the former imposes a new stage of work (change in work rate) usually every 2 to 3 min. This approach typi-

Table 6.3 Ranges for Maximal Aerobic Power ($\dot{V}O_2$) in Selected Trained and Untrained Adults (mL $\cdot$ kg^{-1} $\cdot$ min^{-1})

	Men	Women
Elite Nordic skiers	65-94	60-75
Elite runners	60-85	50-75
Elite cyclists	62-74	47-57
Basketball	40-60	43-60
50-59 yr (50th percentile)	27-40	20-28
Individuals with heart disease	13-25	11-18
Individuals with heart failure	6-23	6-21

Adapted, by permission, from W.L. Kenney, J.H. Wilmore, and D.L. Costill, 2012, *Physiology of sport and exercise*, 5th ed. (Champaign, IL: Human Kinetics), 269; Clinton A. Brawner, PhD, Henry Ford Hospital.

Table 6.4 Commonly Used Treadmill Protocols

Protocol	Stage	Time (min)	Speed in mph (kph)	Grade (%)	Estimated $\dot{V}O_2$ (mL · kg⁻¹ · min⁻¹)	Estimated MET level
Bruce	1	3	1.7 (2.7)	10.0	16.3	4.6
	2	3	2.5 (4.0)	12.0	24.7	7.0
	3	3	3.4 (5.5)	14.0	35.6	10.2
	4	3	4.2 (6.8)	16.0	47.2	13.5
	5	3	5.0 (8.0)	18.0	52.0	14.9
	6	3	5.5 (8.9)	20.0	59.5	17.0
Naughton	1	2	1.0 (1.6)	0.0	8.9	2.5
	2	2	2.0 (3.2)	0.0	14.2	4.1
	3	2	2.0 (3.2)	3.5	15.9	4.5
	4	2	2.0 (3.2)	7.0	17.6	5.0
	5	2	2.0 (3.2)	10.5	19.3	5.5
	6	2	2.0 (3.2)	14.0	21.0	6.0
	7	2	2.0 (3.2)	17.5	22.7	6.5
Athlete	1	3	8 (12.9)	0.0	46.0	13.3
	2	3	8 (12.9)	4.0	54.3	15.5
	3	3	8 (12.9)	6.0	58.0	16.6
	4	3	8 (12.9)	8.0	62.0	17.7
	5	3	8 (12.9)	10.0	65.8	18.8

MET = metabolic equivalent of task

cally allows most people being tested to achieve or nearly achieve a steady-state response during submaximal work levels (data such as HR and blood pressure are measured toward the end of each stage) before the next stage of exercise is imposed. Workloads increase every 2 to 3 min until the patient, client, or athlete reaches fatigue. A steady-state protocol is advantageous if one wishes to compare changes in submaximal responses such as HR and blood pressure over time or in response to an intervention such as a medication or a regular exercise training program.

If, however, only peak or maximal values are of interest, such as maximum HR or maximal $\dot{V}O_2$, then a ramp protocol serves well. Using a ramp testing method involves ramping the work rate up using small, continuous increments throughout the test. For example, when using a cycle ergometer, this might be 25 watts · min⁻¹ or even 1 watt · 6 s⁻¹ up to maximal effort. The ramp protocol may also provide a better description of the relationship between the continuous changes in work rate and the changes in the variables measured during the test

(e.g., HR and O_2 uptake), an approach that works well if one is trying to analyze the data for subtle changes during the course of a test (e.g., determining ventilatory threshold). With respect to measuring max $\dot{V}O_2$ or maximal HR, similar values are usually achieved regardless of whether a steady-state or ramp testing approach is chosen (76).

Measured Maximal O_2 Uptake

Today O_2 uptake or cardiorespiratory fitness are routinely measured using indirect open circuit spirometry, as described earlier in this chapter. It is a variable that is influenced by age and gender.

Among nonexercising individuals, max $\dot{V}O_2$ generally is 15% to 20% higher in males than in females (28,56). This difference between sexes is partly due to the typically 10% higher hemoglobin (Hb) concentration in men compared with women. (The difference between sexes is greater among untrained people than among trained people.) Hb is the molecule responsible for transporting O_2 to the skeletal muscles that are more metabolically

---

Aging and Fitness

It is well known that, after age 30, cardiorespiratory fitness (as measured by max $\dot{V}O_2$) decreases in both men and women at a rate of about 10% to 15% per decade on average (100). The rate of decline with age, however, is not uniform; accelerated declines noted are in the latter decades (3%-6% in the 20s and 30s and >20% per decade in the 70s) (33). What is much less clear are the mechanisms responsible for the age-related decline.

One factor many would suspect to be a primary cause for the decline in fitness is the well-known decline in maximal or peak HR that occurs with age. Because HR contributes to cardiac output, any decline in maximal HR with age would contribute to a reduction in peak or maximal cardiac output. However, the average rate of the decline in HR (approximately 4%-6% per decade) does not sufficiently parallel the decline in fitness. Interestingly, Fleg et al. (33) observed a close relationship between the rate of decline in O_2 pulse (mL of $O_2 \cdot beat^{-1}$) and the decline in peak $\dot{V}O_2$. Because O_2 pulse is the product of stroke volume and a-$\bar{v}O_2$ difference, it is difficult to determine whether age-related declines in central function (stroke volume), peripheral function (a-$\bar{v}O_2$ diff), or both play the major role. Some work suggests that stroke volume may not be altered much in healthy people as they age (34), suggesting that the loss in fitness is mostly attributable to intrinsic changes in peripheral function, such as regional peak blood flow or skeletal muscle function (e.g., muscle atrophy, reduced mitochondrial content and function) (64).

It is important to point out that these declines in fitness can be attenuated with regular exercise training. Specifically, this decrease in max $\dot{V}O_2$ with increasing age generally is reduced to approximately 5% to 6% per decade among individuals who exercise regularly (47). Similarly, regular exercise training may lead to preserved skeletal muscle mass and histology (e.g., proportion of slow-twitch red or Type I fibers) (98) despite increasing age.

---

active during exercise. Another key reason for the difference in max $\dot{V}O_2$ values between men and women is the generally larger stroke volume in men. Because the left ventricle is larger in men, peak cardiac output and the transport of O_2 are greater as well. One might surmise that differences in body fat might also account for the differences in cardiorespiratory fitness between men and women; essential fat is greater in females (approximately 14%) than in males (approximately 6%). However, such differences are minimized when O_2 uptake is expressed in mL of O_2 consumed $\cdot$ kg of nonfat or lean body mass$^{-1} \cdot min^{-1}$ (versus just mL of $O_2 \cdot kg^{-1} \cdot min^{-1}$) (23).

No discussion about max $\dot{V}O_2$ would be complete without mentioning the criteria used to define maximal (versus peak) $\dot{V}O_2$. Although the criteria used for determining max $\dot{V}O_2$ may vary based on the laboratory doing the testing and analysis, maximal values generally are believed to be attained if further increases in work rate result in a plateau or no further increases in $\dot{V}O_2$ (i.e., increase of 150 mL $\cdot$ min^{-1} or less) (10) or if RER exceeds 1.15 during testing. If at the end of the test there is no observed plateau in measured $\dot{V}O_2$, which often is the case when testing a patient with a chronic disease

(e.g., heart disease, kidney disease, cancer), then the highest attained value typically is referred to as *peak* $\dot{V}O_2$. As one might expect, variability in subject effort, testing mode, and testing personnel can all influence when a test is stopped, and such factors might lend themselves to the measurement of a peak $\dot{V}O_2$ instead of a true maximal $\dot{V}O_2$. We know for sure that we should not use achievement of estimated peak HR alone as a criterion for whether a maximal or peak $\dot{V}O_2$ was achieved.

Concept of an MET

Another way to express the energy cost of an activity—a method that is linked to $\dot{V}O_2$—is to report the work effort in terms of an MET (an acronym for "metabolic equivalent of task"). One MET is defined as the energy expended ($\dot{V}O_2$, expressed as mL $\cdot$ kg$^{-1} \cdot$ min^{-1}) while sitting quietly. Simply put, 1 MET is a person's resting $\dot{V}O_2$. For the average adult, 1 MET generally is assumed to be approximately 3.5 mL $\cdot$ kg$^{-1} \cdot$ min^{-1}; however, some researchers have shown this value to be overstated by as much as 20% (86). An estimation that an activity requires 10 METs simply means that the O_2 cost is 10 times resting $\dot{V}O_2$

$(3.5 \text{ mL} \cdot \text{kg}^{-1} \cdot \text{min}^{-1} \times 10)$, which is $35 \text{ mL} \cdot \text{kg}^{-1} \cdot \text{min}^{-1}$. Alternately, 1 MET is also equivalent to expending approximately $1 \text{ kcal} \cdot \text{kg}^{-1} \cdot \text{h}^{-1}$ (5). A 70-kg athlete who performs a 7-MET activity such as Nordic skiing for a period of 1 h would expend approximately 500 kcal $(7 \times 1 \text{ kcal} \cdot \text{kg}^{-1} \cdot \text{h}^{-1} \times 70 \text{ kg})$.

Working the above math in the other direction, given an individual's $\dot{V}O_2$ measured while performing a specific activity, we can define that activity in terms of METs completed. For example, for an athlete with a measured maximal $\dot{V}O_2$ of $60 \text{ mL} \cdot \text{kg}^{-1} \cdot \text{min}^{-1}$, we can estimate peak METs as 17 (60/3.5). Table 6.5 shows the MET levels associated with walking at various velocities and at different elevations or grades (1). Table 6.6 lists the MET level associated with several common occupational and leisure activities.

It is important to again mention that although we assign a resting $\dot{V}O_2$ value of $3.5 \text{ mL} \cdot \text{kg}^{-1} \cdot \text{min}^{-1}$, there

Table 6.5   Approximate Energy Requirements in Metabolic Equivalents of Task (METs) for Various Common Horizontal and Grade Walking

Horizontal walking speed or velocity mi · h⁻¹	PERCENT (%) GRADE								
	0	2	4	6	8	10	12	14	16
1.80 [48.3]	2.4	2.9	3.4	3.9	4.4	4.9	5.4	5.9	6.3
2.00 [53.7]	2.5	3.1	3.6	4.2	4.7	5.3	5.8	6.4	6.9
2.50 [67.1]	2.9	3.6	4.3	5.0	5.7	6.4	7.0	7.7	8.4
2.70 [72.4]	3.1	3.8	4.3	5.3	6.0	6.8	7.5	8.3	9.0
3.00 [80.5]	3.3	4.1	5.0	5.8	6.6	7.4	8.3	9.1	9.9
3.30 [88.5]	3.5	4.4	5.3	6.3	7.2	8.1	9.0	9.9	10.8
3.50 [93.9]	3.7	4.6	5.6	6.6	7.5	8.5	9.5	10.4	11.4
3.70 [99.2]	3.8	4.9	5.9	6.9	7.9	8.9	10.0	11.0	12.0

Energy requirements computed using American College of Sports Medicine equations (1); assumes individual has achieved a steady-state response.

Table 6.6   MET Values Associated With Common Occupational and Leisure Physical Activities

Activity	MET value	Activity	MET value
Assembly line worker	3.5	Mowing lawn with power mower	4.5
Bowling	3.0	Painter	4.5
Bus driver	3.0	Security guard	2.5
Carpenter	6.0	Skiing	
Desk worker	1.5	Downhill, light	5.0
Farmer	5.0	Nordic	7.0
Firefighter	12.0	Snow removal	
Gardening	5.0	Hand shovel	6.0
Golf		Blower	4.5
Carrying clubs	5.5	Steel worker	8.0
With cart	3.5	Swimming laps (slow)	8.0
Hunting	6.0	Tennis (general)	7.0
Mail carrier, with satchel of 35 lb	4.5	Walking (3 mph)	3.3

Did I Pass?

After completing an exercise stress test in a noninvasive clinical laboratory, it is not uncommon for the patient to ask, "How did I do?" The answer to the question depends greatly on why the test was performed. Regardless of the reason, one piece of information that is available from almost all such tests is exercise capacity or fitness, expressed as estimated METs. Given workload (treadmill speed and grade; table 6.5) or work rate (watts from an arm or bike ergometer), we can estimate METs. The MET level associated with the different stages of some common exercise protocols is already calculated in table 6.4. For the patient who asked, "How did I do?" we can use the peak MET level achieved to help characterize fitness level and, in some patients, estimate prognosis (i.e., future risk of mortality).

For example, when describing or categorizing an individual's fitness level, if a 45-yr-old patient completes three stages of the standard Bruce protocol (table 6.4), we would estimate his or her MET level at 10 METs ($\dot{V}O_2$ of approximately 35 mL $\cdot$ kg^{-1} $\cdot$ min^{-1}). Acknowledging that fitness declines with age and that differences in fitness exist based on gender, one can still broadly say that for women over 40 yr of age, a fitness level of average or above represents values above 8.0 METs. Similarly, for men over 40 yr of age, above 10 METs (2) is associated with an average or greater level of fitness. Therefore, broadly, the previously mentioned patient's achieved level of 10 METs is considered average to above average regardless of gender or age.

Relative to prognosis, prior work by Myers et al. (77) and Al-Mallah et al. (6) suggests that MET levels less than 5 or 6 are associated with the highest future risk of mortality. In fact, for every 1-MET increase in achieved fitness, survival improves 12%. The independent relationship between fitness and survival remains regardless of whether the person has other health problems such as diabetes, smoking, obesity, or high blood pressure. Taking this concept one step further, we also know that among those with known heart disease (e.g., heart attack, coronary bypass surgery), the MET level a patient exercises at during a training session is also related to prognosis (17). Specifically, training at more than 4.5 METs in cardiac rehabilitation is associated with a 97% event-free survival rate at 1 yr. In comparison, training at less than 3 METs in cardiac rehabilitation is associated with an 86% event-free survival rate at 1 yr.

Finally, imagine that 4 wk after suffering an uncomplicated heart attack a mail carrier sees his physician and asks if he can return to work. His occupation involves a lot of walking at a mild to moderate pace while carrying a mail satchel that can weigh up to 35 lb (15.9 kg). Assuming that the patient is free of any complications after his heart attack, to answer this inquiry the physician needs to address two issues: the approximate energy expenditure associated with the patient's job (i.e., delivering mail) and the patient's fitness level (i.e., whether he is fit enough to do his job). To answer the first question, we turn to a source (e.g., table 6.6) (5) that lists a variety of occupational, leisure, and recreational tasks and their estimated MET levels. Accordingly, we learn that the task of delivering mail is estimated at 4.5 to 5 METs. To answer the second question, the patient undergoes a symptom-limited exercise stress test, and the test results show an estimated peak MET level of 8 METs. As a general rule, patients are allowed to return to work when the task they wish to perform is less than or equal to 75% of their achieved peak effort. In this example, 5 METs is 63% of 8 METs, so returning to work—perhaps part time at first to progressively regain endurance—could be approved. Other factors, such as job-related stress, whether the patient is engaged in a regular exercise program, self-pacing, and level of assistance provided by coworkers, can also influence decisions about whether a patient can return to work.

is variability among different people at rest. Some will have an actual resting $\dot{V}O_2$ while seated of 3.6 mL $\cdot$ kg^{-1} $\cdot$ min^{-1}, whereas others measure out at 2.9 mL $\cdot$ kg^{-1} $\cdot$ min^{-1}. Applying a constant of 3.5 mL $\cdot$ kg^{-1} $\cdot$ min^{-1} to men and women, girls and boys, older and younger, obese and normal weight, and highly fit and less fit helps the reader appreciate the fact that METs are often, in fact, simply an estimate and should be interpreted as such. If fitness or caloric expenditure needs to be quantified in an athlete, patient, or client, $\dot{V}O_2$ should be measured.

MATCHING TRAINING REGIMEN TO ENERGY AND ORGAN SYSTEMS AND TRAINING TO IMPROVE AEROBIC POWER

When developing a training regimen to improve performance, one must first look at what organ systems and energy systems are involved in the sport or activity and then develop a program that progressively and systematically overloads the organs and energy pathways involved. Doing so will stimulate and elicit a training adaptation. For activities that rely on aerobic power or cardiorespiratory fitness, the concept of *specificity of training* dictates that the training regimen should overload the cardiorespiratory system and the aerobic metabolic pathways in the skeletal muscles (i.e., beta oxidation, Krebs cycle, electron transport system; see chapter 1) that are engaged by the activity (e.g., legs for running and cycling, arms for rowing). Some activities are almost entirely aerobic in nature (e.g., walking), whereas others are quite anaerobic (e.g., gymnastics). Still others involve a constant changing of energy systems (e.g., basketball, soccer); for these sports, the training regimens need to include practice drills and conditioning that aims to adapt more than just one energy pathway.

Two other important issues pertaining to specificity need to be emphasized. First, performance time is related to energy system. For example, although the absolute distances associated with 5,000-m ice speed skating and 3,000-m running are clearly different, performance times are relatively similar—around 6.5 to 8 min. Therefore, a portion (e.g., 30%- 35%) of an athlete's conditioning time during practice should stress the adenosine triphosphate-phosphocreatine (ATP-PC) and anaerobic glycolysis systems, and the balance should stress the aerobic pathways.

Second, sports such as ice hockey and American football involve many short-duration periods of very intense anaerobic activity followed by a relatively longer aerobic period devoted to recovery or rest. However, because these sports rely on the ATP-PC stores for performance and because this energy pathway is replenished using the ATP generated from aerobic pathways during recovery, spending some training time developing the aerobic systems during practice is useful for indirectly improving performance. Specifically, spending some conditioning time targeting the aerobic energy pathways will improve the ability of the skeletal muscles to efficiently replenish ATP-PC stores via aerobic pathways during recovery, thus optimizing the athlete's ability for intense work

during the next anaerobic shift in hockey or snap of the ball in football.

As a patient or client begins a training program, keep in mind the reason for such training. Is it to improve general fitness, assist with weight loss, or improve performance in an athletic endeavor? Knowing the goal will help guide how the overload stimulus is applied and help assess or gauge whether your plan or regimen was successful.

Numerous methods address the second key principle involved with prescribing exercise, which is the principle of *progressive overload*—altering frequency, duration, or intensity of exercise to adjust and progress volume overload. A good acronym to help you remember the important elements of progressive overload is FITT-VP: F = frequency of exercise, I = intensity of exercise, T = time or duration of activity, T = type of activity, V = volume of activity, and P = progression of exercise.

Duration (Time) and Frequency of Training

Duration or training time can refer to minutes or hours of training during a single training session or to the months or years spent training. We focus our discussion on the former. In general, adults interested in garnering most of the health and fitness benefits derived from an aerobic training regimen should accumulate 30 to 60 $min \cdot d^{-1}$ ($>150 \ min \cdot wk^{-1}$) of moderate-intensity exercise or 20 to 60 $min \cdot wk^{-1}$ of vigorous-intensity exercise (see also figure 6.5) (2,39). In some people with initially poor

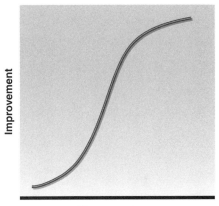

Figure 6.5 Relationships between exercise dose (intensity, duration, or frequency) and improvement (e.g., $\dot{V}O_2$, resting HR). Note that the initial and terminal portions of the curve are relatively flat, suggesting that low and high volumes of exercise are associated with smaller improvements.

fitness levels, as little as 20 min of exercise per session is sufficient to stimulate improvement. Additionally, among people who are overweight or obese and who exercise to assist with long-term weight loss or management, more than $250 \text{ min} \cdot \text{wk}^{-1}$ is recommended (26). In general, the longer the duration of each training session and the longer the training program (months vs. weeks), the greater the gains in aerobic fitness.

Although some people who are poorly fit can experience an improvement in fitness from training only $2 \text{ d} \cdot \text{wk}^{-1}$, most professional organizations establish 3 sessions $\cdot \text{wk}^{-1}$ as the minimum number of bouts needed. For general purposes, this frequency of training may be all that is needed if intensity of effort is vigorous; $5 \text{ d} \cdot \text{wk}^{-1}$ may be needed for a moderate-intensity training regimen. During preseason training, some more highly trained athletes and sport competitors may train 7, 8, 9, or more times $\cdot \text{wk}^{-1}$ (more exercise is associated with continued improvement, but the magnitude of the gains is less; see the upper portion of figure 6.5). In these individuals, the slight increase in risk associated with developing an overuse injury is offset by the hoped-for small gains in performance. These gains might be of sufficient magnitude to improve on one's previous best performance or win the event.

Determining Intensity of Aerobic Training

Possibly the most influential variable associated with developing the greatest gains in fitness is exercise intensity (41). Prior research (95) indicates that guiding intensity using $\dot{V}O_2$ reserve correctly reflects the intensity overload that is imposed on the cardiorespiratory system and the aerobic energy pathways during exercise. The following equation computes target training using $\dot{V}O_2$ reserve:

$$\text{Target training } \dot{V}O_2 \text{reserve}$$
$$= \left(\text{maximum or peak } \dot{V}O_2 - \text{resting } \dot{V}O_2\right)$$
$$\times \text{ \% intensity + resting } \dot{V}O_2 \qquad \text{(Equation 6.10)}$$

Because it is impractical to measure gas exchange and therefore O_2 uptake during every training session, a surrogate is needed that commensurately reflects $\dot{V}O_2$ reserve. The surrogate that works best is HR reserve because the relationship between HR reserve and $\dot{V}O_2$ reserve during cardiorespiratory exercise generally is quite linear. This close relationship allows us to use HR as a method for guiding exercise intensity in athletes, clients, and patients. The formula for prescribing exercise intensity using HR reserve is as follows:

$$\text{Target training HR reserve}$$
$$= \left(\text{HR maximum or peak - resting HR}\right)$$
$$\times \text{ \% intensity + resting HR} \qquad \text{(Equation 6.11)}$$

In the section "Summary of Methods for Prescribing Target Exercise Intensity," note that the HR-based methods rely on either a measured or an estimated maximum or peak HR. In the absence of an exercise test to derive measured maximal HR, several equations exist for estimating peak HR in generally healthy individuals (see "Three Commonly Used Generalized Equations for Estimating Maximal HR in Apparently Healthy Men and

---

Summary of Methods for Prescribing Target Exercise Intensity

HR Methods*

- [(HR maximum or peak − resting HR) × % intensity] + resting HR
- HR maximum or peak × % intensity

$\dot{V}O_2$ Methods

- [($\dot{V}O_2$ maximum or peak − $\dot{V}O_2$ rest) × % intensity] + resting $\dot{V}O_2$
- $\dot{V}O_2$ maximum or peak × % intensity

Subjective Methods

- Rating of perceived exertion: On a scale of 6 to 20 that evaluates overall body fatigue or exertion, training at a level of 11 to 15 (fairly light to hard)
- Talk test: Exercising at the fastest pace possible while retaining the ability to speak comfortably in response to a speech-provoking stimulus (e.g., answering a question).

*Based on measured or estimated maximum or peak heart rate.

Three Commonly Used Generalized Equations for Estimating Maximal HR in Apparently Healthy Men and Women

$$220 - age$$

$$208 - (0.7 \times age)$$

$$207 - (0.7 \times age) \quad \text{(Equation 6.12)}$$

References 37, 97, 40

Women"). In the absence of an exercise test, both rating of perceived exertion (15) and the talk test (104) can be used to guide exercise intensity.

With respect to setting the correct target intensity or target intensity range, most research suggests a minimum threshold intensity of 45% of $\dot{V}O_2$ reserve or HR reserve (2,94). Moderate intensity is described as training up to 60% of HR reserve, whereas vigorous intensity is identified as training between 60% and less than 90% of HR reserve (2). Most general fitness enthusiasts who engage in exercise training do so within an intensity range set between 70% and less than 85% of HR reserve. Keep in mind that, in general, the higher the fitness level of the individual, the greater the level of intensity of effort needed to induce further gains in fitness. Therefore, highly conditioned athletes may need to train at 90% of HR reserve to garner further gains in fitness.

Volume and Progression

Before describing various training regimens designed to impose a sufficient overload stimulus, it is important to introduce the concept of training volume, which is the product of frequency, intensity, and duration of exercise. Training volume usually is expressed as MET-min · wk^{-1} or MET-h · wk^{-1} (kcal · wk^{-1} can also be an expression of exercise volume). Computing volume of exercise allows us to quantify the total load placed on the body from all stimuli (intensity, duration, and frequency) compared with only minutes per week or level of effort. Using MET-h · wk^{-1} allows the athletes, clients, and researchers involved in exercise training programs of different intensities, durations, and frequencies to combine the three elements of overload into a common unit, thus allowing for discussion about observed similarities and differences in response to such training. For example, a 12% improvement in fitness resulting from a training regimen of 500 MET-min · wk^{-1} (200 min · wk^{-1} of slow walking at approximately 2.5 METs) in a group of breast cancer survivors in Germany can be compared and contrasted with

results from another study done in Norway, where similar patients who performed a 5.8-MET activity (e.g., very fast walking for 150 min · wk^{-1}) showed a 25% increase in fitness secondary to training 870 MET-min · wk^{-1}. In this example, one might surmise that a 75% increase in volume doubled the relative improvement in fitness.

Training Methods

Before discussing several of the common training methods used to improve cardiorespiratory fitness, let's discuss the potential benefits of warming up before exercise and cooling down afterward. Exercise certainly can be performed without any prior warm-up. However, a warm-up period before moderately vigorous or vigorous training (8,12,19,53,57,73,74,79,89) is advisable because it increases both whole-body and muscle temperatures and increases both systemic and regional blood flow (and O_2 delivery) to what will soon be more metabolically active tissues. Such a warm-up usually begins by performing 4 to 5 min of sport- or activity-specific exercise at a very slow pace (e.g., very slow swimming for swimmers or slow jogging for runners). This typically is followed by several minutes of static (not ballistic) stretching of the major muscle groups to ensure full range of motion. This is followed by increasing levels of ballistic activity that can begin with general callisthenic-type activities (e.g., push-ups, vertical jumps) and should finish with sport-specific motions, such as throwing for baseball, dribbling to a hard stop and then shooting for basketball, or kicking for football (i.e., American soccer).

Likewise, after a session of moderately vigorous or vigorous exercise, a period of mild exercise (cool-down) is advised to keep the muscle pump (i.e., venous return) active, avoid lightheadedness due to blood pooling in the extremities, and restore body temperature and metabolism to pre-exercise levels. Additionally, an active cool-down or recovery period facilitates removal of exercise-induced waste products (e.g., lactate). Again, sport- or activity-specific exercises (e.g., easy throwing for baseball or jogging after hard running) would apply here. This could be followed by a period of static flexibility exercises.

Long–Slow Duration

The long–slow duration method of endurance training includes minutes to hours of continuous exercise, usually performed at the same pace over long distances. The intensity or pace is submaximal but can vary based on the fitness level and goals of the individual. For example, a competitive runner might train at a pace that is 6 min · mi^{-1}, equivalent to an intensity of 85% of HR reserve. Conversely, a regularly active 60-yr-old individual exercising for general health and fitness might simply walk briskly at a pace of 18 min · mi^{-1} or jog slowly at 10 to 12 min · mi^{-1}, equivalent to intensities of approximately 70% to 80% of HR reserve. In both instances, the pace is usually at or below lactate or ventilatory-derived anaerobic threshold.

Fartlek, Tempo, and Aerobic Interval Training

Fartlek is the Swedish word for "speed play"; such training is said to be the forerunner of interval training. Unlike interval training, the work and relief intervals are unstructured rather than formally timed, and all exercise is continuous. Fartlek training uses varying durations of exercise during a training session that usually is conducted outdoors over natural terrain (flats and hills) and involves changing intensity (e.g., sprint pace, speed pace, long–slow duration pace, recovery pace) for different distances (e.g., 60 m, 400 m, 2 km). There is no set pattern to speak of. Fartlek training is common among runners and cyclists, and its use typically is limited to no more than 1 time/wk. Such training leads to improvement in aerobic capacity and, to some extent, anaerobic capacity.

When training pace is set just below an athlete's lactate threshold or ventilatory-derived anaerobic threshold, it is referred to as *tempo training* (or *pace training* because usually it is set at a pace that is at or near an athlete's race or event pace). Tempo training involves continuous exercise at an intensity that usually equates to between 80% and 90% of HR reserve. For athletes who have plateaued in their training relative to further improvements in maximal $\dot{V}O_2$, incorporating tempo training can be effective relative to improving pace or $\dot{V}O_2$ at lactate threshold.

When tempo training is altered slightly such that intensity is increased to approximately 95% of HR reserve (or just above lactate threshold) and duration of exercise is shortened to 4- to 10-min bouts, it is referred to as *aerobic interval training*. Aerobic interval training is similar to the interval training for anaerobic performance described in chapter 7 in that it leads to some improvement in anaerobic performance. However, it often involves work bouts that are longer in duration (4-10 min vs. 30 s-2 min) and recovery bouts that are shorter in duration.

Other Important Training Topics: Overtraining and Cross-Training

Overtraining and cross-training are issues of importance when discussing training regimens for improving aerobic power, and neither is a training method like those described previously. Overtraining is a rare but important outcome that can result when too much (e.g., months) intense training is not balanced with sufficient periods of rest. Cross-training often is either used as a strategy for increasing the total volume of exercise performed over a period of time or used to diversify the training modes in hopes of avoiding an injury or maintaining fitness while recovering from an injury. Both are important considerations and are described in detail in the following sections.

Overtraining Syndrome

Before going into detail about overtraining syndrome (OTS), it is important to emphasize two key points. First, despite much attention and research on this topic over the past 25 yr, identifying an athlete who is truly suffering from OTS is rare (65). For relatively brief periods of time (days to even weeks) during a season, many athletes are tired and emotionally spent and underperform. These athletes may be experiencing many issues, but true OTS is most likely not one of them. Second, OTS is attributable to a prolonged imbalance between training volume and intensity of work and adequate recovery. Recovering from OTS often takes months once it is formally diagnosed by the trainer and physician.

As mentioned, fatigue, performing below expectations, feeling burned out, and injuries are common among athletes across all sports, and experiencing one or more of these problems does not necessarily mean that OTS is at hand. To place these and other training-related issues into context, let's first review three relevant terms (65).

- *Undertraining* or *insufficient training* refers to a suboptimal training stimulus (overload) that leads to less-than-optimal training responses.

- *Sufficient overload stimulus* indicates that improvements in biological function and performance are observed, but these improvements are not optimal. Changes in intensity, duration, or frequency could lead to even greater biological changes and further increases in performance.

- *Functional overreaching* or *supercompensation* indicates that gains in performance and biological adaptations are optimized. The overload stimulus is great and may result in a very temporary loss of performance, but performance improves to levels that are equal to or greater than previously achieved. In this instance, even though the training overload stimulus is high, it is accompanied by a sufficient amount of rest or recovery (65).

This leaves OTS, which is rare and requires a prolonged recovery of months or possibly years. Table 6.7 summarizes the potential changes and symptoms associated with overtraining. Note that OTS can involve a variety of organs and organ systems—autonomic nervous system, immune system, hormone responses (e.g., cortisol), and behaviors—all of which can influence performance.

It is important to discuss what steps can be taken to recognize and prevent OTS. With respect to the former, two useful strategies—one biological and one affective—can be applied. To screen for possible biological red flags or predictors, consider monitoring an athlete's HR over time, such as upon waking in the morning or at a fixed workload during exercise (e.g., jogging a mile in 6 min or swimming freestyle for 500 m in 6 min). Do this up to 2 times/mo. In each instance, HR should be at or below preseason values. Over time if HR is consistently higher at rest or while running or swimming a fixed distance at a certain pace, this might be a red flag for OTS and needs to be correlated with other factors (see table 6.7).

The second strategy pertains to gathering information on how the athlete feels. Nearly 20 yr ago, Hooper et al. (50) conducted a longitudinal study in which they measured a host of factors in elite Australian swimmers. All athletes also kept detailed records about how they were feeling as a way to evaluate their well-being. Among the three athletes who experienced a measured loss of performance and high levels of fatigue (eventually classified as OTS), logs revealed self-reported descriptions such as "washed out," "can't get going," and "feeling sluggish." The authors concluded that these self-assessment logs were a unique and cost-effective approach to monitoring for potential evidence of OTS.

As long as athletes continue to be obsessed with training and elite performance, OTS will remain a potential problem—one that is confounded by the fact that individual athletes respond to the same intense training regimen in vastly different ways. As a result, identifying, monitoring, and recovering from OTS is not always predictable.

Cross-Training

Training with a focus on more than one sport at the same time or to improve more than one component of fitness at the same time often is referred to as *cross-training*. Cross-training is a must for athletes who participate in multisport events such as the triathlon. Cross-training also might be viewed as a means to transfer the training effects or benefits gained from one mode or type of training to another (96)—for example, using cycling in place of running to improve cardiorespiratory fitness and running performance. In this example, such an approach might increase fitness while running for the nonexerciser who is just getting started. However, among already-trained athletes, any gains in fitness are usually never as great as when the athlete spends the same amount of time performing their specific sport (i.e., the concept of specificity of training). That said, adopting cross-training as an approach to maintain or attenuate loss of one's fitness level while recovering from an injury, as well as to minimize any potential risk for overtraining, is common (52).

With respect to cross-training, the key messages to remember are as follows. First, for the already generally

Table 6.7 Possible Changes and Symptoms Associated With OTS

Physical performance	Biological changes	Mood/other
• Loss of muscle strength or endurance • Decrement in performance times • Decreased aerobic power	• Decreased body mass • Increased risk for infection • Increased HR or blood lactate during standardized submaximal exercise • Increased HR at rest	• Loss of appetite • Sleep disturbance • Inability to concentrate • Anxiety • Easily fatigued • Waking unrefreshed • Sore, tender muscles

OTS and Sport Performance

Developing methods for studying and preventing OTS is difficult because sport psychologists, exercise physiologists, and coaches cannot structure prospective, controlled research experiments that involve training regimens designed to induce OTS in an athlete. Such an approach is unethical, and athletes do not welcome the signs and symptoms associated with OTS. Therefore, much of the information available about OTS so far has been gathered from cross-sectional studies or anecdotes provided by athletes and coaches interested in the topic. That said, some commonsense approaches targeting prevention that involve training should to be considered. The first involves the concept of periodization, or breaking down a year of training into smaller segments called macrocycles (usually several months), then mesocycles (up to 4 wk), and finally microcycles (1 wk).

Various macrocycles may represent the preseason, heavy competition or in-season, and postseason periods. Across each 10- to 15-wk period, the volume and intensity of training are varied. Within each macrocycle, similar attention is given to balancing and varying the volume and intensity of training across smaller segments of time, such as a 4-wk mesocycle and a 1-wk microcycle. This paradigm for training provides a structured opportunity for the coach or trainer to systematically balance the duration and extent of the overload stimulus with rest and recovery in a planned, disciplined manner.

Our discussion about prevention would not be complete if we did not mention the need for athletes to adopt a lifestyle that allows for rest and recovery outside of the sport, which includes proper sleep and nutrition habits. The latter involves ensuring that energy intake meets metabolic expenditure, which for some athletes (e.g., triathletes) can approach 8,000 kcal $\cdot$ d^{-1} during intense training periods. Proper nutrition includes a sufficient intake of complex carbohydrate to replenish carbohydrate stores in muscle and liver (which may equate to 70% or more of all calories consumed in a day) as well as sufficient fluid intake to maintain proper hydration.

Data quantifying the prevalence of OTS among athletes and the frequency at which it occurs are quite varied. Once OTS is identified, treatment for these athletes can include several lines of care. First, any organic diseases (e.g., infection) must be ruled out, and any problems identified should be treated. Second, the best treatment for OTS is rest, which may come in the form of complete rest (i.e., no activity) or relative rest, which can be a marked reduction of exercise to as little as 5 to 10 min $\cdot$ d^{-1}. Which approach is superior is an area in need of additional research. In some instances, treatment might also involve a sport psychologist given that OTS often is associated with underperformance. Finally, and only with proper medical oversight, medications may be used to treat any coexisting depression or profound sleep disturbances (7).

healthy nonathlete participating in multiple activities, cross-training can help maintain interest and adherence to exercise as well as effectively increase overall volume. Second, among elite athletes, the greatest gains in performance are achieved through sport-specific training. Third, in individuals who are injured, cross-training should be considered as a means to help maintain fitness during periods of recovery or rehabilitation.

PHYSIOLOGICAL ADAPTATIONS ATTRIBUTABLE TO AEROBIC TRAINING

Individuals who engage in some type of regular aerobic- or cardiorespiratory-type exercise training regimen

experience firsthand the benefits of such an endeavor. Subjectively, they may notice that hustling up two or three flights of stairs is associated with less leg fatigue or shortness of breath, or perhaps they feel that they recover faster during a break in the action when playing basketball. Objectively, if they record their HR at rest each morning upon waking, they might notice it decrease over several months by as much as 10 beats $\cdot$ min^{-1}. These changes are all part of the training effect one expects—changes that today are still being elucidated at the cellular and subcellular levels.

Our discussion focuses on the chronic adaptations in structure (anatomy) and function (physiology) that result from regular, habitual aerobic- or endurance-type activities or sports. This section does not address the short-term adjustments that occur immediately before, during, or after a single bout of exercise. We categorize the chronic adaptations attributable to endurance- or cardiorespiratory-type exercise training as follows:

- Changes that occur within the skeletal muscle
- Changes that occur systemically, with a major focus on those involving the cardiorespiratory system

Changes that Occur in Skeletal Muscle

A thorough review of all the changes that occur in the skeletal muscle in response to endurance-type exercise training is simply beyond the scope of this chapter. "Selected Histological and Biochemical Changes Induced by Endurance Training" summarizes these adaptations, all of which improve the ability of the skeletal muscles to generate ATP during exercise. Analyzing and assessing changes attributable to an exercise training program typically involves obtaining a sample of the involved skeletal muscles using a bioptome (figure 6.6).

Histochemical Changes

In the skeletal muscle, myoglobin (similar to Hb but not found in the blood) is an O_2-binding globular protein that facilitates the intracellular movement of O_2 from the cell membrane to the mitochondria, where it is used in cellular respiration. During intense exercise, myoglobin also briefly functions as a local source of O_2 reserves. Each molecule of myoglobin contains a single heme group that contains iron and is capable of binding to a single molecule of O_2. The myoglobin content of skeletal muscle may increase in response to an endurance training regimen—in some studies by as much as 75% to 80% (49,75) and typically noted only in those muscles that are engaged during training.

Chronic endurance training also yields a marked increase in the capacity of the skeletal muscles to completely break down carbohydrate in the presence of O_2 (i.e., aerobic glycolysis) and oxidize or break down fat. Utilization of glucose and fats, which are transported in the blood and both stored and utilized in the skeletal muscle, is improved with endurance training due to mitochondrial adaptation and biogenesis (e.g., increase in the number, size, and surface area of mitochondria) and increases in the level of activity or concentration of the enzymes and proteins involved in the oxidative metabolic pathways (e.g., Krebs cycle).

The mitochondria in the skeletal muscle are small organelles (0.5-1 μm) that exist in an interconnected network or reticulum. They mostly are found below the sarcolemma (i.e., subsarcolemmal) or distributed between myofibrils (intermyofibrillar, which accounts for approximately 75% of total cellular mitochondrial volume). Among skeletal muscles recruited for exercise, endurance training improves mitochondrial volume density by as much as 100% in 6 wk. The magnitude of the

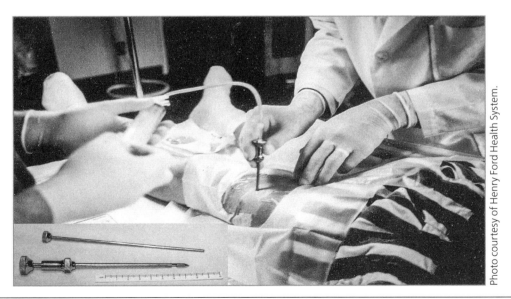

Photo courtesy of Henry Ford Health System.

Figure 6.6 The collection of skeletal muscle tissue using the needle biopsy procedure. This procedure involves the insertion of a pencil-size needle (i.e., bioptome; see insert) through a 1-cm incision in the skin and fascia. The incision is made under local anesthesia. Once the bioptome is inserted into the muscle, tissue is drawn into a window-like opening and an internal cutting blade is used to cleave off the tissue. Approximately 20 to 40 mg of tissue is harvested and then prepared for subsequent analysis.

Selected Histological and Biochemical Changes Induced by Endurance Training

- Increase in myoglobin content (most noted in those muscles engaged during training)
- Increase in mitochondrial volume, possibly in subsarcolemmal mitochondria (vs. intermyofibrillar mitochondria)
- Consistent with the above, increase in activity and concentrations of enzymes (e.g. citrate, synthase, succinate dehydrogenase) and proteins involved in ATP production in the Krebs cycle, beta oxidation, and electron transport system
- Increase in intramuscular storage of carbohydrate and fats
- Slight increase in skeletal muscle size and mass
- Increase in the number of capillaries surrounding each muscle fiber (i.e., capillary density)
- Increase in the cross-sectional area of slow-twitch red Type I fibers
- Mild increase in the percentage of slow-twitch red fibers consistent with a similar decrease in fast-twitch white Type IIx fibers

increase is influenced by the muscle fiber type involved with training, the initial mitochondrial content of the cell, and its location in the cell (intermyofibrillar vs. subsarcolemmal) (16). In response to endurance training, mitochondrial content increases more in Type IIa and possibly Type IIX fibers than in Type I fibers (16). Although further work is needed, in response to training it appears that subsarcolemmal mitochondria adapt to a greater degree and likely sooner compared with intermyofibrillar mitochondria.

Endurance training increases the myoglobin and mitochondria content of skeletal muscles as well as the level of activity or concentration of the enzymes and proteins involved in ATP production via the Krebs cycle, beta oxidation, and the electron transport system. Overall, O_2 consumption by the electron transport system is enhanced, usually to an extent that is greater than the increases in O_2 delivery with training. In humans, activity of key oxidative enzymes in the Krebs cycle (e.g., succinate dehydrogenase, citrate synthase) and proteins in the electron transport system (e.g., cytochrome c) all favorably increase 35% or more with endurance training (48,83).

Similarly, oxidation of fat is enhanced because the complete breakdown of fat (e.g., palmitic acid) to CO_2, H_2O, and ATP also involves the Krebs cycle and the electron transport system. Additionally, a key enzyme marking the beta-oxidation of free fatty acids, 3-hydroxyacyl-CoA dehydrogenase, is increased. A Danish study involving different doses of endurance training in overweight, sedentary young men showed that both a high volume (600 kcal · d^{-1}) and a moderate volume (300 kcal · d^{-1}) of

exercise similarly improved peak fat oxidation compared with the no-exercise control. Both volumes of training were associated with an approximately 45% increase in 3-hydroxyacyl-CoA dehydrogenase (83).

Finally, the ability of the skeletal muscle to store carbohydrate and fat is increased as well. Specifically, with endurance training, muscle glycogen stores increase 2.5-fold (as do the enzymes responsible for muscle glycogen synthesis and breakdown) and intramuscular fat stores increase approximately 1.5 times. Having greater fat stores and concentrations of the enzymes needed to oxidize fat allows for the greater utilization of fat versus carbohydrate during submaximal exercise—a glycogen-sparing effect that contributes to improved performance.

Histological Changes

A discussion about the adaptations that take place in the skeletal muscle with endurance training would not be complete without describing the histological changes (i.e., fiber size, capillary density, muscle fiber type) that occur. Until recently, most exercise scientists believed that aerobic exercise training has little effect on skeletal muscle size (hypertrophy). However, a review by Konopka and Harber (63) suggests that this might not be entirely true. They suggest that sufficient data exist to indicate that skeletal muscle size and mass are increased (approximately 7%) with aerobic training to the extent that such training should be considered a countermeasure for the muscle loss that often accompanies increasing age. The overload stimulus imposed during training must be of sufficient intensity (>70%), duration (>30 min), and frequency (>4 d/wk) in order to achieve the number of contractions needed to induce skeletal muscle hypertrophy.

Capillary density refers to the number of capillaries that surround each skeletal muscle cell or fiber. As you might suspect, long-term endurance training increases the number of capillaries per fiber. Among athletes (18) it is not uncommon to identify upward of six capillaries per muscle fiber, which is some 10% to 30% greater than what is observed in untrained individuals. The increased capillary density that accompanies endurance training provides an important beneficial function at the cellular level. With more capillaries, the diffusion or movement of O_2 (and nutrients) from within the capillaries to the cell membranes that they interface with is improved; therefore, more O_2 and nutrients are made available to the cell and the mitochondria. Likewise, the removal of metabolic waste products and heat is enhanced because of the increased number of capillaries around each fiber. As skeletal muscles hypertrophy with endurance training, the increase in capillary density helps ensure that diffusion time and distance from capillary to inside the cell are at least maintained at pretraining levels or, as described previously, improved beyond pretraining levels.

All skeletal muscles comprise thousands of individual cells or fibers, and these fibers can be categorized based on some common properties. The three types or isoforms (i.e., slightly different forms of the same protein—in this case, myosin—that perform the same function) of skeletal muscle fibers that are most commonly found in humans and the nomenclature we use to present them are as follows:

1. Slow-twitch red, which contain mainly the myosin heavy chain Type I isoforms
2. Fast-twitch red, which contain mainly myosin heavy chain Type IIA isoforms
3. Fast-twitch white, which contain mainly the myosin heavy chain Type IIx isoforms

The muscle fibers engaged with training adapt in response to such a stimulus. For example, the cross-sectional diameter of slow-twitch red fibers is increased (i.e., hypertrophies) because these fibers are preferentially recruited (vs. fast-twitch red or fast-twitch white) during submaximal, lower intensity activities.

Additionally, studies comparing endurance-trained athletes to anaerobically trained athletes (e.g., sprinters in track) show that slow-twitch fibers make up a higher percentage of total skeletal muscle content in the involved muscles of endurance athletes. Conversely, fast-twitch fibers make up a higher percentage of fibers in anaerobi-

cally trained athletes. This suggests some preferential selectivity of the sport performed by the athlete or possibly a certain level of plasticity in the involved skeletal muscles such that individual fibers can be converted or transition from one type of fiber to another. Earlier evidence suggested that fiber transition did not occur; however, gain and loss of Type I fibers with training and detraining has been documented (66). Part of the initial problem relative to resolving the question about transition of fiber types was that the methods used to determine fiber typing were unable to identify changes or shifts due to training. More recent immunohistochemical methods involving monoclonal antibodies and gel electrophoresis indicate that gain and loss of specific fibers does, in fact, occur in a manner that parallels changes in physical activity. An extensive longitudinal study investigating the relationship between variability in responses to training and genetics (HERITAGE Family Study) included a substudy investigating the potential transition of fiber types after a 20-wk program in which participants trained 3 to 5 times/wk on a stationary cycle (82). After 20 wk of training, Type I fibers increased 3.5 percentage points compared with baseline, consistent with a 5.4 percentage point decrease in Type IIx fibers.

The plasticity in the skeletal muscles that occurs with training should cause one to ask whether changes in fiber type distribution also occur with disuse (detraining) or disease. The short answer is *yes*. Several studies involving bed rest or microgravity demonstrate a shift away from slow-twitch red fibers to Type II fibers, and the same is true for patients with disease (27,92,99). Among patients with heart failure, an approximately 10% to 20% decrease occurred in slow-twitch red fibers compared with age-matched controls (27,92). The shift in fiber type distribution may be attributable, in part, to disease- or inactivity-related **apoptosis** (i.e., the death or loss of muscle fibers or cells) (3), which itself is partly linked to the state of mitochondrial biogenesis. In the example of the patient with heart failure, apoptosis may be preferential for the loss of slow-twitch red fibers, thus yielding a shift or a relatively greater percentage of remaining Type II fibers. Several studies have shown that exercise training can partly reverse the initial loss of slow-twitch red fibers that accompanies patients with heart failure (45,59). Specifically, Keteyian et al. (59) observed an approximately 30% increase in slow-twitch red content in male patients who completed 24 wk of supervised exercise training. In summary, the interconversion of fiber type in response to endurance training (and detraining and disease) does seem to occur. Although the magnitude of the changes may vary somewhat, the changes do demonstrate the adaptability and plasticity of skeletal muscle fiber distribution.

Changes in the Cardiorespiratory System

Maximal aerobic capacity (max $\dot{V}O_2$) represents the ability of the heart to transport O_2 to the metabolically more active skeletal muscles and the ability of those muscles to utilize the O_2 that is delivered to them. This section describes the chronic adaptations that develop in the cardiorespiratory system (and to a lesser extent the autonomic nervous system) in response to a regular endurance training program. These include changes that occur at rest, those that occur during a bout of submaximal exercise, and those that occur at peak or maximal exercise. Table 6.8 summarizes the potential important adaptations in the cardiorespiratory system.

Changes at Rest

The following changes occur at rest.

1. *Increase in heart size and dimensions.* An exercise-induced increase in heart size has been appreciated by scientists and clinicians for decades (figure 6.7). Sophisticated noninvasive imaging using echocardiog-raphy and magnetic resonance techniques shows that the supranormal enlarged or hypertrophied heart of the athlete ("athlete's heart") generally is related to the type of sport they participate in (72). Specifically, the size of the left ventricle and, to some extent, the thickness of the walls of the ventricle are influenced by the type of overload stimuli imposed on it: volume overload or pressure overload.

The cardiac hypertrophy of endurance training typically is greater in males than in females and is associated with a mild increase in mean pressure during exercise. This adaptation occurs in response to the high volume (i.e., volume overload) of blood that is constantly moved through the heart during a training session. In this instance, the internal dimension of the left ventricle at the end of filling (i.e., diastole) approaches but usually does not exceed the upper limit of normal—58 mm (figure 6.8b)—along with possibly mild increases in septal and posterior wall thicknesses. This remodeling of the left ventricle often is referred to as *eccentric hypertrophy* (9) (figure 6.7) and likely is attributable to increased contractile proteins and the assembly of new sarcomeres inside the cardiac myocyte in series (end to end). This increase

Table 6.8 Summary of Selected Exercise-Training Changes that Occur at Rest, During Submaximal Exercise, and at Maximal Exercise

Condition	HR	Stroke volume	Cardiac output	Other
Rest	↓↓	↑↑	No change	• Increase in heart size, mass, and dimensions • Increase in blood volume (approximately 25%) and Hb • No marked changes in minute ventilation, tidal volume, breathing rate, residual volume, or vital capacity
Submaximal at fixed work rate	↓↓	↑↑	No change	• No change or slight decrease in O_2 consumption • Decrease in blood lactate • Decrease in blood flow per kilogram of metabolically active tissue
Maximal exercise	No change	↑↑	↑↑	• Increase in peak or maximal $\dot{V}O_2$ (10%-30%) • Increase in peak arteriovenous O_2 difference • Increase in minute ventilation due to increases in breathing frequency and maximal tidal volume

↑ = increase; ↓ = decrease.

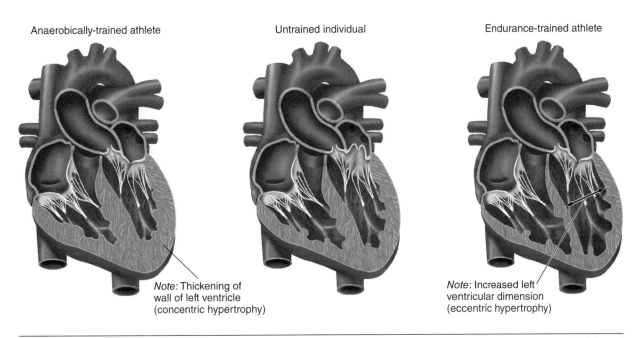

Figure 6.7 Examples of eccentric hypertrophy due to volume overload in an endurance-trained athlete (right) and of concentric hypertrophy due to pressure overload in an anaerobically trained athlete (left) compared with a normal, untrained individual (center).

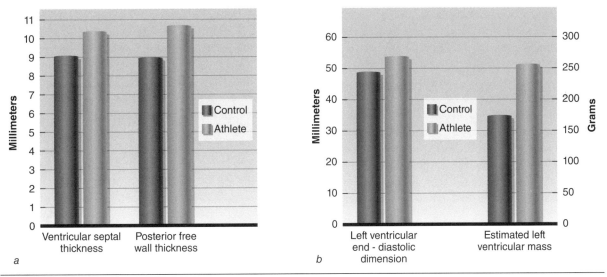

Figure 6.8 Comparison of left ventricular dimensions and mass in athletes and nonathletes as assessed by echocardiography.

Data from Martin et al. (71).

in chamber size contributes to the increase in stroke volume observed in endurance athletes at rest, during submaximal exercise, and at peak exercise.

Using magnetic resonance imaging techniques, figure 6.9 shows the mild enlargement of the ventricular septal wall and posterior wall of the left ventricle in a 20-year-old aerobically trained male athlete, compared to a similar image from a 19-year-old untrained healthy male control. Not shown in these two-dimensional photographs is left ventricular end-diastolic volume, which was calculated to be 90 mL · m^{-2} in the athlete and 81 mL · m^{-2} in the normal control (upper limit of normal = 88 mL · m^{-2}).

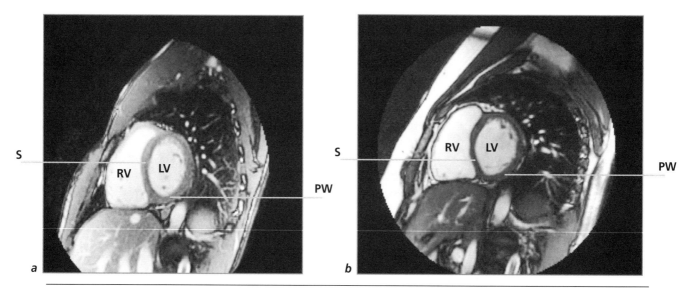

Figure 6.9 Comparison of ventricular wall thicknesses in (a) a 20-year-old aerobically trained athlete (septum = 13 mm and posterior wall = 12 mm; normal < 12 mm) versus (b) an untrained 19-year-old active, healthy control (septum 10 mm and posterior wall 9 mm). LV=left ventricle, RV=right ventricle, S=septum, PW=posterior wall.

Images gathered using magnetic resonance methods, provided courtesy of Khaled Abdul-Nour, MD, Henry Ford Health System.

In general, the cardiac hypertrophy observed in the anaerobic or non–endurance athlete engaged in power sports such as wrestling, weightlifting, and short-distance track sprinting is characterized by a left ventricular end-diastolic dimension that generally is of normal size but with ventricular wall thicknesses that may be at the upper end of or slightly above normal (e.g., posterior wall of the left ventricle approaches but remains less than 12 mm) (figure 6.8a). In contrast to the endurance athlete, the stimulus for the increase in hypertrophy among power athletes is pressure overload (vs. volume overload); systolic and diastolic pressures can exceed 300 and 150 mmHg, respectively, during extreme effort. In these athletes the remodeling of the left ventricle is referred to as *concentric hypertrophy* (9) (figure 6.7) and is due to the acquisition of increased contractile proteins and the assembly of new sarcomeres inside the cardiac myocyte in parallel.

Both types of athletes (endurance trained and non–endurance trained) demonstrate cardiac hypertrophy. The former generally displays an increase in the dimension of the left ventricular cavity (with accompanying greater stroke volume), and the latter generally shows an increase in wall thicknesses. Keep in mind that the hypertrophy (eccentric or concentric) may take months if not years of training to develop and is influenced by individual differences in response to training, differences in the overload stimulus imposed, and initial level of training.

2. *Decrease in HR at rest.* One of the most well-appreciated and consistent responses to an endurance training program among leisure exercisers, patients, and well-conditioned athletes alike is a decrease in HR at rest. This response often results in bradycardia (<60 beats · min⁻¹) and is observed in both longitudinal training studies and cross-sectional studies that compared athletes or exercisers with nonexercisers. The mechanisms responsible for such a reduction include increased parasympathetic nervous system (vagal) input to the sinoatrial node (see chapter 3) and, to a lesser extent, a decrease in efferent sympathetic tone or activity to the myocardium. Additionally, there is likely a partial resetting of the intrinsic rate of the sinoatrial node that is independent of autonomic nervous system input (90). This is possibly attributable to decreased sensitivity of cardiac tissue (β-1 receptors) to catecholamines such as norepinephrine, a mechanical effect influenced by training-induced changes in cardiac dimension (left ventricular cavity dimension and wall thickness), and an increased level of the neurotransmitter acetylcholine found in the atrial tissues.

3. *Increased stroke volume.* As mentioned previously, endurance-type training increases stroke volume at rest. Given the lowering of HR with training, cardiac output (which is the product of HR and stroke volume) essentially is unchanged at rest. The increase in stroke volume is attributable to greater filling of the left ventricle during diastole due to the greater internal dimensions (i.e., increased cavity size) and a greater blood volume. Another factor that may contribute to the increase in stroke volume is an improved inotropic or contractile state (54).

Regardless of the mechanisms, the increase in stroke volume with endurance training is another hallmark feature of the endurance-trained (vs. untrained) individual. Such a change may take years to develop. The lack of an increase in an athlete or client who regularly exercise trains may simply be due to the level or magnitude of overload stimulus imposed on the heart during training.

4. *Increase in blood volume and Hb.* Both blood volume and the total amount of Hb are increased with endurance-type training. Both improve the transport of O_2 to the working skeletal muscles during exercise, and improvements in both are correlated with improvements in max $\dot{V}O_2$. An increase (approximately 20%-25%) in blood volume is an early (about 1 wk) adaptive response (20), whereas a change in Hb mass may take weeks to months to develop, if it develops at all. Green et al. demonstrated a 12% increase in blood volume after only 3 d of cycling for 2 h at an intensity of 65% of max $\dot{V}O_2$ (42). Among both men and women, the total amount (grams) of Hb might be 20% to 30% higher in trained versus untrained individuals (60,61,101), yet Hb concentration (g · 100 mL^{-1} of blood) is changed relatively little or may initially be slightly reduced ("sport anemia") with training as blood volume expands more than red blood cell volume or the amount of Hb.

5. *Resting lung measures and volumes.* Current thinking is that in apparently healthy people, the liters of air exhaled per minute (i.e., minute ventilation) at rest generally is not influenced by exercise habits. Similarly, in healthy people, key lung volumes and measures such as tidal volume, respiratory rate, total lung volume, residual volume, and vital capacity are altered little (if at all) as a result of a regular exercise training program. In general, improvement in cardiorespiratory fitness due to endurance training is not associated with—and typically occurs independent of—changes in resting lung volumes and measures.

Changes During Submaximal Exercise

Several classic adaptations that are induced as a result of regular acerbic training are those that occur during submaximal exercise, when one compares responses several weeks after training with those that were observed before a training regimen began. Such comparisons usually are made at the same work rate, such as cycling at 75 watts

before 10 wk of training and then again at 75 watts after training. Table 6.8 summarizes the common training adaptations that occur during fixed or standardized submaximal exercise (see also figure 6.10*a*).

1. *No change or slight decrease in O_2 uptake.* When the individual is doing the same amount of work (e.g., 75 watts), there is little if any difference in O_2 consumed. Any decrease that might be observed is mostly attributable to improved efficiency or skill of movement, which is usually most pronounced when an untrained individual becomes skilled or efficient over several months or more.

2. *Decrease in blood lactate.* A key adaptation indicative of a training effect is the reduction in blood lactate at a fixed work rate (35,36). Several physiological mechanisms are responsible for this, including a smaller O_2 deficit at the onset of exercise due to the improved ability of O_2 transport and utilization to more quickly match the energy demands at the start of exercise; the improved use of the lactate produced during exercise for a fuel source during exercise; and improved oxidative metabolism in the skeletal muscles due to improvement in the proteins and enzymes involved in beta (fat) oxidation, the Krebs cycle, and oxidative phosphorylation. All of this results in the improved use of the O_2 that is delivered and, therefore, less reliance on anaerobic metabolism and less lactate production. Because of these changes and the resultant decrease in blood lactate at a fixed work rate, the athlete can increase exercise pace or velocity, which results in improved performance (before hitting lactate threshold). In fact, as athletes train over many years, meaningful increases in max $\dot{V}O_2$ become more difficult, whereas continuous improvements in performance velocity or pace at lactate threshold may persist (21,22,56).

3. *Decrease in HR.* Another profound and very consistent training effect is a decrease in HR during submaximal exercise (32,38,103). This decrease in HR is associated with less myocardial O_2 consumption at any given work rate and is due to several factors, including a reduction in sympathetic nervous system efferent activity that results in lower blood plasma norepinephrine levels. Interestingly, the decrease in plasma norepinephrine initially parallels the decrease in HR over several weeks and then begins to plateau, whereas further decreases in HR are observed (103). This suggests that other mechanisms are involved, such as less of a decrease (i.e., less inhibition) in parasympathetic (vagal) tone during exercise or less direct sympathetic efferent input to the heart.

4. *Increase in stroke volume.* Without question, the increase in stroke volume that occurs with exercise is an adaptation that has meaningful implications

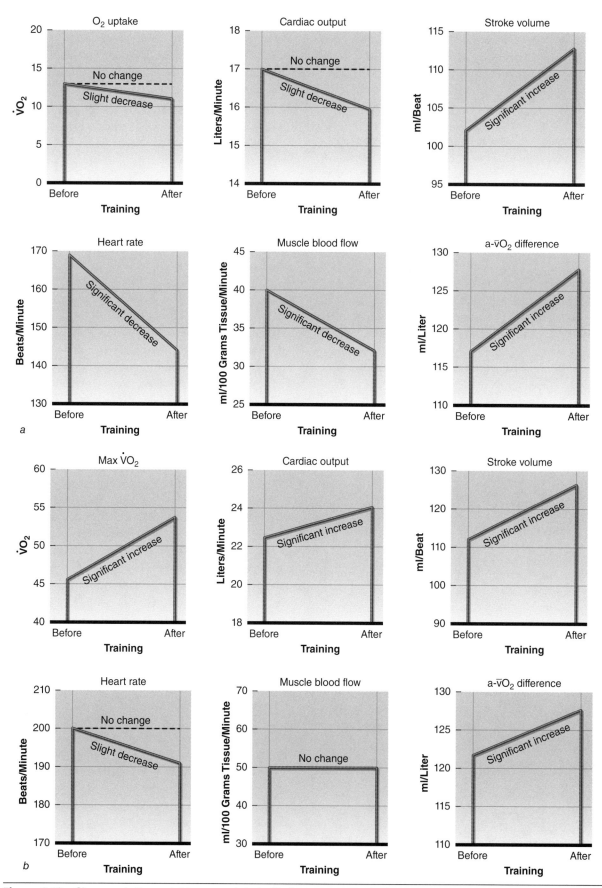

Figure 6.10 Changes in the variables associated with O_2 transport and other related systems during *(a)* steady-state exercise and *(b)* maximal exercise after an endurance-type exercise training regimen.

(13,38,84,101). Similar to what is responsible for the increase in stroke volume at rest, the increase in plasma volume and left ventricular cavity size (i.e., increased preload) and possibly increased inotropicity all contribute to the increase in stroke volume during submaximal exercise (84). The greater the size of the left ventricular cavity, the greater the stroke volume, which results in better forward flow per heartbeat. Given the decrease in resting HR and the increase in stroke volume, what might you predict for the adaptation of cardiac output during submaximal exercise at a fixed work rate after a 3-mo training program? If you said, "the same or no difference," you are correct because changes in cardiac output are closely linked to changes in work rate. Because the work rate at which cardiac output was measured was the same before and after training, cardiac output would be the same.

5. *Decrease in blood flow per kilogram of active muscle tissue.* After weeks of exercise training, blood flow to the metabolically active skeletal muscle is unchanged or reduced somewhat when exercising at a submaximal work rate that matches the work rate exercised at before undergoing training (43,62,81). In some ways this response might seem counterintuitive in that one might think that training improves blood flow. It does at peak exercise, but not at fixed work rates. This reduction in local blood flow likely is tied to both the improved capillarization surrounding muscle fibers and many of the biochemical changes mentioned previously. Specifically, improvements in these factors increase the ability of the skeletal muscle to extract and utilize the O_2 that is provided, meaning that less blood flow is needed to meet matched metabolic demands. At the whole-body level, more blood is made available for distribution to the skin during submaximal exercise, which helps with heat elimination.

Changes During Maximal Exercise

Regular exercise training improves maximal physical working capacity. Although individual responses can vary, an increase in this parameter is almost always observed in response to aerobic-type training. Such an adaptation occurs regardless of whether the person is a breast cancer survivor who engages in daily walking to attenuate the loss in fitness that often accompanies the disease and the drugs used to treat it or the professional basketball player who has been less active during the off-season and begins to train regularly in August before training camp opens in September. Table 6.8 summarizes the changes that occur at maximal exercise as a result of a regular aerobic-type training program (see also figure 6.10*b*).

1. *Peak aerobic power or cardiorespiratory endurance as measured by max $\dot{V}O_2$.* Among both men and women, the increase in max $\dot{V}O_2$ in response to training (13,31,36,84) typically falls between 10% and 25%. The magnitude of the increase depends on a variety of factors, including genetics, mode of training, the volume or training stimulus (including differences in intensity, duration, and frequency), and the nature of the training (e.g., interval, tempo, Fartlek). The initial fitness level of the person who starts a training program is influential as well. For example, a 55-yr-old patient with known heart disease and a peak $\dot{V}O_2$ of 22.4 mL · kg^{-1} · min^{-1} improves his peak capacity by 3.6 mL · kg^{-1} · min^{-1} to 26 mL · kg^{-1} · min^{-1} after 10 wk of training; this equates to a 16% increase. Conversely, a 27-yr-old sedentary individual with a peak $\dot{V}O_2$ of 38 mL · kg^{-1} · min^{-1} trains for 10 wk and improves his peak $\dot{V}O_2$ by 4.2 mL · kg^{-1} · min^{-1}; this equates to an approximately 11% increase. The absolute increase was greater in the younger individual (4.2 vs. 3.6 mL · kg^{-1} · min^{-1}), but on a relative basis when expressed as a percentage of one's initial fitness, greater gains were observed in the older patient with heart disease (16% vs. 11%).

To better appreciate the physiological adaptations that are responsible for the improvement in max $\dot{V}O_2$, one needs to again review the Fick equation (see chapter 3) and rearrange it as follows:

$$\text{Max } \dot{V}O_2 = \text{heart rate} \times \text{stroke volume}$$
$$\times \text{arteriovenous } O_2 \text{diff}$$

(Equation 6.13)

Improvements in exercise tolerance must be due to delivery (stroke volume or HR), improved extraction and utilization, or a combination of both (table 6.9). Now is a good time to address the question "What limits the upper end of human performance?" Is it the ability of the body to transport O_2 to the metabolically active tissues or the ability to extract O_2 once it is delivered? Among healthy people, the general consensus is that peak aerobic capacity typically is limited by O_2 delivery (stroke volume, cardiac output, maximal regional blood flow). However, this answer can be modified as needed to incorporate the concept that other factors should be considered as well, including attenuation of any of the steps associated with movement of O_2 in the body (e.g., gas exchange in the pulmonary system, transporting O_2 to the tissues, and the movement of O_2 into and within the skeletal muscles).

Table 6.9 Exercise Training–Induced Changes in Variables That Contribute to the Increase in Maximal Aerobic Capacity

	Max $\dot{V}O_2$ (L·min⁻¹)	=	Maximal HR (beats·min⁻¹)	×	Maximal stroke volume (mL·beat⁻¹)	×	Arteriovenous O_2 difference (mL·L⁻¹)
Before training	3.16		196		115		140
After training	5.07		193		175		150

Maximal cardiac output (HR × stroke volume): before training = 22.54 L · min⁻¹; after training = 33.78 L · min⁻¹.

2. *Increase in peak cardiac output due to greater stroke volume and no change or slight decrease in peak HR.* At rest, end-diastolic volume increases in response to chronic training, greatly due to the training-related increase in plasma volume. This improved volume of blood pumped per beat of the heart (stroke volume) is evident both at rest and during maximal exercise. Additionally, a training-induced improvement in the ability to maintain preload (venous return), a slight reduction in afterload, and possible improvements in contractility can also contribute to the improved stroke volume (54). This improvement in peak stroke volume (and, subsequently, cardiac output) is a hallmark feature of the highly trained endurance athlete (31,32,84,85).

Although cardiac output is the product of HR and stroke volume, with respect to identifying training adaptations that contribute to improved peak aerobic power, we can essentially ignore any meaningful contribution of a change in peak HR. In healthy people, peak HR is influenced little by training and instead is influenced mostly by increasing age. If anything, peak rate may be slightly lower with training, likely due to a slight downregulation in sympathetic tone at peak exercise. In summary, improved aerobic capacity due to endurance training is greatly due to improved central transport (cardiac output) and delivery (muscle blood flow) and partly due to improved ability to extract O_2 in the metabolically more active skeletal muscles. The improvement in cardiac output is almost totally attributable to improved stroke volume, which itself is mostly attributable to the greater training-induced increase in plasma volume along with some improvement (reduction) in afterload and myocardial contractility (54).

3. *Increase in maximal minute ventilation and greater pulmonary diffusion capacity.* Although the volume of air exhaled per minute at rest is essentially unchanged after weeks or months of training, it is increased greatly with training. Typical increases in maximal ventilation can approach 25% or more, reaching values as high as 120 to 140 L · min⁻¹ among previously untrained healthy people and 80 to 100 L · min⁻¹ in patients with a known health disorder such as heart disease. Among competitive elite athletes, minute ventilation can exceed 200 L · min⁻¹.

The increase in maximal ventilation is attributable to increases in both frequency of breathing (breaths per min) and maximal tidal volume (volume of air per breath). Also, it is not uncommon among very elite athletes who have the ability to generate a very high blood flow (i.e., cardiac output)—so high, in fact, that the amount of time that blood transits the lungs (<0.4 s) may be insufficient to fully saturate Hb (which normally is approximately 98% saturated; see chapter 3). In this instance, a ventilation–perfusion–diffusion mismatch occurs that leads to a decrease in Hb saturation (**exercise-induced hypoxemia**; oxyhemoglobin ≤ 91%) due to the rapid transit of blood through the alveoli–pulmonary capillary interface (80). It is important to note, however, that this mismatch is not due to limitations in minute ventilation. In generally healthy people, maximal ventilatory ability rarely is viewed as a factor that limits max $\dot{V}O_2$. In fact, maximal ventilation should instead be viewed as independent of or secondary to any increase in max $\dot{V}O_2$ with training.

Pulmonary diffusion capacity, or the volume of gas that diffuses through the alveoli–capillary membrane each minute for a pressure difference of 1 mmHg (mL · min⁻¹ · mmHg⁻¹), is higher in trained athletes than in untrained individuals (24,58,70,78). Several mechanisms may be responsible for this increase, including an increase in alveoli–capillary surface area and an increase in the flow of blood to the lungs (70).

4. *Improved blood flow to metabolically active muscle but no change in blood flow when expressed per kilogram of muscle tissue.* Endurance training improves absolute total blood flow to working skeletal muscles during maximal work, and such a response has been observed in healthy people, athletes, and patients with known disease (84,93). It is important to note, however, that because the total amount (kilograms) of muscle mass involved with the exercise may also increase due to training, the relative rate of blood flow ($L \cdot min^{-1} \cdot kg$ of muscle tissue^{-1}) during a bout of maximal exercise is essentially unchanged if one compares values measured before and after a regular training regimen.

Other Adaptations in Response to Aerobic Exercise Training

The previous material detailing chronic adaptations to an aerobic-type training regimen focuses almost exclusively on changes that develop in the cardiac and pulmonary (and autonomic nervous) systems. As one might expect, however, other organs and organ systems adapt as well. A few of these are reviewed in the following sections.

Changes in Connective Tissues

Bones, ligaments, tendons, and cartilage generally all respond favorably to a chronic training program. Bone remodeling (new bone formation via osteoblast cells and resorption via osteoclast cells) is a dynamic process influenced by a variety of factors, including age, a calcium-rich diet, and the imposed exercise load. Bone mineral content or density (bone formation due to osteoblast production of calcium phosphate and the deposition of such into a protein matrix) reaches maximum levels through age 30. Relative to exercise, this influential factor is site specific; this means that the beneficial bone adaptations are unique to those bones undergoing the load during exercise. Therefore, those bones that bear body weight are the ones that adapt. For example, bone mineral density of the femur is greater in basketball players than in swimmers (67). Conversely, upper limb bone mineral content is greater in weightlifters than in runners (46).

As one might expect, ligaments and tendons also adapt favorably to regular physical stress. The breaking strength of both ligaments and tendons is increased and the attachment sites (entheses) of tendons (osteotendinous junctions) and ligaments (osteoligamentous junctions) are both improved with training (11). Conversely, the entheses often are sites of injury due to excessive exercise (e.g., tennis elbow, golfer's elbow, jumper's knee). With respect to regular exercise and cartilage, only in the past 15 yr or so have advanced imaging techniques (e.g., magnetic resonance imaging) allowed researchers the ability to more thoroughly investigate the effects of

regular training on joint health. Before then, most studies of exercise and cartilage involved animal models. Although there is good evidence that cartilage thins when exposed to periods of unloading such as paraplegia or postoperative immobilization, the increased loading on cartilage experienced by athletes does not appear to meaningfully influence cartilage thickness (29).

Changes in Gastrointestinal Function

Although the effect of exercise on other organs and organ systems is well described, the effect of exercise on esophageal, stomach, and intestinal health and function is less clear. Acutely, light and moderate exercise training may improve gastric emptying (14), whereas a bout of severe, heavy exercise may impede gastric emptying and intestinal absorption of nutrients and, in very rare instances, may cause gastrointestinal bleeding. Regular habitual exercise shortens mouth-to-cecum transit time and improves colonic motility and is therefore associated with a reduction in colonic transit time. The latter is one theory about how regular exercise contributes to reducing an individual's overall risk for developing colon cancer.

Brain Health and Function

Over the past decade the chronic or long-term effects of exercise on cognitive function (i.e., memory, attention, processing speed), dementia, and brain health have been actively studied in both animals and humans. *Dementia* is an umbrella terms that describes a variety of diseases and conditions that develop when nerve cells (neurons) die or no longer function properly. A major and common type of dementia is Alzheimer's disease, a disorder that progressively impairs an individual's daily and then eventually bodily functions. Several studies show a reduced future risk for developing dementia and disease-specific mortality among people who exercise. Interestingly, the risk associated with Alzheimer's-related death appears to be dose related such that more exercise was associated with lower (up to 40% reduction) risk (87,102). In patients with dementia, exercise programs of 6 to 12 mo also appear to be associated with better cognitive scores compared with nonexercisers (4,102). How exercise may aid in brain health is not yet defined and is an area of intense scientific investigation. Possible pathways might be new neuron development (neurogenesis), reduction of proinflammatory cytokines, improved cerebral artery vascular health and function, attenuation of neurodegeneration, and decreased loss of gray matter volume. Clearly, more research is needed in this important area, especially because the number of people aged 65 yr and older is expected to increase greatly over the next 20 yr.

Detraining

A large portion of this chapter is devoted to summarizing the benefits and adaptations that occur across many organs and organ systems with regular aerobic-type exercise training. Consider what might happen if the athlete stops exercising. Assume that a period of time passes during which exercise volume and physical activity are markedly reduced or totally eliminated. As a result—and as one might expect—most of the favorable adaptations described previously would be reversed. What is somewhat variable is the time course and magnitude of the decline, both of which are related to a variety of factors such as fitness level and training volume prior to stopping exercise, type of inactivity (bed rest vs. normal ambulation), and duration of inactivity.

It is important to point out that the planned decrease in activity or exercise volume that many elite competitive athletes undertake before a big meet or competition, referred to as the precompetition *taper*, differs from

and does not apply to our discussion about detraining. Concerning tapering, numerous studies (51,52,55,88) indicate that decreasing training duration by up to 90% 5 d to 1 wk or more before sport competition, with or without a corresponding decrease in exercise intensity, may improve athletic performance by up to 3%. To maximize the results from tapering, coaches and athletes typically restrict its use to the most important competitions each year.

Let's get back to our discussion about detraining by reviewing two studies, one involving enforced bed rest and the other involving the discontinuation of exercise. The first study—a landmark trial often referred to as the *Dallas Bed Rest Study*—was conducted in 1996 and involved two active, somewhat trained college-age subjects and three sedentary subjects (84). All five subjects underwent a variety of physiological measurements before and after 20 d of bed rest and then again after completing 8 wk of regular exercise training. Exercise training involved outdoor running using both continuous and interval training methods and consisted of two workouts per day Monday through Friday and one workout on Saturday. As shown in figure 6.11a, mean

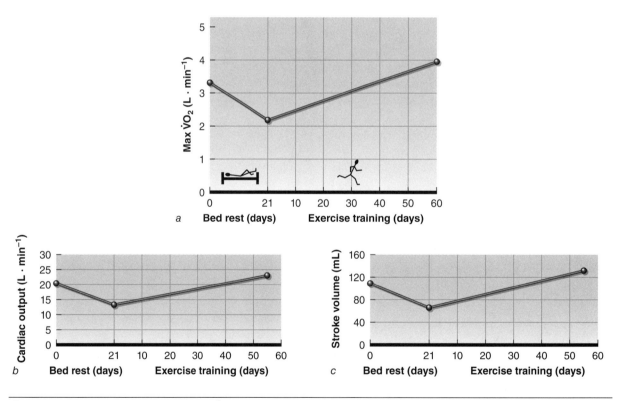

Figure 6.11 Mean changes in *(a)* max V̇O₂, *(b)* cardiac output, and *(c)* stroke volume in subjects who underwent 20 d of bed rest and approximately 8 wk of regular exercise training.

Adapted from Saltin et al. (84)

maximal aerobic power or capacity (as measured by maximal $\dot{V}O_2$) decreased dramatically (26%) after 20 d of bed rest. This was due to a rapid decrease in central transport (i.e., peak cardiac output; figure 6.11b), which itself was due to a commensurate decline (29%) in peak stroke volume (figure 6.11c). The average peak HR achieved for all five athletes was essentially unchanged after bed rest (+4 beats · min⁻¹) and after training (–2 beats · min⁻¹). It is important to note that the magnitude of the declines in fitness, cardiac output, and stroke volume generally were greater in the two more highly trained athletes who enrolled in the study than in the three individuals who were more sedentary at baseline. Of note, the loss in fitness in all five subjects was reversed with retraining. The three sedentary athletes returned to pre-exercise fitness levels within several weeks, and then, due to continued exercise training, they achieved values at 8 wk that were greater than those observed before bed rest (i.e., the training effect).

Table 6.10 presents the results from another classic study involving six highly trained competitive cross country runners who were running as much as 70 mi (112.7 km)/wk before study participation. In this study, measurements of both fitness and cardiac function were gathered at baseline and at several time points over the 3 wk after the athletes stopped running (no bed rest). Again, a decline (10%) was observed in cardiorespiratory fitness in just 7 d. Other declines were observed in cardiac mass (cardiac atrophy was 15% at 4 d and 39% at 3 wk), wall thickness, and left ventricular end-diastolic volume (12% at 1 wk). The latter suggests a loss or decrease in plasma volume and its associated effect on cardiac filling and, subsequently, on stroke volume, cardiac output, and max $\dot{V}O_2$.

Although our discussion about detraining has been limited to changes in fitness and cardiac function, important changes also develop in skeletal muscle function and other organ systems. Concerning skeletal muscle, detraining markedly alters ATP production, mitochondrial size, oxidative enzymes and proteins, and myoglobin (44).

Finally, we would be remiss if we did not mention a unique type of detraining that is rarely experienced but that has a profound effect on the body: detraining due to microgravity or spaceflight, an environment in which astronauts experience major biological adaptations. The primary systems involved include cardiovascular, pulmonary, and skeletal muscle function; bone health; immune function; and appetite and caloric intake. Many changes that manifest in these systems mimic the changes associated with prolonged bed rest. Soon after exposure to microgravity there is a reduction in blood volume due to a loss of plasma volume, which is approximately 10% to 20% (25). Upon the return to Earth, this hypovolemia contributes to reduction in cardiac filling and subsequent stroke volume. As a result, peak cardiac output is reduced up to 25%. Like the effects of bed rest, this maladaptation contributes greatly to a 20% or more reduction in exercise capacity (max $\dot{V}O_2$) measured immediately postflight (68). The hypovolemia, along with cardiac atrophy, likely also contributes to the orthostatic intolerance observed early after return from spaceflight. To attenuate the negative effects of microgravity on the body, countermeasures are used during space flight and remain an area of intense research. One such countermeasure is exercise, such as treadmill walking while in space, as a means to stimulate the cardiovascular

Table 6.10 Changes in Maximal O₂ Uptake and Left Ventricular Characteristics in Runners Undergoing 3 Wk of Detraining

	Before training	DETRAINING			
		Day 4	Week 1	Week 2	Week 3
Max $\dot{V}O_2$ (mL · kg⁻¹ · min⁻¹)	62.3	—	56.4	57.8	56.7
Mass (g · m⁻²)	109.5	92.7	80.2	70.2	67.1
Internal dimension (cm)	5.1	4.9	4.7	4.7	4.6
Posterior wall thickness (mm)	10.7	9.6	9.0	8.2	8.0
End-diastolic volume (mL · m⁻²)	64.5	61.9	56.7	55.2	53.8

Data from Ehsani et al. (30)

system and provide fluid volume control to improve orthostatic tolerance.

Summary

- Aerobic power or capacity can be measured, and a variety of training programs and characteristics can be combined to impose a progressive and sufficient overload stimulus on the body.
- The human body has great plasticity, meaning that it can undergo adaptations or changes in response to the stimuli or stresses placed on it.
- For endurance-type training, the body adapts in a manner that results in an improved ability to tolerate submaximal energy demands using aerobic metabolic pathways for a greater period of time or at a higher intensity. These changes can be observed at rest and during exercise and occur both centrally (the heart) and peripherally.

Definitions

apoptosis—Planned or programmed cell death due to biological changes. For skeletal muscles, excessive apoptosis due to disuse or disease can alter muscle fiber distribution, size, and function.

exercise-induced hypoxemia—Exercise-induced reduction in arterial oxyhemoglobin saturation of less than or equal to 91% (normal = approximately 98%). Not uncommon among elite athletes; due to inequality in ventilation-perfusion or pulmonary diffusion that can limit max $\dot{V}O_2$.

power—The rate at which work is being performed per unit of time. Power = work/time.

respiratory exchange ratio—Measured during indirect, open-circuit spirometry. The ratio between the volume of CO_2 exhaled per a unit of time and O_2 consumed per the same unit of time. RER = $\dot{V}CO_2/\dot{V}O_2$.

work—The product of force (kilograms) times distance (meters). Work = force × distance.

Principles for Testing and Training: Anaerobic Strength, Power, and Range of Motion

We thank George L. Panzak, PhD, RN, USAW-SPC, for his contributions to this chapter.

Skeletal muscle strength, anaerobic power, and flexibility are all crucial components of health-related physical fitness and performance. Many adults would benefit from improved function in each of these physical fitness components. For instance, hustling up a flight of stairs or running to catch the bus requires an element of anaerobic power in that "rush" of a moment. Even simple tasks, such as bending over or squatting down to pick up a dropped item or carrying a load of groceries into the house, can be made easier by improving flexibility and strength, respectively. These are also important, to varying degrees, in almost any athletic activity. Competitive swimming and wrestling are excellent examples of the need for a combination of anaerobic power, strength, and flexibility.

Decrements and improvements in anaerobic power, flexibility, and strength are possible in all individuals regardless of age, sex, or race. A natural reduction in these indices occurs for everyone as they age past about 25 to 30 yr. However, the rate of decline of each fitness parameter can be attenuated—and in most cases improved—through a properly designed exercise training program. This chapter discusses anaerobic power, range of motion (ROM), and skeletal muscle strength and provides advanced-level information with respect to testing and training. This information will help you appreciate each individual's response to activities requiring these components of physical fitness and will help you understand how to develop regimens for improvement.

ANAEROBIC EXERCISE TRAINING PRINCIPLES

There are several important considerations for any type of training program designed to enhance exercise ability, including the following: testing and evaluation; training principles; training methods; periodization, intensity, and volume training phases; warm-up or training-day preparation; anaerobic and aerobic exercise selection; speed, agility, and speed endurance movements; cool-down post training practices; and recovery and restoration for sport or other activities. We first briefly review the training principles. The other listed concepts are then presented in the context of anaerobic, strength, and flexibility training.

Improvement in physical performance (aerobic, anaerobic, strength, and flexibility) as a result of training requires adherence to several basic principles. First, the *principle of specificity* of training states that the design and implementation of a training program must invoke the use of the energy system(s), skeletal muscles and their actions, and ROM that will be used in the performance that one desires to improve. Physiological and other gains from training are specific to the type of training performed. The *principle of overload* refers to the stimulation of the energy system and skeletal muscles beyond that to which they normally are exposed. These systems will then adapt (*principle of adaptation*) over time and result in improved performance (or reduced performance if there is negative adaptation due to inappropriately applied training intensity and/or load). Conversely, according to the *principle of reversibility*, improvements in performance can be lost when a training stimulus is reduced or removed. The *principle of progression* (or progressive overload) states that a gradual and systematic increase in training overload is necessary for optimal rates of improvement in specific fitness variables. If the rate of progression is too slow, fitness will not optimally improve. If the rate of progression is too great, there is a risk of injury or the effects of overreaching and overtraining (i.e., negative adaptation).

The *principle of accommodation* states that improvements may be diminished when proper progressive overload and variability of training do not occur. The *principle of individual differences* means that variations (e.g., muscle fibers, heart and lung capacity, motor skills, sex, previous training experience and injuries, genetics) among individuals before beginning a training program will result in different rates and amounts of adaptation to training. Finally, the *principle of periodization* (or variation) refers to altering one or more components of training (i.e., frequency, intensity, time, and type) for a specific purpose, such as preparing for a competition or recovering from a difficult period of training. These principles must be considered when developing any exercise training program with an expectation of improvement. In this chapter we apply these principles to anaerobic power, resistance, and flexibility training.

Anaerobic Power Testing and Training

Anaerobic power is the rate of energy release by cell metabolic processes. This does not require oxygen for adenosine triphosphate (ATP) generation in the electron transport chain process within the mitochondria. Training programs for improving anaerobic power focus on improving performance of exercise that lasts between a few seconds and several minutes. Improvement in anaerobic performance might also be important for longer duration athletic events that require sprinting at the end of the competition. This type of training focuses on the ATP-phosphocreatine (ATP-PCr) and the anaerobic glycolytic energy systems (see chapter 1).

Exercise testing can be used to assess the rate and capacity of the energy systems to generate ATP. The distinction between the anaerobic and aerobic (i.e., oxidative) metabolic pathways will vary depending on the intensity and duration of a given exercise. Figure 7.1 shows the approximate relative contributions of each of these energy systems during a bout of exercise. This figure demonstrates that use of stored ATP and resynthesis of ATP via the ATP-PCr system provide the greatest percentage of energy use in activities lasting only a few seconds (e.g., jumping, very short sprinting) to about 20 to 30 s. As exercise duration advances, the contribution of the ATP-PCr system to total ATP production and use is greatly reduced. At the same time, anaerobic glycolysis greatly increases its contribution to ATP production and use. If exercise duration continues beyond about 60 to 90 s, the oxidative aerobic energy contribution increases to become the dominant source of ATP. Both the ATP-PCr and anaerobic glycolytic pathways respond to specific training, resulting in performance improvements that are specific to activities lasting a few seconds for the ATP-PCr system and 30-60 s for the glycolytic system.

The general principles of specificity, reversibility, progressive overload, accommodation, individual differences, adaptation, and periodization apply to anaerobic power training. One must also assume that individual training responses will occur due to the design and implementation of a training program as well as the skeletal muscle fiber composition (i.e., percentage of slow-twitch fibers vs. percentage of fast-twitch fibers) due to the genetic makeup of a given individual. Methods of testing anaerobic power are not universally accepted,

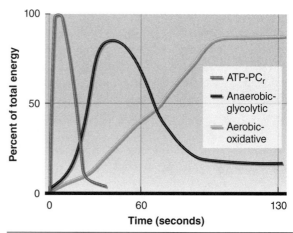

Figure 7.1 Percentage of energy systems used during exercise. The *x*-axis represents total exercise duration, and the *y*-axis represents the percentage of total ATP derived from the specific energy systems.

and many testing procedures exist. The most common are short run-based sprinting, the Wingate Anaerobic Test (WAnT; reviewed later in this chapter), and the oxygen deficit assessment. The Wingate Anaerobic and oxygen deficit tests are nonspecific testing methods. Other tests are designed to assess anaerobic energy necessary during short periods of exercise duration at very high intensities by mimicking sport-specific movements. Some of these tests are the vertical jump (73), the Margaria-Kalamen power test (43), the modified Margaria-Kalamen power test (also known as the football staircase test) (34), and the Modified Box Long Jump test (3). These tests are discussed further in the next section.

Anaerobic Testing

To accurately determine baseline status in order to guide exercise prescription development and determine whether adaptations occur in response to the exercise training stimulus, one needs to be able to precisely measure the variables of interest. Measures must be both repeatable and accurate (i.e., valid). To attain valid measures, the testing methods used must be performed according to specific guidelines with respect to

- the subject's preparation (e.g., meals, rest, prior exercise, time of day),
- equipment use (e.g., appropriateness of equipment for measuring variables of interest, maintenance, calibration),
- implementation according to testing guidelines applied by the individual conducting the test, and
- proper interpretation of the test results.

Typical variables of interest in anaerobic performance assessment are **anaerobic power** and **anaerobic capacity**. Anaerobic power is important for sprint, mid-distance, and distance athletes but for varying reasons. For most athletic events, there is interplay between the aerobic and anaerobic energy systems. Referring to figure 7.1, we see that sprinters (events lasting <1 min) primarily utilize anaerobic energy sources for ATP production, whereas middle-distance athletes (events lasting 1-4 min) utilize both aerobic and anaerobic sources for ATP production and distance athletes (events lasting >4 min) predominantly utilize aerobic metabolic sources. Even aerobic athletes must rely on anaerobic sources at critical times of an event, such as during a hill climb in a bicycle race or the final sprint of a 10K running race.

No direct method of determining anaerobic energy power and capacity currently exists. Although blood lactate accumulation generally is considered to be the product of a high rate of anaerobic glycolysis, blood lactate concentration measurement does not provide specific information about the rate and capacity of the production of ATP. Additionally, the concentration of blood lactate is affected by the rate of its clearance, which differs in individuals due to factors such as—but not limited to—recovery activity status (i.e., active vs. passive); recovery exercise intensity; ratio of individual Type I to Type II skeletal muscle fiber composition; the rate of blood flow to the skeletal muscles, heart, and liver (Cori cycle); and the rate of release of muscle lactate into the blood.

The following is a comprehensive list of common anaerobic tests. The test to be utilized must be specific to the variable being measured. For instance, to measure maximal anaerobic muscle **power**, a test needs to be very short in duration to isolate the ATP-PCr energy system. Maximal muscle power is used in activities such as jumping and throwing. A vertical jump test is a good example of a short-duration muscle power test. **Speed** often is thought of as a useful measure, and the 40-yd (36.6 m) dash is used by the National Football League (NFL) to measure speed and provide a ranking of potential draftees. However, a 40-yd dash technically assesses **acceleration**, **velocity**, and average speed. The top speed is the fastest instantaneous rate of an object in motion while the average speed is the distance divided by the time to cover that distance. Velocity includes the component of direction of movement; thus, speed is the magnitude component of velocity. *NSCA's Guide to Tests and Assessments* (Human Kinetics, 2012) provides an excellent review of the performance of anaerobic tests.

Anaerobic Tests

- Vertical jump test (8)
- Margaria-Kalamen power test (43)
- Modified Margaria-Kalamen power test (also known as the football staircase test) (34)
- Three Modified Box Long Jump (MBLJ) test (3)
- Wingate Anaerobic Test (WAnT) (10)
- Critical power test

Various studies have determined that each test in the list is a valid measure of anaerobic power. In particular, the vertical jump test, standing long jump test, Margaria-Kalamen power test, Modified Margaria-Kalamen power test, and Three Modified Box Long Jump test are considered to be the most sport specific of these tests. The following sections briefly discuss and provide directions for performing several of these anaerobic tests.

PROCEDURES FOR THE VERTICAL JUMP TEST

1. Before the test: To measure the highest standing reach, instruct the subject to stand flat footed and reach with one arm fully extended. This should be the same arm that will extend and displace the vanes. Then measure again during plantar flexion with the same arm extended and during plantar flexion. Measure the subject's reach at the middle finger. This measure will be subtracted from the highest recorded vertical jump value.

2. During the test: Instruct the subject to take a vertical jump from a standing position using a countermovement. The subject should squat down and back while simultaneously using a bilateral backward arm swing. The subject should then explode upward by extending the hips, knees, ankles (also known as the *triple extension*), shoulders, elbows, and wrists and attempt to displace the highest vane with one arm (see figure 7.2). Alternatively, a wall-mounted jump-and-reach board may be used.

3. Obtain the measurement and repeat two more times. Instruct the subject to rest for 30 s between trials.

4. Calculate the maximum vertical jump by subtracting the highest standing reach from the highest vane displaced during the three vertical jumps.

Figure 7.2 Vertical jump test.

Vertical Jump Test

The vertical jump test is commonly used by coaches to measure lower body power and to test an athlete's jumping ability. Previous investigations determined that the vertical jump has high predictive value for measuring power production for athletes participating in such sports as track and field, weightlifting, volleyball, basketball, and football (73). To test the vertical jump, schools, organizations, and sport teams commonly use a commercially designed device consisting of plastic swivel vanes placed in 0.5-in. (12.7 cm) increments. Because the athlete uses countermovement of the lower body and an outstretched arm swing to reach the highest vane, the vertical jump test is also known as the jump and reach test.

As an example of the sophistication that can be developed for simple tests such as the vertical jump, consider the following. Some researchers believe that the gold standard for measuring the vertical jump includes implementing video analysis and using biomechanical reflective markers on the hip to measure the displacement of the center of mass. Although valid and reliable, this method is cost prohibitive for many fitness and sport organizations. A commercial jump mat is another device used to measure the vertical jump. Embedded in the mat are microswitches that determine flight time—that is, the interval between the liftoff and landing of both feet during a countermovement from a flat-footed standing position. The flight time in the air is used to compute the subject's vertical jump. A study by García-López et al. using three countermovement jumps found that jump mats designed with photo cells showed higher correlations with force plate technologies compared with a contact jump mat (26). Similarly, Whitmer et al. determined that practitioners using a contact jump mat may underestimate the vertical jump for high performers (73). A promising new technology for measuring vertical jump is the Myotest (Myotest Inc., Switzerland). The Myotest device is approximately the size of a small smartphone, and the subject wears the device on the hip to determine the displacement of the center of mass. This test does not require the subject to reach and stretch with one arm to displace vanes like when using a vane jump test device. A study by Nuzzo et al. found that the Myotest demonstrated the best intrasession reliability for measuring a countermovement vertical jump when compared with the vane jump test device and jump mat system (48).

PROCEDURES FOR THE MARGARIA-KALAMEN POWER TEST

1. Set up automatic timing devices to measure the time from touching the third step to touching the ninth step.
2. Obtain the subject's body weight in kilograms using a certified scale.
3. Instruct the subject to perform two warm-up sprints at approximately 50% and 80% of maximum volitional effort, respectively.
4. Instruct the subject to use a standing start 6 m from the vertical face of the first step.
5. Instruct the subject to repeat the test after a 15-s rest period. The subject should complete five trials at 100% effort.
6. Determine test scores by entering body mass (in kilograms) and best sprint time from five trials (in seconds) into this formula: power (kg · m · s^{-1}) = [body mass (kg) × distance (m)]/time (s). Multiply the final power value by 9.807 to convert to watts.

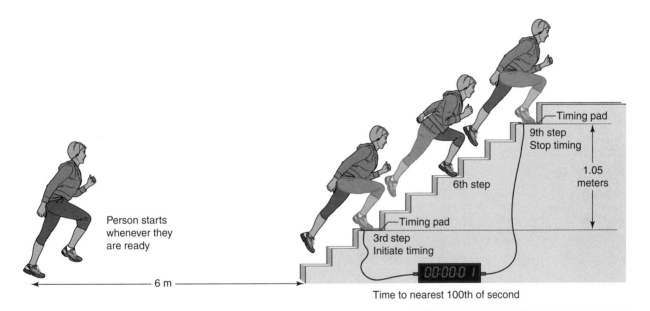

Figure 7.3 Margaria-Kalamen power test.

Based on E. Fox, R. Bowers, and M. Foss, 1993, *The physiological basis for exercise and sport,* 5th ed. (Dubuque, IA: Wm C. Brown), 675.

Margaria-Kalamen Test

Historically, researchers have attempted to assess anaerobic power in sports that rely on the ATP-PCr energy system. The Margaria-Kalamen power test (see figure 7.3) was designed to be of very short duration (<5 s) and to test anaerobic power—specifically, that of the ATP-PCr system. In this test, the subject sprints 6 m to a staircase consisting of nine steps that are 17.5 cm high. The subject sprints three steps per stride, covering 1.05 m of vertical distance (i.e., the distance from the third step to the ninth step). Start and stop times are recorded using commercial electronic switch pads (secured with doubled-sided carpet tape) placed at the third and ninth steps. The time for each trial is recorded to the nearest thousandth of a second and used to calculate power after the completion of five trials (34). This test has a test–retest reliability of 0.85 in the general population, suggesting good reliability (41).

PROCEDURES FOR THE MODIFIED MARGARIA-KALAMEN/ FOOTBALL STAIRCASE TEST (FST)

1. This test requires a 6-m run sprint to a staircase consisting of 20 steps. The subject sprints two steps per stride, covering 3.12 m of vertical distance (i.e., the distance from the first step to the 20th step).

2. Start and stop times are recorded using commercial electronic switch pads (secured with double-sided carpet tape) placed at the second and 20th steps.

3. The time for each trial is recorded to the nearest thousandth of a second and is used to calculate power after the completion of five trials (34).

4. Power is calculated as it is for the Margaria-Kalamen test.

Modified Margaria-Kalamen Test

The modified Margaria-Kalamen test (also referred to as the football staircase test) was developed for several reasons. Previous investigations found that subjects are able to alter Margaria-Kalamen test results by tripping the timing device with the lead leg before the center of gravity reaches the first step and subsequent final step. It was found that this movement could artificially improve the test times by 10% to 24% (34). Additionally, in a study of football players it was found that the Margaria-Kalamen test had poor reproducibility (34). In an attempt to improve the results, a modification was made to ensure that the test required greater than 1.5 s to complete: The vertical distance was increased to 3.12 m over 20 steps. Additionally, a two-step protocol was adopted to prevent errors associated with the three-step approach of the Margaria-Kalamen test. The results determined that for the football skill players and linebackers had a peak power of 1,389 ± 210 and 1,675 ± 301 watts, respectively.

PROCEDURES FOR THE THREE MODIFIED BOX LONG JUMP (MBLJ) TEST

1. Obtain boxes 50 cm wide by 60 cm long.

2. Place boxes 90 cm apart on a level track or surface.

3. Instruct subjects to stand on the first box and perform a three modified long jump as follows: Drop from the first box with both feet and immediately jump with both feet onto the second box, drop from the second box with both feet and immediately jump with both feet onto the third box, and drop from the third box with both feet and immediately perform a long jump using both feet.

4. Measure the distance of the long jump (where the heel closest to the edge of the third box lands after performing the long jump).

5. Complete three trials, resting 10 min between each trial.

6. Measure (to the nearest 0.01 s) the time from when the feet leave the first box to the time they land from the long jump.

Three Modified Box Long Jump (MBLJ) Test

Almuzaini and Fleck (3) noted that horizontal jumping ability demonstrated higher correlation with various types of power sports compared with vertical jumping ability. They found in their study that a three modified box long jump test resulted in stronger correlations with measures of power compared with other horizontal jump tests.

PROCEDURES FOR THE WINGATE ANAEROBIC TEST

Before the test: Each subject should be given full details of the test procedures, including the likelihood of excessive exhaustion and potential symptoms (e.g., vomiting, faintness). There is evidence that time of day can affect test performance; testing later in the day elicits higher peak values. Additionally, the warm-up process can affect performance. Other factors that may affect performance include intrinsic motivation, external motivation (i.e., encouragement), blood glucose status, and recent bouts of high-intensity exercise. Each of these factors should be standardized in the testing facility. An ergometer that is either mechanically or electrically braked should be positioned to the proper seat height and be equipped with pedal stirrups to maintain foot contact with the pedals throughout the full revolution.

Test Procedures for Cycle or Arm Ergometer

1. The subject should warm up at a low intensity on the testing equipment. This includes two or three 15-s bouts of sprinting to replicate the pedal speed desired for the test. The subject then stops pedaling for 1 min before beginning the test.

2. To begin the test, the subject pedals against zero resistance for 10 s. A countdown begins at 5 s; during this time, the subject should begin to pedal at their fastest rate possible.

3. At the beginning of the data collection period, a flywheel resistance of 0.075 kg/kg of body weight is immediately placed against the pedals.

4. The subject is verbally encouraged for 30 s to maintain the fastest pedal speed possible. The resistance remains constant throughout the test.

5. At the end of the 30-s data collection period, the load can be reduced to a very light intensity level and the subject can pedal at a slower, comfortable rate until they feel adequately recovered.

Posttest Data Calculations

Peak anaerobic power, mean anaerobic capacity, and the fatigue index are calculated from the data as follows.

- *Peak anaerobic power.* Peak anaerobic power reflects the phosphagen component of anaerobic energy release. It is the highest workload performed during any 5-s period (kg/5 s) and is calculated using the following formula: 0.075 kg of resistance × kg of body weight × 6 × highest number of flywheel revolutions in a 5-s period (typically the first 5 s of pedaling).

- *Mean anaerobic capacity.* Mean anaerobic capacity reflects the glycolytic component plus the phosphagen component of energy release. It is expressed in total work (kg/30 s) and is calculated using the following formula: 0.075 kg of resistance × kg of body weight × 6 × number of flywheel revolutions in a 30-s period.

- *Fatigue index.* Fatigue index reflects the oxidative capability of the muscles; a higher fatigue index is indicative of a greater proportion of fast-twitch muscle fibers. It is calculated using the following formula: (work performed in initial 5-s period – work performed in final 5-s period)/work performed in initial 5-s period × 100.

Note that there are variations of the Wingate test with respect to equipment used, time of exercise, and workload utilized.

Wingate Anaerobic Test

Common anaerobic assessment tests include the Wingate Anaerobic Test (WAnT), the critical power test, and the maximal accumulated oxygen deficit test. Because each of these tests provides different information, there is no consensus on a single gold standard for measuring a person's anaerobic capabilities. Of these, the Wingate test has become the most often used. Currently, the National Hockey League (NHL) uses the Wingate test in its battery of assessments of selected prospective players entering the annual draft. Most often it is performed using a cycle ergometer, but occasionally it is performed on a treadmill or an upper body ergometer. The critical component of the Wingate test is accurately measuring the work performed over the duration of the test.

Critical Power Test

The critical power test was designed to quantify the amount of work a muscle group could perform before exhaustion during dynamic contractions (45). The critical power test, performed on a cycle ergometer, results in two estimates: critical power (defined as the highest sustainable power output without fatigue occurring) and anaerobic work capacity (which theoretically is correlated to the total available energy contained in the muscle). Initially, this test required three to four exhaustive efforts on a cycle ergometer (46). Since its origin, many variations of this test have attempted to reduce the multiple exhaustive work bouts required. This test can be used to develop and assess the results of training programs.

Anaerobic Training Methods and Adaptations

Think about how many truly anaerobic sprint activities you perform in a typical day. There likely are not many other than going quickly up a flight of stairs or running to catch the bus. Nonathletic individuals do not perform anaerobic activities regularly. Most daily activities are done at an intensity and duration requiring predominantly aerobic metabolism. Because anaerobic ability generally is not related to health with respect to the development of chronic disease, anaerobic training typically is not recommended for the general public. Most anaerobic training programs are designed and implemented for athletes who compete in events requiring significant anaerobic effort (e.g., sprinting in running, swimming, and cycling; team sports such as basketball, football, and baseball).

Many athletic events utilize both the anaerobic and aerobic energy systems. Figure 7.4 categorizes many common sporting activities based on the energy system utilized and the dynamic and static nature of the skeletal muscle movements performed. The balance of utilizing predominantly the aerobic or anaerobic system for energy utilization also correlates to the intensity of muscle tension utilized during isometric static activities and dynamic exercise movements. Isometric or static activities include movements where the muscle length remains unchanged or nearly unchanged during submaximal or maximal exercise intensities while the limb remains in a fixed position. Conversely, movements where limbs are changing positions consist of either dynamic concentric or positive muscle contractions or dynamic eccentric muscle contractions. Muscle fibers shorten during dynamic concentric or positive muscle contraction and lengthen during an eccentric or negative muscle contraction.

In general, when determining the relative energy contribution during either an isometric static activity or during a dynamic concentric or dynamic eccentric muscle contraction will depend on the level of muscle tension developed during submaximal and maximal exercise intensities. For example, when muscle tension is lower (e.g., < 60% peak VO_2, or < 6 on the Omni scale, or < 12 on the Borg RPE scale) then the aerobic system predominates. And this balance switches to more anaerobic at greater muscle tension intensities reflected by increases in these measures. Most activities incorporate a mix of static and dynamic movements and lie somewhere between the two extremes of the continuum. An understanding of these interactions between energy substrate contribution and muscle tension intensity is important when developing training programs that are specific to the energy system utilized and the duration of exercise.

Think about a couple of athletic activities and, without looking at figure 7.4, try to determine their level of static and dynamic movement and whether they are predominantly anaerobic or aerobic with respect to the energy system used. Now look at figure 7.4 to see whether you correctly categorized the activities.

Table 7.1 presents a more specific breakdown of the energy systems used in athletic activities. Three important notes about this table are as follows:

- Performance time is strongly related to the percentage of use of each energy system. For instance, a 100-m swim event and a 400-m track and field event each take 45 to 70 s to complete depending on the athlete's age and skill level. Note how each of these events is very similar in energy system usage.

- As noted in figure 7.1, there is overlap between the anaerobic glycolysis and aerobic energy systems for generating ATP-PCr. This is because these systems are difficult to separate during laboratory assessments and typically supply ATP in unison during physical activity or exercise.

- The percentages of energy system use are specific to certain activities during a performance. For instance, golf has a very high anaerobic component. This is based on the golf swing and not on walking performed between shots.

Understanding the energy system use relationships is key when developing training plans for improving performance in physical activities or athletics. It is also important to understand the physiology of anaerobic training with respect to the type of results desired. Anaerobic training will lead to improvement of these metabolic systems and adaptation of the neuromuscular

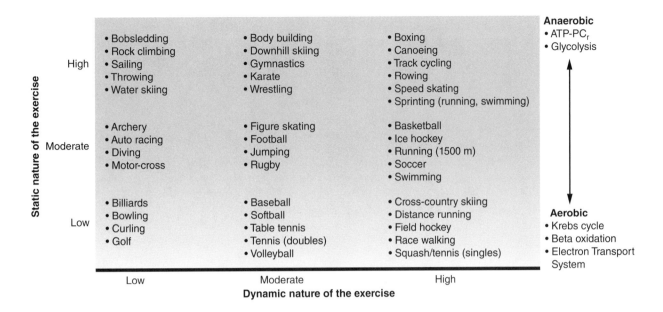

Figure 7.4 Classification of sporting activities based on the static and dynamic movement components and the energy system utilized.

Adapted from *Journal of American College of Cardiology*, Vol. 24, J.H. Mitchell, W.L. Haskell, and P.B. Raven, "Classification of sports, 26th Bethesda Conference," pg. 866, copyright 1994, with permission of Elsevier.

system. With specific anaerobic training, these improvements will be greater in those systems used for short-duration, high-intensity exercise. Crossover improvement (although less) will also occur in the systems used for longer duration, lower intensity exercise. Figure 7.5 summarizes the physiological variables and relationships specific to anaerobic training. The degree of contribution of each variable depends on the type of anaerobic training performed. Notice that in each of the pathways (i.e., metabolic and neuromuscular) some factors are very specific to anaerobic exercise, whereas others are more prominent in aerobic exercise training and performance.

It is important to emphasize that many of these variables are interrelated. For instance, in every skeletal muscle there is a mix of Type I (slow twitch; slow oxidative), Type IIA (intermediate twitch; fast oxidative/glycolytic), and Type IIX (fast twitch; fast glycolytic) fibers. Although these fibers and their motor units are recruited by low, intermediate, and high nerve stimulation frequencies, respectively, the motor units are recruited in the order according to their contraction force. At low exercise intensities (i.e., 30%-40% of peak $\dot{V}O_2$) Type I fibers are predominantly recruited, but there is also some recruitment of Type IIA fibers (29, 69). At intensities well above $\dot{V}O_2$max (i.e., up to 150%), the recruitment pattern is predominantly Type IIX, Type IIA, and then

Type I fibers (29). Another way to look at motor unit recruitment patterns is to use a method that assesses motor unit recruitment at a percentage of the maximal voluntary contraction of a muscle or muscle group. As with the previous example using percentage $\dot{V}O_2$max, motor unit recruitment with increasing percentage of maximal voluntary contraction follows a pattern of Type I, Type IIA, and Type IIX as the percentage increases from 40% to 75% and above. Essentially all motor units are recruited as the percentage of the maximal voluntary contraction increases above 75%.

Anaerobic Training Stimulus

The appropriate stimulus for training is determined by the amount of overload performed by the individual in order to adequately stress the anaerobic metabolic and neuromuscular systems. Overload is a product of the amount of training being performed either during a single session (acute overload) or over a prolonged (i.e., weeks, months) period of training sessions (chronic overload). Overload is a combination of training intensity, duration, and frequency. Of these factors, intensity typically is considered to be the most important. Intensity pertains to the level of effort, relative to peak effort, that is applied in a given training session. Exercise intensity can be guided using several techniques, including

Table 7.1 Energy System Use (%) of Various Athletic Activities

Sport or sport activity	Emphasis (%) by energy system		
	ATP-PCr and anaerobic glycolysis	Anaerobic glycolysis and aerobic	Aerobic
Aerobic dance	5	15-20	75-80
Baseball	80	15	5
Basketball	60	20	20
Fencing	90	10	Negligible
Field hockey	50	20	30
Football	90	10	Negligible
Golf	95	5	Negligible
Gymnastics	80	15	5
Ice hockey			
Forward, defense	60	20	20
Goalie	90	5	5
Ice speed skating			
<500 m	80	10	10
1,000 m	35	55	10
1,500 m	20-30	30	40-50
5,000 m	10	25	65
10,000 m	5	15	80
In-line skating, >10 km	5	25	70
Lacrosse			
Goalie, defense, attacker	50	20	30
Midfielders, man-down	60	20	20
Rowing	20	30	50
Skiing			
Slalom, jumping	80	15	5
Downhill	50	30	20
Cross-country	5	10	85
Recreational	20	40	40
Soccer			
Goalie, wings, strikers	60	30	10
Halfbacks or sweeper	60	20	20
Stepping machine	5	25	70
Swimming and diving			
Diving	98	2	Negligible
50 m	90	5	5
100 m	80	15	5
200 m	30	65	5
400 m	20	40	40
1,500 mi, 1650 yd	10	20	70

Reprinted, by permission, from E.L. Fox and D.K. Mathews, 1974, Interval training conditioning for sports and general fitness (W.B. Saunders).

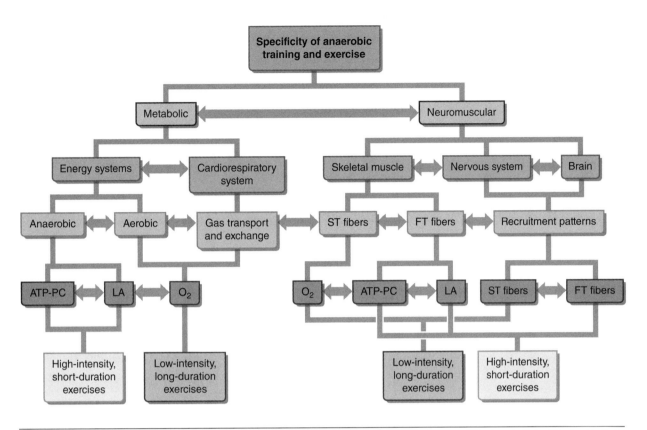

Figure 7.5 Specificity of anaerobic training and exercise. LA=lactic acid; O$_2$=oxygen; ST=slow twitch; FT= fast twitch; ATP= adenosine triphosphate; PC= phosphocreatine.

Reprinted from M.L. Foss, S.J. Keteyian, and E.L. Fox, 1998, *Fox's physiological basis for exercise and sport,* 6th ed. (Pittsburgh, PA: William C. Brown). By permission of the authors.

1. blood lactate

2. heart rate

3. pace, and

4. subjective effort rating.

Although these are very different methods of assessing intensity, each is related.

Because an objective of anaerobic training is to focus the overload primarily on the anaerobic energy pathways, we must be able to determine whether this occurs. Of the methods for assessing anaerobic exercise intensity, blood lactate change is the most directly related to metabolism. Because increases in blood lactate typically result during high levels of ATP production via the glycolytic metabolic pathway (see chapter 4 section on ventilatory derived anaerobic threshold), methods for assessing blood lactate have been developed. This method requires a droplet of blood from an earlobe or fingertip and the use of a rapid handheld analyzer. During an exercise test protocol with progressive small increments in intensity (i.e., ramp protocol), blood is sampled at each workload increase; blood lactate values (in mmol · L^{-1}) are plotted on the y-axis, and work rate

is plotted on the x-axis (see figure 7.6 for an example). Normal resting blood lactate concentrations usually are about 0.5 to 1 mmol · L^{-1} and increase with progressive exercise as the result of intracellular anaerobic metabolism increasing to meet the ATP demand. As the intracellular concentration of lactate increases, the concentration gradient favors its release from the cell, and lactic acid forms in the blood as extracellular H$^+$ combines with lactate. The increase in blood lactic acid concentration is slow and linear at lower exercise intensities. As exercise intensity increases, there is a point at which blood lactate concentration begins to rapidly increase. This point has been termed the *lactate inflection point,* the *onset of blood lactic acid accumulation (OBLA),* and the *anaerobic threshold (AT).* In general, this point has been described to have a value of approximately 4 mmol · L^{-1} and is a highly reproducible value known as a *threshold* (e.g., in trained female runners) (32). However, Maglischo found that up to 40% to 50% of swimmers tested over a 10-yr period attained their lactate threshold either above or below the 4 mmol · L^{-1} value (42). This much variability in a training group could have a significant effect on training

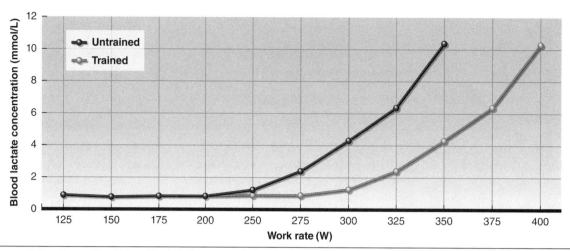

Figure 7.6   Example of blood lactate increase during incremental exercise.

tolerance and progression because some athletes might be training at too high (or too low) of an intensity that was developed around the 4 mmol · L^{-1} blood lactate value. This suggests that, if possible, individuals training at intensities requiring the knowledge of their blood lactic acid threshold should know the exact training intensity at which this threshold occurs. Determining the blood lactic acid threshold routinely is impractical, even in today's age of rapid analyzers that provide blood lactic acid values in seconds and portable handheld devices. Therefore, a surrogate measure that reflects this value would be useful.

As previously listed, heart rate, pace, and subjective rating of effort (e.g., Borg rating of perceived exertion; aka RPE) can be used to guide anaerobic training intensity. Any of these easily measured variables can be collected during a lactate threshold test. Using the value that corresponds to the blood lactic acid threshold and monitoring it during training provides a relatively accurate method for determining whether the athlete is above the threshold. For instance, if an athlete plans to train at an intensity above the anaerobic threshold, she can use a heart rate device to determine whether her rate is at or above the identified threshold heart rate. Alternatively, most individuals can easily be taught to take their pulse at the radial or carotid artery. However, this may decrease accuracy because most individuals must stop to assess their heart rate; thus, the heart rate is not measured during exercise.

Pace can be used in the same manner as heart rate to guide anaerobic exercise intensity. Perceived exertion (refer to the Omni or RPE scales) is a subjective rating compared with heart rate and pace, which are objective. Despite this, research suggests that self-described per-ceived exertion in National Collegiate Athletic Association Division I football players can be used to train near (at, above, or below) the anaerobic threshold (30). Recent investigations have also suggested that significant anaerobic metabolism begins at an average RPE scale rating of 14 (6-20 scale; 14 is between *somewhat hard* and *hard*) (60). Others have suggested that the anaerobic threshold occurs between a 14 and 16 rating on the RPE scale (75). These differences may reflect the previously mentioned inaccuracy of subjective rating.

Frequency and duration of anaerobic training depend on factors such as the type of athlete (aerobic, anaerobic, power and strength), the sport, planning for a single exercise session versus planning over a longer time period (i.e., weeks, months), and the seasonal training phase (off-season, preseason, in-season, peaking). Athletes who are primarily anaerobic performers (e.g., sprinters, football players, power athletes) will spend a large amount of any single training session performing anaerobic activities. However, even these athletes cannot spend 100% of their time above their anaerobic threshold. For any athlete, greater amounts of high-intensity training increase the risk of injury, **overreaching**, **overtraining**, and a delayed recovery response. Spending some training time well below the anaerobic threshold likely has meaningful benefits. For instance, lactate clearance needs to occur to prepare for the next exercise bout; this occurs best with very low intensity exercise (i.e., aerobic exercise) such as walking. A well-developed aerobic energy system likely facilitates the rate of lactate clearance and PCr store regeneration after exhausting anaerobic exercise (68). This can enhance the individual's rate of preparation for another bout of training.

Depending on the sport, it is not uncommon for anaerobic athletes to spend their early season performing some aerobic training. To illustrate why aerobic energy system training might be necessary for the anaerobic athlete you must understand that a 45- to 60-s 400-m runner or a hockey player who has shifts lasting 30 to 60 s during a game will rely on their aerobic energy system for 30% to 40% of their ATP production; their ATP-PCr system supplies about 10% to 15%, and their glycolytic system supplies the remaining ATP generation required. Given this information, it makes sense for these athletes to improve their aerobic energy system. As the duration of an event decreases from 1 min to less than 30 s, the reliance on the aerobic energy system—and thus the need for aerobic training—is reduced. As the season progresses for the 400-m runner or the hockey player, training will switch to less aerobic training and more

energy system–specific anaerobic training. Table 7.2 depicts the general seasonal training plan for a professional NHL team. Note the difference in energy systems between the early off-season (aerobic) and the regular season (mostly anaerobic).

The duration of a single anaerobic training session typically ranges from 1 to 2 h. This provides adequate time for a proper warm-up and cool-down as well as time for the interval training that is common in anaerobic training sessions (note the emphasis on interval training in the energy system row of table 7.2). The percentage of a training session spent below (aerobic), near (anaerobic endurance), or above (anaerobic speed) the anaerobic threshold varies with the duration of the focus event. Table 7.3 provides the relative energy system contribution to ATP production in competitive swimming. Based on this knowledge, single training sessions can be planned

Table 7.2 Detroit Red Wings Seasonal Training Overview

	May: playoffs and early off-season	June: off-season	July: off-season	August: training camp preparation	September: training camp	October-April: regular season and playoffs
Energy system	Aerobic: low to moderate intensity	Interval and sprint training	Interval and sprint training with 3:1 and 4:1 work:rest ratios	Interval and sprint training with 3:1 and 4:1 work:rest ratios	Interval and sprint training	Mostly anaerobic; amount depends on minutes played
Strength	Base strength and corrective exercises	Absolute strength and circuit training	Absolute strength and power circuit training	Absolute strength and power circuit training	Sport-specific strength	Strength maintenance with 1-2 lifting sessions/wk
Speed and agility	Minimal	Coordination and balance	Coordination and balance	Coordination and balance	Decrease volume	Maintain agility
Plyometrics	None	Initiate	Increase intensity	High intensity	High intensity	None
Flexibility	Daily emphasis on improving ROM where limited	Daily emphasis on improving ROM where limited	Daily emphasis on improving ROM where limited	Daily emphasis on improving ROM where limited	Emphasis on injury prevention	Emphasis on injury prevention
Postural corrections	Corrective exercises for imbalances	Corrective exercises for imbalances	As needed	As needed	As needed	As needed
Rehabilitation	Surgeries and incompletely healed	As needed	As needed	As needed	As needed	As needed
Skating (nongame)	None	Skill development only	Skill development and initiate conditioning	Skill development and continue conditioning	Primary conditioning	At practice; varies depending on minutes played

Table 7.3 Relative Energy System Contributions to ATP Production During Freestyle Competitive Swimming

Swim distance	ATP-PCr	Glycolytic	Aerobic (oxidative)
50 m (20-30 s)	15%-80%	Up to 80%	<25%
200 m (105-130 s)	Up to 30%	25%-65%	5%-65%
400 m (225-300 s)	0%-20%	10%-55%	25%-80%+
1,500 m (15-17 m)	<10%	10%-20%	80%-90%

Adapted from Rodriguez and Mader (55).

to focus specific amounts of training on a specific energy system. Combining this with a seasonal plan allows for long-term planning of training sessions. Note that the energy system contributions listed in table 7.3 can be applied to any competitive sport that requires an all-out effort of a similar duration.

Training Sessions

Interval training is a method commonly used in anaerobic training. This type of training consists of a series of repeated bouts of exercise interspersed with periods of active or passive rest. The purpose of interval training is to allow more work to be performed at a high intensity than otherwise would be possible during continuous exercise. This training model has received recent attention as a method of aerobic training in clinical populations, although the concept as it relates to anaerobic training actually dates back to Germany in the 1930s. Around the same time, Swedes were developing a different form of interval training termed *fartlek* ("speed play" in English). Interval training is utilized more by anaerobic athletes than fartlek training, which is more often used by distance runners.

An example of interval training is a set of 6 to 8 × 200 m or 3 to 4 × 800 m running. The length of each bout of exercise and the interval between these repeated bouts are key to this type of energy system training. It is typical for the interval between to be performed actively (i.e., active rest) with a walk or slow jog. This type of recovery period can be shorter or longer in duration. Shorter distance bouts coupled with longer duration recovery intervals is a key factor in training anaerobic athletes who primarily utilize the ATP-PCr and glycolytic systems (i.e., those running 400 m or less). Aerobic athletes who compete at distances requiring about 3 to 4 min and longer tend to run longer bouts and have shorter rest periods. Athletes who rely almost equally on the glycolytic and oxidative

energy systems compete in races of 1 to 3 min duration. See table 7.4 for recommendations of exercise:rest ratios for these different types of athletes.

Another type of anaerobic training is sprint training, which is a common component of training programs for both power and endurance athletes. Sprint training consists of near-maximal aerobic efforts lasting 10 to 30 s (sometimes longer) followed by a short recovery. In the swimming community, a recent type of training known as ultrashort race pace swim training is an example of sprint training. Sessions last in the range of 30 to 60 min, which is only one quarter to one half the duration of the common swimming practice. In this type of training swimming is always performed at the highest intensity possible, and no swimming is performed at a submaximal pace. Recovery between swims is inactive. For most athletes the distances swam at practice are between 25 and 50 yd or m, which is approximately 10 to 35 s for each repetition depending on the stroke performed. Nonaerobic energy systems account for roughly 95% of energy metabolism during a maximal 10-s sprint, and the percentage decreases as the duration of the sprint increases (e.g., 73% during a 30-s sprint). For this reason, it is common for sprint athletes to engage in shorter sprints, whereas middle-distance and endurance athletes perform longer (e.g., >30 s) sprints.

Several metabolic benefits result from sprint training and include adaptations to the ATP-PCr, glycolytic, and oxidative systems. Creatine kinase and adenylate kinase enzyme activities have been shown to increase by 36% and 20%, respectively, after 5-s maximal intervals among sprint athletes (20,67). Interestingly, these adaptations were not seen when intervals were increased to 30 s or when longer intervals were combined with the shorter bouts. Glycolytic adaptations such as increased phosphofructokinase (PFK), lactate dehydrogenase, and glycogen phosphorylase have been shown to increase after longer

Table 7.4 Relative Energy System Contribution To ATP Production

% of maximum power	Primary energy system stressed	Typical exercise interval time	Range of work-to-rest period ratios
90-100	Phosphagen	5-10 s	1:12 to 1:20
75-90	Fast glycolysis	15-30 s	1:3 to 1:5
30-75	Fast glycolysis and oxidative	1-3 min	1:3 to 1:4
20-30	Oxidative	>3 min	1:1 to 1:3

Adapted, by permission, from F.A. Rodriguez and A. Mader, 2010, Energy systems in swimming. In *World Book of swimming: From science to performance,* edited by L. Seifert, D. Chollet, L. Mujika (Hauppauge NY: Nova Science Publishers).

Technology and Athletic Training

Advances in athletic performance often are the result of advances in training techniques. Today, athletes in both team and individual sports generally have outstanding access to professional coaches; excellent training facilities; knowledge about nutrition practices for training, competition, and recovery; and recognition and treatment of injuries. Advances in technology have always been an important part of improving athletic performance. Today, possibly more than ever, the number of technology-based tools for assisting coaches and athletes is growing at a rapid pace.

An example of this rate of development is video. Thirty years ago video typically relied on 8-mm film that had to be developed and played back using a projector and screen. The delay between filming and viewing typically was several days. The advent of VHS and Betamax in the 1980s brought about the ability to film and immediately view the video on a television screen. This brought the time between filming and viewing down to mere minutes. Real-time video feedback is now the norm and is thought to be the ultimate in effectiveness because an athlete can see their performance immediately without leaving the practice or playing field. Also, the devices used to record these videos (i.e., smartphones, tablet computers) are very small and can store enormous amounts of data. Add in low-cost and free apps that can provide frame-by-frame playback, voiceover review, and the ability to draw on the video and detect time (e.g., time to 5 m in the 100-m sprint) and assess technique (e.g., determine the angle and trajectory of a basketball free throw), and the potential for the latest video technologies to provide tremendous benefit is apparent.

Other technological improvements include the field of sensors. For instance, heart rate monitors have provided real-time feedback to athletes at a relatively low price for more than 20 yr. More recently, these devices have included software to assess heart rate variability, which is the microsecond difference between each successive heartbeat. This methodology provides information about the autonomic nervous system's control of heart rate. Although the technology is not yet fully developed, some athletes and coaches are using this information to assess recovery from training sessions and to determine indications of overreaching or overtraining.

On the horizon are items such as smart clothing, which might measure sweat and respiratory rates; stretch sensors that can assess posture, edema, or swelling and motion; wireless insoles for shoes that can measure weight distribution and motion during running or walking; and sensors than can noninvasively provide feedback on individual muscle or muscle group contraction and relaxation speeds and potentially be able to assess indicators of muscle fatigue. Because all new technology is wireless, the data can be sent to a monitoring device (e.g., smartphone) for real-time analysis and feedback from an athlete's coach or trainer. Additionally, much of this type of technology can have benefit in a clinical setting with patients recovering from injury or surgery or who are dealing with a chronic disease.

sprint intervals (e.g., >10 s). Shorter intervals lasting less than 30 s have shown decreases in tricarboxylic acid enzyme levels (20), and the combination of short and long intervals had no effect on oxidative enzymes (24). However, performing longer intervals (>30 s) or multiple sets (more than 10) has been shown to enhance citrate synthase and succinate dehydrogenase levels (40). In summary, metabolic adaptations to sprint training are dictated by the length of the sprint, or number of intervals. Therefore, an aerobic athlete may benefit by performing longer intervals (>30 s), whereas a sprinter may benefit from shorter bouts (<10 s) (57).

RESISTANCE TESTING AND TRAINING

The term *muscular fitness* combines the three primary variables that can be measured: strength, power, and endurance. Resistance training is an important part of a program aimed at improving overall muscular fitness. It can be used either for performance training or to improve or maintain optimal health and wellness. Resistance training can affect health in a number of ways, including enhancing and maintaining bone density and strength, improving body composition, reducing the risk of a musculoskeletal injury, and reducing the risk of heart disease and chance of premature death. In a report from the Aerobics Center Longitudinal Study of more than 7,000 men, low muscle strength was found to be related to the risk of having the constellation of risk factors often termed **metabolic syndrome** which is a risk factor for heart disease (61). This was especially true in young men. This exemplifies the association of muscle strength (a marker for overall muscle fitness) with long-term health. Fortunately, resistance training is associated with improvement of many metabolic syndrome variables, including insulin sensitivity, blood glucose values, and blood pressure. Additionally, improved muscular fitness can reduce the risk of orthopedic injury, arthritis, pain, and other functional limitations and may reduce the number of work days missed due to illness. In general, most individuals who perform regular resistance training have a higher energy level and reduced fatigue.

A primary goal of resistance training is to improve muscular strength and endurance. This can be applied to the whole body (e.g., generally improve strength, mobility, and balance; make it easier to perform activities of daily living, reduce risk of future disease and injury), specifically to areas of need (e.g., recovery from an injury or surgery), or for sport-specific improvements. The American College of Sports Medicine (ACSM) has developed position stands with specific recommendations for resistance training for both younger and elderly individuals (5, 6). Training programs are designed with the general principles of specificity, progressive overload, individual differences, and each of the other training principles. A training program can be designed to focus on improvements in strength (i.e., maximal force generation), power (i.e., maximal rate of force development; force × distance/time), and endurance (i.e., the ability to sustain a contraction force over time), or a combination of these. To put these in perspective, the person who lifts the heaviest weight on a bench press has the most absolute strength because they lifted the most weight. If two individuals have the same strength measure on a bench press, the one who performs a single bench press in the shortest amount of time has the greatest power because she moved the weight through a similar distance faster. If two individuals bench press the same amount of weight relative to their maximum (e.g., 50% of maximum), the one who does the most repetitions has the highest endurance level because he sustained contraction force for the longest period of time. Strength, power, and endurance each can be improved with resistance training. Many other physiological factors can affect skeletal muscle performance variables, including (but not limited to) fiber size and recruitment pattern, fiber type, neural input, local circulation, and metabolism.

Before developing and implementing a resistance training program, it is generally useful to evaluate the goals of the program and muscular performance ability. With respect to goals, athletic individuals typically desire to increase specific types of strength. Types of strength are specific to the type of contraction performed. The general types of skeletal muscle contraction are isometric (i.e., static contraction without a change in muscle length) and dynamic (i.e., contraction with a change in muscle length). There are many methods for testing skeletal muscle fitness. In a typical isometric strength assessment, a cable tensiometer or a dynamometer is used to measure static strength at a specific joint angle. The most commonly used dynamic strength assessment is the one-repetition maximum (1RM), which is defined as the maximal amount of weight an individual can lift with proper form only once. Submaximal estimates can be made to predict one's 1RM. Training programs that stress improvements in strength and endurance can be designed around an individual's initial strength assessment results. The 1RM test and isometric strength assessment are also simple to use to assess training progress because each can be performed

Assessing Skeletal Muscle Activity Using Electromyography

The surface electromyogram (EMG) is a noninvasive method that assesses muscle function by recording the electrical impulses produced when skeletal muscles contract using electrodes placed on the skin over the muscles to be assessed. As a muscle cell (fiber) is stimulated by an action potential from the central nervous system via motor neurons, the fiber twitches or contracts. The signal recorded by the EMG is a combination of all the motor units, afferent nerves, and muscle cell depolarizations in the area of electrode placement. The central nervous system can control coordinated movement and force of contraction by stimulating the proper muscles and by summating twitches, respectively. The EMG indirectly provides information about motor unit recruitment, and the number of muscle fibers depolarized as the EMG signal obtained is related to the amount of action potential activity in that region. The pattern of the EMG differs depending on the type of contraction (e.g., static vs. dynamic or eccentric vs. concentric) and the duration of the contraction. Changes in fiber conduction velocity, frequency of action potentials, and the amplitude of the action potentials occur in these differing scenarios and can reflect the type of fiber activated, the rate of muscle fatigue, and changes in motor unit recruitment, respectively.

A surface EMG can provide useful information for a variety of medical purposes, including diagnosing various neuromuscular diseases and assessing pain, among others. It may also provide useful information for athletes who are interested in specific muscle activation during a particular activity and in whether a muscle is being activated (or overactivated) to compensate for a limitation in another muscle. Observation of the EMG waves during a specific contraction can provide valuable real-time biofeedback regarding both activation and deactivation of motor units. At times, athletes attempt to change the pattern of muscle use and rest during a specific activity. The surface EMG can provide feedback to help the athlete achieve this aim.

The surface EMG might also be used to assess for neuromuscular asymmetry. An example of this is from a study that used EMG in patients who had a recent anterior cruciate ligament (ACL) injury (27). The goal was to determine whether the skeletal muscles played a role in the gait (walking) adaptations that occur after injury. In particular, the researchers were interested in the flexor and extensor forces for each limb. They hypothesized that the group would have less knee flexion than normal in the injured knee and that the extensor and flexor forces would be similarly increased compared with normal. By using EMG and comparing with the noninjured leg, they reported that the patients' gait actually used less muscle force for both extending and flexing the knee during walking. This likely was due to a loss of strength in the muscles surrounding the injured knee. Understanding this type of muscle use strategy in patients with an ACL injury can help guide necessary rehabilitation strategies to strengthen these muscles and eliminate any asymmetry.

on a regular basis. For many of these tests, normative data tables can be used to compare results by sex and age and determine whether an individual's values are adequate or need improvement. See *ACSM's Guidelines for Exercise Testing and Prescription* (50) for many of these normative data sets. Results can be used to counsel an individual regarding their muscle fitness level and design or alter an existing resistance training program. When comparing changes in training status in an individual, it is typical to use absolute values such as pounds, kilograms, and number of repetitions. When comparing changes among individuals, these values should be reported relative to the individual's body weight or fat-free mass (e.g., kilograms lifted per kilogram of body weight) to adjust for differences in the size of individuals.

When performing resistance training, the principle of specificity must be adhered to with respect to the muscle groups trained and types of contractions performed. Let's

first discuss the types of contractions that can be part of a resistance training program. There are two main types of skeletal muscle contractions: isometric (or static) and dynamic. Isometric contraction occurs when the force of contraction is equal to or greater than the resistance. No change in fiber length occurs, but the fibers can either resist or apply force. A change occurs in fiber length during a dynamic contraction. The muscular force produced must be greater or less than the resistance to result in a shortening (concentric) or lengthening (eccentric) contraction, respectively. The resistance to movement for dynamic contractions can be isotonic (same force with changing fiber length), isokinetic (constant speed of contraction), auxotonic (continual changes in joint angle and speed), and plyometric (contraction begins in a stretched and loaded fiber).

In order to develop an effective training program, one must be able to identify several important variables:

- The types of contraction used for a specific movement, activity, or sport
- The muscle groups used
- The specific improvements desired (e.g., speed, power, static strength)
- The energy system(s) used (Type I vs. Type II fibers)

For example, with respect to the type of contraction performed, and reflecting back on figure 7.4, which provided an appraisal of the relative dynamic and static movements of various activities, we see that wrestling is an athletic event that might use quite a bit of isometric contraction. This can occur when the two opponents are applying equal force to each other, such as when engaged in a tie-up maneuver. Enhanced static strength can help a wrestler better resist and absorb the opponent's moves and attacks. The muscle groups of the arms, shoulders, and upper back as well as the core muscles are important for this purpose. A wrestler might also wish to develop general strength and power, which can be useful in throwing an opponent. On the other end of the spectrum, activities such as walking, running, and cycling use primarily dynamic contractions. These are general assessments of isometric and dynamic contractions that can be refined much more specifically. For instance, walking and running utilize both concentric (hamstrings during knee flexion) and eccentric (quadriceps just after heel strike) contractions. When comparing walking and running, it is apparent that there exists a difference in the ROM of these contractions due to differences in stride length; the contraction is applied in a smaller ROM for walking than for running. By evaluating a movement and determining this type information, a resistance training program can be developed to target these specific movements.

Testing Muscular Strength, Power, and Endurance

Assessments of muscle strength, power, and endurance are performed to provide information that can be used to develop a training program and to assess changes over time due to training or detraining. Unfortunately, no single test allows for the assessment of overall body strength, power, or endurance. Isometric grip strength assessed using a hand dynamometer has been shown by some to be related to overall muscle health. For instance, one study showed a moderate relationship between grip strength and elbow flexion, knee extension, and trunk extension strength in older individuals (52). Another study suggested a relationship between grip strength and health and mortality (51). However, a similar relationship has not been established in younger individuals. In college football players, the relationship between grip strength and maximal bench press strength is very weak (2). This shows that testing results are specific to the muscle group assessed. In accordance with the principles of exercise prescription, the testing methodology should mimic the activity or performance conditions as closely as possible. This means that the muscle or muscle groups involved, the function of interest (i.e., strength, power, endurance), and the contraction type (e.g., dynamic, static, eccentric) should be considered when determining the testing methodology for an individual.

Assessing Strength

It is likely you learned the 1RM strength test in an undergraduate class to measure isotonic strength (sometimes also referred to as *isoinertial strength* to reflect the constant resistance to the work being performed during contraction) (5). This is the most commonly used strength assessment. The 1RM ends by determining the greatest amount of weight that can be lifted once and not more. It typically is performed using free weights or universal machine weights, and the bench press, military press, leg extension, and leg curl are the most commonly performed assessments. The number of trials performed to progress to the 1RM can negatively affect the final value, likely due to fatigue. Therefore, performing fewer trials is likely to provide a more valid assessment than performing many trials. In weaker, diseased populations, variations such as a 5RM (a weight that can be lifted five times but not six) can be used to predict the 1RM. As for any evaluation of physical status, the individual being tested should have adequate familiarity and practice with the equipment to be used. This practice period can also be used as a warm-up period to prepare for maximal effort. However, the individual being tested must be careful to avoid excessive fatigue and should have adequate time to recover after the warm-up period to allow for optimal test results.

Advanced strength assessments often use an isokinetic dynamometer (figure 7.7). Several companies make these sophisticated machines. They typically are quite expensive, and a practitioner must have special training to use them properly. These machines often are used in physical therapy and athletic training settings and tend to be a reliable form of strength assessment. One benefit of using this device is the ability to assess bilateral (right vs. left side) and ipsilateral (same side agonist vs. antagonist)

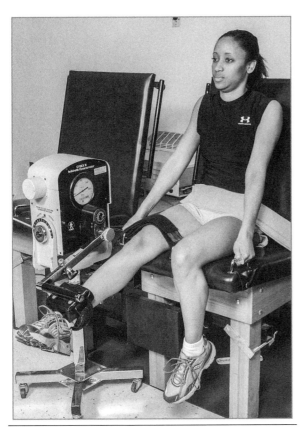

Figure 7.7 Example of isokinetic dynamometry assessment.

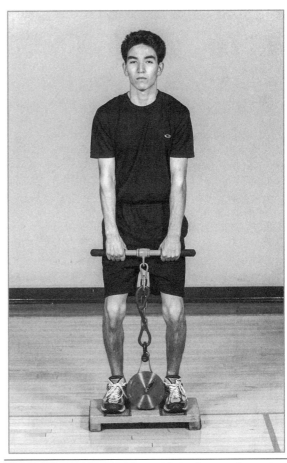

Figure 7.8 Example of isometric testing equipment.

imbalances because each limb can be tested separately. Normative population data exist with which comparisons can be made. An isokinetic dynamometer can assess both the concentric and eccentric strength of a single joint at a constant speed of contraction. The resistance of the dynamometer is equal to the muscular force applied through the ROM tested, thus allowing the limb to move only as fast as the machine's preset speed. It should be noted that the movement allowed is always angular movement around an axis of rotation. Therefore, the device actually measures torque, or the force used to rotate the dynamometer arm around its axis. This resistance is controlled by either an electric or hydraulic controller. It is important to set up the equipment so that the joint moved is aligned with the axis of rotation. Any misalignment will result in testing error. In general, the setup involves adjusting the seat and the machine arm length and using straps to hold the limb in place so that the joint stays aligned with the axis of rotation. These devices can alter the velocity of limb movement (e.g., 60, 180, or 300° per s) as well as the ROM allowed. However, a drawback is that the machines used to assess strength in this manner bear little resemblance

to daily activities or athletic moves and thus violate (to a degree) the principle of specificity. There are also many factors to consider for accurate measurement, including compensating for the effect of gravity, lever arm oscillation, and impact artifact as the lever arm reaches the end of its range. Muscle power and endurance or fatigue can also be assessed using this device.

When performing a static strength test, it is important to determine the appropriate joint angle to assess. Considering the specificity principle, the joint angle tested should closely mimic the angle used during a task of interest. Because joint angle is fixed during isometric testing, it does not reflect strength throughout a full ROM. Testing is performed by having the subject complete a maximal contraction against a resistance that is greater than that which he or she can generate. Common devices used for this type of testing are handheld dynamometers, cable tensiometers, and force platforms (figure 7.8). Many of these devices either have a self-contained computer or are wirelessly connected to another computer with software.

They can easily provide reliable data (e.g., average of several trials; total force in newtons, kilograms, or pounds) as long as the testing technique is performed properly by an individual who is well trained in using the device and performing muscular fitness tests. An isokinetic dynamometer can also be used to assess static strength. These devices can be specialized to measure the force, torque, or power exerted by an individual. It is common to perform two to four fast maximal voluntary contraction exertions with adequate rest between each. The rest period should be long enough for a perceived full recovery from the effort.

Assessing Power

Muscle power assessments determine the rate at which maximal work can be performed. Power is the product of muscle force and velocity of movement. In athletics, power assessment is interested in the explosiveness of strength.

$$\text{Power (watts)} = \text{force} \times \text{distance/time}$$

(Equation 7.1)

where force = strength, and distance/time = speed.

Power assessments generally have a stronger relationship to athletic movements than do isotonic or isometric strength assessments. Simple determinations of power include vertical and horizontal jump tests for distance and the seated medicine ball put. Advanced assessments can be made on specialized equipment. In the case of power assessment, a constant force is applied and the subject moves against the force as fast as possible. Peak muscle power occurs at about 70% of the 1RM load. At loads less than 70% of the 1RM the velocity increases, and at loads greater than 70% the velocity slows.

One factor that can affect both strength and power test results is the timing of the test with respect to previous rest, training, and competition. An individual may require several days to completely recover from a recent strenuous training or competition bout. This includes not only recent muscular strength and power activities but also recent aerobic-based activities. The general recommendation is to limit physical activity for 48 to 72 h before maximal strength and power testing. Proper setup, administration of rest, and calibration of the testing equipment are also important for valid results.

Assessing Muscular Endurance

Muscular endurance assessment tests an individual's ability to repeatedly develop and sustain a force that is less than their 1RM. A great example is the bench press test used in the NFL combine assessments. All players bench press a 225-lb (102 kg) barbell as many times as possible with a spotter in place for safety. There certainly are limitations to this test, including direct comparisons between players with differing 1RM values and the fact that no football player will ever have to provide a similar level of exertion during a game. To compare between individuals, it might be better to initially determine each person's 1RM and then perform the endurance test at a specified percentage (e.g., 75%) of 1RM. The NFL compares players in the same position (e.g., linemen, linebackers). Other examples of muscle endurance tests include the total number of sit-ups or push-ups that can be performed.

Muscle Length–Tension Relationship and Angle of Pull

An isolated muscle exerts its maximal force or tension while in a lengthened position. However, at some point the lengthening is too much and the ability to develop force is reduced. The range for peak force development is slightly more than the resting length of a muscle. The same is true for a muscle that is shortened with respect to force production. In fact, as the muscle length approaches 40% to 60% of resting length, the ability to produce force approaches zero. The physiological rationale for this effect on force production is described in the ROM section (see figure 7.17). To review, there is an optimal muscle length at which the most actin–myosin cross-bridges and subsequent level of excitation-contraction coupling necessary for force production. For most individuals, this lies in the range of 80% to 120% of resting length (figure 7.9). Shorter than 80% of resting length results in overlap of the potential cross-bridges, and longer than 120% can result in some of the filaments no longer being in contact. In each of these instances a loss of the possible maximum number of cross-bridges occurs.

The angle of pull of a muscle relative to the load exerted against is an important biomechanical consideration. Although we might conclude that a person can make a maximal lift of a weight as long as the muscle length is optimal, the angle of the joint that the muscle spans is also very important. This is because the bones act as a lever for the muscles, which act as a pulley to move a bone through the ROM of a joint. Figure 7.10 depicts several bone, joint, and muscle relationships and the force that can be produced. Note that some joints, such as the elbow, have a curvilinear response related to the joint angle. Also notice that in a single joint a vastly different force–angle relationship can occur depending on whether flexion or extension is performed.

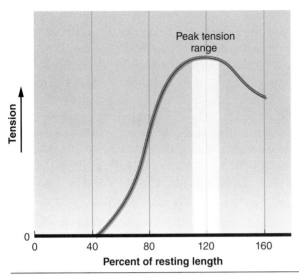

Figure 7.9 Length–tension relationship of muscle force generation.

Reprinted from M.L. Foss, S.J. Keteyian, and E.L. Fox, 1998, *Fox's physiological basis for exercise and sport,* 6th ed. (Pittsburgh, PA: William C. Brown). By permission of the authors.

These relationships (length–tension and angle of pull) are important to understand when considering a resistance training program. When performing a dynamic contraction through an ROM, the heaviest weight that can be used is one that is equal to the weight that can be lifted at the optimal length–tension and joint angle equal to the lightest weight throughout the full range. For instance, when performing flexion of the forearm (see the example in figure 7.10), this would be about 45 lb (20.4 kg) because this is the lightest weight in the angle of pull range. As noted in the same figure, this occurs at an angle of about 40°. When lifting free weights with correct form, this relationship will always be true. Resistance equipment that allows for variable resistance will permit a near-maximal force to be applied throughout the ROM of a joint. However, this type of equipment is very sophisticated and expensive because it relies on changing the angle of the resistance with respect to the changing angle of a joint and the subsequent changing maximal force production.

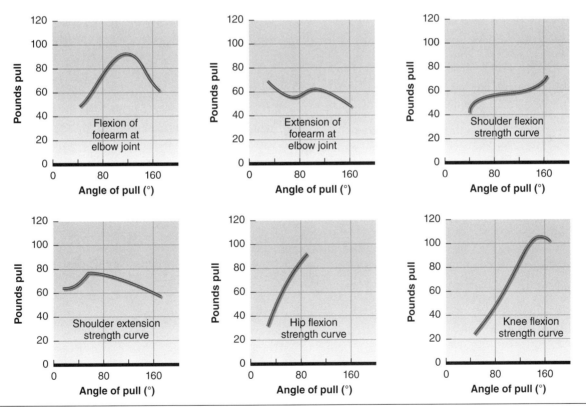

Figure 7.10 Change in maximal force generation with differing angles of pull.

Reprinted from M.L. Foss, S.J. Keteyian, and E.L. Fox, 1998, *Fox's physiological basis for exercise and sport,* 6th ed. (Pittsburgh, PA: William C. Brown). By permission of the authors.

Resistance Training Prescription

A properly designed resistance training program takes into account an individual's limitations, deficits, and individual training purpose and goals and is adjusted as needed as the individual progresses over time. Therefore, these aspects should be assessed before initiating a training program. This will allow for training to start at the proper point and for proper application of the training principles (e.g., frequency, intensity, duration, progression, individual differences) to ensure the best chance of optimal gains.

As with any exercise training program, each individual should have some level of clearance to participate. It is important to understand chronic diseases that might be acutely affected by resistance training. For example, a patient for whom blood pressure control is important (e.g., history of hypertension, previous stroke or transient ischemic attack) must be closely monitored. Those with a pre-existing injury or recovering from surgery likely can improve with a program that takes these limitations into account and that is properly designed by an exercise professional trained to deal with these types of concerns. Assessment of these issues, along with the goals of the program, can be gathered in a needs analysis (39). It is likely that several goals will be identified during a needs analysis. Some of these include increases in strength and power. Others might include improved body composition, improved flexibility, and better muscle coordination for specific activities or rehabilitation from an injury or surgery.

Lifting routines, equipment used, strategies, and so on can be quite varied. A creative strength and conditioning specialist or other exercise professional can develop a very unique program. Over the past 10 to 20 yr, options for exercises and equipment have expanded, and many books have been written on resistance and strength training. Additionally, there are countless online blogs about resistance training and many videos on YouTube demonstrating a huge variety of resistance training exercises and techniques. Although it often is useful to have this information readily available, one must be able to evaluate the safety and effectiveness of various training strategies. While creativity can be important in keeping an individual engaged and interested in a resistance training program, all exercises and routines must still conform to the principles of exercise training as previously reviewed. The following variables must be considered when developing a resistance training program to develop muscular strength.

- *Muscle actions.* Muscle actions can be dynamic (concentric, eccentric) or isometric. The ACSM position stand on resistance training provides the highest level recommendation (level A) for including all types of muscle action in a general resistance training program for any level of exerciser (i.e., novice, intermediate, or advanced) (53). Those desiring specific improvements (usually athletes) may require more of one type of muscle action. Most training should involve dynamic contractions. Isometric training is beneficial, but only in select individuals.

- *Load.* Load or resistance can be constant or variable. When considering strength gains for novice and intermediate exercisers, a beginning load of 60% to 70% of 1RM is recommended for muscular strength benefits. Experienced resistance trainers can increase this to 80% to 100% of 1RM. Determining training load requires assessment of the 1RM.

- *Repetitions.* The number of repetitions performed in a single set is directly related to the desired improvement. Typically, loads of fewer repetitions (1-6RM) are associated with strength improvement, whereas loads of more repetitions (≥12-15 RM) are associated with endurance improvement. A range of 8 to 12 repetitions generally is recommended for most individuals with a goal of general strength and endurance benefits.

- *Volume.* The total volume of training is a combination of repetitions, sets, and frequency of performance. Substantial evidence shows that a single set can improve strength, particularly in novice exercisers (15). Although most strength gains occur with a single set, adding one or two more sets (maximum of three sets) will improve strength beyond a single set, but to a diminishing degree. The frequency of training depends on many factors, including individual goals, current level of conditioning, sport-specific season (e.g., off-season, training camp, in-season, championship preparation), type of muscle action used, and the individual's ability to recover from a previous resistance training session. It typically is recommended that a muscle group not be trained on consecutive days of a week. The novice trainer should not perform a resistance training routine more than 2 or 3 d/wk. The intermediate trainer can progress to 3 or 4 d/wk. These individuals can begin to perform split routines in which a particular training bout is focused on a different area than the next training bout. This will allow more overall resistance training bouts per week while minimizing the risk of overreaching and overtraining. Using the split routine approach, an advanced individual can train 4 to 6 d/wk. Elite trainers (powerlifters, Olympic lifters, bodybuilders) may even perform more than one resistance training session on a day.

• *Exercise selection.* An important issue with exercise selection is the equipment to be used. A variety of weight equipment is available, and it generally falls into the free weight and resistance machine categories. According to ACSM recommendations, novice and intermediate individuals should use a combination of free and machine weights, whereas advanced trainers should focus more on free weights. Machine weights typically are safer to use than free weights due to less likelihood of losing control during a lift. Control is an important safety criterion, especially in individuals who are weakened or who have little experience in the control necessary to safely use free weights. Weighted devices (e.g., vests, medicine balls, kettlebells, shoes) can also be used. Additionally, exercises using a person's body weight with (e.g., suspension straps) or without (e.g., push-ups, unweighted squats) devices can be selected. The number of joints involved in an exercise (e.g., single vs. multiple) should also be considered. Both single-joint and multijoint exercises are associated with strength improvement. Single-joint exercises can better isolate a muscle group, whereas multijoint exercises may better mimic a desired movement to improve strength. Note that performing multijoint exercises with textbook technique often requires more skill. Unilateral (one side of the body) and bilateral (both sides of the body) exercises both improve strength and can be used to vary training.

• *Structure of routine.* In general, large muscle groups should be trained before smaller muscle groups, and multijoint exercises (e.g., bench press, leg press, classical Olympic lifts) should be performed before single-joint exercises (e.g., knee extension, biceps curl, triceps curl). This is based on research that suggests that the performance of large muscle groups is reduced when multijoint exercises are done later in a workout. Split routines can be used with more advanced individuals who perform more sessions per week. Additionally, higher intensity exercises should be performed before lower intensity exercises.

• *Intervals between sets.* For most individuals, regardless of training experience, a minimum of 2 to 3 min of rest is recommended when using heavy loads and primary exercises (e.g., bench press, squat, Olympic lift, and leg press). Other types of exercises with lighter loads can have a shorter rest period (1-2 min) between sets. Research suggests that the rest time between sets needs to be adequate to restore the muscle to near **homeostasis** in order to perform the next set at the intensity and pace required. In some cases, the rest interval can be an active period of movement (e.g., walking) or even another set using a different muscle group.

• *Speed and velocity.* Novice healthy populations should train at a low to moderate rate (180-240 °/s with a 1-2 s:1-2 s concentric:eccentric contraction rate ratio). The rate can be increased as an individual moves to the intermediate level and then to the advanced level. Super slow ($\geq$ 5:5 s) and fast (<1:1 s) repetitions are not as effective for increasing strength but may have a role in very specific circumstances. The ultimate rate for any individual should correspond closely to the desired contraction velocity of any activity in which there is a desired increase in strength. An isokinetic dynamometer is best for controlling repetition velocity, but repetition velocity can also be controlled during the use of free weights, machine weights, body weight, and other devices.

• *Progression.* Once an individual can easily perform one or two repetitions beyond their current recommended number of repetitions, the load and resistance can be increased by 2% to 10% of the current load. The load for smaller muscle groups should be increased at the lower end of this range, and the load for larger muscle groups can move toward the middle to upper end of this range. Because there exists a risk of overreaching or overtraining with resistance training, large increases in total volume should be avoided.

This list focuses on enhancing muscular strength with resistance training. Varying these factors can result in other types of muscular improvement, including enhancement of hypertrophy, endurance, and power. These types of training programs typically are used by those interested in bodybuilding and competition lifting and traditionally have not been used for healthy individuals interested in improving muscular health and performance. Table 7.5 compares the primary variables that make up an exercise prescription for resistance training with respect to the specific muscle characteristic (i.e., strength, hypertrophy, power, endurance) affected. The differences between these characteristics have been shown to affect the muscle in specific ways to enhance a specific physiologic response. These training variables include the physiologic effect on neural, hormonal, metabolic, and hypertrophic factors of skeletal muscle that can respond to enhance muscle adaptations.

The ACSM position stand on exercise training for healthy adults (53) states that resistance training may result in an improvement in several sport-related activities, including sprint speed, vertical and horizontal jump distance, sport-specific motor performance, agility, kicking, and throwing. It is also likely that resistance training can play a role in preventing athletic and nonathletic injuries. For instance, women have a 2- to 10-fold higher risk of ACL injury compared with men (13) due to factors such

Table 7.5 Comparing Aspects of Resistance Training for Strength, Hypertrophy, Power, and Endurance

	Strength	Hypertrophy	Power	Endurance
Muscle action	Concentric, eccentric, isometric	Concentric, eccentric, isometric	Concentric	Concentric, eccentric
Load/resistance and volume	60%-70% to 80%-100% of 1RM; 1-3 sets; 8-12 to 1-12 reps	70%-85% to 70%-100% of 1RM; 1-3 sets; 8-12 to 1-12 reps; advanced focused on 6-12RM vs. 1-6RM	Incorporate with strength training 1-3 sets at 30%-60% 1RM upper body and 0%-60% 1RM lower body for 3-6 reps; add heavy loading (85%-100% 1RM) for force increase and light loading (see above) for explosive increase	Light loads of 10-15 reps ranging up to 25 reps
Exercises	Combined unilateral and bilateral single and multiple joint	Single and multiple joint using free or machine weights	Primarily multiple joint	Variety of unilateral and bilateral exercises using single or multiple joints; multiple or large muscle groups for maximal metabolic demand
Exercise order	Split routines; large before small muscles; multiple before single joint	Similar to strength training	Similar to strength training	Vary sequences depending on level of training
Rest periods	2-3 min between sets; shorter for nonprimary exercises	1-2 min between sets for novice and noncore exercises; 2-3 min for heavier core exercises	2-3 min between core exercise sets	1-2 min for high reps (15-20) and <1 min for moderate reps (10-15); in general a reduced recovery compared with other forms of resistance training
Movement velocity	Slow to moderate progressing to moderate rate; simulate desired activity velocity	Slow to moderate for beginners and variable (slow to fast) for advanced exercisers	Fast with submaximal loading to maximize power development; max speed for highest intensity work; consider matching speed of desired movement	Moderate rate with lower reps increasing to moderate to fast rate with higher number of repsw
Frequency	2-3 d · wk⁻¹ progressing to 3-4 d · wk⁻¹; never same muscle group 2 d in a row	2-3 d · wk⁻¹ progressing to 3-4 d · wk⁻¹	2-3 d · wk⁻¹ progressing to 3-4 d · wk⁻¹ and 4-5 d · wk⁻¹	2-3 d · wk⁻¹ progressing to 3-4 d · wk⁻¹ and 4-6 d · wk⁻¹

Reprinted, by permission, from NSCA, 2015, Bioenergetics of exercise and training, T.J. Herda, and J.T. Cramer. In *Essentials of strength training and conditioning*, 4th ed., edited by G.G. Haff and N.T Triplett (Champaign, IL: Human Kinetics), 60.

as ACL size and tensile strength, influences of hormones and menstruation, greater joint laxity, and differences in quadriceps angle and posterior tibial slope. Landing with little knee flexion can increase the risk of ACL injury. During a game that includes jumping (e.g., soccer, volleyball, basketball), this type of landing pattern may be more likely to occur as proprioception is blunted during heavy fatigue. Injury prevention focused on resistance, proprioception, and neuromuscular training can result in a 50% to 60% reduced risk of subsequent noncontact ACL injury (1). USA Volleyball uses a comprehensive program of screening, training, and re-evaluation with their female players to reduce the risk of ACL injury. Their specific focus is on strength training, flexibility enhancement, neuromuscular and proprioceptive control, tissue recovery, assessment of training patterns, and biomechanical analysis. This is just one of many examples of how resistance training is used to reduce the risk of injury as well as improve athletic performance.

In general, men are stronger and more powerful than women. Much of this difference is related to the greater amount of muscle mass in men compared with women. The amount of potential maximal force generated by male and female skeletal muscle fiber contraction is actually very similar; when expressed per unit of muscle cross-sectional area, there is less than a 2% difference in strength. The difference in absolute strength is greater in the upper body than in the lower body, likely because total muscle mass is more similar in the lower body between the sexes. Compared with women, men have about 40% more muscle in the upper body and only 33% more muscle in the lower body (37). Although this difference is greatly explained by variances in height and weight, hormonal influences (particularly testosterone in men) play an important role in the amount of difference in muscle mass between the sexes. Both men and women can significantly increase overall strength and improve body composition with a regular strength- and hypertrophy-based resistance training program.

Skeletal muscle oxidative adaptations to resistance training have been greatly understudied compared with endurance training. Research has shown that 8 wk of resistance training enhanced electron transport chain enzyme messenger ribonucleic acid expression in skeletal muscle and increased the rate of fat oxidation (4). Further, mitochondrial density and tricarboxylic acid enzyme citrate synthase have been shown to increase after resistance training (64,65). However, these increases may be attributable to the increase in muscle hypertrophy and not necessarily to enhanced enzyme expression.

Table 7.6 Metabolic Adaptations to Resistance Training

Metabolic marker	Resistance training response
Citrate synthase	Increase or no change
Succinate dehydrogenase	Increase or no change
Mitochondrial density	No change
Capillary density	Increase or no change
Pyruvate dehydrogenase	Increase or no change
Hexokinase	Increase
Phosphofructokinase	Increase
Lactate dehydrogenase	Increase
Glycogen storage	Increase
Glycogen sparing	No change
Fiber type shift	Type IIx <<< Type I

Resistance training has been shown to increase glycolytic enzyme levels and activity (table 7.6). Both phosphofructokinase and hexokinase have been shown to increase after 12 wk of weight training (64). This indicates an improvement in glucose uptake and breakdown. This is a major reason why resistance training is an extremely important treatment for metabolic diseases. The ATP-PCr system may increase activity, but the adaptation is due to the increase in muscle size.

Resistance Training Techniques

Reviewing all of the exercises and training program possibilities is beyond the scope of this chapter given the many variables and many types of individuals who seek assistance with resistance training. The following paragraphs review the basics of developing a resistance training routine and provide some specific practical examples and general discussion items.

When one is considering developing a resistance training program, an initial question might be "What type of equipment should be used?" The typical comparison is free weights versus machines; however, one should also consider using other types of devices (e.g., medicine balls, suspension bands, elastic cords) and even body weight. Plenty of research data show that any of these modes of resistance can produce beneficial results. However, each type of resistance equipment has specific advantages and disadvantages that should be considered for each individual who begins resistance training (see table 7.7).

Table 7.7 Comparison of Types of Resistance Training Equipment

Equipment	Resistance	ROM	Safety	Benefits
Free weights (barbell, dumbbell, kettlebell)	• Constant within a lift; ability to alter • Maximum weight limited by weakest part of ROM of movement	Limited to range of individual	• Under individual's control unless a spotter is used to assist • Can result in injury	• Required control of weight lifted can also train supportive or postural muscles • May provide more strength gain in a short period compared with machine weights
Machine weights	Depends on machine type: • Constant resistance within a lift; ability to alter • Variable resistance on specific machines • Variable rate of movement on specific machines (i.e., isokinetic)	Limited to range of machine	Little risk of mishandling weight resulting in injury	• Provides stabilization to reduce the need to control the resistance • Can be safely used in novice and weakened individuals • Easy to change resistance
Suspension training	Limited by body weight, but resistance can vary by the angle of muscle pull (i.e., contraction)	Limited to range of individual	Generally safe, but extreme exercises may increase risk of joint injury	• Excellent for emphasizing core stability • Brings into use supporting muscles required for balance and coordination of this type of training
Calisthenics	Limited by body weight; little ability to alter	Limited to range of individual	Generally safe, but some exercises (e.g., plyometric jumping) may increase injury risk (e.g., ACL tear), particularly in those not prepared to perform	• Plyometric exercises can improve power and ability to apply force when muscle is lengthened • No cost for equipment
Elastic bands	Variable by band thickness and distance stretched during movement	Limited to range of individual	Generally safe; minor risk of impact injury if elastic breaks or is released while stretched	• Ease of use and inexpensive equipment • Lots of variety of exercises

The type of equipment to be used depends on the goals of the individual. Those desiring to increase general strength and gain the health benefits of resistance training likely can use machine weights, free weights, or a combination to perform an exercise session. This type of equipment is easy to use and can effectively be used for all major muscle groups. In those with specific goals, such as rehabilitating from an injury or surgery or improving athletic performance, other types of equipment often can be better used for a specific movement or ROM and can isolate specific muscles or muscle groups to a greater degree. For instance, a baseball pitcher with a desire to

throw at high velocities with a reduced risk of injury likely would seek out a resistance training program. Rotator cuff injuries are common in baseball pitchers, and one exercise often used is internal and external rotations. These tend to be difficult to do with a resistance exercise machine. Dumbbells can be used, but elastic resistance bands often are the equipment of choice because they easily provide varying amounts of resistance throughout the entire sports-specific movement. They are also very convenient to use and are inexpensive.

The training status of an individual must be considered. In this chapter we use the terms *novice, intermediate,* and *advanced* with respect to resistance training. Those at a novice level likely have little experience with any type of equipment and may not even understand the terms *set* and *rep*. They likely do not know proper technique for the basic resistance training movements of push, pull, bend, and squat. Exercises with several of these movements (e.g., snatch lift, squat press) will add complexity that might become overwhelming. The novice will almost certainly experience delayed-onset muscle soreness (DOMS) within several hours to days after their initial several training sessions. These are important concerns that must be part of their resistance training program design. Those with more resistance training experience might be able to perform more complex exercises and rou-

tines. Although these individuals likely are better trained and have more relative strength, power, and endurance, they will progress at a slower rate than the novice individual. The novice individual might expect up to a 40% to 50% improvement in strength, whereas intermediate and advanced individuals may have less than a 10% to 20% improvement (53). Much of this improvement will be realized in the first several of months of training.

The development of power for sport is a critical element for enhancing performance. The Olympic weightlifting movements, consisting of the two-hand snatch and two-handed clean and jerk (figures 7.11 and 7.12), have a long history. Previous reports demonstrate the ability of these two types of lifts, and their variations and

Figure 7.11　The two-hand snatch.

assistance exercises consisting of the power clean (figure 7.13), to create and exert high forces into the ground with a rapid rate of acceleration (35). These exercises greatly help athletes in developing power. The power clean is an assistance exercise used by high school and college strength coaches as well as Olympic and professional athletes for developing power in their respective sports. This variation of the clean and jerk allows athletes to learn how to accelerate a loaded barbell by driving through their individual ROM. Individuals also experience the effects of gravity once they complete the movement with the weight across their chest and thereby learn to receive

a load and react to its deceleration. The power clean, used in conjunction with the squat exercise during a synchronized strength and power training program, can improve performance in explosive lower body movements (35, 49).

Another consideration related to training status is the periodization of training. Periodization is not for everyone who performs resistance training. The person who desires a general benefit from resistance training may not vary their training program for long periods of time, particularly when they are past the progression phase and simply are working to maintain their current fitness level. Therefore, periodization typically

Figure 7.12 The clean and jerk.

Figure 7.13 The power clean.

is not important for the general public. Many athletes require a sustained level of performance throughout a competitive season (e.g., football, basketball, soccer, and hockey) that can last up to 6 mo. Other types of athletes, however, have a planned period of desired peak performance. Often these are runners or swimmers or other racing athletes who have a championship meet at the end of the season for which they desire to be at their best. In this scenario, the resistance training program will vary throughout the year with a goal of maximizing training potential leading up to the targeted competition. There may be periods of strength gain, gains in general and explosive power, recovery, maintenance, rehabilitation, posture correction, and peak performance. Each of these periods should be considered and carefully planned to coincide with an athlete's training phase in their particular sport. To plan the timing of the periods, a coach should work backward from the date of the competition to develop changes in volume, intensity, and technique. These periods are known as *cycles* and are based on Selye's general adaptation syndrome, which defines the general biological response to stress (alarm, resistance, exhaustion, or recovery) leading to adaptation or maladaptation. Figure 7.14*a* demonstrates a simple periodization model by Matveyev that is ideal for beginners or novice athletes. The term *macrocycle* refers to an annualized plan of training. Within the macrocycle are *mesocycles*, which can last from as little as 2 wk up to 2 mo. Figure 7.14*b* demonstrates a modification of the simple periodization model to include cycles of training volume and intensity. The goal of this type of periodization is to apply different resistance training

stressors at desired times to invoke adequate adaptation and subsequent improvements in muscular strength, power, or endurance.

Although isotonic concentric contraction is the most common type of muscle action performed during resistance training, other types of training can be considered. These include eccentric contractions as well as variable resistance and isokinetic (accommodating) training. One might ask whether eccentric dynamic resistance training provides any benefit beyond traditional concentric training. Eccentric contractions have different physiological characteristics than concentric contractions, including cortical (brain) activity, motor unit activation pattern, sympathetic nerve activity, less fatigability, electromyography wave amplitude at an equivalent force of contraction intensity, and the ability to generate a greater absolute force during maximal contraction. Eccentric contractions appear to require a lower metabolic demand than concentric contractions. Eccentric exercise also results in a greater level of DOMS, muscle damage, and subsequent limitations in contractile ability compared with concentric exercise. Untrained and older individuals are especially vulnerable to these negative factors. There is evidence that eccentric training improves eccentric and total strength (i.e., eccentric + concentric + isometric torque) and enhances muscle mass to a greater degree than concentric training (56). This may be why eccentric training is popular with bodybuilders. However, as the principle of specificity states, adaptations from eccentric training are very specific to the type and velocity of training contractions. Therefore, eccentric training might not be best for the majority of individuals.

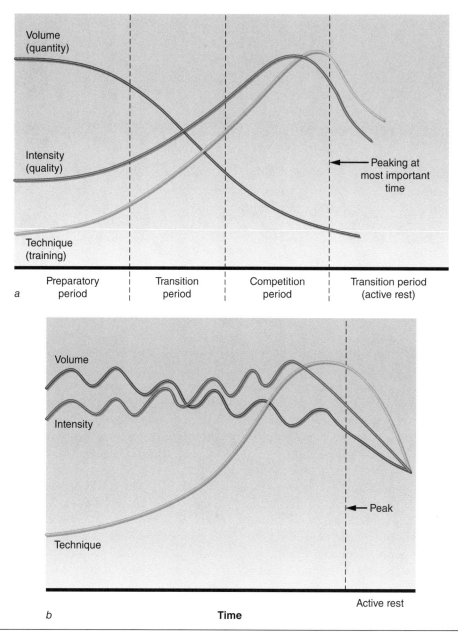

Figure 7.14 Example of the linear prioritization model for periodization.

Adapted from M.H. Stone and H.S. O'Bryant, 1987, *Weight training: A scientific approach* (Boston, MA: Pearson Custom Publishing), 124. By permission of M.H. Stone.

In specialized (and expensive) equipment for variable resistance training, the resistance is reduced at the weakest point of an ROM of a particular exercise (e.g., biceps curl) and increased at the stronger points. Equipment has been designed with lever and cam systems that change the length and radius of the lever and cam, respectively, which varies the resistance in an attempt to mimic the musculoskeletal system's ability to produce torque at various joint angles. A drawback of this type of system is that not every individual has the same biomechanics, and

thus the resistance may not mimic the ability of all individuals. Because this is different from constant resistance training, which requires the weight lifted throughout the ROM to be no greater than the weakest point, the theory is that maximal resistance can be applied throughout the range. There is evidence that this type of training might provide enhanced benefits in specific circumstances. For instance, a study of men over 65 yr of age reported that variable resistance training resulted in strength gains similar to those made using a constant-resistance training program (70). However, the variable resistance trained group had a significantly improved fatigue test result, which may be an important factor in the functional capacity of the elderly. Chains and elastic bands are another method of variable resistance. However, unlike machines with levers and cams, the resistance with these devices increases throughout the ROM of the movement. Figure 7.15 provides an example of elastic bands in use.

Isokinetic resistance training uses special (and expensive) equipment that maintains a desired constant speed throughout the ROM of an exercise. The speed is measured in degrees of movement per second and can range from 0 to 360 °/s. The amount of force applied, no matter how little or great, does not make the lever arm of the machine move faster. Theoretically, a motivated individual applying constant maximal force will provide maximal resistance throughout the ROM of the muscle group and thus develop strength throughout the entire range. This type of training can be used for specific athletic movements that require a desired speed of contraction, such as throwing and kicking movements. However, as with most movements, the rate of contraction varies throughout the range of movement. Therefore, holding the speed constant throughout the range may not provide the specificity needed to mimic the desired movement.

In some cases, a constant-load resistance that allows the speed of contraction to more closely mimic the real-life action may provide benefit. Some evidence also shows that ligament and tendon strength may not be adequately developed with any type of machine training due to the stability the machine provides compared with free weights, bands, or suspension straps.

Core strength and stability are important facets of an overall resistance training program. Although the core typically is thought of as the abdominal muscles, it actually refers to a group of muscles in the trunk area that act as a foundation for limb movement to achieve maximal strength and stability. These include the abdominals (rectus and transverse abdominis), the multifidus, the external and internal obliques, the diaphragm, the erector spinae, and the muscles of the pelvic floor (levator ani, coccygeus). The latissimus dorsi, gluteus maximus, and trapezius muscles also have a supporting role in core strength and stability. The core also stabilizes the spine, which is important for proper biomechanics and for reducing the risk of injury. Anyone wishing to improve their health or performance likely would benefit from focusing on core muscle training. Many core exercises can be performed without any equipment. These include planks (normal, modified, side), abdominal crunches, and

Figure 7.15   Elastic band variable resistance exercises.

bridges, among many others. Core training exercises can be performed as often as daily and should be incorporated into an overall resistance training program.

The principle of plyometric exercises is that the maximal concentric contraction is stronger when it is preceded by an eccentric contraction of the same muscle, thus utilizing elastic energy stored from the eccentric contraction. Plyometric training is performed primarily to improve explosive concentric contractile strength of the legs. The arms can also be trained using the plyometric technique. Athletes are interested in this type of explosiveness for such events as sprint running, jumping, weightlifting, and starting in both running and swimming. This method of training was developed in the 1960s in Russia by Dr. Yuri Verkhoshansky, who called it "shock training." Fred Wilt, a U.S. track coach in the 1960s, coined the term *plyometrics*. Lower body plyometric training involves primarily an exercise known as the *depth jump*, where upon landing the elastic energy developed by eccentric extension of the quadriceps is followed by an immediate maximal concentric contraction, resulting in a jump for maximal

height and distance. The key of the depth jump is that the elastic energy stored during the eccentric contraction is not dissipated but rather stored and used in the explosive and immediate concentric contraction. Due to the high forces of landing before the explosive contraction there is a risk of injury, including ACL rupture, ankle sprains, and severe DOMS. Injury risk can be reduced by making sure that a proper progression of jump exercises and increasing volume of training are performed. Good quadriceps and hamstrings strength is important. Nonplyometric jumping exercises can be used to prepare novice athletes for plyometric training. Many different types of jump training exist (see "Example of Progression of Jump Training in Preparation for Plyometric Depth Jumps"). Working through these exercises over several months and performing 5 to 10 maximal repetitions of each for 1 to 2 sets about twice per week will prepare an athlete for true plyometric depth jump training (see "Example of Depth Jump Training").

In addition to traditional resistance training methods, there exist a multitude of advanced or alternative

Example of Progression of Jump Training in Preparation for Plyometric Depth Jumps

Key point: Use maximal effort for maximal recruitment for all exercises. Perform 5 to 10 repetitions (depending on the athlete's condition), and travel no more than 15 yd (13.7 m) for exercises performed over a distance.

1. *Skipping:* Keep the body low to the ground. Skip slowly at first and then increase speed.
2. *Power skips:* Focus on a combination of maximal vertical height and forward distance; distance and speed are not important.
3. *Double-leg jumps in place:* Make sure that the legs are straight and that the toes are pointed on takeoff. Land close to the arch of the foot on the ball of the foot and the heel simultaneously.
4. *Double-leg jumps for height and distance:* Same as number 3, but direct forces slightly forward.
5. *Tuck jumps:* Jump up as high as possible while keeping the knees up as high as possible in the tuck position.
6. *Single-leg jumps in place:* Extend the ankle joint as much as possible during jump. Other leg is slightly flexed at the knee so it is off the ground.
7. *Single-leg jumps with forward movement:* Same as number 6, but direct jumping forces slightly forward. You should be able to land 12 to 18 in. (30.5-45.7 cm) in front of the previous foot plant position.
8. *Ankle jumps:* Similar to number 3, except fully extend the legs on takeoff and do not bend the knees more than 10° to 20° on landing.
9. *Double-leg bounding:* Keep the body vertical and the center of gravity low to the ground.
10. *Leaping using alternating legs:* Keep the body low to the ground, unlike in bounding.

EXAMPLE OF DEPTH JUMP TRAINING

Depth jump training requires a platform from which to step off. For the novice the height should be only 8 to 12 in. (20.3-30.5 cm) from the floor. As training advances the height can be gradually raised to as much as 30 to 40 in. (76.2-101.6 cm). The following are step-by-step procedures for performing a depth jump.

1. Stand on the platform with the feet together and the toes over the edge of the platform.
2. Step off the platform and land toward the forefoot.
3. As the feet hit the floor, immediately counteract the force with a maximal explosive vertical jump.
4. The arms may be used to propel the body upward.
5. Perform three to six jumps initially and stop immediately upon fatigue.
6. Build to two to three sets of ten jumps

---

training methods. Some of these are more popular than others. Most have not been heavily researched, and thus much of the information about them is anecdotal and elicited from those who have utilized them by trial and error in a training setting. The following is a list of some of these methods.

- *Super slow:* This requires the individual to perform lifts at a much slower speed than normal. An example for a bench press is a 10-s concentric lift followed by a 4-s eccentric return of the weight. The theory is that longer tension times for the muscle may enhance strength and endurance. This method has shown an improvement of about 50% over the traditional method (72). One criticism is that the movements are not speed specific to most daily or athletic events.

- *Maximal effort:* The individual performs basic lifts (e.g., clean and jerk, snatch, bench press, squat, deadlift) with only 1 to 3 repetitions per set. This method is popular among power and Olympic-style weightlifters. It can improve strength but has not been shown to be effective for producing hypertrophy. It is not recommended for novice or intermediate individuals, and if used excessively it may increase the risk of injury and plateau (accommodation) or overtraining.

- *Dynamic effort:* The individual moves very quickly from lifting light resistance to lifting moderate to heavy resistance. The theory is that a sudden increase in force development may enhance strength benefits.

- *Super sets:* This method typically involves performing two or three different exercise sets in a row. It often utilizes the same muscle group or may involve the antagonist muscle group.

- *Rest–pause:* The individual performs as many repetitions as possible until exhaustion, rests only a few seconds, and then again performs as many repetitions as possible. One variation is to switch to a lower weight during the pause.

- *Partials:* The individual performs only a portion of the full ROM with the goal of focusing on that portion of the movement. This technique has been proposed for competitive lifters who have a sticking point that they would like to improve. It also may benefit those with an injury or pain in a particular part of their ROM. An isokinetic dynamometer can be useful for this method.

- *Forced repetitions:* A spotter is required for this method. The spotter assists the lifter with the last one to several lifts during an exhaustive set to failure. This can be done with a resistance that is heavier than typical. Data show that this method may not be useful for strength gains (22). Some have argued that using a resistance that is heavier than normal may invoke a Golgi tendon reflex response and inhibit voluntary muscle activation.

There certainly are many other methods of training, including routines using alternative devices such as dumbbells and kettlebells, suspension training, core strength and stability training, balance training, and agility training. Additionally, there are specific workouts with catchy or branded names (e.g., Bootcamp,

CrossFit®, Insanity®) that use a variety of exercises in specific routines and that may include aerobic training practices. It is unlikely that any single routine is perfect for all the benefits any single individual desires. Specificity is vitally important. However, variety may be as important for avoiding accommodation, or a lack of continuous adaptation.

Expected Benefits of Resistance Training

Because the types and goals of resistance training are quite varied, it makes sense that the physiological changes associated with this type of training also vary. Although all nonathletic individuals will likely improve any component of muscular fitness with a regular training program, the training status of an individual at any time will affect the rate of progression. Those who have little previous resistance training experience typically have a very fast rate of improvement. The following lists the potential rate of increase in muscular strength for individuals at different training levels.

- Novice: 40%
- Intermediate: 20%
- Advanced: 16%
- Very advanced: 10%
- Elite: 2%

Note that the total time of training in these groups ranged from as little as 4 wk to as long as 2 yr (39). The better trained an individual is, the less absolute change in strength, power, and endurance can be expected. Also, everyone has a genetic, physiological, and biomechanical limit to their ability to improve. Improvement beyond 4 to 12 mo of regular resistance training will occur at a significantly slower rate than improvement with initial training. However, in advanced and elite competitive athletes, these smaller changes may translate to significant athletic improvement, particularly in events in which small increments in

---

Resistance Training

Although resistance training typically is thought to be primarily for those who are young and athletic and attempting to improve athletic performance or look very fit, resistance training is also increasingly recognized as an excellent prevention strategy and treatment for many types of chronic disease. Resistance training lowers blood pressure values in those with hypertension, can maintain bone integrity in those prone to osteoporosis, improves health outcomes in those with chronic kidney disease, and helps maintain lean tissue in those undergoing weight loss, among many other benefits.

Progressive resistance training gained traction as an adjunct treatment in those treated for cancer or who are cancer survivors. A review and meta-analysis demonstrated the following benefits of resistance training in those with breast cancer (18):

- Reduced risk of breast cancer–related lymphedema
- No change in arm volume in those with established lymphedema
- Improved upper and lower body strength
- Positive effect on health-related quality of life in those not currently on cancer therapy

Interestingly, until as recent as 2003 there were no trials assessing resistance training in these patients. This was largely due to the belief that the threat of immune system dysfunction from potential fatigue and exhaustion were too great a risk in these patients coupled with the idea that resistance training might negatively affect those who had lymph tissue removed during surgery and were at risk for lymphedema (fluid buildup) in their arms. In just over a decade, resistance training has become widely recommended as a standard part of exercise therapy in the breast cancer population (50).

performance might mean the difference between first and second place.

Gains in muscle fitness occur through a variety of mechanisms. Table 7.8 lists potential physiological adaptations and benefits that result from resistance training.

RANGE OF MOTION

Flexibility is synonymous with the ability to move a joint or series of joints through an ROM. The ability to move a joint through its potential normal ROM has benefits for both health (e.g., mobility and balance, pain control, injury prevention) and performance aspects of life. Static flex-ibility is the amount of ROM of a joint held in an extended position. Static flexibility can be assessed passively when the person stretching (or an assisting person or object) holds the joint in an extended position. It can also be assessed actively when the person's own muscles (agonists and synergists) hold the joint in an extended position. Dynamic flexibility is defined as the resistance of a joint during a controlled movement. Resistance is applied by a combination of relaxation of the extended (antagonist) muscle and contraction of the agonist muscle. The ROM of a joint can be readily measured using several types of tools (e.g., sit-and-reach box, tape measure, goniometer, inclinometer, Leighton flexometer; see figure 7.16), and a

Table 7.8 Chronic Physiological Adaptations and Benefits Resulting From Resistance Training

Adaptations	Benefits
Enhanced muscle mass and cross-sectional area	Effect mostly on Type II fibers with potential hyperplasia; may have Type I effect depending on program regimen Increased force and power Increased lean:fat body composition ratio Improved body image Enhanced muscle tone Elevated resting metabolic rate Reduced risk of metabolic syndrome and hypertension
Enhanced motor unit recruitment, firing rate, and synchronization	Fast-twitch selective recruitment, improved force production, and rate of force development Improved movement coordination
Enhanced ATP and glycogen storage	Enhanced energy use for increased force and power
Elevated anaerobic muscle endurance	Note that oxidative (aerobic) capacity becomes reduced (mitochondria and capillary, myoglobin decreases)
Improved blood vessel tone	Reduced hypertension risk Enhanced blood supply
Improved activity performance	Easier to perform activities of daily living (ADLs) Improved athletic performance
Biochemical changes	Possible increases in testosterone and growth hormone, enhanced insulin sensitivity, and reduced secretion Enhanced glucose tolerance
Reduced non-HDL blood lipid values	Lower LDL
Improved endurance	Training type specific
Connective tissue improvement	Enhanced bone mineral density and strength, tendon and ligament size and strength
Immune system effects	Increased white blood cells (leukocytes) and cytokines (e.g., TNF)
Other possible effects	Improved performance of activities of daily living and sport, reduced low back pain, elevated resting metabolic rate, reduced osteoporosis and sarcopenia risk, enhanced flexibility
Improved mortality	Live longer

HDL = high density lipoprotein; LDL = low density lipoprotein; TNF = tumor necrosis factor

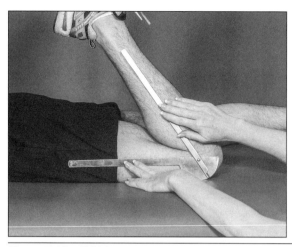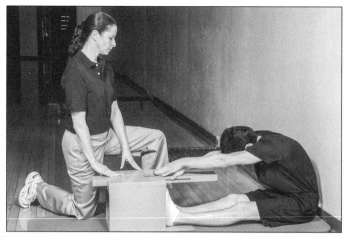

Figure 7.16 ROM measurement devices (goniometer and sit-and-reach box).

training program can be designed when deficits are determined. Deficits may be identified by comparing results against published normal standards (50). ROM can be measured actively (i.e., subject uses own muscle to move the joint) or passively (i.e., another person moves the joint without the subject's assistance). As with any other form of exercise training (e.g., aerobic, anaerobic power, strength), the principles of specificity, progressive overload, reversibility, individual differences, and training variation apply. A typical general flexibility training program might begin with 5 to 10 min of static or dynamic stretching followed by some low-level aerobic exercise. Most people would benefit from daily flexibility training, even in joints that are not deficient.

Some believe that stretching before exercise training might put an individual at risk for musculoskeletal injury and instead suggest performing stretching exercises only after an exercise training session. Additionally, some forms of stretching (e.g., ballistic stretching) and poor stretching technique may increase the risk of injury in select individuals. Factors that influence the ability to move through an ROM include the following:

- Individual bone and ligament structure
- Previous injury and scar tissue
- Subcutaneous fat
- Sex (women typically are more flexible than men)

Although it is universally accepted that ROM training to improve or maintain flexibility is important for both performance and general health, this topic tends to get

little attention in exercise physiology texts and courses. Although no data apparently exist, anecdotal evidence suggests that flexibility training often is overlooked in performance training despite its potential positive influence.

Over the past several years, considerable data have been published on "sitting disease." Much of this information has focused on the associated risk of disease and death from too much sitting, likely due to the high rate of an overall sedentary lifestyle. A study commissioned by the Chiropractors' Association of Australia found that office workers in Australia sit for about 15 h/d when considering their time spent commuting, performing office work, using a computer at home, eating, watching TV, and so on. They also found that most stand or walk for only about 73 min/d; the remainder of the day was devoted to sleeping. Poor sitting posture of the back and hips as well as prolonged hip flexion and ankle dorsiflexion can add to the normal loss of ROM and function associated with aging. Aging-related loss of flexibility likely is attributable to multiple factors, including a loss of elasticity of connective tissue. Although loss of ROM typically is accepted as a natural process of aging, data suggest that normal ROM can be maintained in a healthy aging population and that any loss in ROM should be considered abnormal and treated as such (54). This recommendation is based on the assessment of active ROM measures in a diverse group of 1,892 healthy individuals. For all goniometer measures except hip extension, there was a less than 20% difference in ROM values among the 25 to 39 yr, 40 to 59 yr, and 60 to 74 yr age groups. This is important because

Common ROM Assessments

Joint	Assessment method
Shoulder girdle	Goniometry
Elbow	Goniometry
Wrist	Sit-and-reach box, tape measure
Trunk and back	Goniometry, lift-and raise test
Hip	Goniometry
Knee	Goniometry
Ankle	Goniometry

poor ROM contributes to the loss of mobility that occurs with aging.

Posture certainly can affect health and can be related to increased soreness, headaches, back and limb pain, and fatigue. Some also believe that posture can negatively affect digestion and breathing and can be related to the development of high blood pressure. Exercise routines that focus on ROM movements, such as yoga and tai chi, have been noted to increase flexibility measures and other functional fitness measures in a wide range of individuals. Normal or excessive flexibility (i.e., hypermobility) generally has positive implications for those in athletics or the performing arts (9). This section focuses on issues related to flexibility and ROM.

Normal Versus Abnormal Flexibility

When considering flexibility as a potential hindrance to mobility and performance, one first has to define normal and abnormal flexibility. See the "Common ROM Assessments" box for a list of joints that commonly are assessed for ROM. See also the Human Kinetics text by Norkin for specific instructions on flexibility assessment methods (47).

Each of the joints listed in "Common ROM Assessments" moves in multiple (two or more) directions, which may include extension, flexion, adduction, abduction, rotation, and circumduction. Although some may visually determine the level of ROM of a joint, it is most accurate to measure the ROM. Following well-established standardized procedures, one can assess the ROM of moveable joints and compare the results with normative data. However, even when established criteria are followed, assessments can be spurious when the performer of the assessment either does not follow the criteria or is inexperienced at applying the measurement (58). It is important for the individual making the ROM measurement to be well versed in muscle, bone, and joint anatomy; to be well practiced; and to carefully apply specific assessment technique (47). The gold standard for ROM assessment is radiographic images (i.e., X-ray). However, most normative data sets have not used this technique as a standard assessment or to compare against the results of the individual performing the ROM measurement (23). Goniometry is the most commonly used method of ROM assessment. It has a high intrarater and interrater reliability when used with standardized methods by those who are well trained (54). This means that measuring ROM two or more times provides consistently similar values. However, these measurements are not exactly the same between assessments. Boone and Azen (14) suggest that there is a standard deviation of 4° between tests. Based on this measurement error, he states that a change of at least 6° between measures is required to accurately detect a change in ROM. When using any other device for ROM measurement, proper procedures must be followed to reduce the risk of measurement error associated with the device.

When determining normal and abnormal ROM, it is important to understand the population in which the determination will be made. There are many normative data sets published for ROM (50). These data are designed to provide ROM values that are considered the usual or ideal standard. However, there are anticipated differences in the population based on race and sex, and specific data sets that present these differences might

Table 7.9 Factors Related to Poor Joint ROM

Acutely occurring	Developing over time	Disorders	Structural
Dislocation	Ankylosing spondylitis	Cerebral palsy	Fat tissue
Fracture	Past fracture	Congenital torticollis	Swelling
Infection	Osteoarthritis	Muscular dystrophy	Adhesions or fibrosis
Legg-Calve-Perthes disease	Rheumatoid arthritis	Stroke	Myofascial tightness
Brain injury	Age	Spasms	Tissue extensibility
Pain	Poor posture	Loss of strength	
Tendon rupture	Joint effusion	Multiple sclerosis	
Cramping	Bony block		

be important. Additionally, for athletes there may be sport-specific ROM values that are associated with performance level and risk of injury.

Limits of Flexibility

Certain individuals may have very limited ROM in a specific joint or systemically. A variety of factors are associated with or directly result in a poor ROM. Table 7.9 lists many of the common causes of poor joint ROM. Those listed in the *acute* column are considered factors with a sudden onset. Those listed in the *developing* column are current or previous issues that lead to the development of reduced ROM over time. Those listed in the *disorders* column are particular issues that are related to a loss of ROM. Finally, those in the *structural* column are specific joint-related anatomical limitations that affect ROM.

One might also consider the active (i.e., muscles, tendons), passive (i.e., ligaments), and neural factors that affect the ability to move through an ROM as well as the **distensibility** of the skin tissue. At the tendon and muscle levels are multiple competing properties such as connective tissue **elasticity** and **plasticity** (or **extensibility**) as well as **proprioceptors** in the skeletal muscle that sense and react to changes in muscle and tendon length. When a muscle is stretched, the protein filament titin is important in returning the muscle to a normal length. This occurs at the Z disc (also known as the Z *line*) and the M line, which anchor the thick (myosin) filaments within muscle fiber. Within connective tissue, including tendons, are elastin fibers that provide the elastic ability for a stretched tendon to return to its normal length. If connective tissue is stretched beyond the capability of the

elastin fibers, it can remain permanently elongated; this can reduce the stability of a joint by adversely increasing its ROM. When a muscle is properly stretched, the complex plasticity regulation of skeletal muscle uses intracellular signaling pathways that allow the muscle to permanently change size in response to a demand. This is the goal of flexibility exercises designed to increase an individual's ROM.

When a skeletal muscle is stretched, several proprioceptors monitor muscle length (muscle spindles) and muscle tension (Golgi tendon organs). The muscle spindles activate the stretch reflex, which is an involuntary response to an external stimulus stretching the muscle beyond its capacity. This reflex, resulting in muscle contraction, is activated only during dynamic movements and is inhibited during static stretching. Thus, static stretching allows the achievement of a more elongated muscle. However, static stretching may not result in proper specificity to muscle length during movement. The Golgi tendon organs, which are in close proximity to the musculotendinous junction, can cause a reflexive relaxation (autogenic inhibition) of the stretched muscle. Reciprocal inhibition can occur in the muscle opposite of that being stretched through the same Golgi tendon organ–mediated reflexive relaxation by maximally contracting the opposing muscle. These types of inhibition can be used in specific types of stretching (i.e., proprioceptive neuromuscular facilitation and contract–relax–antagonist contract).

The continuum of skeletal muscle damage can range from DOMS to strain injury and potentially to complete rupture. The hallmark of these injuries is an eccentric contraction that creates muscle tension while the muscle is lengthening. In a laboratory setting using a rabbit model, a combination of muscle stretch and contraction with a force several times greater than that used during a maximal isometric contraction was required to obtain a muscle strain failure (28). With DOMS, microscopic damage occurs to the contractile elements of the muscle fiber, particularly in individuals who are unaccustomed to exercise. Muscle strain can be caused by a stretch that is beyond the plastic capability of the muscle. The result is partial-tear, likely microscopic disruption of the muscle–tendon location. There are levels of strain, and a minor strain often precedes a more disabling injury. This type of injury has been demonstrated in water skiers who, during the start position, straighten their knees too soon and are pulled into an excessive hip flexion position with extended knees. The hamstrings contract in an extreme stretched position, causing an excessive eccentric contraction; this can often cause hamstring strain injury. Injuries such as this are more likely to occur when an individual is fatigued or if they have recently overused the muscle. Prevention strategies, which include performing a dynamic warm-up before an activity and enhancing overall flexibility, can help reduce the likelihood of injury. The purpose of the warm-up is to increase body temperature and blood flow and activate normal ROM. Regarding stretching to enhance flexibility, it is important to understand that excessive forces generated during stretching (i.e., beyond 70% of maximal isometric force capability) may lead to a greater risk of strain injury compared with stretching at a reduced force (e.g., 50% or less) (66).

There certainly is evidence that stretching may not reduce the risk of muscle strain (74) or DOMS (33). However, this may depend on the type of sport or physical activity being performed. In activities that require a high amount and intensity of stretch–shortening cycles such as jumping and landing (e.g., basketball, soccer, gymnastics, dancing), a stretching program initiated with a goal of increasing tendon viscosity and compliance reduces the risk of injury. In other activities such as running, cycling, and swimming, which do not involve high-intensity stretch–shortening cycles, a regular stretching routine does not appear to reduce the risk of injury as much. Additionally, multiple studies have suggested that acute static stretching before a resistance training bout can negatively affect the ability to generate maximal force and thus can affect performance. For instance, Fowles et al. (25) had subjects perform static stretching of their ankle plantar flexors and reported a 28% reduction in their ability to produce maximal force immediately afterward. Serra et al. (63) reported a negative effect of static stretching that was similar in the lower and upper body for both experienced and inexperienced resistance trainers. This negative effect may occur if the elongation of the muscle fibers results in a loss of the optimal length–tension relationship and thus fewer potential actin–myosin cross-bridges, resulting in a reduced strength of contraction termed the ***stretching-induced force deficit*** (figure 7.17). Dynamic stretching has been recommended to promote strength and power performance. Ryan et al. (59) reported a

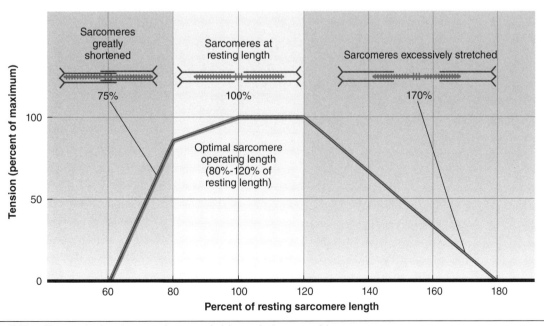

Figure 7.17 Theoretical actin–myosin cross-bridges during stretching.

beneficial effect of 6 to 12 min of dynamic stretching on flexibility and performance measures of vertical jump height and velocity. However, force reductions have also been reported during dynamic stretching (19) and may result in greater poststretching force losses (concentric and eccentric hamstrings contraction) in those with poor flexibility compared with those with better flexibility (7).

In practice, it is almost universally recommended that all individuals—whether competing in athletic events or simply desiring to remain healthy, mobile, independent, and injury free—perform a regular flexibility enhancement (i.e., stretching) routine. Despite the previously reported potential negative effects, overall the evidence points to a reduced risk of injury and the promotion of performance in those who stretch regularly. However, the implementation of a stretching routine with respect to type of stretching, forces generated, joints and muscle groups emphasized, and timing is specific to each individual. It is likely that the type, timing, and duration of stretching for maximal benefit varies among individuals on the basis of age, sex, previous stretching experience, current flexibility level, and other factors.

Although no specific public health or professional society statements on flexibility and stretching currently exist, the topic is presented briefly in the general recommendation for exercise and physical activity for healthy adults and for older adults (50) (table 7.10). Recommendations for athletes must be sport and individual specific. Sport-specific studies of stretching before practice or competition have not yet been performed. Anecdotally, many coaches have begun recommending that athletes refrain from static stretching or reduce the stretching time to less than 90 s directly before competing in sports such as swimming. Others promote 5 to 10 min of dynamic warm-up and stretching before a practice or competition.

Hypermobility: The Good and Bad

Some individuals demonstrate excellent natural flexibility and can move joints such as the shoulders, elbows, wrists, and knees through a very large ROM. A portion of these individuals are considered hypermobile, meaning than their ROM is excessive during normal joint

Table 7.10 ACSM Recommendations for Flexibility Training in Nonathletic Individuals

Who	Frequency	Intensity	Type	Muscles	Volume	Other
≤65 yr	≥2-3 times/wk	Feeling of tightness or slight discomfort	Static holding 30-60 s Proprioceptive neuromuscular facilitation for 3-6 s contraction at 20%-75% maximum voluntary contraction followed by 10-30 s assisted (passive) stretch	Each major muscle–tendon group	60 s for each exercise; 2-4 repetitions of each exercise	Best performed when muscle is warmed up with prior aerobic exercise or use of heat packs or hot bath; progression process is unknown
>65 yr	≥2 times/wk	5-6 on 0-10 subjective rating scale	Static (avoid ballistic) holding 30-60 s	All major muscle groups	Up to 60 s for each exercise; repeat up to 2 times	Can also perform physical activities that maintain or increase flexibility in as many muscle groups as possible

Based on Pescatello (50).

Criteria for Joint Hypermobility

This five-step assessment was developed by Carter and Wilkinson (17). One point is awarded for each item that is positive, and a score of 4 points is considered positive for joint hypermobility. They found that 7% of a group of 300 children aged 6 to 11 yr scored positive for joint hypermobility.

- Passive hyperextension of the fingers so that they lie parallel to the extensor aspect of the forearm
- Passive apposition of the thumbs to the flexor aspects of the forearms
- Passive elbow hyperextension beyond 10°
- Passive knee hyperextension beyond 10°
- Excessive range of passive dorsiflexion of ankle and eversion of foot

Beighton and Horan (12) modified this test by changing the first and last items, respectively, to the following:

- Passive extension of the fifth (little) finger beyond 90°
- Forward trunk flexion with the knees fully extended and palms of the hands resting flat on the floor

movement or that they are able to move their joints in an abnormal direction (or both). This may occur in a single joint or in many joints. This type of ability typically is genetic, and for many it does not relate to an increased risk of injury. Women are up to three times more likely than men to have joint hypermobility, and as much as 10% of the general population has some degree of hypermobility. Some pathologic etiologies can cause joint hypermobility, including (but not limited to) **Marfan syndrome**, **Ehlers-Danlos syndrome**, **Achard syndrome**, **osteogenesis imperfecta**, **homocystinuria**, and **hyperlysinemia**.

Normal ROM and hypermobility can be important factors of injury prevention for the general population who performs exercise for health or recreational reasons. Good joint flexibility also is thought to play an important role in reducing the risk of injury and enhancing performance for participants in many different sporting activities. This includes those who require sudden high-velocity, high-force (i.e., high power) movements such as jumping and throwing as well as those who apply force through a long ROM, including swimmers and runners. There is, however, evidence that negative consequences may be associated with hypermobility, including an increased risk of trauma, recurrent dislocations, effusions, and possibly premature osteoarthritis. Additionally, enhancement of ligament length such as with a sprain can lead to joint instability; likewise, a ligamentous injury (i.e., partial/second degree or complete/third degree

rupture) can lead to excessive joint **laxity** and instability, including potential hypermobility. Injury to a ligament is the result of a disruption of the ligament matrix. This includes platelet interaction with matrix components, which changes their shape and can result in the formation of a blood clot. Through healing, the ligament matrix is remodeled but can lack its original shape and strength and is subject to future injury. It is rare that a stretching routine designed to enhance flexibility results in a ligament injury; these injuries typically occur during a specific movement such as landing after a jump or a slip and fall. However, there is a small risk of ligament injury in a stretch that moves beyond the limit of a joint (see "Criteria for Joint Hypermobility").

Flexibility Training Practices and Theory

Assuming that increased ROM is desired, one might question the best method for enhancing ROM. It is intuitive to believe that a regular stretching routine would result in a greater ROM specific to the joint(s) involved. But does this effect really occur? In general, studies that have assessed the effects of stretching have not been of high quality (e.g., lack of a control group, nonrandomized, poor adherence). Several authors have summarized various types of stretching-related studies. For instance, Decoster et al. (21) evaluated 28 studies when investigating the effect of stretching on hamstring flexibility and found that only 22% of the studies were performed with appropriate research methods. Harvey et al. (31) reviewed 13 studies and deemed that only four were of good enough quality to evaluate the lasting effect of static stretching. They concluded that a regular static stretching routine resulted in improvement in ROM of at least 8° that lasted for at least 24 h after stretching ceased. One question about this ability to increase ROM is the mechanism by which it occurs. The answer is complex and likely involves multiple physiological adaptations (71). Currently, several mechanisms for increasing muscle extensibility are proposed; these include viscoelastic deformation, plastic deformation, increased sarcomeres in series, and neuromuscular relaxation. Table 7.11 provides a brief review of each. As noted in the table, these mechanical theories have been debated. The sensory theory, which suggests a reduction of perceived discomfort that might limit ROM, currently is the most oft-cited mechanism for increased ability to move a joint through an increased ROM as the result of stretching.

A variety of types of stretches can be performed for training. To review, those that are held at a point at which no joint movement or change in muscle length occurs are termed *static*. Those that involve motion throughout the stretch are termed *dynamic*. Table 7.12 lists the common types of stretching methods along with a brief description.

Considerations for using a stretching program are as varied as the types of stretches available. There is considerable controversy regarding the need for a stretching program. The literature varies with respect to the lasting effects on increasing ROM, reducing injury risk, and improving performance. For instance, some have advised avoiding all types of static stretching due to the possibility of permanent ligament and tendon laxity or **viscoelastic creep** (where over time these structures lose their elastic properties and remain elongated and possibly destabilize the joint; e.g., knee or shoulder) (62). To illustrate this concept, Baumgart et al. assessed soccer players who regularly performed a series of lower body static stretching (11). They reported a significant increase in anterior tibial translation. It is known that an increase in anterior tibial translation places an individual at risk for injury due to reduced passive and active knee stability. The authors suggest that this was a consequence of the response of the ligaments to stretching, resulting in reduced elasticity due to viscoelastic creep. Common daily activities including prolonged sitting (e.g., deskwork) may also result in a loss of viscoelasticity (36).

However, it is generally accepted that each of these types of stretching increases ROM. Specificity is important when designing a flexibility program, and variables to consider include age, training status, injury (e.g., muscle imbalances, recent fractures, joint replacement) and disease (e.g., arthritis, osteoporosis) status, goals of increased ROM, timing with respect to activity and exercise, and the current ROM of all affected joints. As with any training program, a flexibility program must consider the FITT principle: frequency, intensity, time, and type. Other considerations include timing of the routine with respect to activity and performance, if applicable, and

Table 7.11 Mechanical Theories of Enhancing Muscle Extensibility

Mechanical theory	Brief review
Viscoelastic deformation	Both muscles and tendons have elastic and viscous (liquid) properties. Studies have refuted this theory as muscle and tendon length continues to increase as stretch duration increases.
Plastic deformation	Suggests that muscle and tendons change length permanently due to stretching when a stretch is sufficient to pull muscle and connective tissue past its elastic limit and that the muscle does not return to its prestretch length when the stretch force is ended. However, many studies have contradicted this theory.
Increased sarcomeres in series	Based on animal studies of muscle immobilization in an extremely lengthened position that resulted in more sarcomeres in series; however, it does not appear that these studies actually increased muscle length and thus ROM. A loss of sarcomeres occurs during shortened muscle length immobilization, suggesting that an adaptation of sarcomeres affects the amount of force development that can occur. No studies have been conducted in humans.
Neuromuscular relaxation	A process known as the stretch reflex suggests that muscle length elongation is limited by neural control. This theory suggests that slow, long-duration stretching can overcome this reflex and invoke relaxation. Studies have not supported this theory primarily based on a lack of EMG activity suggesting neural activity.

Table 7.12 Types of Stretches

Type	Method	Advantages	Disadvantages
Static	The individual moves a joint to a position of stretch and holds for a period of time (typically 10-30 s). Active, passive, and isometric are all types of static stretching.	Effective for increasing ROM of most joints	May increase risk of injury when done before strength or endurance exercise
Static–active	The individual uses their agonist muscle to move a joint and hold the position with a stretch of the antagonist muscle. Reciprocal inhibition, by stretch of the agonist, works to relax the antagonist.	Increases ROM and strengthens the agonist muscle	May be difficult for the elderly or others who have limited strength
Static–passive	The individual assumes a stretched position, and the position is held in place with another body part, a device (e.g., towel), or another person.	Useful in those with limited strength or recovering from injury; excellent for cool-down after fatiguing exercise; may reduce postexercise muscle soreness	Potential for injury if someone assisting with the stretch moves the joint beyond capable ROM
Static–isometric	Once in a static stretch position, the individual contracts the stretched muscle against stationary resistance for 5-15 s and then relaxes for a minimum of 20 s. Can be repeated several times; some may advise different contract and rest durations.	Quick method for increasing static–active flexibility and strength in contracted muscle	May result in injury to children who are still growing; may increase risk of injury if performed too often (recommend 36-48 h between sessions)
Dynamic	The individual moves body parts through an ROM with control. Speed and reach through range can vary and typically increase throughout performance.	Reduced risk of injury compared with ballistic method; useful for sport-specific movements to increase ROM; excellent specific movement for warm-up; may increase maximum strength after performance	May not maximally increase ROM
Dynamic–ballistic	The individual uses the momentum of the moving body or limb to force a joint beyond its normal ROM. Fast and powerful movements are required; bouncing is a common method. Maximum muscle length is attained only momentarily.	May be useful in athletes who often perform ballistic movements	May invoke stretch reflex and ultimately reduce ROM; can cause injury (strain) in the unaccustomed; can reduce maximum strength after performing
Proprioceptive neuromuscular facilitation	The individual performs passive stretching with a partner for 10-30 s; then, the partner provides resistance so the individual can contract the stretched muscle at up to 75% of maximal intensity for 3-6 s. Once the individual relaxes the muscle, the partner moves the stretch further and holds for up to 30 s. This can be repeated several times. Agonist contraction activates reflexive relaxation via stretch of the Golgi tendons, whereas antagonist contraction results in reciprocal inhibition of the agonist muscle.	Can produce the largest increases in ROM of any stretching technique and produces results in a short period of time (as little as a single session). May produce increased results with stretch–contract–antagonist contract process, which produces reciprocal inhibition of agonist via the stretch reflex	Cannot be performed alone (requires a partner); reduces maximum strength immediately after stretching

progression of the stretches with respect to FITT. Table 7.13 reviews general suggestions for developing a flexibility training program.

PUTTING IT ALL TOGETHER

Developing a training plan for high-level anaerobic athletes requires combining anaerobic speed and speed endurance, flexibility, resistance, and even aerobic training. Many factors must be considered when developing the training plan, including individual versus team training; the specific sport and event; the need for coordination, technique, and balance training; previous training experience; current training status; schedule of off-season, in-season, and peak performance competitions; and the time course of adaptation of tissues being trained and their subsequent recovery. Portions of three training plans for elite-level anaerobic-based sports are listed in tables 7.2, 7.14 and 7.15 as examples of the complexity of various training plans.

Competitive swimming is a sport that requires all facets of anaerobic training. It is unique because the amount of in-water training (in particular, the distances swum) often is far greater than any pool swim event (1,650 yd or 1,500 m freestyle). This is quite controversial among swim coaches, and some discount the excessive distances of most programs in favor of shorter training distance repetitions and overall training session distance. The most radical approach is the implementation of a training process known as *ultrashort race-pace training*. Currently, there are no research data on this technique, and very few elite-level individuals use it. In this type of training, distances per repetition are kept at 25 yd or m and occasionally a single 50- or 100-yd or m swim, and full training sessions are only a fraction of the typical swim training distances. A criticism of this type of training is that it is not specific to distances swum during races. There are very few races of 50 yd or m and none at 25 yd or m. The majority are 100 to 200 yd or m, and several at are 400+ yd or m. Most programs incorporate flexibility and resistance training in a portion of the training program known as *dryland training*. Resistance training attempts to be specific to the details of each of the four competitive strokes. Controversy exists about whether land-based resistance training translates to increases in strength and power in the water. Good evidence shows that resistance training benefits sprinters (i.e., events of ≤100 yd or m lasting under 1 min) and provides

diminishing but positive benefits for middle-distance and distance swimmers. Table 7.14 is an example of in-season resistance training at Purdue University. Note that the exercises vary throughout the week and that all are performed for 3 sets, typically in the range of 8 to 15 repetitions, and are focused on strength, endurance, and ROM improvements. This particular example was for a swimmer who had a significant back injury (L5 pars fracture) within the past 6 mo. The program was altered to protect the athlete from excessive lumbar extension and loading. This program will transition to more power and explosive power training as the season progresses toward the primary competition (e.g., Big 10 championship or National Collegiate Athletic Association championship).

Professional anaerobic athletes, such as hockey players, often work with a training staff year round. This includes the off-season, training camp, the regular season (which can be up to 6 mo long), and the playoffs. In this type of situation, each portion of the year is divided into specialized training periods. Table 7.2 is an annual training plan for an NHL team. Professional players have contract stipulations (e.g., individual player and player association) and outside influences (e.g., home during the off-season, other trainers) that must be considered when outlining and planning a training routine. The routine initially should be broad enough to incorporate the entire team. This can be observed in the example provided (table 7.2), where there are general types of training for energy system, strength, speed, agility, and so on depending on the time of the season. Individual plans certainly are developed depending on each player's specific concerns, such as postural correction and rehabilitation. Even in season, the general categories are tailored as needed to suit any individual player, such as those demonstrating excessive fatigue, those with injuries that are not severe enough to prevent them from playing, and those who might not play as often or as many minutes per game as other players.

Training for amateur athletics at the elite level can be as demanding as training for many professional sports. Unlike professionals, the amateur athlete likely has little outside influence (e.g., personal trainer, agent) that will supersede the team or program's training plan. Table 7.15 provides an example of the training principles of the USA women's volleyball team. This is led by a certified athletic trainer who focuses on current training in an attempt to reduce and treat injuries. This emphasis is noted in the major areas of focus, including core corrective exercises,

Table 7.13 Flexibility Training Program Development

	Frequency	Intensity	Time	Type	Other
Static	Daily; no less than 3 times/wk	Pain tolerance	Up to 30 s per joint stretched Best performed after exercise	May incorporate passive and active	Can perform 2-3 times per joint per session; avoid before aerobic or resistance exercise
Dynamic	Low intensity daily; 36-48 h between moderate- or high-intensity (i.e., ballistic) sessions	Controlled movement to pain tolerance	5-10 movements throughout ROM	General dynamic ROM recommended vs. ballistic	Suggested low level for warm-up before aerobic or resistance training
Proprioceptive neuromuscular facilitation	36-48 h between sessions	Contractions at 50%-75% of maximum	10-30 s passive stretch followed by 3-6 s contraction	Passive stretch–agonist contract–relax or passive stretch–agonist contract–relax–antagonist contract–relax	May promote compliance due to quick, substantial possible gains

Anaerobic Training Across the Life Span: Yes or No?

Large amounts of information link chronic aerobic exercise training to long-term health benefits, including reductions in morbidity and mortality. For this reason alone, everyone should perform aerobic training as they age. But do people need to perform anaerobic-based training as they age? With respect to flexibility and resistance training, the answer is an easy *yes*. As people age, they naturally lose strength and ROM of their joints. This loss is enhanced if people do not address these changes and if they simply become increasingly inactive. The effects of poor flexibility and strength, especially of the legs, can lead to less independence, frailty, a loss of lean mass (i.e., sarcopenia), a weakened skeleton (i.e., osteoporosis), and less mobility and functionality (e.g., more difficulty getting out of a chair or into and out of a car), and eventually succumbing to the use of walkers, wheel chairs, or even bed restriction. Evidence for performing anaerobic resistance and flexibility training throughout the life span is compelling.

What about anaerobic sprint training? Previously, we mentioned that few moments in typical daily life require sprinting. However, all initial movement (e.g., getting up from a chair or bed, taking a step, picking something up) is initiated using ATP from the anaerobic energy system (stored ATP or ATP generated from anaerobic glycolysis) (16). The anaerobic system is utilized any time a quick response is required. As with all other physiological processes, the ability to utilize the anaerobic energy system is reduced with aging. This can be due to the natural loss of Type II muscle fibers (i.e., atrophy), a reduced rate of ATP resynthesis, or a reduced neural ability to activate motor units specific to Type II muscle fibers. Additionally, aging increases the risk of loss of balance and falling. Although both flexibility and strength training can be helpful for improving balance, the initial corrective response of the body when off balance requires quick movement of both the upper and lower body. Adding some anaerobic sprint training to the exercise regimen of the aging can possibly help individuals react to situations that require a very quick muscular movement or reduce injury from falling. However, because anaerobic training often requires near-maximal effort and is performed nearly to exhaustion and because assessing improvement requires exhaustive or high-intensity movements (e.g., Wingate test, vertical jump), questions exist about the safety of performing this type of training and testing in an older population. The answer is likely *yes,* all types of anaerobic training should be performed throughout the life span. However, safe methods of training and testing anaerobic sprints still need to be developed and assessed.

Table 7.14 Sample Early-Season Resistance Training Program for Purdue University's Men's Swim Team

Exercises	Set 1		Set 2		Set 3	
MONDAY	REPS	LOAD	REPS	LOAD	REPS	LOAD
Balance routine						
Reach, roll, lift						
Split squat with weight vest						
Machine overhead press						
Normal-grip pull-up						
PB back extension						
Push-up with hold (2 s)						
Half-kneeling face pull						
Goblet squat with hold (2 s)						
Prone support plank						
Side support plank						
Glute bridge with hold (2 s)						
WEDNESDAY	REPS	LOAD	REPS	LOAD	REPS	LOAD
X-band walk						
Cable pull-through						
DB eccentric bend (5 s down)						
Suspension inverted row						
Single-leg box squat						
DB scaption						
One-arm hammer row						
Split squat with hold						
Keiser antirotational press (5-s pause)						
PB mini rollout						
FRIDAY	REPS	LOAD	REPS	LOAD	REPS	LOAD
One-leg balance matrix						
Reach, roll, lift						
Plate press-out vest squat						
DB alternate incline						
Normal-grip pull-up						
PB leg curl						
Push-up						
One-arm squat and row (5-s hold)						
Reverse hyper with 3-s hold						
Walking lunge with weight vest						
Side support plank						

DB = dumbbell; PB = physioball.

Table 7.15 USA Women's Volleyball Guiding Principles of Training

Energy system development	• Strategize for both aerobic and anaerobic demands • Pin hitters: specific transition footwork, positive shin angles, balance and sprint efficiency • Middles: effective lateral movements, transition horizontal to vertical power • Liberos and setters: ability to run a large portion of the court with speed and stamina
Core corrective exercises	The core is not only abdominals but also the back musculature, posterior shoulder, and gluteal complex. The core is the foundation on which all movement is built. When the core is properly aligned, the athlete can transfer energy throughout the body more effectively, so they'll produce more strength and power with less fatigue. Poor core strength equates to an increased potential for injury.
Flexibility and mobility	At times volleyball can put the body in vulnerable positions (e.g., sprawling for a ball, landing inefficiently on one leg, or requiring additional torque to hit a not-so-perfect set). It is key to have awareness and be able to train athletes in biomechanically advantageous positions. Dynamic stretch: The goal is to utilize momentum to emphasize a stretch on the muscle fibers throughout a full ROM. This will challenge tissue tensility, activate muscles, and increase overall body temperature.
Movement prep: proprioception and neuromuscular activation	• Movement-focused training • Push–pull movement patterns • Multidirectional movements in a full ROM • Movement prep: a series of exercises intended to increase heart rate, blood flow to muscles, and core temperature; prepares the athlete for volleyball-specific explosive movement patterns required on the court • Note: Research suggests that neuromuscular activation and proprioceptive skills are overlooked in novice athletes, thus leading to increased lower extremity injuries.
Strength principles	• Upper body strength: scapula, thoracic spine stability, plyometrics, proprioception, push–pull patterns in both horizontal and vertical planes • Core: rotary (concentric and eccentric), antirotary and isostabilization exercises • Lower body strength: two-leg and one-leg push–pull movements in both horizontal and vertical planes, emphasis on posterior chain, loaded and unloaded, proprioception and plyometrics • Triple extension movement patterns • Eccentric loading • Shoulder-, knee-, and ankle-specific proprioception and corrective exercises
Power principles	Strategies on speed of movement; analysis of power:work ratios and overall peak power
Speed and agility	Position-specific work: foot work, wall drills, functional patterns (loaded and unloaded)
Biomechanical considerations	Analysis of the following: cervical spine mobility and posturing, scapular movement, thoracic spine mobility, core, lumbar spine mobility, hip mobility, gluteal activations, quadriceps:hamstrings strength ratio, ankle joint mobility, proprioception, valgus knee mechanics, force distribution in the foot

(continued)

Table 7.15   USA Women's Volleyball Guiding Principles of Training *(continued)*

Injury rehabilitation principles	Includes the following: understanding the mechanism of injury, related injury history, selective functional movement assessment correlations, specific manual therapy goals, stability and mobility goals, pain and inflammation management, therapeutic modality utilization, nutritional plan, strength plan, regeneration plan, volleyball skill and movement plan, mental training plan, and short- and long-term goals
Regeneration principles	• Proper nutrition (fueling appropriately before, during, and after workouts) • Implementing a proper warm-up and corrective exercise • Sleep habits • Hydration analysis • Active rest workouts on days off • Complete (or passive) rest and meditation • Lymphatic flow enhancement • Massage or self-myofascial work • Hydrotherapy • Flexibility: static, dynamic, and proprioceptive neuromuscular facilitation • Yoga • Mental training • Manual therapy specific to injury: active release technique, muscle energy technique, soft tissue mobilization • Low-impact cardio • Whole-body vibration

movement preparation, and injury rehabilitation and regeneration. Volleyball players, particularly females, are at risk of ACL injury from repetitive jumping and from fatigue leading to poor proprioception, and often a jump landing or change of direction produces enough force on the ACL to cause rupture. Thus, a strong focus on flexibility, mobility, proprioception, strength, biomechanical analysis, and regeneration is aimed at reducing ACL and other injury.

Summary

- Development of anaerobic power, enhanced muscular strength and power, and adequate ROM are important for all individuals to perform better and with less risk of injury.
- Training programs should be specific to the desired benefit (e.g., general health, ease of movement in performing ADLs, athletic performance, reduction of the physiological effects of aging).

- Anaerobic power testing and training focus on assessing and improving activities that are performed for less than approximately 90 s.
- Resistance training aims to improve skeletal muscle strength and power.
- ROM training to improve or maintain flexibility can be beneficial for improving balance and coordination and reducing pain and injury risk and can help improve performance.
- Improvements from each of these types of training (anaerobic power, resistance, ROM) can be noted in individuals of any age and should be considered as part of a well-rounded exercise program.

Definitions

acceleration—The rate of change of velocity per unit of time.

Achard syndrome—Consists of a receding lower jaw and joint laxity in the hands and feet.

adaptation—Changes occurring in response to interval and other types of anaerobic training occur within the muscle fibers and within the energy system (i.e., ATP-PCr and glycolytic systems).

anaerobic capacity—The total amount of energy from the anaerobic energy systems (i.e., the combined amount of output from the ATP, PCr, and lactic acid systems).

anaerobic power—The ability of the ATP-PCr energy pathways to produce energy for muscle contraction.

distensibility—Capable of being distended or stretched.

Ehlers-Danlos syndrome—A group of disorders that affect the connective tissues that support the skin, bones, blood vessels, and many other organs and tissues.

elasticity—The ability to return to the original resting length after a passive stretch.

extensibility—Ability of skeletal muscles to be stretched to their normal resting length and beyond to a limited degree.

homeostasis—The process through which bodily equilibrium is maintained.

homocystinuria—An inherited disorder of the metabolism of the amino acid methionine, often involving cystathionine beta synthase.

Hyperlysinemia—An inherited condition characterized by elevated blood levels of the amino acid lysine.

individual differences—Differences in the rate of training effect are due to many variables, including previous training experience, quantity of the types of muscle fiber in a trained limb, and recovery between training sessions.

laxity—Lacking in rigor or firmness; not taut; loose and not easily retained or controlled.

Marfan syndrome—A hereditary disorder of connective tissue that results in abnormally long, thin digits and frequently in optical and cardiovascular defects.

metabolic syndrome—A risk factor for heart disease.

osteogenesis imperfecta—An inherited disorder characterized by extreme fragility of the bones.

overreaching—Short-term accumulation of training stress resulting in performance decrement.

overtraining—Long-term accumulation of training stress with decrements in performance.

plasticity—The tendency to assume a new and greater length after a passive stretch.

power—The rate at which work is done as the amount of work per unit of time (i.e., watt, horsepower).

proprioceptor—A sensory receptor found in muscles, tendons, joints, and the inner ear that detects motion or body and limb position.

reversibility—Adaptations gained during training that are lost over a period of reduced training (i.e., detraining) or the complete cessation of training.

speed—Distance traveled divided by the time of travel.

stretching-induced force deficit—Reduction in muscle force capacity after a static stretch.

velocity—A vector quantity with a magnitude of speed and direction.

viscoelastic creep—The ability of a tissue to elongate over time.

Chapter 8

Body Composition and Weight Management

For most of its history, the field of body composition has been limited mainly to either laboratory measurements, which make assumptions based on a small sample cadaver analysis (11,27,28,72), or field tests, which through mathematical equations make additional assumptions based on the aforementioned laboratory measurements. In short, the dependence on indirect measurements based on assumptions constructed from limited sample sizes has contributed to a degree of uncertainty when measuring body composition, especially in populations not included in some of the original cadaver studies (e.g., non-Caucasians, children, diseased populations) (11,27,28,72). Perhaps for this reason the utilization of body composition has been limited mainly to athletics and research.

However, with the advent of more precise methods of body composition measurement instruments such as dual-energy X-ray absorptiometry (DEXA), magnetic resonance imaging (MRI), or computed tomography (CT), we now have the ability to measure in vivo components that were previously assumed. These advances have assisted in providing new reference models, which has allowed better, more specific population prediction equations. Some of these technologies have become commonplace, such as DEXA, which is used to measure bone mineral density for conditions such as osteoporosis. Other areas show potential but remain underutilized due to prohibitive costs and lack of awareness. One example of this is the use of CT imaging in oncology to measure skeletal muscle loss—an important clinical marker related to disease progression and early death—in patients undergoing chemotherapy.

Body composition measures have applications when it comes to one of the most prevalent and costly health conditions in the United States: obesity. According to the most recent statistics, more than 35% of males and females in the United States are considered obese, and nearly 70% of all Americans are categorized as overweight (24). With a preponderance of evidence linking obesity to several chronic health conditions, there is much focus on developing and improving weight-loss techniques and strategies. When successful, weight loss has been shown to prevent, improve, or even resolve certain chronic health conditions (e.g., diabetes, hypertension).

BODY COMPOSITION MODELS AND THEORIES

The basic assumptions of many body composition models still used today originate from dissections of three Caucasian male cadavers aged 25, 35, and 46 yr. From the analysis of these cadavers, Behnke et al. (4) then later Brozek et al. (6) and Siri (62) developed the two-component model, which separates the body into the sum of fat mass (FM) and fat-free mass (FFM) (4,6,62). FFM is defined as nonfat tissue (e.g., bone, muscle, organs, and connective tissue), whereas FM consists of energy reserves (e.g., visceral fatty depots and subcutaneous fat) as well as **essential fat**.

The models of Behnke, Brozek, and Siri each assume that density for FM and FFM are constant and that proportions of protein, water, and bone mineral are stable. Therefore, any change in body density (as measured by either underwater weighing or air displacement plethysmography) is attributed to changes in FM. Interestingly, both the Brozek and Siri equations, still used today, yield similar percentage body fat values for individuals less than 30% (36). However, as mentioned, systemic errors occur with populations found to have different FFM densities (e.g., African Americans, Asians, athletes). Based on more recent works, population-specific equations should be used in place of the Brozek and Siri equations (table 8.1).

Traditional body composition methods such as underwater weighing use a two-component model, whereas contemporary techniques have the ability to measure individual elements at the atomic (e.g., carbon), molecular (e.g., protein), cellular (e.g., cell mass), and tissue (e.g., bone) levels (57). Combining different body composition methods, one can expand the two-component model into a multicomponent model. In coming sections we discuss DEXA as using a three-component model. An example of a four-component model from Withers et al. (74) would be the use of underwater weighing, isotopic dilution, and DEXA (i.e., body density, total body water, bone mineral density, and other) (74). The main advantage for using a multicomponent model is that components such as total body water or bone mineral, which can vary from the assumed two-component model, can be accounted for. The disadvantage is that each method in a multicomponent model has individual measurement error that is additive for each measurement used; this is known as the *propagation of errors* (36).

Reference Methods

Although no gold standard for measuring body composition currently exists, the three commonly used methods that are considered most accurate and are used as **reference methods** are hydrodensitometry (i.e., underwater weighing), air displacement plethysmography (commercially known as the Bod Pod), and DEXA.

Hydrodensitometry

The origins of hydrodensitometry (also known as underwater weighing or hydrostatic weighing) date back to the Greek mathematician Archimedes, who was given the task of determining whether a king's crown was made of pure gold. His breakthrough was when he realized that the volume of an object was equal to the volume of

Table 8.1 Ethnicity-Specific Two-Component Equations for Converting Body Density (Db) to Percentage Body Fat

Population	Age (yr)	Conversion formula Males	Females
African American	9-17	–	(5.24/Db) − 4.82
	19-45	(4.86/Db) − 4.39	–
	24-79	–	(4.86/Db) − 4.39
American Indian	18-62	(4.97/Db) − 4.52	(4.81/Db) − 4.34
Japanese	18-48	(4.97/Db) − 4.34	(4.76/Db) − 4.28
	61-78	(4.87/Db) − 4.41	(4.95/Db) − 4.50
Singaporean	61-78	(4.94/Db) − 4.48	(4.84/Db) − 4.37
Caucasian	8-12	(5.27/Db) − 4.85	(5.27/Db) − 4.85
	13-17	(5.12/Db) − 4.69	(5.19/Db) − 4.76
	18-59	(4.95/Db) − 4.50	(4.96/Db) − 4.51
	60-90	(4.97/Db) − 4.52	(4.97/Db) − 4.52
Hispanic	20-40	(4.87/Db) − 4.41	–

Adapted, by permission, from V. Heyward and D. Wagner, 2004, *Applied body composition assessment*, 2nd ed. (Champaign, IL: Human Kinetics), 9.

displaced water when that object was placed in it. Think about how your favorite drink rises in a glass when you add ice. If the crown were made entirely of gold, its density would be the same as that of pure gold. Unfortunately for the crown maker, this was not the case! Centuries later, Behnke et al. (4) developed a method for measuring density in humans using the same principle.

Underwater weighing (figure 8.1) is a two-component model that utilizes Archimedes' principle of water displacement to determine a person's FFM and FM. This principle asserts that the weight of an individual outside of water subtracted by the weight of an individual in water is equal to the volume of water that is displaced. (Later we discuss the need to correct for the density of water as well as trapped air, which can introduce error.)

$$Db = m/V \qquad \text{(Equation 8.1a)}$$

Where

Db = body density

m = weight in air

V = Volume

For hydrodensitometry body density is calculated below:

$$\text{Body density} = \text{weight in air} /$$

$$(\text{weight in air - weight in water}) \qquad \text{(Equation 8.1b)}$$

Body density then can be used to calculate FFM and FM. This is because of the known density of fat (0.9007 g/cc) and fat-free body density (FFBd) (1.1000 g/cc). Differences in the assumed FFBd account for the several population-specific equations (table 8.1). When compared with the density of water, which varies with temperature but is close to 1.0 g/cc, it becomes clear why individuals with a greater percentage of FM float and those with more muscle mass do not. However, body density must be converted to FFM using one of the known population-specific equations discussed previously, such as the ones developed by Siri (62).

When performing the underwater weighing, corrections are also needed for the pockets of air remaining in the lungs and gastrointestinal tract. Because air trapped in the gastrointestinal tract is much smaller (compared

Figure 8.1 Underwater weighing.

with air in the lungs) and cannot be measured directly, it is assumed to be 0.1 L (see equation 8.2). Air in the lungs is considerable and can lead to large error if a subject does not exhale completely. Motivating a subject to completely exhale is both a challenge and a limitation of the hydrostatic method. However, even when a subject successfully exhales completely, an amount of air known as the *residual volume* (RV) cannot be exhaled. The RV must be calculated, preferably using dilution methods such as helium dilution. If dilution methods are not available, RV can also be calculated by measuring the vital capacity and multiplying by 0.34 in males and 0.28 in females. Additionally, prediction equations for RV based on age, gender, and weight exist; however, these can contain significant error (up to 3.9%). Conversely, when RV is measured, the range for a true value is around ±1.2%. This could mean the difference in confidently saying a person's body fat is between 20% and 22.4% with measured RV or between 20% and 27.8% when RV is estimated!

Once corrections for trapped air are made, water density must be corrected for (unless you can find a subject who will enter a 4 °C water tank!) because water density at 4 °C is 1 g/cm³. In comparison, the water density at 25 °C is 0.997071 g/cm³. Once trapped air and water density are accounted for, body density can be calculated using equation 8.2. Once body density is calculated, percentage fat can be calculated using one of the two component equations (e.g., Siri).

$$Db = Wa/\left[\left(Wa - Ww\right)/Dw - \left(RV - 0.1\right)L\right]$$

(Equation 8.2)

Where

Db = body density

Wa = weight in air

Ww = weight in water

Dw = density of water

RV = residual water

0.1 = correction for air trapped in the gastrointestinal tract

Air Displacement Plethysmography

Compared with underwater weighing, air displacement plethysmography (commercially known as the Bod Pod,

figure 8.2) yields similar estimates of percentage body fat (23). However, unlike underwater weighing, one advantage of air displacement plethysmography is that it requires very little effort from the subject or tester. Collins et al. (12) reported that the **technical error of measurement** (i.e., the difference in measurement from one tester to another) is essentially negligible at only 0.448%. Another advantage of air displacement plethysmography is the time required for testing. In a group of children aged 10 to 18 yr, Lockner et al. (47) found that the Bod Pod took on average 1 hr less to measure body fat compared with underwater weighing. Because of these advantages the use of the Bod Pod has become popular across many populations, including professional football prospects who participate in the annual NFL Scouting Combine.

Air displacement plethysmography is also similar to underwater weighing in that it determines body volume to calculate FM and FFM from body density. However, as the name suggests, air displacement plethysmography measures the volume of air displaced as a person sits inside a chamber that contains a known volume of air. This is done in part by applying Boyle's law ($P_1V_1 = P_2V_2$), which states that at a constant temperature, the change in pressure is inversely related to the change in volume. An example of Boyle's law is how air bubbles from a scuba diver get larger as they approach the surface. The volume of the bubble expands due to the decrease in water pressure as the bubble approaches the surface. (This is also what causes "the bends.") Thus, by knowing the volume of air in the empty Bod Pod and measuring the change in pressure as a person sits inside, you can solve for volume of the person. However, because a person is not isothermic and because Boyle's law requires a constant temperature, the Bod Pod also needs to adjust for changes in temperature and humidity generated by a person (22). As with underwater weighing, once a person's body volume is known, this can be converted into body density (see equation 8.1a) and then percentage body fat using the same equations used with underwater weighing.

Despite these advantages, careful attention to procedures is necessary to avoid prediction errors. As mentioned previously, the Bod Pod measures volume based on changes in pressure when a person is placed in the chamber. However, because the chamber temperature is adiabatic (i.e., the temperature and pressure change because body heat is released), Poisson's law of pressure (a modification of Boyle's law) is used to measure volume

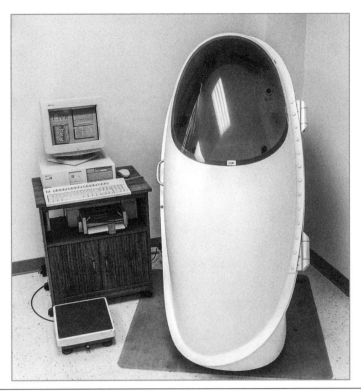

Figure 8.2 The Bod Pod.

Adapted, by permission, from V. Heyward and D. Wagner, 2004, *Applied body composition assessment*, 2nd ed. (Champaign, IL: Human Kinetics), 34.

when temperature is not constant (equation 8.3) (22). A common error associated with the Bod Pod occurs when nonisothermic items (e.g., clothing, hair, jewelry) and air in the lungs artificially reduce body volume, which leads to an underestimation of body fat (36). To minimize these errors, tight spandex shorts or swimsuits and a swim cap must be worn. To account for the volume of air in the lungs during normal tidal breathing, the Bod Pod measures **thoracic gas volume**. When thoracic gas volume is measured, a subject breathes normally into a tube for a few seconds. Then, the tube is temporarily blocked and the subject is asked to give two or three gentle puffs. If the calculation of thoracic gas volume is not within the acceptable range after three attempts, the Bod Pod software instead uses an estimated measure of thoracic gas volume. Interestingly, studies in both children and athletes found that the measured body density was very similar (i.e., ±0.002 g/cc) when using the estimated thoracic gas volume versus the predicted thoracic gas volume (47).

Poisson's law

$$P_1/P_2 = (V_2/V_1)\gamma \qquad \text{(Equation 8.3)}$$

Where

P = pressure

V = volume

γ = ratio of the specific heat of the gas at a constant pressure to that at constant volume

Dual-Energy X-Ray Absorptiometry

It can be argued that DEXA (figure 8.3) is not really a three-component model but rather is two separate two-component models. This is because DEXA calculates separately portions of the body containing only soft tissue mass (i.e., fat and lean) and portions containing both soft tissue and bone (table 8.2). These calculations are based on the absorption of two distinct X-ray beams that vary depending on the manufacturer (e.g., 40 and

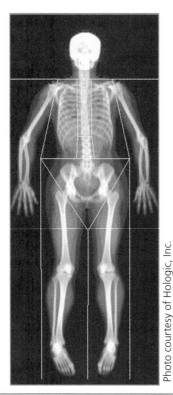

Photo courtesy of Hologic, Inc.

Figure 8.3 Dual-energy X-ray absorptiometry.

70 keV). The two beams are necessary when separating tissue that is inhomogeneous (i.e., comprising different constituents such as bone and soft tissue). Based on known absorptions of lean and fat tissue, DEXA first measures and calculates percentage fat and lean tissue in the portions devoid of bone. Then, based on the known absorptions of bone and soft tissue, DEXA does the same in the regions containing bone and soft tissue. Because fat and lean tissue are not directly measured in the regions containing bone, DEXA assumes the proportion of fat and lean tissue in the soft tissue mass around the bone to be the same as what was calculated in the regions without bone.

Compared with underwater weighing, DEXA requires little effort from subjects and takes only a few minutes to perform. Another advantage is that DEXA can measure fat and muscle distribution in localized areas of the body (e.g., the trunk and extremities). Although radiation exposure is a drawback, the amount is minimal—about equivalent to that of a cross-continental airplane flight.

DEXA is recognized as a reference model because of its precision compared with underwater weighing and multicomponent models (58). Sources of error include incorrect body positioning, hydration status, disagreement between measured weight and DEXA weight, and food intake (57). Another drawback of DEXA is the inability to measure individuals with a large body mass index (BMI). Early-model DEXAs had a weight limit of only 118 kg, whereas newer models have limits up to 180 kg. Problems arise if a person does not completely fit within the scanning field (34). Although not optimal, offset scanning can be done; this is where nonscanned areas are "mirrored" by the computer to calculate the missing portions (34).

Dilution Techniques for Estimating Body Water

As mentioned earlier, FFM in the human adult contains approximately 73% water. Interestingly, this proportion of water to FFM is strikingly similar not only across humans but also in other animal species (69). **Hydrometry** is another two-component model that measures total body water. FM and FFM can be estimated based

Table 8.2 Separate Components Calculated From DEXA

Soft tissue	Bone + soft tissue	
Fat + lean tissue mass	Fat + lean tissue mass (composition assumed based on bone-free scan)*	= Total tissue mass (i.e., fat + lean + bone)
	Bone mineral mass	

*Estimated from a scan of soft tissue only.

on the known relationship between total body water and FFM (equation 8.4).

$$FFM = TBW/0.732$$
$$\%BF = \left[BM - (TBW/0.732)\right]/BM \times 100$$
(Equation 8.4)

Where

FFM = fat-free mass

TBW = total body water

BM = body mass

%BF = percentage body fat

Hydrometry is predicated on the dilution principle that utilizes an isotopic tracer (e.g., tritium, deuterium) and measures the aforementioned tracer in a diluted solvent (i.e., total body water). As long as the concentration of the tracer in the solvent is known and the original amount of tracer is known, then the volume of the solvent (i.e., total body water) can be determined. As with other two-component models, hydrometry involves assumptions that can lead to error. These assumptions are as follows:

1. Lean body mass (LBM) contains 73.2% water. However, deviations can occur in some individuals, particularly in diseased populations (e.g., heart failure, end-stage kidney disease), who can have great variations in total body water.

2. The tracer diffuses evenly and completely through the total body water. Although this is mainly true, some amounts of tracer can be absorbed in body tissue (e.g., adipose tissue).

3. The tracer is not metabolized. Again, this is mainly true for common tracers used today.

4. The procedure for hydrometry involves an overnight fast, a baseline body fluid sample (saliva, urine, or blood), administration of a known dose of an isotopic tracer, an equilibrium period (typically 3-4 h), a second body fluid sample, and measurement of the isotope concentration in the body fluid (e.g., mass spectroscopy). Because hydrometry measures total body water, it is often used as part of a multicomponent model.

Advanced Body Composition Methods

The ability to measure fat and muscle distribution in distinct regions of the body can provide valuable information about health risk beyond whole-body composition techniques. Three methods that have changed the field of body composition research are CT, MRI, and magnetic resonance spectroscopy. Although these methods are expensive to administer, each is highly reliable and reproducible (30).

Computed Tomography

When an X-ray passes through a portion of the body, a two-dimensional image of that X-ray will display as either dark or light depending on the attenuation of the object. For example, bone is denser than lung tissue. When taking a chest X-ray, more of the X-ray is absorbed at the ribs, causing them to appear white. Conversely, the lungs appear black because more of the photon particles pass through them. However, a limitation with a standard X-ray is the overlapping of different tissues and organs, which prevents the ability to differentiate these components.

In the 1970s, British scientist Godfrey Hounsfield developed a way to use a computer to reconstruct multiple images taken 180° around an object. The reconstructed images provided high-contrast pixel attenuations known as **Hounsfield units**. Because X-rays pass through body tissue differently depending on the density of that tissue, the CT scan can easily separate between fat and muscle tissue. The first total-body composition measure taken was done in 1984 at the University of Gothenburg (59). The induction of CT has brought about greater understanding of fat distribution, specifically the importance of abdominal fat. Both the CT and MRI have shown how increased fat in the viscera contributes to diabetes, heart disease, and liver disease (30). One example of where research has led to clinical application is the diagnosis of fatty liver disease. CT scans of the liver give an attenuation value; the less attenuation of the liver (measured in Hounsfield units), the greater the amount of fat in the liver.

Magnetic Resonance Imaging

One of the major drawbacks of the CT scan is exposure to ionizing radiation. Because of this, the CT scan may not be the best option when needing to perform multiple tests or when testing children. MRIs, on the other hand, do not use ionizing radiation and instead are based on the interaction of hydrogen protons with a magnetic field controlled by the MRI (58). Similar to CT, MRIs produce whole-body images that display the distribution of adipose tissue throughout the body (see figure 8.4). MRIs can also be used regionally and have been shown to be sensitive enough to detect body composition changes (e.g., increased muscle mass with training) over time. MRIs have also provided accurate measures of extramuscular adipose tissue (EMAT), which is the fat that is visible on MRI images between muscle groups and beneath the muscle fascia. The amount of EMAT in an individual has been found to be a better predictor of insulin resistance than subcutaneous fat.

Magnetic Resonance Spectroscopy

Magnetic resonance spectroscopy (MRS) has the ability to measure concentrations of atomic and metabolic components in vivo; these components are not measurable by any other means (46). MRS measures the resonance (typically of hydrogen atoms) and is conducted using the same machine as the MRI. Similarly to MRI and CT scans, MRS has many applications outside of body composition (e.g., quantifying neurotransmitters and metabolites in the brain). A specific area of interest in the field of body composition is the presence of ectopic fat. Ectopic fat is defined as

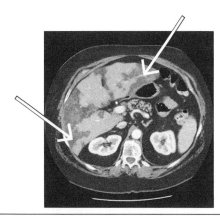

Figure 8.4 An MRI image of fatty liver disease. The arrows point to dark regions of the liver where fat has accumulated.

Photo courtesy of Henry Ford Health System.

deposits of triglycerides not located in adipose tissue (e.g., skeletal muscle, liver, heart) (38). The presence of ectopic fat inside the myocyte (i.e., intramuscular adipose tissue) is associated with insulin resistance and diabetes (38).

Field Methods of Body Composition

Due to the expense and technical support of the previously mentioned reference models, many professionals rely on the use of field tests to assess body composition. Typically, these field tests use regression equations based on the relationship of the field test measures to a reference model (e.g., hydrodensitometry). The two field tests most commonly used are the skinfold method and bioelectric impedance.

Skinfold Method

The skinfold method involves the use of calipers to measure subcutaneous fat directly beneath the skin. Subcutaneous fat composes 40% to 60% of total body fat and is positively correlated with the body's internal fat stores (i.e., the greater subcutaneous fat, the greater fat throughout the body). Once the skinfold measures are taken, these are entered into a prediction equation to calculate FM and FFM.

Although the skinfold technique offers a valid and inexpensive option for estimating percentage body fat, a major disadvantage is an estimated error of 3% to 11% (36). This error mainly is due to the skill (or lack thereof) of the individual administering the skinfold test but can also be attributable to the use of inappropriate prediction equations (e.g., using an adult prediction equation when testing adolescents). Out of the many equations developed, the Jackson and Pollock generalized equation (equation 8.5) is still one of the most widely cited. Developed in the 1970s, the generalized equations for men and women used a quadratic regression equation based on underwater weighing (39,40).

Females: $Db = 1.099421 - 0.0009929 (X_1) + 0.0000023 (X_1)^2 - 0.0001392 (X_2)$

Males: $Db = 1.10938 - 0.0008267 (X_1) + 0.0000016 (X_1)^2 - 0.0002574 (X_2)$

(Equation 8.5)

Using Body Composition Methods to Identify Cachexia in Cancer Patients

Cachexia is not simply weight loss. In some cancer populations, such as patients with early-stage breast cancer, weight gain is actually a more common occurrence that is associated with poor prognosis. Cachexia is a metabolic syndrome linked to an underlying condition (e.g., cancer, heart failure) that results in the loss of muscle with or without fat that impairs function (41). Although the mechanisms that lead to cachexia are not fully understood, one factor that differentiates itself from sarcopenia (see "Body Composition and Weight Management Across the Life Span") is an increase in metabolic rate. This increased metabolic rate may be attributable to the cancer tumor itself, especially when the cancer cells spread to metabolically active tissue. An example of this was seen in patients with colorectal cancer, where an increase in liver mass due to metastatic cancer cells was associated with a concurrent decrease in total muscle mass (41). Regardless of what triggers the increased metabolic rate, the end result is increased muscle catabolism, which places the cancer patient at higher risk of early death (41).

One misconception about cachexia is that obese individuals are somehow protected because of increased energy reserves. Although the obesity paradox suggests a protective effect with increases in certain cancer populations (e.g., individuals undergoing chemotherapy), in general BMI does a poor job of classifying those patients with a high mortality risk (29). The obvious reason why is the inability of BMI to differentiate FFM from FM. Using a bioelectrical impedance analysis, Gonzalez et al. (29) showed that FFM index (i.e., FFM divided by the square of height) was a better predictor of mortality regardless of how much FM was present. In other words, it did not matter if a patient's BMI was higher or lower; if the patient had a reduced amount of muscle mass, the risk of death was increased.

The clinical implications for measuring FM and FFM become evident if you compare two obese patients, one with cachexia and one without. Both patients lose 20 lb (9.1 kg) after a diagnosis of cancer. Using body composition methods, a clinician can identify the higher risk patient. Methods such as the CT measurement of lumbar skeletal muscle currently are being used, although unfortunately not widely. Regardless, the identification of cachexia is one potential area where knowledge of body composition methods can be invaluable.

Where

Db = body density (percentage body fat can be converted using a two-component equation—e.g., Siri)

X_1 = sum of triceps, iliac crest, and midthigh skinfolds (females) or sum of chest, abdominal, and midthigh skinfolds (males)

X_2 = age (rounded to the nearest year)

More recently, studies have shown that the generalized equation is still a valid measurement that shows a high correlation with DEXA-derived percentage body fat ($r = .87$ for women; $r = .95$ for men) (53). The major drawback for the generalized equation, however, was the cohort, which consisted mainly of non-Hispanic Caucasians. This led to systematic error in some ethnic groups (e.g., percentage body fat was underestimated in African American men and overestimated in Hispanic men and women) (55). Because of this, O'Connor et al. (55) developed newer sum of skinfold equations based on a diverse group (55). Instead of comparing the sum of skinfolds to underwater weighing, the more recent equations use DEXA as a reference model and incorporate BMI (equation 8.6) (55). The advantages of these equations is they are ethnic specific, have good agreement with DEXA (3.6% in females and 3.1% in males), and do not involve the use of another equation to convert body density into percentage body fat. The drawback is that the population used was between the ages of 18 and

35 yr, and thus the equations are not suitable for older populations. Another drawback when using generalized skinfold equations is that percentage body fat is greatly underestimated in obese populations when the sum of skinfolds is greater than 120 mm (53).

Caucasian females: %BF = $(0.169 \times X_1)$ − $(0.0007 \times X_1^2)$ + $(0.849 \times BMI)$ + 1.260

Hispanic females: %BF = $(0.169 \times X_1)$ − $(0.0007 \times X_1^2)$ + $(0.849 \times BMI)$ + 3.146

African American females: %BF = $(0.169 \times X_1)$ − $(0.0007 \times X_1^2)$ + $(0.849 \times BMI)$ − 0.078

Caucasian and Hispanic males: %BF = $(0.190 \times X_1)$ − $(0.0005 \times X_1^2)$ + $(0.604 \times BMI)$ − 5.377

African American males: %BF = $(0.206 \times X_1)$ − $(0.0005 \times X_1^2)$ + $(0.604 \times BMI)$ − 1.987

(Equation 8.6)

Where

%BF = percentage body fat

X_1 = sum of triceps, iliac crest, and midthigh skinfolds (females) or sum of chest, abdominal, and midthigh skinfolds (males)

BMI = body mass index

Bioelectric Impedance

Bioelectric impedance is a quick and technically simple method of estimating total body water by measuring the **impedance** (Z) of a small current as it passes through the body. Impedance is a function of both resistance (R), which is the amount a current is stopped as it attempts to pass through an object, and reactance (X_c), which is the amount a current is slowed as it passes through an object: (Z = R + X_c). You could use impedance to measure the volume of a tube filled with salt solution using the following equation: V = pL^2/Z, where V is volume, p is the specific resistivity of a solution, and L is the length of the tube (i.e., cm) where the electrodes are placed.

The same principles and formula are also used to determine total body water in humans. However, the human body is not a perfect cylinder but rather is five different-shaped cylinders: the arms, the legs, and the trunk. Bioelectrical impedance analysis (BIA) utilizes prediction equations to determine total body water. One example of this is a simple linear regression equation

developed by Lukaski et al.(48) that used deuterium dilution as the reference model (total body water = 0.63 × height²/R + 2.03; r = .95) (68).

Because FFM contains the vast majority of body water, the impedance measured in a lean individual will be much less than in someone with a higher percentage of body fat. (Remember that water is a good conductor of electricity, whereas fat is an insulator.) Once total body water is known, FFM can be calculated using the following equation: FFM (kg) = total body water (kg)/0.73. Thus, a major assumption used in BIA is that FFM contains 73% water.

Traditionally, BIA is measured at the whole-body level using a single frequency (50 kHz) that travels between four electrodes placed on the right side around the wrist and ankles (figure 8.5). More recently, some BIA models measure multifrequency impedances that can differentiate intracellular water and extracellular water in addition to total body water (45). This is because at a frequency of 50 kHz the electricity travels through the path of least resistance (i.e., around cell membranes), which in theory would be the extracellular water, but at higher frequencies the current travels through cell membranes, in theory yielding the intracellular water. In addition to the electrode-based BIAs, commercial BIA models are available that use electroplates that measure foot to foot, hand to hand, or foot to hand impedance. Although some of these electroplate models have shown some agreement compared with reference models in healthy populations, the results are mixed at best and warrant more research (76).

In general, if the basic assumptions are met, BIA can be a good predictor of body fat. As with the other

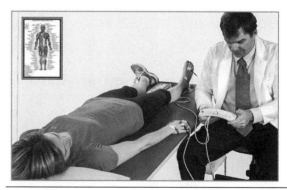

Figure 8.5 Single-frequency BIA. Traditional single-frequency BIA utilizes four surface electrodes at a frequency of 50 kHz.

Photo courtesy of RJL Sciences.

body composition methods, it is important to identify potential sources of error and, when available, use equations that account for differences in gender, age, athletic status, and ethnicity. When it comes to the use of BIA, the number one assumption is the euhydration of subjects. For this reason, individuals with electrolyte abnormalities or fluid imbalances (e.g., congestive heart failure, end-stage kidney disease) may have spurious results. However, even with euhydrated subjects, the assumption that lean body mass contains 73.2% water can vary individually. Das et al. (17) showed that individuals with grade 3 obesity (BMI >40 kg/m^2) had a greater percentage of lean body mass water (i.e., 75.6%) compared with the reference value. Although this explains why obese individuals systematically have a lower BIA-derived percentage body fat compared with reference models, the use of BIA to show changes in percentage body fat seems to be comparable with that of DEXA (64,66).

The use of BIA in individuals with internal devices (e.g., pacemaker, artificial hip) is an interesting subject. Although the small current generated through the body in BIA does not appear to interfere with pacemaker function, to date the U.S. Food and Drug Administration has not approved its use in these individuals. Also not known is the effect of the internal devices themselves, which may cause error in estimating total body water (7).

Anthropometry

Anthropometry—the measurement of the size and proportion of the human body (36)—often is used to estimate body fatness or risk associated with excess body fat. The most common measurement used is BMI (table 8.3).

Body Mass Index

BMI is calculated by dividing body weight in kilograms by meters squared: BMI = kg/m^2. BMI is easy to use, is inexpensive, and can be measured across large populations. An example of how BMI can provide quick, simple measures across many demographics is the **Behavioral Risk Factor Surveillance System**, an ongoing telephone survey conducted by the Centers for Disease Control and Prevention. Information gathered by the survey and other similar studies has provided valuable insight into trends in obesity as well as differences between gender, ethnicity, and demographic location. However, although BMI is highly correlated with adipose tissue (r = .82), it is not a perfect surrogate for body fatness (44). This is made clear when considering that NBA basketball star LeBron James, who had 6.7% body fat as a rookie, would otherwise be categorized as overweight with a BMI of 27 kg/m^2 [height of 80 in. (2 m), weight of 245 lb (111 kg)]. BMI may misclassify other populations (e.g., the elderly) and certain clinical populations (e.g., patients with human immunodeficiency virus).

It may also partially explain the phenomenon known as the **obesity paradox**. Across the general population, a BMI greater than or equal to 30 kg/m^2 is associated with higher mortality (26). However, mortality follows a U-shaped curve, with the highest risk of mortality in individuals with BMIs lower than 18.5 kg/m^2 and higher than 35 kg/m^2. The reasons why individuals in the overweight (i.e., BMI of 25-29.9 kg/m^2) and grade I obesity (i.e., BMI of 30-34.9 kg/m^2) categories do not fall into a higher risk category is unclear, although it may either suggest a protective effect with some excess adipose tissue or simply be a limitation of using BMI.

Table 8.3 Body Mass Index Classifications

Classification	Value (kg/m²)
Underweight	<18.5
Normal weight	18.5-24.9
Overweight	25-29.9
Grade 1 obesity	30-34.9
Grade 2 obesity	35-39.9
Grade 3 obesity	>40

Flegal et al. (26)

Circumference Measures

Circumference measures have been proposed as an alternative or adjunct to BMI. Circumference measures have good reliability (better than the skinfold method), can be used on very large individuals, and are quick and inexpensive to perform. However, perhaps the most important advantage of circumference measures, especially compared with using BMI, is that they can provide information on fat distribution. Mounting evidence shows that individuals who are shaped like an apple (i.e., store more fat in the intra-abdominal region) are at greater risk for chronic diseases such as diabetes, hypertension, and heart disease. It is perhaps for this reason that out of the 17 or more reported circumference measurement sites (table 8.4), waist circumference is the most often used. In fact, reductions in waist circumference are associated with improved insulin sensitivity, reduced visceral fat (i.e., intra-abdominal fat), and decreased risk of diabetes (44).

Waist circumference should be measured during exhalation with a flexible, inelastic tape measure perpendicular to the floor. The anatomical sites commonly used to measure waist circumference are as follows:

- At the umbilicus
- Above the iliac crest
- At the narrowest point between the last rib and the iliac crest
- Below the lowest rib
- At the midpoint between the last rib and the iliac crest

Wang et al. (69) compared four waist circumference sites (the umbilicus was not tested) and found that all sites had good repeatability (intraclass correlation = 0.996) and were related to body fat. However, the site just above the iliac crest had the highest correlation to body fat ($r = .89$ for females; $r = .60$ for males) compared with the other sites.

WEIGHT MANAGEMENT PRINCIPLES

Just as an automobile traveling 50 miles · d^{-1} requires more gasoline than does one traveling 20 miles · d^{-1}, a person running 5 miles · d^{-1} will require more food than someone who is sedentary. However, unlike an automobile, a person can vastly increase the amount of

fuel storage in the form of triglycerides located inside fat cells known as *adipocytes*. This is evident when comparing an average lean adult with roughly 35 billion adipocytes and around 0.5 µg of triglycerides against an obese individual with approximately 140 billion adipocytes and 1.0 µg of triglycerides (32).

Many contributing factors determine whether someone will be at an energy surplus (leading to weight gain) or an energy deficit (resulting in weight loss). In general, the components that contribute to weight gain or loss can be defined as follows:

$$E_s = E_I - (REE + TEF + AEE) \qquad \text{(Equation 8.7)}$$

Where

E_s = energy balance

E_I = energy intake

REE = resting energy expenditure

TEF = thermic effect of food

AEE = activity energy expenditure

Most likely, you learned that you need to produce a caloric deficit (through reduced food intake, increased activity, or a combination of both) of 3,500 kcal in order to lose 1 lb (0.45 kg) of fat. Therefore, based on the amount of daily calories reduced or expended, you should be able to predict how long it would take for a 180-lb (81.6 kg) nonathletic male to lose a pound of fat when his daily caloric requirement is 2,500 kcal. Suppose that this individual reduced his caloric total to 2,000 kcal. Based on the 3,500-kcal estimate, it would take approximately 7 d for him to lose 1 lb of fat (i.e., 500-kcal deficit × 7 d = 3,500 kcal). Extending this logic further, if he continued at a 500-kcal deficit, he would lose 52 lb (23.6 kg) in 1 yr and vanish from the face of the earth in 180 wk! Clearly, there are flaws in the "3,500 kcal/1 lb" argument, which is why individuals lose weight at different rates and why the rate of weight loss slows over time. (This is discussed in more detail later.)

Etiology of Obesity

We all have preconceived notions about what causes obesity. Surprisingly, many of those beliefs are not strongly supported by research. Examples of these are the common beliefs that breast feeding and eating breakfast lowers risk of obesity; however, when tested under well-controlled or randomized conditions, both of these presumptions are not supported (9). What cannot be denied are the laws of thermodynamics, which state that energy cannot be

Table 8.4 Standardized Sites for Circumference Measurements

Site	Anatomical reference	Position	Measurement
Neck	Laryngeal prominence ("Adam's apple")	Perpendicular to long axis of neck	Apply tape with minimal pressure just inferior to the Adam's apple.
Shoulder	Deltoid muscles and acromion processes of scapula	Horizontal	Apply tape snugly over maximum bulges of the deltoid muscles, inferior to acromion processes. Record measurement at end of normal expiration.
Chest	Fourth costosternal joints	Horizontal	Apply tape snugly around the torso at level of fourth costosternal joints. Record at end of normal expiration.
Waist	Narrowest part of torso, level of the "natural" waist between ribs and iliac crest	Horizontal	Apply tape snugly around the waist at level of narrowest part of torso. An assistant is needed to position tape behind the client. Take measurement at end of normal expiration.
Abdominal	Maximum anterior protuberance of abdomen, usually at umbilicus	Horizontal	Apply tape snugly around the abdomen at level of greatest anterior protuberance. An assistant is needed to position tape behind the client. Take measurement at end of normal expiration.
Hip (buttocks)	Maximum posterior extension of buttocks	Horizontal	Apply tape snugly around the buttocks. An assistant is needed to position tape on opposite side of body.
Thigh (proximal)	Gluteal fold	Horizontal	Apply tape snugly around thigh, just distal to the gluteal fold.
Thigh (mid)	Inguinal crease and proximal border of patella	Horizontal	With client's knee flexed 90° (right foot on bench), apply tape at level midway between inguinal crease and proximal border of patella.
Thigh (distal)	Femoral epicondyles	Horizontal	Apply tape just proximal to the femoral epicondyles.
Knee	Patella	Horizontal	Apply tape around the knee at midpatellar level with knee relaxed in slight flexion.
Calf	Maximum girth of calf muscle	Perpendicular to long axis of leg	With client sitting on end of table and legs hanging freely, apply tape horizontally around the maximum girth of the calf.
Ankle	Malleoli of tibia and fibula	Perpendicular to long axis of leg	Apply tape snugly around minimum circumference of leg, just proximal to the malleoli.
Arm (biceps)	Acromion process of scapula and olecranon process of ulna	Perpendicular to long axis of arm	With client's arms hanging freely at sides and palms facing thighs, apply tape snugly around the arm at level midway between the acromion process of scapula and olecranon process of ulna (as marked for triceps and biceps skinfolds).
Forearm	Maximum girth of forearm	Perpendicular to long axis of forearm	With client's arms hanging down and away from trunk and forearm supinated, apply tape snugly around the maximum girth of the proximal part of the forearm.
Wrist	Styloid processes of radius and ulna	Perpendicular to long axis of forearm	With client's elbow flexed and forearm supinated, apply tape snugly around wrist, just distal to the syloid processes of the radius and ulna.

Adapted, by permission, from V.H. Heyward and Ann L. Gibson, 2014, *Advanced fitness assessment and exercise prescription*, 7th ed. (Champaign, IL: Human Kinetics), 425; Adapted from Callaway et al. 1988.

Cutting Down on Unsafe Weight-Loss Practices in Wrestling

Weight cutting is a method of rapidly losing weight when an athlete must achieve a certain body weight to participate in a sport competition. Often, this rapid weight loss is attributable to the loss of body fluids, which can impair performance and pose a serious health risk (71). The sport often associated with these unsafe weight-loss practices is wrestling. In the 1970s Thorland et al. (67) performed the Iowa studies, which first brought attention to weight loss in high school wrestlers. They reported that a large number of wrestlers lose an excessive amount of weight in a short period of time prior to the date of certification (i.e., the last official weigh-in before a match). Additionally, they reported that the methods of weight loss (e.g., fluid restrictions, wearing excessive clothing during exercise) usually were suggested by either a coach or a teammate.

In the span of about 1 month in the fall of 1997, three collegiate wrestlers died while performing unsafe weight-loss methods. Each death occurred under the supervision of their coaches. Later it was discovered that all three athletes were attempting to lose an average of 8 lb (3.6 kg) within a few hours through vigorous exercise while wearing a rubber suit inside an excessively heated gymnasium (52). Unfortunately, according to the American College of Sports Medicine, these types of methods for rapidly losing weight are still practiced. (Sadly, there are even websites and YouTube videos that misinform young athletes on these practices.)

However, as a result of these tragedies (as well as earlier work by Tipton and others), governing bodies that oversee high school and college wrestling have instituted guidelines for appropriate weight loss. The National Collegiate Athletic Association (NCAA) has banned using the following to induce rapid weight loss:

- Saunas or steam rooms (on or off campus)
- Vapor-impermeable suits (e.g., rubber suits)
- A practice room over 75 °F
- Laxatives
- Excessive restriction of food or fluid
- Diuretics

Additionally, the NCAA and many high school athletic associations have developed protocols to determine the lowest acceptable weight a participant can achieve (5% body fat for the NCAA) and the fastest rate of acceptable weight loss (no greater than 1.5% of initial body weight/wk). To determine how much and how fast weight loss can occur, the NCAA uses a baseline body fat measurement. The three acceptable methods used by the NCAA to measure percentage body fat are as follows:

- Skinfolds using the sum of the triceps, subscapular, and abdomen. Body density is determined using an equation from Thorland et al. (67) and converted to percentage body fat using the Brozek equation (6).
- Hydrostatic weighing using a direct measure of RV. Body density is converted into percentage body fat using the Brozek equation (6).
- Bod Pod. Body density is converted into percentage body fat using the Brozek equation (6).

Before performing the baseline body fat measure, athletes need to be in a euhydrated state. To determine whether an athlete is hydrated adequately, a urine sample is taken to measure specific gravity. If an athlete is found to have a urine-specific gravity greater than the acceptable threshold (i.e., >1.020), then they must wait 24 hours before attempting another body fat test.

created or destroyed but rather is changed from one form (food) to another (adipose tissue or heat as a byproduct of metabolism). Research from the 1950s showed that humans and animals alike seem to adjust ad libitum eating based on the amount of daily activity (50). Meyer et al. (50) reported that most individuals adjusted their caloric intake depending on their occupation (i.e., the less physically demanding a job, the less the intake of calories). The one exception to the rule was the group with the most sedentary jobs, who actually consumed more than any other group outside of the most active workers. Parenthetically, the sedentary group also weighed the most.

Much has changed over the past 50 yr, and many of those changes have contributed to an **obesogenic environment**. Food is now relatively cheaper (except for fruits and vegetables) and more available than it was a generation ago. Dietary intake surveys indicate a per capita increase of 168 kcal/d for men and 335 kcal/d for women between the years 1971 and 2000. Fewer jobs are physically demanding: The mean daily expenditure during work is 142 cal/d less than in 1962 (10). It should come as no surprise that the human genome pool, selected for energy efficiency over centuries of food scarcity, should now find itself in this dilemma.

Genetic Etiology of Obesity

Evidence from twin studies as well as parent–child studies suggests that obesity has a very strong genetic component (75). One early example is an adoptee study that found a strong relationship between the BMI of adoptees and that of their biological parents but no relationship between the BMI of adoptees and that of their adoptive parents (63). Although twin studies have provided evidence of a genetic link, identifying the genes specific to the obesity phenotype has mostly remained elusive.

One breakthrough was the discovery of the ob/ob gene in mice. Mice with mutations to this gene were three times heavier than normal mice. Later it was discovered that the ob/ob gene manufactures a protein that produces the hormone **leptin**. Leptin is released by adipocytes and regulates food intake and metabolism at least partially through its effects on the hypothalamus, which contains a high number of leptin receptors. There was much excitement when leptin was first discovered in 1994, especially when administration of leptin to obese mice with the ob/ob gene showed a dose–response relationship with weight loss. However, the administration of leptin in obese humans who have normal leptin levels does not appear to be effective for weight loss.

More recently, genome-wide association studies have led to the discovery of the FM and obesity-associated gene FTO; this is the most robust genetic–obesity link to date (21). Interestingly, the FTO gene was originally examined as a potential type 2 diabetes gene (75). Although the mechanism for how the FTO gene causes obesity is unclear (FTO is not associated with leptin in humans), the specific allele (i.e., gene variant) associated with obesity has been found across ethnic populations (e.g., European, African).

Often gene studies attempt to find a link with the hormones responsible for the regulation of appetite and metabolism. Previously we discussed the role of the hormone leptin, which is a proposed regulatory mechanism contributing to energy balance. Another regulatory hormone, mainly secreted by the stomach, is **ghrelin**. Ghrelin has been found to directly influence appetite (61). Weight loss has a direct relationship with ghrelin: The greater the weight loss, the greater the amount of ghrelin. It does not seem to matter whether the weight loss is exercise induced or diet induced (49). Although no clear genetic links have been made between ghrelin and obesity, high levels of ghrelin are found in individuals with **Prader-Willi syndrome**, a genetic disorder of the 15th chromosome that is associated with delays in growth and development as well as an insatiable appetite that leads to obesity. Pharmaceutical companies currently are investigating agents targeting ghrelin receptors with the hope of helping obese individuals both with and without Prader-Willi syndrome curb their appetite and lose weight.

In addition to genetics and lifestyle, other potential factors contributing to obesity have been proposed (e.g., Bisphenol A, lack of sleep, constant ambient temperature). Although studies have found some provocative links to obesity [e.g., high levels of Bisphenol A have been detected in obese children (20)], the evidence is based on correlative data rather than causal data, and more studies are need before any definitive statements can be made.

Weight-Loss Strategies

The public health burden of obesity is great. Deaths directly related to obesity were estimated to be around 3 to 4 million worldwide in 2010 (54). Currently, obesity ranks as the number two cause of preventable death (smoking is the leading cause). Obesity trends have risen sharply worldwide over the past few decades. Since the late 1980s, the percentage of obese individuals in the United States has increased from 23% to 34% (25).

Treatment options for weight loss vary in effectiveness and often are used in combination with behavioral strategies that have been shown to increase effectiveness and prevent relapse (table 8.5). Unfortunately, obesity in general is refractory to many conventional weight-loss methods. Some reports state that by 3 yr posttreatment nearly 100% of patients regain all weight lost (15). However, some studies, such as the Look AHEAD trial, do not paint as bleak of an outlook and rather suggest that long-term weight maintenance is possible. According to the National Weight Control Registry, predictors of long-term success (i.e., 10 yr after initial weight loss) include maintaining leisure-time activities, dietary restraint, frequency of self-weighing, consuming a greater percentage of energy from fat, and disinhibition (65).

Traditional Weight-Loss Strategies

Reducing food intake alone or increasing exercise alone typically are not successful, especially in the long term. However, modest but clinically significant weight loss of greater than 10% can be possible and sustainable if both are done together and behavioral strategies are implemented. Current recommendations include daily dietary restrictions of 500 to 1,000 kcal below resting metabolic rate and 60 to 90 min of aerobic exercise 5 d/wk. Although the concept of creating an energy deficit by increasing energy expenditure and decreasing energy intake seems straightforward, individuals attempting weight loss contend with several barriers, including the body's compensatory measures in response to an energy deficit, environmental cues, and a lack of understanding regarding the requirements needed to achieve weight loss.

When individuals restrict energy intake, the body takes compensatory measures to decrease energy expenditure and increase hunger. Similarly, as a person loses more weight, the effects of weight-bearing exercise are lessened (e.g., fewer calories are burned while walking because the person has less weight to support). These changes in energy requirements are contrary to the static 3,500 kcal = 1 lb rule. This means that instead of a weight loss of roughly 10 lb (4.5 kg)/yr through modest reductions of 100 kcal/d, the actual weight loss would be much less (37). Hall et al. (33) created a computer simulation program to better estimate the

Table 8.5 Treatment Options for Weight Loss

Treatment	Description	Effectiveness	Recidivism
Traditional reduced-calorie diet	Intake is reduced by 500-1,000 cal/d.	Low-moderate	High
Exercise	Calorie expenditure is increased mainly through aerobic exercise (250-300 min/wk).	Low	Moderate-high
Partial meal replacement diet	Prepackaged foods fortified with vitamins and minerals replace traditional meals.	Moderate	Moderate-high
Very-low-calorie diet	Medically supervised diets use only partial meal replacements and provide ≤ 800 cal/d.	Moderate-high	High
Pharmacotherapy	Mechanism depends on pharmacological agent (e.g., sympathomimetic, lipase inhibitor, serotonergic) used.	Low-moderate	High
Weight-loss surgery	The stomach volume and/or the absorption of food are reduced through surgery.	High	Low-moderate

Based on Casazza et al. (9).

dynamic energy balance changes that occur over time and the expected weight loss based on calories restricted and expended.

These changes do not diminish the roles of diet and exercise; if anything, they emphasize them. However, because these compensatory mechanisms can lead to discouragement in individuals, it is important to provide realistic expectations and behavioral strategies that have been shown to be effective. Out of the known behavioral strategies, perhaps the one most often associated with weight-loss success is self-monitoring (8,19,31,56,73). A study by Guare et al. (31) found that individuals who kept food diaries more frequently maintained weight loss better (−18 kg) at 1 yr compared with those who did not (−5 kg). Another study by Carels et al. (8) on 40 middle-aged males and females found that the use of an exercise diary was associated with fewer reported difficulties in exercise ($r = −.48$), greater weight loss ($r = .44$), and greater amounts of weekly exercise. Another mode of self-monitoring is simply recording a daily weight (42). Although frequent weighing sometimes is discouraged by behaviorists who are afraid clients will be discouraged by stepping on the scale too much, it is actually one of the strongest predictors of success (42).

Looking Back at Two Lifestyle Weight-Loss Trials

A criticism of lifestyle interventions (i.e., nutrition, physical activity, and behavioral strategies) for weight loss is that long-term adherence is poor. Although recidivism and weight regain are challenges for lifestyle interventions, the Diabetes Prevention Trial and Look Ahead Trial both demonstrated that long-term weight loss and clinical improvements are feasible (1,2).

The Diabetes Prevention Trial examined the incidence of new diabetes cases by comparing the effects of lifestyle changes with those of metformin (a common diabetic medication) over a period of nearly 3 yr. The lifestyle intervention included a 16-wk individualized program covering diet, exercise, and behavioral modification with the goal of more than 7% weight loss and 150 min/wk of exercise. Compared with the metformin group and placebo group, the intervention group lost more weight, exercised more, and reported eating fewer calories. However, the most striking finding of the study was the incidence of diabetes, which was 58% lower in the lifestyle group compared with the placebo group but only 31% lower in the

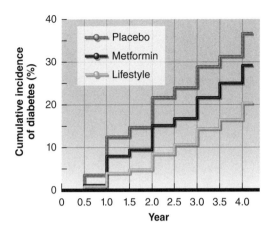

Figure 8.6 Results from the Diabetes Prevention Trial. Compared with a placebo and the common diabetes drug metformin, lifestyle changes (which included exercise and a weight loss of 7%) most lowered the risk of developing diabetes.

metformin group compared with the placebo group (figure 8.6).

The Look Ahead Trial showed that a lifestyle intervention program consisting of reduced caloric intake and increased physical activity had important health benefits in overweight individuals with type 2 diabetes. Specifically, those in the intervention group showed greater weight loss, greater fitness, and lower levels of glycated hemoglobin compared with the control group. With respect to weight loss, the intervention group averaged around 6% total weight loss at 4 yr, and greater than 40% of subjects maintained at least 10% weight loss. Both trials support the fact that changes in diet and exercise habits can be sustainable and lead to important improvements in health.

Very-Low-Calorie and Partial Meal Replacement Diets

Very-low-calorie diets (VLCD) typically are given under medical supervision and consist of supplements fortified with vitamins and minerals (e.g., shakes, bars). The most well-known example of a celebrity who used a VLCD plan is Oprah Winfrey, who lost 67 lb (30.4 kg) in 1988.

The total daily caloric intake for a person on a VLCD may vary depending on their estimated resting metabolic rate but is roughly 800 kcal/d (35). These supplements typically are high in protein (70-100 g/d) and low in carbohydrate (<80 g/d) (68). Criticisms of VLCD plans

Body Composition and Weight Management Across the Life Span

Childhood and Adolescence

The number of children and adolescents who are overweight or obese is at an all-time high (i.e., approximately 20%) and continues to increase at an alarming rate (32). The rise of childhood obesity has led to the presence of adulthood diseases such as type 2 diabetes and hypertension. Because weight changes during childhood and adolescence are also attributable to growth, the Centers for Disease Control and Prevention defines *overweight* as a BMI equal to or greater than the 85% age percentile and defines *obese* as a BMI equal to or greater than the 95% age percentile. Correlates that place a child at a higher risk for obesity are time spent indoors, time spent using electrical devices (e.g., video games, tablets, TV), and the amount of sugary beverages consumed. Because of this, public health programs have focused on reducing sugary drinks and screen time while promoting the consumption of more fruits and vegetables and increasing physical activity to 60 min/d.

On the opposite side of the spectrum, many adolescents, especially those involved in sport, are at risk of being too lean. Essential body fat is defined as the minimal amount of fat needed to maintain physiological processes and protect vital organs. The amount of essential body fat differs between males and females (i.e., approximately 5% vs. 12%, respectively); females' greater needs are related to reproductive function. Many female-dominated sports (e.g., dancing, gymnastics) have a high incidence of unsafe weight practices. The American College of Sports Medicine has produced two position statements on the female athlete triad, which involves components of energy availability, menstrual function, and bone mineral density (3). Because of the increased energy expenditure of a sport, many female athletes are in a low or negative energy balance. In some cases, increased pressure to achieve or maintain leanness can lead to disordered eating or excessive exercise, which can exacerbate the female athlete triad. The end result of low energy availability is amenorrhea (i.e., the absence of three or more consecutive menstrual periods) and the loss of bone mineral density. If the problem is not addressed appropriately through education and proper nutrition, long-term health consequences (e.g., increased stress fractures, osteoporosis) may occur.

Young Adult to Elderly

Cardiovascular fitness as well as muscle strength and size typically peak somewhere between the ages of 20 and 30 yr before gradually declining. Height peaks and begins to decline somewhere between 30 and 40 yr of age as a result of increased compression on the intervertebral discs and, later, the loss of bone mineral density. Although exercise can attenuate these declines, studies in master athletes show that these functional declines are inevitable. Likewise, body composition changes throughout adulthood. Although weight may be stable or gradually increase between ages 20 and 50 yr, there is typically a concurrent process of reduced muscle mass, known as sarcopenia, and increased body fat. These changes are associated with hormonal changes (e.g., reductions in insulin growth factor-1 and dehydroepiandrosterone), but the influence of environment versus aging is still debatable. Regardless of how much of this change is not controllable through exercise or diet, favorable changes in body composition can still occur throughout the aging process.

mainly involve safety and weight regain (2 yr after losing 67 pounds, Oprah Winfrey regained her weight).

In the 1970s, there were reported fatalities of individuals who experienced substantial weight loss while on VLCDs. Factors that contributed to these deaths were likely the poor-quality protein (i.e., hydrolyzed collagen) and incomplete supplementation of vitamins and minerals (67). Although contemporary VLCD plans do slightly increase the risk of dehydration, hair loss, and cholelithiasis, the risk of serious side effects is very low, especially when individuals follow the plan under medical supervision (68). Similar to traditional low-calorie and exercise approaches to weight loss, weight regain is an issue with VLCD plans. However, contrary to a commonly held notion, individuals who lose weight rapidly through a VLCD do not end up weighing more (i.e., are heavier than baseline) than those who lose through more conventional methods (9). Data from the Henry Ford Clinical Weight Management Program show that although patients who are on a VLCD plan start to regain weight between 6 and 12 mo, the absolute weight loss at 12 mo is still greater than that of individuals who were on a traditional low-calorie diet with regular food.

A meta-analysis by Johansson et al. (43) reported that weight regain can be attenuated after a VLCD plan with partial meal replacements. Essentially the same as VLCD supplements, partial meal replacements are supplements that are fortified with vitamins and minerals and typically are high in protein. However, unlike VCLDs, which are consumed exclusively, partial meal replacements typically are given once or twice daily along with regular food. Partial meal replacements were used during the Look Ahead Trial and were found to be one of the strongest predictors of long-term weight-loss success (35).

Pharmacological Options for Weight Loss

Despite what advertisements claim regarding weight-loss products (especially supplements, which are not regulated by the FDA), no pill or product available can produce substantial weight loss (i.e., >10%). If there was such a drug, it would likely be the most prescribed drug in the United States. Pharmaceutical development for weight loss is an area of much interest. As discussed earlier, investigational drugs targeting leptin and ghrelin are currently being developed along with other potential agents, all in the hope of finding a more effective and long-

term solution to obesity. Currently there are a handful of FDA-approved drugs for weight loss. Although these drugs have shown to lead to modest weight loss (2%-10%), their use typically is recommended in concert with a low-calorie diet and an exercise program for enhanced weight loss or weight maintenance (43).

Orlistat, which has been available since 1999, is a lipase inhibitor. Because orlistat blocks the absorption of fat into the intestines, it is taken with meals and is not effective for individuals who consume a low-fat diet. Compared with a placebo, orlistat has shown to produce a 2.9-kg weight loss; 21% of users achieved at least a 5% weight loss, and 12% achieved a 10% weight loss (60). Side effects include malabsorption of fat-soluble vitamins and gastrointestinal symptoms such as flatulence, steatorrhoea, and fecal incontinence. For that reason, compliance with orlistat can be poor.

Phentermine-ER topiramate is a newer weight-loss drug that contains a sympathomimetic (i.e., phentermine) as well as an antiepileptic agent (i.e., topiramate). Although the use of topiramate as a weight-loss drug is new, phentermine has been used as an antiobesity drug since 1959. Phentermine is believed to suppress appetite through direct action on the hypothalamus. A randomized trial of phentermine-ER topiramate showed about a 10-kg weight loss compared with a placebo. However, similar to orlistat, there was a high rate of attrition for those on the drug.

Sibutramine is another sympathomimetic drug with mechanisms similar to those of phentermine. However, in 2010 it was pulled off the market because of an increased risk of stroke and heart attack. Likewise, because phentermine can act as a sympathetic agonist and lead to increases in heart rate and blood pressure, caution is recommended for individuals with hypertension or potential heart conditions.

Weight-Loss Surgery

When it comes to weight-loss efficacy and long-term maintenance, bariatric surgery has the strongest evidence of all current treatments. A meta-analysis of studies reporting long-term follow-up (i.e., ≥2 yr) reported more than 50% excess weight loss in most bariatric studies. Percentage excess weight is the method of reporting weight loss in surgical patients and is based on how much weight a patient has lost compared with their ideal weight (equation 8.8) (18).

$$[(OPW - FW)/(OPW - IDW)] \times 100$$

(Equation 8.8)

Where

OPW = operative weight

FW = follow-up weight

IDW = ideal weight (based on the 1983 Metropolitan Insurance height and weight tables)

In the early 1990s, only approximately 16,000 weight-loss surgeries were performed annually. In contrast, by the mid-2000s, the number of surgeries in the United States grew to more than 200,000. The most common type of weight-loss surgery in the United States is the Roux-en-Y gastric bypass (RYGB) surgery, a procedure that replaces the stomach [which has a typical volume of 20 oz (0.59 L)] by partitioning a 1-oz (29.6 mL) pouch from the native stomach. This aids in restricting the amount of volume—and thus calories—an individual can consume in one meal. Additionally, the new 1-oz pouch is attached farther downstream the intestines, which bypasses some digestive enzymes and reduces overall absorption. Another more recent type of surgery, known as *sleeve gastrectomy,* reduces the stomach to approximately the size of a banana.

In addition to better long-term weight loss, a reason for the surge in bariatric procedures has been the improvement—or in some cases the complete resolution—of cardiovascular risk factors (e.g., hypertension, diabetes). The remission of diabetes, which can be measured by how many diabetic patients no longer require insulin or oral diabetes medications, often is a purported benefit of weight-loss surgery. A randomized study comparing RYGB surgery against lifestyle interventions for weight loss found that individuals with type 2 diabetes had either partial (50% of RYGB patients) or complete (17% of RYGB patients) remission of diabetes compared with no remission (partial or complete) in individuals who were randomized to the intensive lifestyle intervention (16).

Despite the benefits, as with any surgery, there are risks involved with bariatric surgery. Replacing open surgical methods with less invasive laparoscopic techniques has reduced serious adverse events and improved time to recovery. The risk of death is very low (<1%). According to a recent Cochrane review, side effects specific to weight-loss surgery (e.g., leakage, infection, obstruction, pouch dilatation) vary depending on the type of surgery as well as the risk profile of the patient undergoing the procedure (13). Interestingly, a study by McCullough et al. (51) found that presurgical fitness (i.e., a peak $\dot{V}O_2$ >15.8 mL · kg^{-1} · min^{-1}) was associated with fewer complications, reduced hospital stay, and reduced 30-d readmissions (51).

Undergoing weight-loss surgery does not diminish the importance of exercise, healthy eating, and behavioral changes because weight regain can occur in this population. Exercise after bariatric surgery has shown to improve fat oxidation and prevent weight regain (5,14). One of the possible mechanisms behind the purported benefits of exercise in this population is the loss of muscle that occurs. Research done at Henry Ford Hospital and elsewhere has shown substantial losses of muscle mass and strength after weight-loss surgery. To attenuate some of the muscle loss, resistance training 2 to 3 d/wk along with more than 250 min of aerobic exercise is recommended. However, it is unlikely that even the most rigorous resistance training program can completely prevent this loss.

Summary

- Body composition measurement can be a very helpful tool across athletic, clinical, and apparently healthy populations. Understanding the assumptions and limitations of a particular body composition method can be valuable when selecting the appropriate test and interpreting results. An excellent tutorial of some of the methods described in this chapter can be found at nutrition.uvm.edu/bodycomp/.

- Weight recidivism rates are high for many individuals who attempt to lose weight. Regular exercise and increased physical activity can play a central role in weight maintenance and overall health. Understanding the environmental, genetic, and behavioral etiologies of weight gain can lead to better understanding and expectations for the individual.

- Modest weight-loss changes (i.e. 5-10% of initial body weight) can lead to improvements for several chronic health conditions (e.g., diabetes, hypertension).

Definitions

Behavioral Risk Factor Surveillance System—A national telephone survey conducted by the Centers for Disease Control and Prevention with the purpose of collecting data regarding health-related risk behaviors, chronic health conditions, and use of preventive services.

essential fat—The minimum amount of fat needed to maintain physiological processes and protect vital organs.

ghrelin—A regulatory hormone mainly secreted by the stomach that has been found to directly influence appetite.

Hounsfield unit—Developed by and named after Sir Godfrey Hounsfield; a quantitative scale used for CT to measure the amount of attenuation of substances in the body.

hydrometry—The measure of body water via isotope dilution methods.

impedance—A function of both resistance (the amount a current is stopped as it attempts to pass through an object) and reactance (the amount a current is slowed as it passes through an object). It is used in BIA to estimate total body water and extrapolate percentage fat and FFM.

leptin—A hormone released by adipocytes that regulates food intake and metabolism at least partially through its effects on the hypothalamus.

obesogenic environment—A surrounding environment that promotes the development of obesity for both individuals and populations.

obesity paradox—Theory that states that extra fat stores provide a protective effect for mortality, as demonstrated by the U-shaped mortality curves associating BMI and death.

Prader-Willi syndrome—A genetic disorder of the 15th chromosome that is associated with growth and developmental delays as well as an insatiable appetite that leads to obesity.

reference method—In body composition, considered an accurate test or model that often is used to cross-validate other body composition models.

technical error of measurement—The difference in measurement from one tester to another.

thoracic gas volume—A measure used during air displacement plethysmography to control for air in the lung; equal to functional residual capacity plus half of tidal breathing.

weight cutting—A method of rapidly losing weight when an athlete must achieve a certain body weight to participate in a sport competition.

Chapter 9

Performance: Environmental Stressors, Genetics, Nutrition, and Ergogenic Aids

We thank Christopher Womack, PhD, Department of Kinesiology, James Madison University, Harrisonburg, VA, for his contributions to the Exercise and the Environment section of this chapter.

Performance is an important aspect of exercise physiology. Regardless of whether the focus is clinical, athletic, or simply improved health and disease prevention, exercise training is targeted toward improving performance. This might be an increase in peak $\dot{V}O_2$ to reduce the risk that a patient with heart failure will require a heart transplant. It might be improving an athlete's ability to tolerate hot and humid conditions during a long-distance event. It might be an individual desiring to move with more ease and perform daily tasks that currently are difficult. Understanding the exercise physiology related to environmental stressors and understanding nutrition methods that may aid performance are important when counseling individuals about performance improvement. This chapter first examines environmental conditions in which performance may be negatively affected and methods for combating performance decrements. This is followed by a discussion of the effects of genetics, nutrition, and the use of nutritional ergogenic aids to enhance performance.

EXERCISE AND THE ENVIRONMENT

Exercise and physical activity are controlled stressors, and the setting in which these activities are carried out can create additional stress. The physiological responses to environmental stressors (e.g., **ambient** temperature, humidity, or atmospheric pressure) can affect exercise performance and increase the risk of illness, injury, or death. Therefore, it is important to have a working knowledge of these responses and their inherent risks as well as information about **acclimation** and **acclimatization**, which are important for maintaining and improving performance in extreme environments (50).

Heat

Imagine a runner attempting to run at her normal pace during the first 90 °F (32 °C) day of summer. By the end of the run, she is extremely fatigued despite running her 5-mi (8 km) loop more than 5 min slower than normal. Heat can take a toll on any athlete, especially when they are not accustomed to this type of environmental condition.

Heat Exchange

The two major challenges imposed by heat are an increase in **core body temperature** and fluid loss. Although the body can maintain a normal core temperature at a range of ambient temperatures and humidity, core temperature increases during exercise in hot and humid environments and as metabolic heat production increases during vigorous exercise intensities. This increase is somewhat—but not completely—offset by the removal of heat by the processes of **convection, conduction, radiation,** and **evaporation** (figure 9.1). Convection occurs when heat is transferred from one place to another by the motion of a substance. A common example in exercise is a breeze blowing across the skin. Conduction is the transfer of heat via direct contact between two objects that are different temperatures. We utilize conduction whenever we cool the body with ice. Radiation is the transfer of heat between objects through electromagnetic

activity. Radiation from the sun causes us to gain heat when we are outside—a fact we are aware of when exercising outdoors in the summer. Evaporative heat loss occurs via evaporation of sweat. This is the main response to an increased core body temperature, which may occur during exercise in humans. It is also the source of many of the physiological consequences that come from exercising in the heat. These heat exchange processes act to maintain heat balance (heat balance = heat gain − heat loss). When heat gain increases beyond the abilities of the body to regulate heat balance (up to approximately

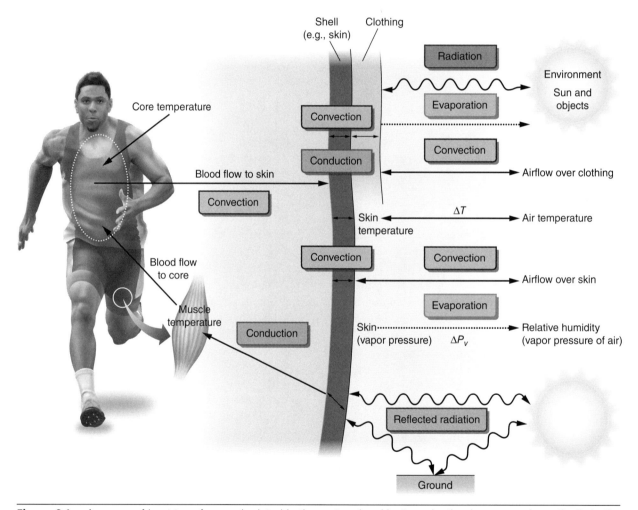

Figure 9.1 Avenues of heat transfer are depicted in the rectangles. Heat production increases dramatically in the muscle with exercise. Internally, conduction and convection are important in the transfer of heat both from muscle to core to skin and from muscle to skin. Externally, during heavy exercise, evaporation accounts for practically all of the heat transfer. Other external mechanisms important at rest include radiation, convection, and conduction. Important factors include delta T (temperature difference between the skin and air) and delta Pv (vapor pressure difference between the skin and air).

Reprinted from M.L. Foss, S.J. Keteyian, and E.L. Fox, 1998, *Fox's physiological basis for exercise and sport,* 6th ed. (Pittsburgh, PA: William C. Brown). By permission of the authors.

Measuring Body Temperature

Because body temperature is important for both performance and health during exercise, its accurate measurement is important. This measurement can be influenced by the environmental setting (e.g., indoors vs. outdoors; humid vs. dry conditions). The core, or internal, temperature of the body best reflects the stressful influence of heat accumulation. However, assessment using the gold standard rectal temperature typically is performed only in controlled environments during research experiments. The use of an external means of temperature assessment that accurately reflects the core temperature would be useful for assessing the effects of sweating, fluid intake, and environmental conditions on performance during practice and competition.

Possible methods of external temperature measurement include forehead, oral, axillary, aural (ear canal), and temporal assessments. Some have speculated that environmental conditions (e.g., indoors vs. outdoors; clouds vs. sunlight; wind, rain, and humidity) may lead to inaccuracies of temperature measurement. Two separate investigations were performed to assess the **validity** and reliability of various body temperature assessments performed during indoor and outdoor exercise (14,29). In a controlled indoor environment (36 °C and 52% humidity) during 90 min of walking and 60 min of postexercise rest, none of the temperature measurements were considered valid compared with the rectal temperature, although all were considered reliable. A similar finding was reported for the uncontrolled outdoor setting. These findings led the investigators to recommend that all meaningful measures of core temperature be performed rectally (e.g., by the medical staff of a large sporting event such as a running race or triathlon). They also suggested that if a clinician needs to assess core temperature in an individual who might be in danger of heat illness and the means of assessing temperature rectally are not available, the clinician should consider using clinical assessments (e.g., central nervous function, assessment of sweating) to make a clinical decision.

40 °C) an imbalance and subsequent increase in core temperatures occurs.

Temperature Regulation

The body works to maintain core temperature at approximately 37 ± 1 °C (98.6 °F; i.e., reference temperature). Input from various thermal receptors and subsequent responses to changes in body temperature occur in the thermal regulatory center (i.e., the hypothalamus), located in a subcortical area of the brain. Body temperature is continuously monitored by central thermal receptors in the anterior hypothalamus, peripheral receptors located in the skin, and deep body receptors located in the spinal cord, abdominal viscera, and larger veins. Sensory input from all of these receptors is transmitted to the posterior hypothalamus, which triggers reflexes in an effort to regulate core temperature (figure 9.2).

Physiological Responses

During exercise in the heat, the thermal gradient between core temperature and skin temperature and between skin temperature and environmental temperature is decreased, resulting in a reduced rate of heat loss. Furthermore, if humidity is high, the evaporative thermal gradient between the skin surface and the ambient air is reduced, thus decreasing the ability to lose heat via evaporation. As a result, core temperature increases and the ability to perform prolonged exercise is reduced. For the runner who was mentioned at the beginning of this section, jogging at 6.0 (9.7 kmh) mph in higher ambient heat and humidity is much more **thermodynamically** stressful than jogging at that same velocity in a **thermoneutral** (i.e., cooler and less humid) environment. To find out why, let's take a look at the major physiological responses affected by heat.

Cardiopulmonary Responses

In a thermoneutral environment, cardiac output is elevated during exercise, and this increased blood flow is preferentially directed to the working skeletal muscles (and diverted from the visceral region). However, during exercise in the heat, skin blood flow increases to

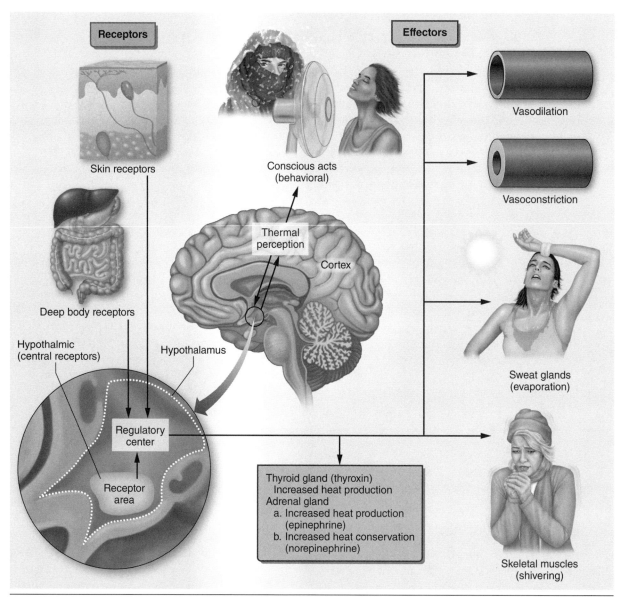

Figure 9.2 A summary of the thermoregulatory system. The internal body temperature is measured by receptor organs and compared to a set point (37 °C; 98.6 °F). If the temperature deviates from the set point, the hypothalamic center automatically relays information to the effector organs, which correct the temperature to the set point value through the mechanisms shown on the right. The return to the set point value then automatically shuts off the effector system. The cortical connections provide a means for voluntary regulation of body temperature.

Reprinted from M.L. Foss, S.J. Keteyian, and E.L. Fox, 1998, *Fox's physiological basis for exercise and sport,* 6th ed. (Pittsburgh, PA: William C. Brown). By permission of the authors.

as high as 8 L · min⁻¹ to enhance cooling (76). Although the increase in skin blood flow is beneficial, it comes at a physiological price. Because muscle blood flow needs to be maintained during exercise to preserve performance, the increase in blood flow to the skin requires the cardiac output to increase to substantially higher levels. In addition, blood flow to the splanchnic region (locations such as the liver, intestines, and kidneys) and inactive skeletal muscle decreases due to local vasocon-

striction—more so than typically occurs in a cooler and drier environment (76). The consequences of these responses are discussed in the section on metabolism and exercise in the heat.

Under any environmental circumstances, plasma volume decreases as a response to acute exercise. Because of sweat loss, which can increase to between 1 and 3 L · h⁻¹ in hot or humid environments, the plasma volume reduction occurs to an even greater extent when exercise is performed in the heat (table 9.1) (34). The loss of water from sweating results in a reduction of plasma volume. For instance, in a person with a 5-L blood volume and a hematocrit of 45%, the normal plasma volume is approximately 2.75 L. Of course, a decrease in plasma volume also means a decrease in blood volume. Even at a lower sweat rate, it is apparent that plasma and blood volumes can become significantly reduced in a short period of exercise (i.e., approximately 60 min). This reduced blood volume results in a decrease in the amount of venous return to the heart (reduced preload), which in turn reduces stroke volume and cardiac output. It should also be emphasized that the reduced blood volume is only a portion of the cause of the reduced stroke volume during exercise in the heat. Traditional theory holds that the increase in blood flow to the skin results in increased venous pooling in the cutaneous veins and can compromise venous return. The combination of reduced blood volume and venous pooling can result in a reduction in stroke volume, primarily because of a decrease in end-diastolic volume. As a result, heart rate increases in order to maintain cardiac output at an appropriate level. However, more recently, it has been suggested that increases in cutaneous circulation and declines in stroke volume are not linked and that heart rate is increased because of increased sympathetic nervous system stimulation. As a result of the increase in heart rate, there is less time for diastolic filling; this results in a lowered stroke volume (22). Therefore, increased heart rate causes the decrease in stroke volume (instead of the other way around). The combination of lowered plasma volume and increased heart rate is the main cause of the decline in stroke volume that is observed during exercise in the heat.

Regardless of the mechanism, a lower peak exercise stroke volume means that $\dot{V}O_2$max will be reduced. Recall that $\dot{V}O_2$max is determined primarily by maximal cardiac output and maximal arteriovenous oxygen (O_2) difference. Although heart rate is elevated during submaximal exercise in the heat, heart rate during maximal exercise cannot go higher than during thermoneutral exercise. Therefore, a lower maximal stroke volume with no change in maximal heart rate will result in a decrease in maximal cardiac output and thus $\dot{V}O_2$max. As dehydration and core temperature increase during exercise in a hot environment, neural reflexes enhance the vasoconstriction of the splanchnic and skin vasculature in an attempt to maintain central blood volume and blood pressure. As this occurs, skin temperature becomes cooler due to evaporative, convective, and radiative heat transfer while core temperature may continue to increase.

To review, exercise in the heat causes a need for increased cardiac output due to increased blood flow to the skin as well as a reduction in stroke volume due to a reduction in plasma volume, venous return, and decreased diastolic filling time. Furthermore, heart rate is increased because of a higher degree of sympathetic stimulation. From these observations, it should be clear that the cardiovascular system is stressed to a great degree during exercise in the heat. Table 9.2 and figure 9.3 summarize the cardiovascular changes that occur during exercise in the heat.

Metabolic Responses

Exercise in the heat not only strains the O_2 delivery system but also independently alters the metabolism of the working muscles. One of these metabolic alterations is an increase in the $\dot{V}O_2$ at a given absolute submaximal exercise intensity. The exact cause of this increased caloric expenditure is unknown, but there are two possible explanations. First, motor unit recruitment increases to perform a given amount of work when exercise is performed in the heat. Because exercise in the heat hastens fatigue of the recruited motor units, more motor units must be recruited in order to maintain the appropriate skeletal muscle contractile force and rate and, thus, exercise intensity. A second possible cause is that more O_2 is needed for oxidative phosphorylation in the electron transport chain for enhanced adenosine triphosphate (ATP) production. This

Table 9.1 Effects of Environmental Heat Loads on Sweat Rate Responses During 15 Min of Moderate Work

Dry bulb temperature (°C)	Wet bulb temperature (°C)	Relative humidity (%)	Sweat rate (L·hr⁻¹)	Heart rate (beats·min⁻¹)
22	14.7	45	0.4	150
35	26.0	50	1.0	155
35	33.4	90	1.6	165

See text for definition of wet bulb temperature.

Reprinted from M.L. Foss, S.J. Keteyian, and E.L. Fox, 1998, *Fox's physiological basis for exercise and sport*, 6th ed. (Pittsburgh, PA: William C. Brown). By permission of the authors.

Table 9.2 Cardiovascular Changes That Occur During Exercise in Heat Compared to a Thermoneutral Environment

	Submaximal exercise	Maximal exercise
$\dot{V}O_2$	$\longleftrightarrow\uparrow$	$\downarrow$
Heart rate	$\uparrow$	$\longleftrightarrow$
Stroke volume	$\downarrow$	$\downarrow$
Cardiac output	$\uparrow$	$\downarrow$
a-vO_2 difference	$\longleftrightarrow$	$\uparrow$
Venous return	$\longleftrightarrow$	$\downarrow$

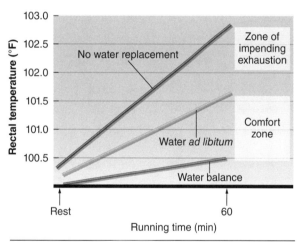

Figure 9.3 Schematic representation of progressive water loss during 1 hr of exercise in a hot and moderately humid environment. When water consumption equals sweat loss, rectal temperature is lowest, compared with no water replacement or water consumed *ad libitum*.

Reprinted from M.L. Foss, S.J. Keteyian, and E.L. Fox, 1998, *Fox's physiological basis for exercise and sport*, 6th ed. (Pittsburgh, PA: William C. Brown). By permission of the authors.

may be due to an increase in core temperature resulting in inefficiency of the electron transport chain; as a result, more O_2 is needed to rephosphorylate a given amount of ATP (13). This increase in cellular O_2 requirements ultimately could be large enough to produce an increase in whole-body $\dot{V}O_2$.

As mentioned previously, $\dot{V}O_2$max is reduced during exercise in the heat. Because of this reduction in $\dot{V}O_2$max and increase in submaximal $\dot{V}O_2$, relative exercise intensity can be increased. For example, assume that running at 6

mph (9.7 kmh) in cool conditions requires a $\dot{V}O_2$ of 40 mL · kg^{-1} · min^{-1}. If a runner's $\dot{V}O_2$max is 60 mL · kg^{-1} · min^{-1}, the relative exercise intensity of a 6-mph (9.7 kmh) jog is about 67% of $\dot{V}O_2$max (40 mL · kg^{-1} · min^{-1} divided by 60 mL · kg^{-1} · min^{-1}). If we examine responses to that same absolute velocity (6 mph [9.7 kmh]) in the heat, we likely would measure the $\dot{V}O_2$ as slightly elevated (e.g., 43 mL · kg^{-1} · min^{-1}). However, the potential $\dot{V}O_2$max in the heat is reduced, for example, to 55 mL · kg^{-1} · min^{-1}. Now the relative exercise intensity is about 78% of $\dot{V}O_2$max (43 mL · kg^{-1} · min^{-1} divided by 55 mL · kg^{-1} · min^{-1}). Thus, the relative and absolute metabolic costs of the activity have increased substantially even though the absolute velocity (6 mph [9.7 kmh]) has not changed.

Exercise is more demanding at a given intensity when performed in a hot environment. Furthermore, some research suggests that the anaerobic contribution to total ATP production increases in the heat as blood lactate levels increase (54). However, the accumulation of lactate is a balance between lactate production and removal. The combination of an increased reliance on Type II (fast twitch) muscle fibers and a reduced blood lactate removal rate due to a reduced oxidative metabolic rate may result in the increase in blood lactate during exercise in the heat. It is also possible that an increase in sympathetic nervous system activity (i.e., increased plasma catecholamines) contributes to the increased blood lactate accumulation rate. Increased sympathetic neural output increases circulating epinephrine, one of the main hormones that stimulate muscle **glycogenolysis**. Furthermore, the glycolytic enzymes (e.g., phosphofructokinase) may increase in activity due to the Q10 effect, which states that the rate of change of a biological or chemical system is related to the increase of temperature by 10 °C. Therefore, if skeletal muscle temperature is elevated during exercise in the heat, it is possible that the rate of glycolytic reactions will increase and that the rate of lactate appearance will increase as a result. Note that the Q10 effect is established only in laboratory conditions. Humans can tolerate an increase in core body temperature of only about 5 °C. Although a larger increase in temperature may theoretically increase the rate of enzymatic reactions, it would also compromise the other physiological systems to a point where the temperature increase could lead to exhaustion or even prove fatal. The provision of water during exercise—and more specifically the use of water to completely replace the fluid lost during exercise—is the best practice to keep core body temperature in a desirable range (figure 9.4).

Additionally, altered blood flow dynamics can cause a reduction in lactate clearance. Recall that the increase

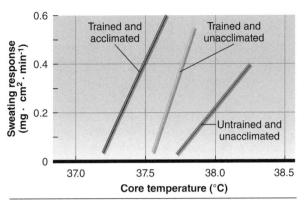

Figure 9.4 In untrained and unacclimatized individuals, the sweating response starts at a relatively high body temperature (37.7 °C; 99.9 °F). When trained the onset of sweating occurs at a lower temperature (37.5 °C; 99.5 °F), and when subjects are both trained and acclimatized sweating begins at an even lower body temperature (37.2 °C; 99.0 °F). Not only is the onset of sweating at a lower body temperature, but its rate of increase is also greater as noted by the steeper slopes of the lines.

Reprinted from M.L. Foss, S.J. Keteyian, and E.L. Fox, 1998, *Fox's physiological basis for exercise and sport*, 6th ed. (Pittsburgh, PA: William C. Brown). By permission of the authors.

in skin (cutaneous) blood flow results in the diversion of blood flow away from major lactate clearance sites, including the liver, kidney, and inactive muscle (primarily slow-twitch Type I fibers). One potential factor that would cause both an increase in blood lactate appearance and a decrease in lactate clearance is an increase in fast-twitch motor unit recruitment. Whether exercise in the heat causes an increase in fast-twitch fiber recruitment has not been definitively proven. However, it is known that fatigue is enhanced during exercise in the heat, which typically results in an increase in motor unit recruitment and a relatively larger reliance on fast-twitch motor units. Furthermore, lactate levels increase to a greater extent in hot conditions in individuals with a higher proportion of fast-twitch muscle fibers to slow-twitch fibers (100).

Neuromuscular Responses

Traditionally, the cardiovascular and metabolic changes that occur during exercise in the heat are implicated as the main reasons for decreased performance observed in hot or humid conditions. However, recent research suggests that the central nervous system may have an important

Heat Production

The calorie is the unit most commonly used to reflect heat energy with respect to metabolism. More than 70% of metabolic energy is converted to heat during movement. The calorie reflects the amount of heat energy required to increase the temperature of 1 g of water by 1 °C. The kilocalorie (kcal), as the prefix implies, is 1,000 times a calorie. The specific heat of the body (taking into account all types of tissue such as bone, muscle, and blood) is 0.83 kcal · kg^{-1} · °C^{-1}. An average 70-kg person would require 58 kcal of heat (i.e., 0.83 × 70) to increase core temperature by 1 °C. The amount of potential temperature increase of the body during exercise can be calculated using the following assumptions:

- For each liter of O_2 consumed, between 4.69 (fat metabolism) and 5.05 kcal (carbohydrate metabolism) is used for metabolism to generate ATP.

- The amount of kilocalories used per liter of O_2 depends on the percentage of fat and carbohydrate being used for metabolism.

- Potential degrees gained per hour = resting energy expenditure/(body weight × specific heat of body).

For example:

$$\frac{580 \, kcal \cdot h^{-1}}{70 \, kg \times 0.83 \, kcal \cdot kg^{-1} \cdot °C^{-1}}$$

$$= 10 \, °C \cdot h^{-1} \text{ potential body heat degree increase}$$
$$\text{(Equation 9.1)}$$

Note that 580 kcal · h^{-1} was calculated by assuming that an individual was consuming 2 L of O_2 · min^{-1} for 60 min at a rate of 4.83 kcal · L^{-1} (2 × 60 × 4.83 = 579.6 kcal · h^{-1}). If the body could not reduce or regulate its temperature, this type of heat production and core temperature increase would quickly lead to heat stroke and death.

role as well. During exercise in the heat, brain serotonin increases, which can produce feelings of fatigue and drowsiness (66). This may be a mechanism for central fatigue and diminished psychological drive to continue exercise, particularly prolonged exercise. Additionally, a study by a Danish laboratory demonstrated that hot conditions can decrease force-generating capacity in both active and inactive muscle groups. This diminished capacity returned to normal when these muscle groups were electrically stimulated (66). Combined, these studies suggest that the muscles could have generated more force but that inhibition of force generation occurred in the central nervous system. Future research is needed to understand exactly how the brain inhibits performance during exercise in the heat, but for the time being it must be considered as a mechanism for impaired exercise function in hot conditions.

Heat Acclimation and Acclimatization

Excessive heat substantially increases the physiologic and metabolic work of exercise. Repeated exposure to a hot environment results in several adaptations that reduce the magnitude of this increased work. These adaptations are referred to as *acclimation* and *acclimatization*. The term *acclimation* is used specifically for observed physiological adaptations to heat or humidity that occur in controlled conditions (e.g., a lab, gymnasium, or home). In contrast, the term *acclimatization* is used for physiological adaptations that occur due to repeated exposure to ambient conditions that are hot and possibly humid. The physiological adaptations presented in this section are based on both acclimation and acclimatization research and are summarized in table 9.3. Individuals who have attained a high degree of fitness before heat acclimatization will

Table 9.3 Physiological Adjustments While Working in the Heat Following Acclimatization

Physiological mechanism	Physiological adjustments
CIRCULATORY SYSTEM	
Pulse rate	Decrease

Skin blood flow
  Time to onset
  Skin blood flow (dry heat)
  Skin blood flow (humid heat)
	
	Decrease
	Increase
	No change

Blood volume	Increase
Blood pressure	No change
SWEATING MECHANISM	

Sweat rate
  Time to onset
  Sweat volume
	
	Decrease
	Increase

Evaporation	Increase
Salt loss in sweat	Decrease
METABOLIC	
Muscle glycogen utilization	Decrease
Blood lactate	Decrease
Oxygen consumption	No change or Increase
SUBJECTIVE SYMPTOMS	
Nausea	Decrease
General discomfort	Decrease

be less impaired when exposed to a hot environment and will acclimate faster than an individual who is less fit. Estimates are that exercise training produces at least 50% to 65% of the total physiological adjustment resulting from heat acclimatization and that the remainder of the acclimatization occurs simply due to heat exposure (31,67). As shown in figure 9.5, the sweating response does not begin until a high internal temperature is reached in individuals who are neither well trained nor acclimatized. Endurance-trained individuals have a lower threshold for initiating the sweat response, whereas individuals who are both endurance trained and acclimatized exhibit the lowest threshold (62). A lower threshold for sweating allows for better body temperature control in hot and humid environments, which is important for both safety (i.e., avoidance of heat injury and illness) and performance aspects of exercise.

Thermal Adaptations

Perhaps the most important adaptation that occurs with repeated bouts of exercise in the heat is a decreased susceptibility to heat-related illnesses (20). Total sweat output increases to allow for more efficient cooling (75). Exercise training can increase maximal sweat rate from 1 to 1.5 $L \cdot h^{-1}$ to more than 2 $L \cdot h^{-1}$. In addition, the threshold temperature for stimulating sweating is lowered so that there is a hastened response to heat stress (figure 9.4), including during activities of daily living and exercise performed for both health and athletic purposes (62). Sweat in acclimatized individuals is also more dilute due to less sodium (42). This effect minimizes the negative consequences attributed to electrolyte changes. These adaptations, combined with an increased ability to distribute blood flow to the skin, cause a decrease in core temperature in both resting and exercise conditions (17,79).

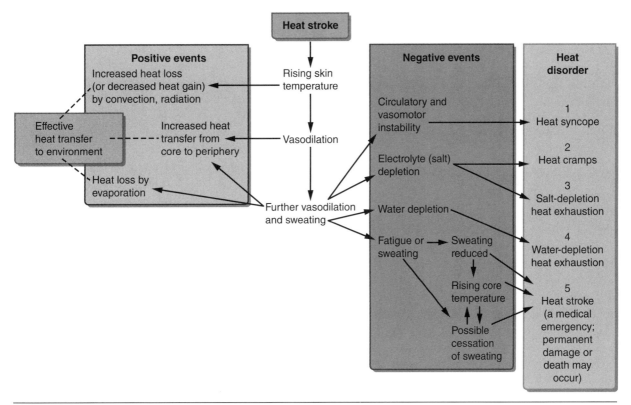

Figure 9.5 When an athlete is exposed to a heat stress during exercise, a number of events occur including an elevated skin temperature, vasodilation, and sweating. Generally, these are positive and facilitate effective heat transfer from the body to the environment and helps to minimize increases in core body temperature (left side of figure). Prolonged exercise in extreme conditions can lead to a variety of heat related disorders (right side of figure).

Reprinted from M.L. Foss, S.J. Keteyian, and E.L. Fox, 1998, *Fox's physiological basis for exercise and sport,* 6th ed. (Pittsburgh, PA: William C. Brown). By permission of the authors.

All of these adaptations typically are evident in as little as 5 to 10 d of exercise in hot conditions. Those with excessive body fat have difficulty with temperature regulation and may not acclimate to the same degree as someone with less body fat. Compared with males, females may have a reduced ability to acclimate to warmer environments due to a higher percentage of body fat and because their maximal sweat rate is lower.

Cardiovascular Adaptations

Because of the elevated cardiac output during submaximal exercise, the cardiovascular system is under a great deal of strain during exercise in the heat. However, within a 5-d acclimatization period, adaptations occur to reduce the degree of cardiovascular stress (98). With repeated exposure to exercise and heat, there is an increase in sodium retention by the kidneys and an increase in plasma proteins (53,83). Both of these responses result in expansion of plasma volume during exercise because both the sodium and plasma proteins act as osmotic attractors that retain plasma in the blood vessels. In some cases, resting plasma volume can increase up to 30% during acclimatization, and this response can occur in as little as 3 d of training (80,83). As a result of this increased plasma volume, venous return is not compromised as early in the sweating process. Also, as stated previously, an important adaptation occurs when blood flow to the skin increases during exercise in the heat after acclimatization (64,90).

Metabolic Adaptations

Much of the cellular metabolic effects that can occur during acute heat exposure during exercise (e.g., greater reliance on anaerobic metabolism, decreased muscle glycogen utilization) are partially corrected with acclimatization. Although the mechanism is not clear, submaximal $\dot{V}O_2$ can decrease slightly (78,99). This adaptation may depend on the mode of exercise because the reduction in submaximal $\dot{V}O_2$ appears to be greater for stair climbing than for cycling or running (98). What causes these differences across modes is unknown.

Muscle and blood lactate are substantially elevated during exercise in the heat (100). Although acclimatization does not reduce lactate levels to values observed during exercise in a cool environment, significant reductions can be measured. At least one research study suggests that acclimation reduces blood lactate during exercise in both hot and cool conditions (101). Although the mechanism for lactate reductions has not been identified, it is possible that acute alterations such

as sympathetic drive and glycogenolysis are reduced once acclimatization has occurred. Finally, if the **Q10 effect** increases glycolysis rate and subsequent lactate formation after heat acclimation, it would be expected that this effect would be reduced due to reductions in core temperature.

Time Course of Adaptations and Other Considerations

The best plan for acclimating or acclimatizing to the heat is not the same for everyone. Consideration needs to be given to how well an athlete tolerates hot conditions before exposure (if known through previous experience), the amount of time needed to acclimate and acclimatize, and the facilities and environment in which the acclimatization and acclimation process will take place. Most individuals can become acclimated within 2 wk of daily heat exposure, and many adaptations begin within 4 to 6 d of beginning the acclimation regimen (68). Research does not support the concept that continued exposure to a hot environment after the exercise session has finished or withholding fluids speed the acclimatization process or provide any further benefit. Therefore, fluids should be provided before, during (**ad libitum**), and after (enough to return to near pre-exercise body weight) the exercise session, and the athlete should be allowed to relax in a cool environment (e.g., shade, an air-conditioned room, or in front of a fan) when not participating in exercise.

If exposure to a warmer temperature via natural or artificial means is not feasible, research has shown that some adaptations can occur from wearing excess clothing (long sleeves and pants, thicker materials), which mimics a hot environment (25). However, this practice should be followed only in cool ambient conditions and should never include nonbreathable materials such as plastic or impermeable protective fabrics, as this can elevate the risk of heat-related illness. Most research suggests that the best acclimatization includes daily exercise of 120 min in the heat, although this can be broken into two 1-h sessions with similar results (82). Because the desired adaptations that produce better heat tolerance are not exercise specific (e.g., increased plasma volume, lowered core temperature, global changes in sweating rate), the mode of exercise can vary.

Thermal Injury

A very important consideration with respect to exercise in the heat is the development of heat disorders or thermal injury. These include syncope (fainting), cramps, exhaustion, and stroke. The physiological responses to exercise in the heat can be categorized as positive (e.g., transfer of excess heat to the environment) and negative processes.

The most common of these conditions is the muscle cramp, which is linked rather closely with dehydration and changes in plasma electrolyte concentrations, particularly the loss of sodium and fluids through sweating (7). Traditional thought is that the dehydration and electrolyte alterations actually cause the cramp, which results in intense pain from uncontrolled persistent muscle contractions in working muscles. However, more recently it has been suggested that the increased stress of heat causes muscle fatigue, increases muscle spindle sensitivity, and decreases Golgi tendon organ sensitivity (81). As a result, small stimuli can activate the muscle spindles and induce a cramp. In this scenario, the heat does not directly cause the cramp but rather hastens muscle fatigue, which subsequently causes the cramp.

Because of fluid loss via sweating, the risk of heat syncope increases if reductions in venous return become large. Heat syncope is characterized by dizziness, pallor of the skin, hypotension, and the risk of fainting. This is a significant concern, not only because syncope reflects heat-related strain but also because fainting can result in serious injury, as many people fall from a standing position. Heat exhaustion can result during strenuous exercise in the heat due to depletion of either water or sodium. Individuals with a high body mass are at an elevated risk. Symptoms of heat exhaustion include headache, dizziness, fatigue, irritability, tachycardia, hyperventilation, diarrhea, nausea, profuse sweating, **ataxia**, coordination problems, and vomiting.

Exertional heat stroke is a life-threatening condition in which the sweating mechanism becomes fatigued and, as a result, both skin and core temperatures become substantially elevated. Core temperature may exceed 105 °F or 40 °C. This can result in muscle flaccidity, involuntary limb movement, seizures, coma, tachycardia, and possibly death. Table 9.4 reviews heat illnesses, including predisposing factors and suggested treatment.

- Exertional heatstroke = hyperthermia (core body temperature 40 °C) associated with central nervous system disturbances and multiple organ system failure.

- Stimulants (e.g., amphetamines, Ritalin, ephedra, and alpha agonists) increase heat production; anticholinergic drugs (e.g., antidepressants, antipsychotics, and antihistamines) inhibit sweating; cardio-

vascular drugs (e.g., calcium channel blockers, beta blockers, diuretics, monoamine oxidase inhibitors) alter the cardiovascular response to heat storage; cocaine increases heat production and reduces heat loss by decreasing cutaneous blood flow.

- Children and elderly persons are particularly susceptible to heat accumulation due to decreased sweating ability, increased metabolic heat production, greater surface area:body mass ratio, decreased thirst response, decreased mobility, decreased vasodilatory response, and chronic medical conditions or medication effects (or both).

- Exertional heat exhaustion = inability to continue to exercise; may or may not be associated with physical collapse. Generally, this does not involve severe hyperthermia (>40 °C) or severe central nervous system dysfunction.

In general, immediate corrective action for the first four disorders includes rest, removing excess clothing, transferring the victim to a cooler environment, and replacing fluids and electrolytes. In the case of heat stroke, emergency medical attention should be obtained in addition to aggressive attempts to lower skin and core temperatures through ice packs or cold water (around 12 °C). Even athletes or individuals who recover from heat stroke can experience permanent damage to the thermoregulatory center in the hypothalamus and a higher risk of heat-related illness in the future (84).

Because of the progressively threatening nature of heat illnesses, it is important to aggressively prevent dehydration. During exercise in hot conditions, athletes should ingest at least 150 to 300 mL of fluid every 15 to 20 min, although this amount could increase with the degree of heat stress imposed. Furthermore, after exercise all athletes should ingest a fluid volume corresponding to 150% of the weight they lost during exercise. This is important because most individuals drinking ad libitum rarely drink adequately to replace the fluid they lost from sweating. Finally, all coaches, trainers, and affiliated parties should be educated about heat illness and be able to closely monitor for its signs and symptoms.

Fortunately, heat acclimatization improves exercise tolerance in the heat and can reduce the risk of developing all of these heat-related illnesses (78). Note that all of the stated adaptations that occur with heat acclimation or acclimatization occur with repeated sessions of exercise in the heat. Some acclimatization can occur if exercise is

Table 9.4 Predisposing Factors to Heat Illness During Exercise

Heat illness	Primary symptoms	Predisposing factors	Treatment
Exertional heat stroke	• Central nervous system dysfunction (disorientation, convulsions, coma) • Severe (>40 °C) hyperthermia • Nausea, vomiting, diarrhea • Severe dizziness and weakness • Hot and wet or dry skin • Increased heart rate and respiratory rate • Decreased blood pressure • Extreme thirst and dehydration	• Obesity • Low physical fitness level • Dehydration • Lack of heat acclimatization • Previous history of heat illness • Sleep deprivation • Sweat gland dysfunction • Sunburn • Viral illness • Diarrhea • Medications (e.g., stimulants, anticholinergic and cardiovascular drugs, cocaine) • Extremes of age • Uncontrolled diabetes or hypertension, cardiac disease	• Institute aggressive and immediate whole-body cooling with water as cold as practical within minutes of diagnosis until temperature is less than 38.3 °C. • Remove equipment and monitor rectal temperature. • Monitor vital medical signs (airway, breathing, circulation, central nervous system status) continuously. • Perform intravenous saline infusion if feasible.
Exertional heat exhaustion	• Fatigue and inability to continue exercise • Ataxia, dizziness, and coordination problems • Profuse sweating • Headache, nausea, vomiting, diarrhea	• Dehydration • High body mass	• Remove victim from practice or competition and move to a shaded or air-conditioned area. • Remove excess clothing and equipment. • Lay victim down with legs above heart level. • Rehydrate using chilled fluids or normal intravenous saline. • Cool athlete to less than 38.3 °C; aggressive cold-water immersion may not be required. • Monitor status and transport to emergency facility if needed.
Exertional heat cramp	• Intense pain • Persistent muscle contractions in working muscles during prolonged exercise	• Dehydration • Exercise-induced muscle fatigue • Large sodium loss through sweat	• Rehydrate and provide sodium replenishment. • Prevent by sodium loading (e.g., 0.5 g in 1 L of sport drink) in cases of heavy or "salty" sweaters. • Use light stretching and massage.

Symptoms and treatment summarized from the 2003 National Athletic Trainers' Association "Inter-association task force on exertional heat illnesses consensus statement."

Reprinted, by permission, from S.S. Cheung, 2010, *Advanced environmental exercise physiology* (Champaign, IL: Human Kinetics), 63.

performed in cooler conditions and the heat exposure is provided at rest by sauna (82,87). However, these adaptations are not nearly as effective as those observed for exercise performed in the heat. The best approach for any athlete preparing for competition or extended training in the heat is to follow a prudent schedule of exercise training in the heat and to work up to the point where the exercise volume normally achieved in a thermoneutral environment can be achieved in the hot conditions with the understanding that this will take up to 2 wk. Those with a higher conditioning status at the beginning of the acclimation process or who have had previous acclimation experience may need less time to acclimate than those with lower fitness levels or who are inexperienced with heat acclimation.

Sport Drinks and Other Coping Strategies

Apart from acclimatization, several strategies can reduce the risk of heat illness and aid performance in the heat. One of the most important strategies is adequate hydration. At a minimum, water should be available and ingested as needed during any exercise session performed in a hot environment. However, during profuse sweating, electrolytes are lost in addition to water. Therefore, as exercise duration progresses, it may become more important to consider some sort of commercially available sport drink that replaces lost fluid and electrolytes. These drinks also are more palatable than water; therefore, the athlete will be more apt to ingest a larger amount of fluid. One common fallacy is that heat-acclimated individuals do not need to consume as much fluid. In fact, because of higher sweat rates occurring earlier in the exercise bout, acclimatized athletes may need to ingest more fluids than nonacclimated individuals. The composition of the ingested fluid is important with respect to electrolytes and carbohydrate content. A recent trend involves the addition of protein to the recovery fluid intake immediately after exercise. These factors and others should be considered with respect to the rate of gastric emptying and the absorption of fluid. Table 9.5 presents several factors that are important when selecting the composition of the fluid.

An important part of the hydration process is the assessment of individual hydration status (table 9.6). Although measuring body weight before and after an acute exercise bout is an excellent method for determining initial fluid replacement, a variety of other assessments may be used to further refine acute hydration changes. These include bioelectrical impedance, urine color, specific gravity and osmolality, plasma osmolality,

and hemoglobin and hematocrit concentrations. Thirst perception, although useful as a stimulus to drink, often is not sensitive enough to rely on for adequate rehydration, particularly in those with significant dehydration. One potential problem with using thirst perception is that it may be diminished once an initial volume of liquid is consumed.

Recently, cooling vests have been shown to be effective in lowering skin and core temperatures and improving performance during exercise in the heat. In individuals wearing a cooling vest before activity (i.e., precooling), both time to exhaustion at 95% of $\dot{V}O_2$max and 5K running time showed improvement—the latter by an average of 13 s (4). Use of these vests may be a low-cost, effective method for decreasing the strain of exercise in the heat. Another cooling target includes helmets that have been used to induce brain hypothermia. Although little data exist on the effectiveness of this technique in athletes, it has been used in individuals during cardiac arrest to reduce the amount of neural damage that might occur. Intermittent cooling techniques also can be used during athletics; currently, the most visible technique is misting fans used on the sidelines of football games.

Cold

The effects of exercising in the heat are less vague than those of exercising in the cold. This is primarily because athletes usually wear adequate amounts of clothing, so the skin and core temperatures are not compromised. Additionally, dressing in layers allows for the removal of clothing as core body temperature increases. Therefore, the core temperature and skin temperature during exercise in the cold is a balance between the environmental temperature, the heat generated by the athlete, and the clothing that the athlete is wearing. In addition, the rate of heat loss may vary across individuals. Leaner and taller individuals lose heat faster due to an increased body surface area. Heat loss is also more effective in an aquatic environment (e.g., during swimming or surfing) than in an air environment because water is 20-fold more conductive than air. Furthermore, moving through water causes a faster rate of heat loss compared with passive submersion. Passive submersion causes the layer of water immediately surrounding the body to serve as an insulating layer, whereas movement of water over the skin disrupts this insulating layer and constantly replaces it with cooler water, thus hastening body cooling.

Table 9.5 Factors Affecting Gastric Emptying and Absorption of Fluid

Factor	Effect	Notes
Volume	• Increasing volumes of ingested fluid increase rate of GE, at least up to an ingested volume of 600 mL. • Increasing volume also increases net fluid absorption.	As gastric volume increases, distension of the stomach increases the activity of the stretch receptors in the gastric mucosa, thereby increasing gastric motility and speeding GE.
Temperature	Increasing fluid temperature (warmer beverages) decreases rate of GE.	Colder drinks can reduce gastric temperature, which may increase gastric motility.
Composition Carbohydrate Na$^+$ K$^+$	• Increasing carbohydrate concentrations generally slows the rate of GE. • High osmolality of a carbohydrate solution further delays GE, whereas low osmolality of a carbohydrate solution leads to a greater rate of fluid absorption. • Isotonic Na$^+$ solution increases GE compared with water. • Addition of NaCl to glucose solution increases intestinal absorption of glucose and water. • K$^+$ has no stimulatory effect on GE and may inhibit GE of carbohydrate-containing fluids. • K$^+$ has an osmotic effect that decreases GE (similar to that of carbohydrate).	Both carbohydrate concentration and osmolality influence rate of GE; however, carbohydrate content likely has a greater effect. To balance carbohydrate needs with fluid replacement, carbohydrate concentrations should not exceed 8%.
Exercise intensity	• Low-intensity exercise has little effect on GE. • Exercise at intensities >70% $\dot{V}O_2$max decreases rate of GE. • Intermittent high-intensity exercise at overall intensities <70% $\dot{V}O_2$max decreases GE relative to continuous exercise. • Exercise at moderate to high intensities may decrease rate of fluid absorption.	Mechanisms behind exercise effects may include exercise-induced release of vasoactive hormones, which may inhibit gastric contractility and slow GE; reduced splanchnic blood flow, which slows GE or reduces the rate of intestinal absorption (or both); and mechanical changes caused by upper body movements, increased respiration, or both, which influence gastric motility. The majority of studies examining GE and exercise use cycling as the mode of exercise. Field studies in competitive arenas also show exercise effects: The rate of GE was reduced during 30 min of competitive soccer (low to moderate overall intensity with intermittent high-intensity sprinting) compared with 30 min of low-intensity walking.
Exercise mode	Treadmill exercise (walking, running) at low to moderate intensities (<70% $\dot{V}O_2$max) has been shown to enhance the rate of GE compared with resting conditions.	Treadmill exercise may increase intragastric pressure via the contractile activity of the abdominal muscles, thereby influencing GE.
Hypohydration	Exercising in a warm environment when hypohydrated may decrease the rate of GE.	

GE = gastric emptying; Na$^+$ = sodium ion; K$^+$ = potassium ion; NaCl = sodium chloride

Reprinted, by permission, from S.S. Cheung, 2010, *Advanced environmental exercise physiology* (Champaign, IL: Human Kinetics), 63.

Table 9.6 Summary of Major Considerations in the Development of a Hydration Strategy for Exercising Individuals

Factors	Current consensus	Developing an individual hydration strategy
Sweat rates and composition	Large interindividual variability exists in sweat rates and composition. Women, children, and elderly persons generally have lower sweating rates than younger men. Increased fitness and heat acclimatization generally increase sweating rates. Hotter environments and the nature of activity (e.g., whether the activity involves protective clothing) can increase sweating rate.	Rely on individual determination of sweat rate and composition rather than general guidelines.
Hydration assessment		• Body weight, taken before and after exercise and over the course of days at consistent times, can give a general indication of sweating rates and fluid requirements. • Urine color (clear to straw color) and specific gravity <1.020 are simple complements to body weight measurement. • Visual indicators (e.g., high level of salt encrusted on clothes) can be used as a general guide to individual saltiness of sweat and electrolyte requirements.
Hydration effects	• Dehydration results in progressive decreases in plasma volume, thus increasing cardiovascular and thermal strain. • Activities reliant on maximal strength, power, or anaerobic metabolism appear to be unaffected by 3% to 5% body weight dehydration. • Aerobic capacity and submaximal exercise may become progressively impaired beyond 2% body weight dehydration, especially with exercise in the heat.	Athletes and workers should strive to drink sufficient amounts to avoid dehydration beyond 2% body weight over prolonged exercise. Even with minimal anaerobic impairment, cognitive impairment may increase the risk of errors and accidents.
Rehydration fluid composition	• Carbohydrate delivery of approximately 60 $g \cdot h^{-1}$ promotes carbohydrate availability and spares glycogen stores, enhancing submaximal exercise. Carbohydrate may have nonmetabolic benefits for high-intensity exercise lasting <1 h. • Carbohydrate concentrations of 4% to 8% do not impair gastric emptying. Simple sugars (e.g., glucose, fructose) in combination appear to empty most rapidly, but fructose >3% can decrease gastrointestinal absorption rate. • Small amounts (0.3-0.7 $g \cdot L^{-1}$) of sodium can increase palatability and drinking via thirst stimulation and decrease the risk for hyponatremia.	• Palatability of fluid should be a key consideration in promoting voluntary fluid consumption during exercise. This varies between individuals and conditions (e.g., tolerance for sweetness decreases in the heat). • Carbohydrate fluids of 6% to 8% concentration should be consumed for prolonged exercise lasting >1 h and possibly for high-intensity exercise lasting <1 h. • 0.3 to 0.7 $g \cdot L^{-1}$ of sodium can be added to fluid. • Individuals exercising >4 h should moderate their fluid intake to below sweating rates and avoid drinking solely water or low-solute fluids to minimize risk of hyponatremia.

Reprinted, by permission, from S.S. Cheung, 2010, *Advanced environmental exercise physiology* (Champaign, IL: Human Kinetics), 56, 57.

Shivering

If heat loss is great enough, some altered exercise responses can develop. Shivering, a process driven primarily by a reduction of skin temperature, involves muscle contractions that lead to an increased muscle metabolism and the production of heat with minimal movement. The result is a protective effect that serves to maintain core body temperature. If body temperature drops enough during exercise to cause shivering, the effect can be an increased submaximal $\dot{V}O_2$ and decreased exercise economy. Because shivering relies on carbohydrate stores, glycogen depletion can reduce or even eliminate shivering. This effect is not beneficial because if shivering ceases, the ability to maintain core body temperature is reduced. However, if an individual is exercising, it is less likely that core temperature would be substantially affected. This is one of the reasons why athletes who finish endurance races such as marathons are given thermal blankets. The athletes are in a glycogen-depleted condition and are in danger of rapid heat loss, and their shivering ability may be inhibited. Alcohol has a similar effect on shivering, leading some people to mistakenly surmise that alcohol ingestion warms the body by suppressing shivering.

Cardiovascular Effects

In contrast to cardiovascular responses in the heat, heart rate can be lower and stroke volume higher at a given cardiac output in cold conditions (57,69). This is because circulation to the skin and periphery is reduced in an effort to protect core body temperature, resulting in an enhanced venous return. It appears that cardiorespiratory capacity, as measured by $\dot{V}O_2$max, is not compromised unless core temperature is reduced.

In general, a 5% to 6% reduction in $\dot{V}O_2$max occurs with each 1 °C decrease in core temperature (8). In these cases, maximal heart rate and stroke volume can decline, the latter due to a reduction in cardiac muscle temperature resulting in a reduced contractility of the left ventricle. Bronchospasm is another possible way that $\dot{V}O_2$max is compromised during exercise in cold environments. If an athlete is susceptible to asthma or has a hyperreactive airway, inhaling cold air may cause a bronchospasm and significantly reduce ventilation to the point where O_2 intake is significantly compromised.

Furthermore, if skeletal muscle temperature drops, it may impair the activity of key metabolic enzymes and mechanical efficiency of the muscle via the Q10 effect.

Decreasing muscle temperature may also cause an increase in the recruitment of fast-twitch motor units, resulting in increased muscle and blood lactate concentrations and leading to fatigue (11). Other factors that could increase lactate accumulation include an increase in plasma catecholamine concentration and the aforementioned increased reliance on carbohydrate metabolism during land exercise in the cold. However, increased glycogen utilization appears to be more pronounced during lower intensity exercise, where the thermogenesis caused by exercise is not as great (100). Another hormonal change is an increase in insulin sensitivity during acute cold exposure (94). This could be due to an increase in skeletal muscle activity via shivering or an increased reliance on carbohydrate as the immediate fuel source during exercise in the cold. Despite these potential occurrences, exercising in the cold does not have to impair performance as long as the individual is prudent about protecting him- or herself with adequate clothing and exhibits caution if susceptible to bronchospasms when breathing cold air.

Health Risks of Cold Exposure

The two main concerns about exercising in cold conditions are frostbite and hypothermia.

Frostbite

Frostbite is a cold–dry thermal injury that occurs when skin temperature reaches between −2 and −6 °C (28.4-21.2 °F). This would require an environmental temperature lower than −29 °C (−20 °F) to freeze exposed areas, resulting in a crystallization of the skin cells. However, as table 9.7 illustrates, risk of frostbite is related to time of exposure and wind velocity in addition to ambient temperature (i.e., wind chill; see table 9.8). Someone who is exercising and generating body heat is able to tolerate lower temperatures and higher winds before experiencing frostbite. If only the epidermal or superficial skin layer freezes, the term used is *frostnip*. Although frostnip is painful, it has no long-term effects. Exercise can also cause an increase in effective wind velocity and thus affect the wind chill. Consider a runner moving at 8 mph in an environment with a wind velocity at 12 mph and a temperature of 10° F (−12.2 °C). When the runner has a tailwind, the effective wind velocity is 4 mph (12 mph − 8 mph) and the runner, if properly clothed, tolerates the cold well. However, when the runner turns around and runs into the headwind, the effective wind velocity becomes 20 mph (12 mph + 8 mph), and the runner is at an increased risk for frostbite of exposed areas.

Table 9.7 Minutes to Frostbite

Wind speed in miles per hour	Air temperature in °F								
	10	**5**	**0**	**-5**	**-10**	**-15**	**-20**	**-25**	**-30**
5	>2 h	>2 h	>2 h	>2 h	31	22	17	14	12
10	>2 h	>2 h	>2 h	28	19	15	12	10	9
15	>2 h	>2 h	33	20	15	12	9	8	7
20	>2 h	>2 h	23	16	12	9	8	8	6
25	>2 h	42	19	13	10	8	7	6	5
30	>2 h	28	16	12	9	7	6	5	4

Wind speed at kilometers per hour	Air temperature in °C								
	-12	**-15**	**-18**	**-21**	**-23**	**-26**	**-29**	**-32**	**-35**
8	>2 h	>2 h	>2 h	>2 h	31	22	17	14	12
16	>2 h	>2 h	>2 h	28	19	15	12	10	9
24	>2 h	>2 h	33	20	15	12	9	8	7
32	>2 h	>2 h	23	16	12	9	8	8	6
40	>2 h	42	19	13	10	8	7	6	5
48	>2 h	28	16	12	9	7	6	5	4

Reprinted from M.L. Foss, S.J. Keteyian, and E.L. Fox, 1998, *Fox's physiological basis for exercise and sport*, 6th ed. (Pittsburgh, PA: William C. Brown). By permission of the authors.

Table 9.8 The Wind Chill Index

	Ambient temperature										
Wind velocity (mpd)	50	40	30	20	10	0	-10	-20	-30	-4	-50
	(Equivalent temperature [°F])										
5	48	37	27	16	0	-5	-13	-26	-36	-47	-57
10	46	28	16	4	-9	-24	-33	-46	-58	-70	-83
15	36	22	9	-5	-18	-32	-45	-58	-72	-85	-99
20	32	18	6	-10	-25	-39	-53	-67	-83	-90	-110
25	30	16	0	-15	-29	-44	-59	-74	-88	-103	-118
30	28	13	-2	-18	-33	-48	-63	-79	-96	-105	-125
35	27	11	-4	-20	-35	-51	-67	-82	-98	-113	-129
40	26	10	-6	-21	-37	-53	-69	-85	-100	-115	-132
		Minimal risk				Increased risk				Great risk	

Reprinted from M.L. Foss, S.J. Keteyian, and E.L. Fox, 1998, *Fox's physiological basis for exercise and sport,* 6th ed. (Pittsburgh, PA: William C. Brown). By permission of the authors.

Hypothermia

Hypothermia is diagnosed when core temperature falls to 95 °F (35 °C) or less. This typically happens only with prolonged exposure to extremely cold air or water. Heat transfer is four to five times greater in the water than in the air. Paradoxically, it is possible to experience hypothermia without being in cold conditions. For example, 75 runners were treated for hypothermia during the 1985 Boston Marathon, a race run at an ambient temperature of 76 °F (24.4 °C) and a moderately high humidity level. Conversely, 250 runners were treated for hyperthermia that day. How could hypothermia possibly have occurred under these conditions? The affected individuals consisted predominantly of slower runners, and their rate of heat production during exercise was not as high as that of faster runners. Furthermore, most were dehydrated, resulting in a lower blood volume. During the race, heat loss via convection, conduction, and evaporation and sudden cooling by being sprayed with cold water resulted in a decrease in core temperature and hypothermia. The runners were treated via intravenous fluids and water ingestion, and the hypothermia abated within 1 h.

Considerations for Exercise Training in the Cold

The primary concerns for exercising in cold environments are maintaining a comfortable microenvironment close to the body (i.e., under clothing) and maintaining core body temperature. Clothing is the primary method of addressing these issues. When dressed properly, individuals can withstand extremely cold environments with little effect on exercise ability. Dressing properly for these extreme conditions is much easier now than in the recent past due to the invention of lightweight materials that wick water from the skin surface and provide an insulating environment. Dressing in layers is important for prolonged exercise so that a layer of clothing can be removed to reduce the amount of insulation when core temperature increases. Three layers typically can be used during exercise in the cold:

1. Base layer. This layer touches the skin and includes undergarments, socks, hat, gloves, and so on. The material should have the ability to pull moisture away from the skin (i.e., wicking)—for example, polyester or Lycra blends.

2. Insulating layer. This middle layer traps warm air. One common material is fleece, which maintains its insulating ability even when wet.

3. Protective layer. This is the outer shell that is in contact with the environment. It keeps wind, rain, and snow out of the other layers and from connecting with the skin. This layer should be water resistant rather than waterproof; this will allow the fabric to breathe, which can help keep underlayers dry.

Altitude

Exercising at high altitude presents a challenge due to a reduction of the partial pressure of O_2 (PO_2) in the blood. The cause of this hypoxic environment is low atmospheric pressure creating a reduced PO_2 relative to sea level. The percentage of O_2 (20.93%) in the atmosphere is the same everywhere on Earth, regardless of whether an individual is at sea level or at an altitude that is considered to impose a physiological stress. Consider exercising in Washington, D.C. versus Denver, Colorado. In Washington, which is nearly at sea level, barometric pressure would be around 760 mmHg (may fluctuate slightly due to the weather). Therefore, PO_2 would equal 760 mmHg × 0.2093, or about 159 mmHg. In Denver, barometric pressure would be about 650 mmHg. Therefore, PO_2 would decrease to around 136 mmHg (650 mmHg × 0.2093). Why is this significant? The PO_2 of the atmospheric air inhaled determines the PO_2 in the alveoli (table 9.9). O_2 diffusion occurs down the pressure gradient, from a high PO_2 in the alveoli to the lower PO_2 in the pulmonary capillaries. If alveolar PO_2 is reduced, the pressure gradient between the alveoli and pulmonary capillaries is also reduced. This ultimately results in a lower PO_2 in the arterial blood, as the PO_2 in arterial blood is approximately equal to the alveoli. See chapter 4 for more information regarding gas exchange physiology.

Furthermore, the reduced alveolar PO_2 reduces the diffusion gradient rate of O_2 between the alveoli and pulmonary capillaries. At rest, this doesn't present much of a problem because the system is overbuilt in that only 25% of the O_2 loaded onto hemoglobin is extracted by the tissues, and the transit time of a red blood cell through the pulmonary capillaries is three times as long as needed for full oxygenation (0.75 vs. 0.25 s). However, during exercise when capillary transit time is much faster, time for O_2 diffusion is reduced. Combined with a reduced P_AO_2 (partial pressure of O_2 in alveolus), the PaO_2 (partial pres-

Table 9.9 Effects of Increasing Altitude on the Partial Pressure of Oxygen in the Lung Alveoli (P_AO_2)

Altitude	Barometric pressure (mmHg)	PA – 47 (mmHg)	Alveolar oxygen extraction	PaO_2 (mmHg)
Sea level	760	713	.1475	105.0
5,000 ft	638	591	.1475	87.0
7,200 ft	586	539	.1475	79.5
10,000 ft	530	483	.1475	73.2

Reprinted from M.L. Foss, S.J. Keteyian, and E.L. Fox, 1998, *Fox's physiological basis for exercise and sport*, 6th ed. (Pittsburgh, PA: William C. Brown). By permission of the authors.

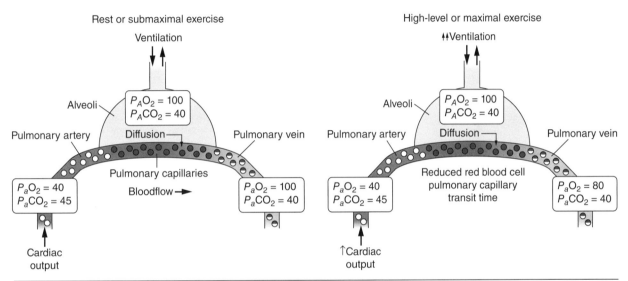

Figure 9.6 Movement of red blood cells through the structures of pulmonary gas exchange. The process involves ventilation, blood flow, and gas diffusion. An increased red blood cell transit may negatively affect oxygen uptake.

sure of O_2 in arterial blood) is reduced and O_2 delivery to exercising skeletal muscle may be impaired and adversely affect the ability to perform exercise. Figure 9.6 provides a schematic of this process.

Cardiopulmonary Responses

Imagine an individual cycling at 300 watts at altitude. This workload would require the same rate of O_2 uptake as it would at sea level. Although the absolute metabolic rate is not different, delivering the required amount of O_2 to the working muscles clearly is a greater physiological challenge. To compensate for this, two physiological adaptations occur very quickly. First, an increase in pulmonary ventilation (known as *hyperventilation*) occurs.

The response is an increase in PaO_2 (discussed further in the next paragraph). Second, cardiac output increases via an elevated heart rate, but no change occurs in stroke volume (73). Imagine that same individual performing a $\dot{V}O_2$max test at altitude. At some point, ventilation and cardiac output and thus O_2 delivery reach the maximum level and cannot fully compensate for the reduced PaO_2. A consistent finding is that $\dot{V}O_2$max is reduced at altitude compared with altitudes below about 5,000 ft (1,524 m). In fact, a strong inverse correlation exists between $\dot{V}O_2$max and altitude, as illustrated in figure 9.7 (28). At altitudes similar to what would be experienced at Mt. Everest, $\dot{V}O_2$max would be only 20% of the value at sea level (56).

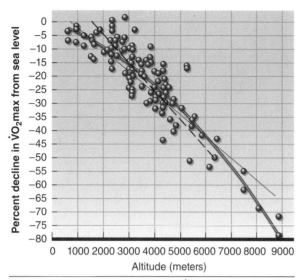

Figure 9.7 Percent decline in $\dot{V}O_2$ max as elevation increases.

Reprinted, by permission, from C.S. Fulco, P.B. Rock, and A. Cymerman, 1998, " Maximal and submaximal exercise performance at altitude," *Aviation Space Environmental Medicine* 69: 793-801.

Regarding elevated ventilation, although PaO_2 is sensed via chemoreceptors in the carotid arteries, this does not have much of an effect on regulating ventilation during exercise at sea level because PaO_2 rarely changes. However, because of the aforementioned decrease in PaO_2 at altitude, these chemoreceptors are triggered, which leads to a substantial increase in ventilation. The increase in ventilation occurs not only during exercise but also at rest (56). This causes more CO_2 to be exhaled. You may remember that CO_2 and pH are inversely proportional to each other. As a result, the lowered CO_2 causes a temporary increase in blood pH (i.e., less acidic) (56). After a few days, pH declines back to normal levels due to the increased excretion of bicarbonate in the urine. A temporary reduction in plasma volume also occurs (56) due to water vapor loss via the increased respiration and because the lowered humidity at altitude tends to cause a great deal of sweat evaporation.

Metabolic Responses

Because of the decline in $\dot{V}O_2$max, exercise at any given submaximal intensity will be performed at a higher percentage of maximum. This is a reason why submaximal exercise is not always well tolerated at altitude. The other reason is that lactate accumulation is higher at altitude (55), mainly because lactate production is accelerated due to a higher reliance on anaerobic ATP production in hypoxic conditions. Interestingly, peak lactate values at the end of a maximal exercise test will be lower at altitude than at sea level; this occurs to an even greater extent in individuals acclimatized to altitude (35). This is known as the *lactate paradox*.

Performance at Altitude

One of the best examples of the effect of altitude on performance is the results from the track and field competition at the 1968 Summer Olympics in Mexico City (7,382 ft or 2,250 m above sea level). In all the distance events (1,500 m to marathon), the winning times were on average about 5% slower than the world record times for those distances. In contrast, Olympic and world records were either tied or broken in all sprint, jump, and throwing events except one. For the distance events, it is clear that the reduced capability for O_2 delivery had a negative effect on performances. In contrast, the "thinner air" due to lower atmospheric pressure actually created a favorable environment for the sprints. Because sprinters, jumpers, and throwers do not rely heavily on aerobic metabolism for their performances, they were able to take advantage of the lower air resistance without any decrement due to reduced O_2. Additionally, there may be a difference in the effect of altitude on performance. Table 9.10 provides estimates of the negative effect (i.e., added total time) of elevation on various race distances at an altitude of 5,000 ft (1,524 m).

Table 9.10 Estimates of the Negative Effect of Elevation on Various Race Distances

	1,500 m	3,000 m	Steeple 5,000 m	10,000 m
Men	5.6 s	15.3 s	25.3 s	63.7 s
Women	6.6 s	18.1 s	30.6 s	76.2 s

Provided by USA Track and Field.

Acclimatization to Altitude

The vast majority of the distance running events in the Mexico City Olympic Games were won by athletes who were born and trained in high-altitude environments. Kipchoge Keino was raised in Eldoret, Kenya, which is more than 6,800 ft (2,073 m) above sea level. He won gold in the 1,500 m and silver in the 5,000 m. Similarly, Mamo Wolde, raised near Addis Ababa, Ethiopia (about 7,500 ft or 2,286 m above sea level), won gold in the marathon and silver in the 10,000 m.

The body can make adaptations to altitude that minimize the effect of successive exposures. Much research has been performed on the training processes of individuals at altitude. The "live high, train low" mantra for runners, swimmers, and cyclists has been around for a while, and many elite athletes spend time at altitude-based training camps (e.g., Colorado Springs, Colorado; elevation = 6,035 ft or 1,839 m) at different times in their training sequence. What types of physiological changes might they expect?

Physiological Adaptations

Most changes that occur with chronic exposure to altitude involve adaptations that aid in O_2 delivery. The hypoxia that occurs with altitude leads to an increase in the expression of hypoxia inducible factor-1 (36). This subsequently triggers the kidneys to produce erythropoietin, a hormone that stimulates red blood cell production from the bone marrow. An increase in red blood cell number along with a decrease in plasma volume creates a significant increase in hematocrit. This also results in an increase in hemoglobin concentration and blood viscosity (95). The effect of hypoxia inducible factor-1 also increases the transcription of vascular endothelial growth factor, which acts to increase capillary density in skeletal muscle, resulting in a greater surface area for O_2 diffusion (90). Furthermore, the muscle myoglobin content and mitochondrial density increase to allow greater utilization of O_2. Finally, 2,3-diphosphoglycerate levels increase, resulting in a lower affinity of hemoglobin for O_2 and enhancing O_2 uptake in muscle (43). The resultant effect is a right shift of the oxyhemoglobin dissociation curve (see chapter 4). Each of these adaptations results in improved O_2 delivery in an attempt to offset the reduced atmospheric PO_2.

Furthermore, the increased blood lactate accumulation evident during submaximal exercise at altitude is reduced; consequently, submaximal exercise is less fatiguing (54). It is interesting to note that the decline in $\dot{V}O_2$max that occurs at altitude is not markedly reversed

with altitude exposure (2). Therefore, the adaptations to altitude exposure allow submaximal exercise to occur with less fatigue and discomfort but do not substantially increase $\dot{V}O_2$max at altitude toward values achieved nearer to sea level.

Altitude Training and Sea-Level Performance

Logically, increased O_2 carrying and delivery capacity with altitude exposure should translate to improved endurance performance once the person returns to sea level. Interestingly, early research showed minimal benefits of altitude training as it pertained to improving sea-level performance (2,48). Anecdotally, many athletes and coaches attest to the benefits, so the findings were difficult to explain. Several athletic governing bodies such as USA Swimming (www.usaswiming.org) and USA Track and Field (www.usatf.org) endorse training at altitude. One potential cause of the negative findings in previous studies was the high degree of variability across athletes (15). It is likely that altitude training may be very effective for athletes who can train at altitude with only minimal effect on their total training volume. However, for athletes who must reduce their training due to the added stress of altitude, the physiological gains from altitude exposure likely are offset by the reduction in training, and the athlete returns to sea level in the same (or worse) physical shape. USA Track and Field recommends that an athlete should simply stay at their typical training location if they cannot spend a minimum of 14 days at altitude for acclimation.

The question then becomes whether there are methods that allow everyone to benefit from altitude training. A definitive answer is still largely elusive. Recent research has evaluated the concept of "live high, train low." In this scenario, the athlete sleeps and sometimes performs daily activities in an altitude environment but trains at sea level. Although this strategy often is impractical, it has been made easier with the advent of altitude chambers that mimic hypobaric conditions (i.e., a lower PO_2). These chambers can be used in daily resting conditions such as sleeping. Other, more sophisticated devices have been constructed to use in exercise situations, including a chamber that fits over the entire length of a 50-m pool lane!

Furthermore, an early study showed the benefits of alternating altitude and sea-level training (23). A group of six elite runners (which included Jim Ryan, who was the first high school athlete to run 1 mi [1.6 km] in under 4 min) trained in alternating 2-wk blocks at altitude and at sea level. At the end of the study, $\dot{V}O_2$max was significantly increased in all subjects, and five of the six subjects achieved personal best times, including world records in the mile and 1500 m. Interestingly, the runner who did not improve was also the one who was most physiologically compromised when exposed to altitude. However, this study also is considered flawed because it did not include controls. Therefore, however compelling the "live high, train low" concept may be, the research is still inconclusive and the findings are mixed (51). This is in part due to a lack of well-controlled studies that eliminate the placebo effect or the athletes' expectation that hypoxia will benefit sea-level performance.

Healthy individuals who are not athletes should have no issue with spending time at increased elevations where others commonly live. These include places such as Denver, Colorado, which sits just above 5,000 ft (1,524 m), or peaks of mountains in the range of 10,000 to 15,000 ft (3,048-4,572 m). Although healthy nonathletes may be fatigued more easily at these altitudes, particularly if they are deconditioned or obese, there generally is no danger. Some individuals should avoid or adequately prepare for visits to places of altitude, including those with coronary artery disease and various lung diseases that affect gas exchange. The reduced PO_2 at altitude for these individuals can lead to a deoxygenation of the blood and can result in angina (chest pain) or severe dyspnea (shortness of breath). Individuals with these diseases who decide to visit these altitudes should be counseled to limit their physical activity to avoid the need for significant increases in O_2 delivery to the heart and skeletal muscles. They also may use supplemental O_2 to maintain PaO_2 at an adequate level.

Underwater

Performing activities underwater presents two specific challenges: high atmospheric pressure (hyperbaria) and an atmosphere in which humans cannot breathe. To properly understand the challenges, risks, and physiology of underwater activities, it is important to recall the concept of partial pressure (discussed briefly in the section on altitude).

Gas diffusion takes place in the body on the basis of partial pressure. In the lungs, the rate of O_2 diffusion from the alveolus (P_AO_2) to the pulmonary capillaries (PaO_2) depends on the difference in PO_2. A large gradient will result in fast diffusion and vice versa. Also, partial pressure is determined by the total pressure of a mixed gas

and the percentage of a specific gas that makes up the air. Finally, it is important to understand that an inverse relationship exists between pressure and gas volume. For example, decreasing pressure causes an increase in the volume of a gas, whereas increasing pressure has the opposite effect.

Scuba Diving

Scuba is an acronym for "self-contained underwater breathing apparatus." This device allows individuals to descend underwater for prolonged periods of time because they can breathe. The scuba diver breathes pressurized gas through a valve that contains mixed air similar to atmospheric air (20.93% O_2, 0.03% CO_2, 79% N_2). The high pressure of the air overcomes the pressure exerted by the water on the body and lungs and helps the air enter the lungs, where it can be diffused. Because atmospheric pressure increases relative to diving depth, P_AO_2 is actually substantially higher than at sea level. At a typical atmospheric pressure, P_AO_2 is about 105 mmHg. However, this will increase during diving because the total air pressure in the alveolus increases. Although P_AO_2 depends on the depth of the dive, values greater than 140 mmHg have been recorded in deep-water divers who do not use breathing devices (72). The specific physiology of scuba diving poses inherent risks, which are discussed in the following sections.

Air Embolus

Consider a situation in which a scuba diver inhales while underwater and then rapidly surfaces while holding their breath. Because the volume of a gas is inversely proportional to pressure, the volume of air in the lungs increases while the diver is ascending. In the extreme, the increased volume can expand the alveolar sacs to the point where they rupture (figure 9.8a). This is why all scuba divers are trained to exhale as they ascend. Although the example used here is extreme, overdistention of the lungs can occur during scuba diving in as little as 5 ft (1.5 m) of water. As a result of alveolar rupture or damage, free air may be forced into the capillaries, forming air and blood emboli, which subsequently can restrict blood flow to the heart or brain.

Spontaneous Pneumothorax

The aforementioned alveolar rupture can also cause an accumulation of air in the intrathoracic space and result in a collapsed lung, or pneumothorax (figure 9.8b). This situation can be painful, and the affected lung usually needs surgical intervention.

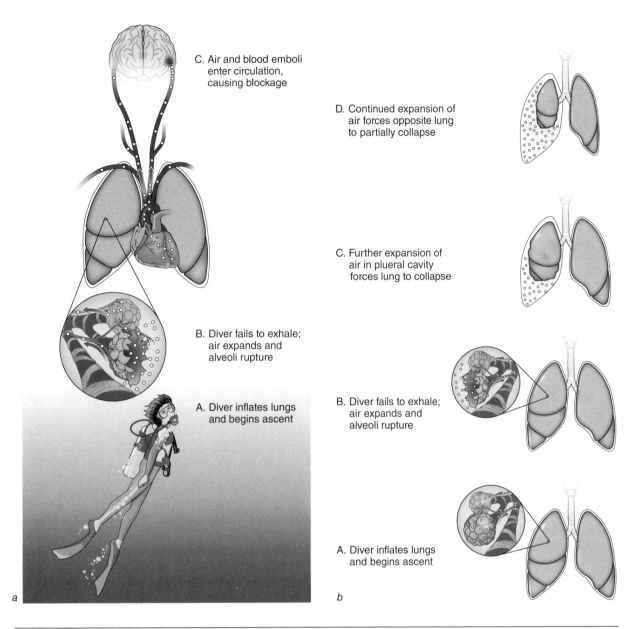

Figure 9.8   The first image depicts (a) the formation of an emboli as a diver ascends without exhaling. The second shows (b) the formation of a spontaneous pneumothorax which also can be caused by a diver's failure to exhale during an ascent.

Reprinted from M.L. Foss, S.J. Keteyian, and E.L. Fox, 1998, *Fox's physiological basis for exercise and sport,* 6th ed. (Pittsburgh, PA: William C. Brown). By permission of the authors.

Nitrogen Narcosis

If you shake a carbonated beverage, bubbles do not appear as long as the top remains securely closed on the bottle. However, once the top is removed, CO_2 bubbles can rapidly form to the degree that they quickly escape the container. Gases in solution can become liberated in the form of bubbles when pressure is suddenly decreased. This can happen in the human body if a scuba diver ascends too rapidly. Nitrogen, despite being an inert gas, is the most common gas in atmospheric air. During scuba diving, the increased atmospheric pressure that occurs while descending into deep water dramatically increases the partial pressure of nitrogen and forces it into solution in the blood. If the subsequent ascent is too rapid, pressure decreases rapidly and the nitrogen forms bubbles. As a result, gas emboli can form and cause blockages and subsequent tissue damage. The condition, which is quite painful, is known as *the bends.* Symptoms include pain in the legs and arms, dizziness, paralysis, shortness of breath, fatigue, and collapse with unconsciousness. Once a diver gets the bends, they need to be treated via recompression. The diver is placed in a chamber that artificially elevates atmospheric pressure so that the nitrogen bubbles are forced back in solution. After that point, pressure is slowly decreased until the individual has been successfully brought back down to standard sea-level atmospheric pressure.

Free Diving

Free diving involves submersion and exploration without the use of scuba and can involve underwater descents to very deep levels. The current world record stands at 100 m (328 ft) and took a little more than 4 min to accomplish. How is it possible to sustain oneself underwater for that long? First, let's look at the factors that limit our ability to hold our breath. If you try to hold your breath as long as possible, you shut off your ability to take in O_2 and to expel CO_2. As a result, PO_2 in the arterial bloodstream decreases, PCO_2 increases, and pH decreases (the latter because of the increase in CO_2). All of these changes trigger chemoreceptors, which subsequently stimulate the respiratory center in the brain stem. Thus, the urge to breathe becomes larger and larger, ultimately causing you to break your conscious decision to avoid breathing. Now imagine a scenario in which you try to hold your breath after hyperventilating. The increased breathing will cause you to expel a lot of CO_2, and the PCO_2 in your blood would be lower when you begin to hold your breath. Therefore, it would take a longer time for the increase in PCO_2 and decrease in pH to occur to the degree that you'd feel like you need to breathe again. As a result, the time that you can hold your breath increases. Finally, imagine someone hyperventilating and then descending into water. PCO_2 would start at a lower value and pH would start at a higher value due to the increased breathing. However, the increased atmospheric pressure creates a situation in which PO_2 is maintained at a high level. Therefore, even though the individual is underwater, these factors have been altered to the point that the urge to breathe is not the same as it would be at sea level. This can be manipulated even further by hyperventilating and breathing hyperoxic gas before submersion. This serves to artificially elevate

PO_2 before submersion and decreases the stimulus for breathing even further. The world record for breath holding is 19 min 21 s, and it occurred underwater after the individual hyperventilated pure O_2.

Of course, all of these manipulations are very dangerous, and experienced free divers do not recommend extreme hyperventilation or hyperoxic rebreathing before unaided dives. In this scenario, it is very easy to black out while ascending back to the surface. Although the increased pressure keeps PO_2 high while submerging, a corresponding decrease in atmospheric pressure occurs while ascending. This causes a rapid decrease in PO_2, and the diver risks PO_2 decreasing to levels that cause fainting while still being submerged. This has resulted in some fatalities among divers.

Microgravity

The influence that minimal gravity has on human physiology when exiting the Earth's atmosphere has been studied, from the first manned space flights in the 1960s to the current plans for people to inhabit Mars. Three major challenges have emerged with respect to microgravity: bone demineralization, muscle atrophy, and cardiovascular deconditioning. Bone demineralization occurs primarily as a result of the decreased stress on bone tissue in space. Muscular atrophy and decreases in strength have been reported during prolonged space flights, and the degree of atrophy is associated with the length of exposure to microgravity. The major results of cardiovascular deconditioning are lowered $\dot{V}O_2max$, alterations in orthostatic tolerance, and a decrease in fluid volume. The latter leads to reduced stroke volume and increased heart rate (99). Orthostatic tolerance refers to the inability to maintain blood pressure with changes in posture (e.g., moving from the supine position to the erect position).

Exercise Performance After Exposure to Microgravity

It has been suggested that $\dot{V}O_2max$ decreases upon return to Earth after a space flight. However, this is difficult to determine conclusively because all postflight studies of $\dot{V}O_2max$ have been confounded by the use of countermeasures designed to minimize the aforementioned physical changes that come from microgravity exposure. However, many studies on prolonged bed rest have observed approximately 5% to 10% reductions in $\dot{V}O_2max$ (19). It would be reasonable to assume that similar changes could be expected with space flight. Because of muscle atrophy, strength has been shown to decline with both simulated and true microgravity environments (19). In many of the Russian space flights, where exercise was an integral part of the daily routine, the aforementioned declines in muscular strength were minimal. This suggests that exercise can serve to minimize or eliminate the consequences of prolonged stays in space.

PERFORMANCE EXERCISE PHYSIOLOGY

Performance exercise physiology (also known as *human performance* or *sport performance*) focuses on helping individuals perform at their best in any situation. This is the true application of the knowledge of physiology as it pertains to exercise and performance. This application is used in athletics, in helping individuals perform physically demanding jobs, or even in helping those who simply want to move around easier in their daily lives. Careers that require an understanding of performance exercise physiology along with the ability to properly assess an individual and develop an effective training plan include athletic training, exercise physiology, clinical exercise physiology, occupational therapy, physical therapy, strength and conditioning, corporate wellness, and personal training.

Body Structure, Genetics, Physiology, and Performance

Is it possible to identify elite athletic talent potential at a very young age? Does genetics play a role in one's ability to excel at a sport? Plenty of evidence suggests that the answer to both questions is *yes*. But is it a "one size fits all" proposition? Because physical outliers exist (e.g., 5 ft 6 in. [1.4 m] basketball player Spud Webb, who won the National Basketball Association dunk competition at the 1986 All-Star game; find it on YouTube), it should be clear that any talent evaluation that uses physical characteristics cannot be 100% accurate. Figure 9.9 shows what can occur when assessing talent and comparing it with competition (40). There will be false positive (i.e., identified as talented but poor in competition) and false negative (i.e., identified as having poor talent but excellent in competition) outcomes. The goal of talent identification is to increase the success rate (identified in figure 9.9 by the oval area) so that the true positive and negative rates are as high as possible.

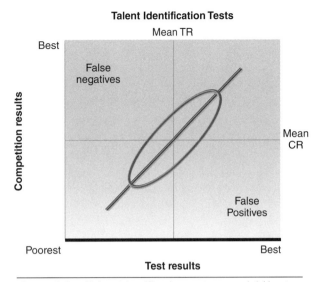

Figure 9.9 Talent identification system model (Australia). Possible outcomes of talent identification assessments compared with competition results.

Body Structure

In general, elite athletes fit the mold with respect to their body structure. In basketball, height typically is of benefit. In the 2004-2005 National Basketball Association season, the league reported the average player height as 6 ft 7.3 in. (2 m). There were 47 players listed as over 7 ft (2.13 m) and only 10 listed as under 6 ft (1.82 m). For comparison, the average American male stands just over 5 ft 9 in. (1.75 m) and weighs about 190 lb (86.2 kg). Among the shortest and lightest athletes are female gymnasts, who average under 5 ft 4 in. (1.63 m) and under 125 lb (56.7 kg). The heaviest American athletes are football players on the offensive and defensive lines. They average more than 300 lb (136 kg). Japanese sumo wrestlers are of similar size, but it is not uncommon for them to weigh more than 500 lb (226.8 kg). It is important to note that for most sports both height and weight are not at the extremes. However, for sports such as ice hockey, baseball, soccer (European football), and rugby, the average height is above that of the average male or female, and in some cases the average weight is higher. Based on this observational data, it would appear that for many sports a higher height and greater weight allow for an advantage over those of average stature.

It commonly is believed that taller individuals are less coordinated than shorter individuals. Other factors may play a role also. For instance, Newton's second law states that acceleration of a body is proportional to the force produced but inversely proportional to the mass, implying that athletes such as taller (and thus likely heavier) runners may be at a disadvantage because they have to produce more force to overcome their higher mass compared with a shorter (likely lighter) runner. In track and field, male sprinters are often around 6 ft (1.8 m). The recent exploits of Jamaican sprinter Usain Bolt, who stands 6 ft 5 in. (1.96 m), run counter to the norm. It is possible that other factors—including **biomechanics**, muscle mass, and muscle type (e.g., torso:leg length ratio, body composition, and muscle fiber composition, respectively)—that typically do not occur in tall individuals may have contributed to his success.

Counter to the premise that taller individuals have poorer coordination than shorter individuals is a recent study from China that examined the effect of height on motor coordination in individuals training in a university-based dance sport program (49). An assessment of basic dance skills was performed both before and after a 16-wk (32 h) training program. The 686 participants were divided by height using 5-cm increment groupings. The authors reported that the tallest individuals had the best dance skill scores, whereas the shortest individuals had the worst scores. Another study looked at a battery of tests as possible predictors of future elite-level female volleyball players (70). Although body height and jump height are predictors of volleyball success, particularly for the skills of spiking, blocking, and attacking, they did not distinguish between elite and subelite players. The authors concluded that the additional measures of motor coordination may help identify future elite-level players.

Talent identification programs have been used in various countries to attempt to determine future athletic potential. Both Germany and Australia have implemented national talent identification programs to assess swimmers and have had good long-term success. China has implemented talent identification programs for diving and has won 29 medals (27 gold) in the Olympics since 1988. Fédération Internationale de Natation (the international governing body for swimming) has reported on anthropometric assessments that relate to predicting future swimming success (58). They state that the following are important factors:

- Height: Although all swimmers have relatively tall stature, this is more important in sprint swimmers.

- Arm span: All swimmers need long upper extremities to achieve a better drive.

- Body fat: Buoyancy in water requires an optimum body fat content. The reported mean percentage body fat for male and female swimmers is 8.8 % and 16.5 %, respectively.

- Somatoype: Swimmers have an ectomorphic mesomorph body type. The muscular component of the body is dominant, and the swimmer has considerably lower fat content than the general population.

- Proportionality: The relationship between the lengths of the limbs is an important factor in some sports, including swimming. The brachial index expresses the ratio of forearm length to upper arm length and directly affects the mechanical advantage of the athletes' upper extremity for producing force and achieving higher power and velocities. This index is relatively higher in sprint swimmers compared with those competing in middle-distance and distance events. Sprint swimmers who have a higher crural index (the ratio between lower leg length and thigh length) have the same advantages.

Although it appears that anthropometric variables are important in many sports with respect to ability, it is likely that a multivariable model might be a better method. This type of assessment might include **kinanthropometric** measures, physiological assessment, psychosocial assessments, biomechanical measurements, and perceptual assessments, among others.

Another aspect of body structure is body composition, or the fat and nonfat makeup of an individual. General information about body composition, including assessment methods, can be found in chapter 8. With respect to performance, body composition can be very important. For instance, those who compete in sports in which weight classes exist (e.g., boxing, wrestling, rowing) may have to lose weight at times to "make weight." This also is often true for athletes competing in sports in which a lean or thin appearance is important (e.g., gymnastics, figure skating, dance). For both of these general types of athletes, body weight alone may not provide sufficient information when attempting to lose or gain weight. Also, the loss or gain of weight may hamper performance. For instance, those who lose weight quickly may become dehydrated, which can negatively affect performance. Those who restrict specific macronutrients (e.g., carbohydrate) may reduce the fuel sources available for training and competition. Specifics about nutrition for changing body composition are beyond the scope of this chapter. However, when the goal is a change in body composition, a proper diet is vitally important and specific to a sport.

Some athletes were famously lean. Charles White, a Heisman Trophy–winning running back for the University of Southern California in the 1970s, stood at 6 ft (1.8 m), weighed 185 lb (84 kg), and had a measured body fat of less than 2%! Compare that with studies that assessed college football linemen, in which the average body fat was 24.8% (30,89). Although athletic performance cannot be precisely predicted based on body composition, a certain body composition in athletics undoubtedly is important (38). Typically, a lean body (i.e., less fat and more nonfat tissue) leads to advantages in speed and power development. A higher body fat percentage may be advantageous for bulk and possibly for buoyancy in swimming, which can help with biomechanics in certain strokes. Table 9.11 provides a range of body fat percentages for males and females in 16 common sports. Factors that affect body composition include age, sex, and genetics (1). One might wonder whether a top-ranked athlete selects a sport or whether the sport ultimately selects the athlete. Because body composition appears to be important for some sports, the latter might be partially true for the high level athlete.

Genetics

What about the role of genetics? Some say that great athletes are born, not made. Chapter 11 discusses the link between genetics and general health and its relationship with fitness and exercise. An individual's genetic makeup certainly has an influence on performance measures such as strength and power, muscle fiber composition, lung capacity, and flexibility, among other attributes, but the amount of influence is difficult to quantify. Early studies involving monozygotic (identical) and dizygotic (fraternal) twins suggested that the heritability (i.e., genetic) level may account for 25% to 50% of the determination of $\dot{V}O_2$max (44,45). The same may be true for muscle strength and power.

Are there genes that can predict athletic performance? It undoubtedly is true that genetic factors play a role in athletic performance (33). In fact, more than 200 gene variants (of 20,000 genes in the human genome) currently are associated with elite athletic performance (12). The challenge of this type of assessment is matching the gene to the specific physical and mental requirements of a particular sport. Because these requirements vary among

Table 9.11 Common Body Fat Percentages of College Student-Athletes

Sport	Male	Female
Baseball	12-15	12-18
Basketball	6-12	20-27
Distance running	5-11	10-15
Football (backs)	9-12	No data
Football (linemen)	15-19	No data
Gymnastics	5-12	10-16
High/long jump	7-12	10-18
Ice or field hockey	8-15	12-18
Rowing	6-14	12-18
Shot put	16-20	20-28
Soccer	10-18	13-18
Sprinting	8-10	12-20
Swimming	9-12	14-24
Tennis	12-16	16-24
Volleyball	11-14	16-25
Wrestling	5-16	No data

Adapted, by permission, from M. Rockwell, 2015, *Body composition: what are athletes made of?* (Indianapolis, IN: NCAA), © 2015 National Collegiate Athletic Association.

sporting events, no single gene makeup will provide all the desired answers. A vast array of **phenotypes** such as body morphology (e.g., muscle fiber type composition, limb length, strength, power) and composition (e.g., body fat, leanness), among other traits, intermingle in a complex interaction to generate sport performance. Much of what determines these variables is related to an individual's gene makeup.

Two gene variants consistently have been associated with performance: ACEI/D (*angiotensin converting enzyme gene polymorphism*) for endurance events, and ACTN3 (*actinin alpha 3 gene*) for speed and power events.

1. ACE I/D. This gene is coded for angiotensin-1 converting enzyme and was the first gene variant associated with human performance. This gene is involved in the regulation of a portion of the renin-angiotensin system. In those with high blood pressure, the use of an angiotensin-1 blocking medication (i.e., angiotensin converting enzyme inhibitor) can help reduce systolic blood pressure because this system is involved in blood pressure control. Specific polymorphisms of the ACE gene (i.e., ACE insertion/deletion [I/D] polymorphism) have been associated with human performance in a variety of populations. It initially was connected to performance in a group of military personnel before and after 10 wk of military training (i.e., basic training). Investigators found the I/I **genotype** to be very strongly associated with exercise endurance, whereas the I/D genotype was less associated with a moderate endurance training effect. The D/D genotype was associated with a nonsignificant change in endurance. The I allele consistently has been associated with endurance performance in men and women and seems to be expressed to a greater degree as training for competitive distances increases (e.g., 10K vs. 20K swimming events or 400 m vs. 5,000 m running events) (61,97). The D allele is shown to be associated with strength and power activities and performance. For instance, an excess of the D allele was expressed in swimmers competing over shorter distances of less than 400 m (96) and in events of less than 200 m (93). This genotype has been assessed in at least 40 studies; approximately 75% reported a significant association of ACE I/D with performance.

2. ACTN3 X/R. This gene codes for the protein alpha-actinin-3 and is found only in Type II muscle fibers. A polymorphism in position 577 can lead to a termination codon (X) rather than an arginine (R). A study of elite European athletes demonstrated that power and strength athletes were less likely to have the ACTN3 XX genotype, whereas endurance athletes were 1.9 times more likely to have the XX versus the RR genotype (26). The more elite the level of athlete, the greater the likelihood of having the XX genotype. Those with the XX genotype lack alpha-actinin-3 in their muscle fibers.

The identification of potential elite athletes at a young age would allow for earlier specialization of training. (Of course, this leads to a discussion of whether it is more important to specialize at an early age or to become a well-rounded athlete, which is outside the scope of this chapter.) With respect to genetics, this raises the question of whether genetic testing at an early age might be useful. The DNA sequence does not change, so when a young athlete is tested, the DNA sequence is the same as it will be when the athlete is physically and mentally mature. A quick search of the Internet produces several companies that market genetic testing, particularly (but not exclusively) of the ACTN3 genotype. In their marketing materials, they state that they can provide information that will suggest whether an individual has a genetic advantage for endurance or power events. Because these tests are relatively new (they were first developed in 2004), it is still too early to understand whether genetic testing provides an advantage that leads to improved athletic performance. At this point, it is likely that the marketing for these types of tests—which are directly aimed at coaches, parents, and athletes—is ahead of the research findings.

Genetic understanding has other potential uses. For example, it may provide information about the risk of certain types of injuries, cardiac arrhythmias or hypertrophy, and concussion recovery rate. Despite the promise of genetic testing, as with testing for diseases such as cancer, many other environmental factors can play a role in the expression of a gene. These factors include skill level, access to training facilities and adequate coaching, support in other aspects of life (e.g., academics, finances), and an individual's work ethic and desire to optimize their athletic potential.

Finally, a new concept in enhancing performance is whether an alteration in one's genetic profile can occur and, if so, whether it can enhance performance. This concept has roots in the gene therapy research now happening in medicine; the goal of such research is to treat or cure diseases. However, to date no medical gene therapy treatments are approved by the U.S. Food and

Drug Administration. As well, no known sport-related gene therapy exists. Clearly, using this type of therapy for an athletic advantage poses an ethical concern that includes consent for the injection of genetic material, as it may be performed in the very young; safety of the process during the acute injection and in terms of short- and long-term health consequences; and whether the therapy is accessible to a wide range of individuals throughout the world. Elite athletes being assessed for genetic manipulation and other performance-enhancing drugs or procedures should have a genetic or biological passport (i.e., an electronic record of a variety of biological markers, including genes, that could be compared with future samples to assess for variances outside of a normal limit). To date, the biological passport process has been used in a number of cycling and track and field athletes resulting in a ban from future sporting participation.

Designing Exercise Training Programs for Performance Improvement

Chapters 6 and 7 present the general aerobic, anaerobic, resistance, and flexibility training principles and discuss them with respect to improving general fitness and health. These same training principles can be adapted and applied to individuals who wish to compete in athletic events with a goal of maximizing performance. In this section we discuss some specific alterations that are designed to improve athletic performance.

Training for peak performance can be considered with respect to the current training and life status (e.g., age, schooling) of an individual and the ultimate competition that will take place in the individual's lifetime. For instance, a 10- or 12-yr-old can train for peak performance at the end-of-season championship, whereas a collegiate or postcollegiate athlete may train for an Olympic trials competition. A great example is provided by Bob Bowman, the current head swimming coach at Arizona State University and better known as the coach of Michael Phelps, the greatest swimmer of all time in terms of world records and Olympic medals. Bowman became Phelps' coach when Phelps was about 11 yr old. Very quickly, Bowman noticed Phelps' long-term potential and began a course of training aimed at the Olympics. Although Phelps was an above-average age-group swimmer and set

a few records, it was not until a little later that his performances became exceptional. At just 15 yr of age Phelps made the 2000 Olympic team that competed in Sydney, Australia; this was ahead of Bowman's timeline targeting the 2004 Olympic Games in Athens, Greece. It illustrates the difference between training for peak performance immediately versus training for peak performance in the long term. Bowman often speaks of training in terms of "capacity" and "utilization"; *capacity* refers to developing a base of fitness and skill over years of training, whereas *utilization* refers to acutely using the fitness capacity for a peak performance.

With any type of training, the principles of specificity and overload are important. The greatest benefit from training occurs when the athlete performs sport-specific training (e.g., swimmers swimming, runners running, baseball players hitting and throwing). However, it should be noted that adjunct training that is designed to enhance a specific type of performance can be performed for most team and individual athletic events. For instance, a football player likely will benefit from a strength and power training program implemented in a resistance training environment. Strength and power training can assist the football player in many ways, including enhancement of speed, agility, and endurance. However, this type of training likely would have little effect on a skill such as catching a football while running at full speed.

Specificity training for sport also should focus on adequately stressing the fuel systems (i.e., phosphocreatine, anaerobic glycolysis, and mitochondrial respiration) to produce ATP at a sufficient rate. Table 9.12 depicts the metabolic systems along with the types of activities that rely on each. It also provides the likely physiological cause of fatigue for individuals relying on each of these metabolic systems. Understanding these relationships is useful for the development of specific training programs for improving performance in each system. However, some types of fitness can transfer from one mode to another. To accomplish a training mode transfer, the training stimulus and neuromuscular patterns used in an activity need to closely mimic the other activity for which a carryover benefit is desired. Table 9.13 lists the energy systems and the objectives for improving the metabolic functions of skeletal muscle. Training should be specific to each of these listed objectives. Finally, interval training allows for race specificity and the ability to stress the different energy systems. When developing an interval training session, the following questions are important to consider: What energy system is being focused on? What is intensity based on (e.g., heart rate, subjective effort, race pace)? What is the distance or time of the interval? What is the work:rest ratio? What type of recovery will be performed (i.e., rest or active)?

Creatine Phosphate–Specific Training

As discussed in chapter 1, creatine phosphate (CP, or phosphocreatine) is a molecule that serves as a substrate to quickly donate a phosphate group to adenosine diphosphate (ADP) to reform ATP. Within the first 2 to 7 s after a skeletal muscle contraction, ATP is reduced to ADP by breaking a high-energy chemical bond that provides the skeletal muscle with the needed energy ($7.3 \text{ kcal} \cdot \text{mol}^{-1}$ of ATP). Because of the principle of specificity, any training that attempts to enhance the CP rephosphorylation process needs to focus on movements that are very high in intensity and very short in duration.

A goal is to enhance the rate of rephosphorylation of ADP. This relies on the availability of substrate (i.e., ADP and CP) and the concentration and activity of creatine kinase. Stores of CP reflect the amount of creatine available. Creatine is produced endogenously by the liver ($1\text{-}2 \text{ g} \cdot \text{d}^{-1}$) as well as obtained exogenously through dietary consumption (nutrition is discussed further later in this chapter).

The rate of ATP generation can be affected by the concentrations of ADP and ATP (stored ATP concentrations in muscle are always low) and temperature. The ADP:ATP ratio cannot be manipulated; however, the process of warming up before intense exercise training or competition may positively affect the rate of the chemical reaction CP + ADP + H → ATP + creatine. The concentration of the enzyme creatine kinase can also affect this rate of reaction. Creatine kinase has been shown to have increased activity after a bout of resistance training (52). This increased creatine kinase activity was independent of creatine supplementation. Training can also be specific

Table 9.12 Metabolic Systems and Their Causes of Fatigue

System	Activity	Cause of fatigue
Creatine phosphate (CP)	Throwing, jumping, kicking, short sprints, single duration or intermittent work (<10s)	↓ CP, ↓ ATP/ADP, ↑ Pi
Glycolysis	Longer sprints, ice hockey, climbing, single duration or intermittent work (<3-4 min)	↓ CP, ↓ ATP/ADP, ↑ Pi, ↑ metabolic acidosis, membrane potential
Mitochondrial respiration (aerobic)	Anything continuous and >5 min duration	Substrate availability, dehydration, hyperthermia, muscle damage, membrane potential

ATP = adenosine triphosphate; ADP = adenosine diphosphate; Pi = inorganic phosphate.

Table 9.13 Objectives for Improving the Metabolic Functions of Skeletal Muscle for the Three Different Energy Systems

Energy system	Objectives	Mechanisms
Creatine phosphase (CP)	↑ Potential ATP regeneration from CP ↑ Rate of CP regeneration during recovery	↑ Muscle CP concentration ↑ Muscle fiber sizes
Glycolysis	↑ Potential ATP regeneration from glycolysis	↑ Muscle glycogen concentration ↑ Glucose uptake ↑ Muscle fiber sizes ↑ Blood and muscle buffer capacity
Mitochondrial respiration (aerobic)	↑ Maximal oxygen delivery ↑ The exercise intensity that can be sustained from mitochondrial respiration ↑ Producing the exercise duration that can be sustained from mitochondrial respiration	↑ Maximal cardiorespiratory function ↑ Heart volume or size ↑ Blood volume ↑ Muscle capillary density ↑ Muscle mitochondrial density ↑ Muscle glycogen store ↑ Muscle damage

ATP = adenosine triphosphate; ADD = adenosine diphosphate.

Are Football Players Fit?

Determining the fitness of a professional athlete sometimes depends on the component of fitness being assessed. Fitness often is described as having five domains: aerobic or cardiopulmonary fitness, musculoskeletal strength, musculoskeletal endurance, flexibility, and body composition. Let's use professional football players as an example. In 1991 we measured the peak $\dot{V}O_2$ of a National Football League (NFL) All-Pro running back to be 45.2 mL $\cdot$ kg^{-1} $\cdot$ min^{-1}. According to the American College of Sports Medicine (76), this places this early-20s running back in the 40th percentile ("fair" rating) for cardiorespiratory fitness among Americans. This man is considered by some to be the greatest college and NFL running back of all time, yet his peak $\dot{V}O_2$ was average or slightly below average compared with the general American public. However, cardiopulmonary endurance was not necessary for him, and he spent the majority of his training time on anaerobic exercises designed to enhance his strength, speed, and power. Thus, he could run a 4.37 s 40-yd (36.6 m) dash and had the ability to quickly move through an opening in the offensive line. This 40-yd time would have ranked him fifth in the 2017 NFL Combine predraft assessment. He was able to do all of this with a body mass index (BMI) of 30.9 kg $\cdot$ m^{-2}. Although this BMI would be considered obese in the general population, his body fat percentage was below 10%.

Based on this information, how would you rate this Hall of Fame player's fitness level at the prime of his career? He is slightly below average for cardiorespiratory fitness and obese based on his BMI. Each of these variables is a known risk factor for the development of cardiovascular disease, and this player's values would place him at risk; however, their effect on a professional athlete is not well understood. Although conflicting information exists, it appears that the life expectancy of an NFL player is less than that of the general population. Most NFL players have peak $\dot{V}O_2$ values similar to those of our All-Pro example. Shields et al. (86) reported that the average peak $\dot{V}O_2$ was 43.0 mL $\cdot$ kg^{-1} $\cdot$ min^{-1} for linemen, 47.2 mL $\cdot$ kg^{-1} $\cdot$ min^{-1} for quarterbacks, and 50.1 mL $\cdot$ kg^{-1} $\cdot$ min^{-1} for receivers and defensive backs. None of these values rise above the 60th percentile for the population and may be related to the development of heart disease, hypertension, and diabetes. Regarding BMI, it has been reported that despite an increase in BMI in NFL players since the 1970s, overall height and body fat percentages have remained essentially unchanged. Kramer et al. thus suggest that BMI is not a proper indicator of health status in football players because it often overestimates disease risk compared with body composition analysis (46).

to the ATP-PC process. This training tends to consist of short-term bouts and can involve resistance or locomotion (e.g., running, cycling, swimming, jumping). A comprehensive review of potential training protocols specific to ATP-PC enhancement is beyond the scope of this chapter. However, here are a few guidelines to follow:

- Total interval duration: 1-30 s

- Intensity: 95% to 100% of peak effort; close to, or at, race pace

- 1-3 d $\cdot$ wk^{-1}; never consecutive days

- 1:3 work:recovery ratio

- Recovery type: typically rest with a goal of fully resynthesizing ATP

  - The rest interval can affect the physiological response.

- A shorter rest interval (e.g., 1 min) versus a longer rest interval (e.g., 4-5 min) results in a possible increase in $\dot{V}O_2$max.

- Shorter rest periods require the athlete to slightly decrease the intensity of the sprint.

- Distances are the same as or less than competitive distance.

- Drills and strength training are important for optimal performance.

Anaerobic Glycolysis and Lactic Acid–Specific Training

As exercise duration progresses beyond about 10 to 15 s, the anaerobic glycolysis and lactic acid metabolic process becomes dominant for producing the required ATP until about 90 to 120 s. Those training for peak performance in this range of exercise duration (e.g., 200-800 m sprinters, hockey players, and many swimmers) require both a good $\dot{V}O_2$max and the ability to exercise at a high percentage of their $\dot{V}O_2$max. The lactate (also known as *ventilatory*) threshold is the intensity corresponding to an abrupt increase in muscle and blood lactate (figure 9.10). Exercising at intensities above this threshold quickly can result in fatigue due to a number of factors, including a reduction in muscle and blood pH from a doubling of [H⁺] (i.e., hydrogen ion concentration) compared with rest, of which up to 94% of the H⁺ comes from lactic acid (39). Two goals of exercise training in these situations are to elevate the $\dot{V}O_2$ at which the threshold occurs (figure 9.10) and to increase the muscle and blood buffering capacity to counteract the reduction in pH. An increase in mitochondrial volume (more enzymes, including those for carbohydrate metabolism such as cytochrome oxidase, succinate dehydrogenase, and citrate synthase) and mass (i.e., increase in total size and surface area) will achieve both of these goals. However, the increase in peak $\dot{V}O_2$ and lactate threshold are independent mechanisms because an individual can improve lactate threshold without improving peak $\dot{V}O_2$ (24). Training to enhance exercise in the 1- to 3-min range also enhances muscle buffering capacity through an increase in the concentration of carnosine.

Training to enhance these aforementioned processes should focus on intervals of 30 to 180 s at intensities at or above the anaerobic or lactate threshold (i.e., race pace). With respect to heart rate, this often is in the range of 90% to 100% of the maximum heart rate or heart rate reserve. If measuring lactate, values would be well above the onset of blood lactate accumulation of about 4 mmol · L⁻¹, often increasing to values of 10 mmol · L⁻¹ or greater. For running, this often will be at a pace faster than the distance to be focused on. For instance, a 1,500-m runner might focus on 400- or 800-m repeat intervals that are 2 to 8 s faster than the pace to be maintained during the race. The work:rest ratio often is 1:2, with the recovery period being active; this will enhance the rate of buffering and metabolism of lactic acid. If the recovery period is too intense or not long enough, then lactic acid levels may remain elevated and blood pH may remain lower (i.e., more acidic). In some types of training, this might be considered advantageous. In swimming, for instance, a lactate set often involves repeats in which the rest period is inactive (the athlete is seated in a chair) in order to maintain higher blood lactate levels for the next repeat interval. The goal is to enhance the physiological stimulus in order to improve buffering capacity. Another tactic in threshold training is to

perform an active recovery between intervals to allow blood lactate levels to return as close to resting levels as possible. This allows the athlete to understand how much active recovery is necessary, allows for training of the physiological processes of lactate clearance, and allows for a likely faster next interval compared with a rest recovery.

A comprehensive review of training protocols specific to anaerobic glycolysis and lactate threshold training enhancement is beyond the scope of this chapter. However, following are a few examples from different disciplines.

Football

Although much of football training focuses on technique (e.g., blocking, tackling, throwing, catching), it is important to focus on anaerobic training. Players must perform at a maximal level for 3 to 7 s, recover, and repeat many times over the course of a game. These athletes often perform a series of sprint repeats lasting as little as 2 to 5 s (speed training) and as long as 10 to 20 s (speed endurance training) with recovery times of 10 and 5 times the exercise time, respectively.

Running

Individuals who run distances of up to 800 m are fueled predominantly by anaerobic glycolysis. Those who run longer distances (1,500-10,000 m) also insert lactate

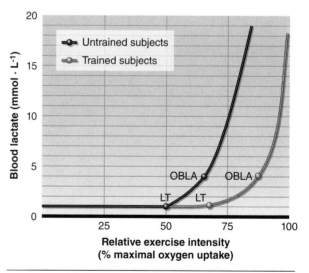

Figure 9.10 The change in the onset of blood lactate accumulation (OBLA) and peak blood lactate levels from the untrained to the trained state. LT=lactate threshold.

Reprinted, by permission, from NSCA, 2008, Bioenergetics of exercise and training, J.T. Cramer. In *Essentials of strength training and conditioning*. 3rd ed., edited by T.R. Baechle and R.W. Earle (Champaign, IL: Human Kinetics), 28.

threshold workouts into their training regimens, although to a lesser degree. The thought process is that the final 400 m of these types of races often require sprinting and thus a reliance on anaerobic metabolism and an ability to generate and tolerate a high blood lactate level. Following are training examples.

1,500-m Competitor

- Repeat interval: 400 m
- Pace: 1 to 4 s faster than average 400 m during 1,500-m race
- Repetitions: 5
- Work:rest ratio: 1:2
- Recovery: active with light aerobic exercise (jog or walk)

5,000- to 10,000-m Competitor

- Repeat: continuous (tempo run) or cruise intervals
- Pace
  - Tempo: 30 to 40 s slower than 5,000 pace; 15 to 20 s slower than 10,000 pace
  - Interval: mile repeats at near 5,000 or 10,000 race pace
- Repetitions
  - Tempo: not applicable
  - Interval: 2 to 4 min (total should be no more than 8% of total weekly training volume to avoid overreaching or overtraining)
- Work:rest ratio
  - Tempo: not applicable
  - Interval: short (only 30-60 s between intervals)
- Recovery
  - Tempo: 5 to 10 min of easy running after tempo run
  - Interval: active (walk or jog)

Aerobic and V̇O₂max–Specific Training

Recall that the cardiorespiratory system, the metabolic capability of the skeletal muscles, and the neurologic system work together as an integrated system. These systems can be trained to greatly enhance endurance performance to allow very high power outputs for long periods of time. Although a high $\dot{V}O_2$max is strongly related to endurance performance, it alone does not guarantee success. Other factors such as resistance to

movement (i.e., drag), velocity at the lactate threshold, economy of movement (i.e., technique), and fuel supply and use are also important. For instance, Derek Clayton from Australia was the first runner to break 2:10 for the marathon; he did so in 1967. His measured $\dot{V}O_2$max was 69.7 mL · kg⁻¹ · min⁻¹ (5.09 L · min⁻¹), which is 10 to 15 mL· kg⁻¹ · min⁻¹ lower than that of his predecessors. How did he run so fast? It was reported that he could race at 86% of his $\dot{V}O_2$max, meaning that his lactate threshold was very high (21). Costill et al.'s conclusion was that Clayton's running success depended on his excellent running economy (i.e., economy of movement) and his ability to utilize a large fraction of his well-developed aerobic capacity (i.e., lactate threshold).

Training for endurance racing–based athletics (e.g., running, cycling, swimming, rowing) often focuses on increasing $\dot{V}O_2$max. There certainly is a place for lower intensity, long-duration training for those who compete in long-duration events (e.g., marathon or ultramarathon running, triathlons of various lengths). In the 1950s through the 1970s, this type of training was very popular and often called *long slow distance (LSD) training*. For instance, Derek Clayton regularly was running 170 to 250 mi (273.6-402.3 km) per wk. Since then, there has been a decreased emphasis on distance and an increased emphasis on quality (vs. quantity) and higher intensity training. The recent marathon world record holder, Haile Gebrselassie of Ethiopia, has a maximum weekly distance of 156 mi (251.1 km). Today, many runners, cyclists, and swimmers train at much lower volumes but incorporate a variety of interval work, allowing them to train at higher intensities. The following are brief descriptions of several types of training that are specific to aerobic performance.

- Long duration, moderate intensity: Performed 1 to 2 times/wk, typically at race distance or longer (30-120 min) at an intensity under the lactate threshold (e.g., approximately 70% $\dot{V}O_2$max).
- Moderate duration, high intensity (pace/tempo): Performed 1 to 2 times/wk for 20 to 30 min at an intensity at or slightly above the lactate threshold.
- Fartlek or speed play: Typically performed 1 time/wk for 20 to 60 min with variable intensities during a continuous bout of exercise. Intensity may vary from approximately 70% $\dot{V}O_2$max to at or above the lactate threshold. An example would be running slower for 1,000 m and then at a higher intensity for 400 m and then repeating.

- Cross-training (not to be confused with the commercial program CrossFit®): Involves using a different mode of exercise training with the goal of transferring the training effects gained from the different mode to a desired exercise mode. An example might be someone who cycles as part of training for long-track speed skating. This occurs frequently; the logic is based partially on the use of the upper leg muscles in both sports. Another example is someone who performs water-based exercise, such as a running motion, to facilitate recovery from an injury suffered during running. Using resistance training is another example of using an alternative exercise mode to gain strength and power that might transfer to another sport (e.g., wrestling, basketball, swimming). Cross-training also allows an athlete to increase overall training volume with a reduced risk of overuse injury compared with performing only one type of exercise.

Numerous studies have assessed the effects of cross-training. Beattie et al. (6) reviewed studies that assessed the effects of resistance training on endurance running. They reported a positive effect on time trial performance at various distances and running economy. In a study assessing increases in training volume (running and swimming) on $\dot{V}O_2$max and a 2-mi (3.2 km) run time trial, Foster et al. (27) reported that mode did not affect $\dot{V}O_2$max in a group of trained runners. Both groups improved their 2-mi time trial, but the running group improved by twice as much as the swim group. These example studies demonstrate that cross-training can enhance training volume using an alternative modality. This can occur while enhancing both strength (if resistance training is used) and aerobic or endurance performance. Unfortunately, although cross-training is used in almost all competitive sports, little scientific data on its effects are available. Despite some research evidence and an abundance of anecdotal evidence, some believe that cross-training does not adhere to the principle of training specificity with respect to cardiorespiratory, neuromuscular, and skeletal muscle exercise. Although in general this is true, it seems that gains can translate to a different modality. However, those gains will likely never exceed those achieved through sport-specific training.

These types of athletes might also incorporate some of the lactate threshold training methods previously reviewed, including various types of interval training. However, intervals for these individuals are of longer duration than are intervals for those training for lactate threshold improvement. Table 9.14 provides typical training practices of elite-class runners, swimmers, and cyclists. These data are generalizations of today's training for these sports. Through the years, training volumes, percentage of time spent training the various metabolic systems, the use of cross-training, and the amount of time spent on technique work have changed. For example, in the 1970s swimming was focused on very high volumes of training (12,000-20,000 m/d), whereas today's top-level swimmers train from 7,000 to 15,000 m/d.

Taper

When a racing or other non-team-oriented athlete is focusing on a particular event, usually a championship (e.g., conference, state, national, Olympics), a reduction in training volume occurs along with an increase in training intensity. This commonly is known as *tapering*. This important concept typically is not used in team sports such as basketball, football, or baseball. In those sports, it is important to be at a high level for each competition during both the regular season and playoffs.

The taper typically involves a process lasting 2 to 3 wk in which training volume is progressively reduced. During this period, however, it is very important to maintain or even elevate the intensity of the training. The goal is to correctly reduce volume and increase intensity so that conditioning does not decrease. Volume can be reduced by exercising less frequently for less time. The reduction in volume should be graded over the length of the taper. Reductions in training volume of 60% to 90% have been reported with successful tapers (37). Training frequency should be reduced by no more than 50%. Some coaches do not reduce training frequency during the taper training period as they attempt to regulate training to the time of day and number of sessions to be performed at the competition.

If performed properly, the taper will allow for skeletal muscle recovery and power enhancement and improved psychological status (i.e., improved mood and reduced perception of effort). Additionally, in swimmers it has been demonstrated that a taper after an intense 22-wk training period positively influences the hemoglobin and hematocrit of highly trained swimmers (59). A study by Zarkadas et al. (102) evaluating two taper

Table 9.14 Training Practices Typical of National and Elite-Class Runners, Swimmers, and Cyclists

	Volume of training	Aerobic training	Anaerobic training	Training at race pace or faster	Seasonal variations
Runners	80-100 mi/wk	1-3 hr runs	Fartlek or longer intervals 1-2 times/wk	1 time/wk	• Greatest training volume achieved during off-season and preseason • Peak 1-2 times/yr
Swimmers	7,000-12,000 m/day	Frequent, longer intervals with very short rest periods	At least 1-2 shorter distance interval sets/day (approximately 500-1,000 m)	Generally performed only during 2-3 wk taper period or 1-2 times/wk during month prior to taper	• Peak 2 times/yr • Shortened off-season (2-3 wk/yr)
Cyclists	350-500 mi/wk	2-6 hr rides are common	Interval work or Fartlek 1 time/wk	Time trial usually 1 time/wk or every other week	• Mileage increased as season progresses • May peak several time/yr

Reprinted from M.L. Foss, S.J. Keteyian, and E.L. Fox, 1998, *Fox's physiological basis for exercise and sport*, 6th ed. (Pittsburgh, PA: William C. Brown). By permission of the authors

durations in triathletes training for an Ironman competition demonstrated a 4% to 6% improvement in 5K run time that occurred along with an improved (from 71% to 75% of $\dot{V}O_2$max) anaerobic threshold. Neary et al. (63) evaluated 11 cyclists who performed a 7-d taper. They showed that a 4.5% improvement in race time during a 20K time trial correlated with a 34% reduction in tissue oxygenation (i.e., ability to extract more O_2 vs. pretaper) as assessed using noninvasive near-infrared spectroscopy and a 4.5% improvement in $\dot{V}O_2$max. Finally, the amount of glycogen stored in skeletal muscle and the liver can be enhanced after a taper compared with the intensive training period. This can be desirable for endurance athletes and those competing in multiple daily sessions or over a period of several days. A successful taper, whether for running, swimming, cycling, or triathlon, should produce a minimum improvement of 2% to 3% in race performance (60). Figure 9.11 depicts the anticipated direction of several physiological variables.

Nutrition and Ergogenic Aids

A sound nutrition strategy can improve human performance; likewise, poor nutrition strategies—in particular, the failure to maintain high glycogen stores—will degrade performance. This effect was investigated when research administered three different diets (low carbohydrate and high fat, mixed, and high carbohydrate and low fat) to men who then performed a submaximal cycle ergometer exercise bout to exhaustion (10). The effect was a progressively higher muscle glycogen content as the amount of carbohydrate in the diet increased; exercise duration was threefold longer in the high-carbohydrate, low-fat group than in the low-carbohydrate, high-fat group.

How can an individual maintain this type of diet in real life? The temptation to eat junk food is everywhere, can be difficult to resist, and can have a profound negative effect (e.g., low glycogen stores, weight gain, increased ratio of fat mass to lean mass) on performance. Despite an

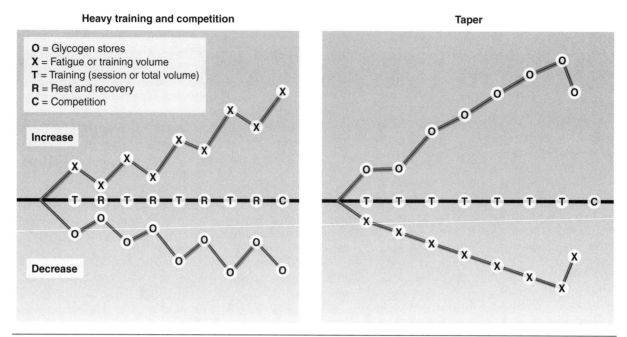

Figure 9.11 The direction of improvement of glycogen content and fatigue during intensive training and their subsequent improvement during the taper process.

abundance of research, not all athletes eat for the purpose of improved performance, and many athletes do not have adequate resources for guidance. For example, a recent report from the United Kingdom found that netball and hockey coaches who responded to a survey could correctly answer only 60% of the sport nutrition questions (16). This may not be completely surprising because only 25% of the coaches reported having any formal nutrition training. There is much to know about sport nutrition, including (but not limited to) the following:

- Macro- and micronutrient and energy needs
- Fluid needs
- Adequate recovery
- Weight control and body composition
- Use of supplements or ergogenic aids

Similarly, a study of National Collegiate Athletic Association Division I, II, and III athletics indicated that only 36% of coaches and 9% of athletes had adequate knowledge of sport nutrition (92). However, a key finding of this study was that both athletic trainers and strength and conditioning specialists often had good knowledge of sport nutrition (78% and 82%, respectively). It likely would be beneficial for almost any competitive sport team to have access to a qualified registered dietitian nutritionist or nutritionist with a specialization in sport nutrition (e.g., Board Certified Specialist in Sports Dietetics from the Academy of Nutrition and Dietetics or a Certified Sports Nutritionist from the International Society of Sports Nutrition). These professionals are specifically trained to work with coaches and athletes to determine whether any nutrition-related issues may negatively affect performance. A nutrition professional could also work with the coaching and training staff to provide general nutrition education to be disseminated to the athletes over the course of a competitive season.

Other issues pertaining to nutrition are timing and frequency of meals with respect to training and competition, athletes who do not eat enough calories to maintain total and lean body weight during periods of heavy training, and athletes who eat intentionally (e.g., vegetarian) or unintentionally (e.g., junk food or food aversions) limited diets that do not provide balanced nutrition.

Meal and Nutrient Timing

Research shows that the timing of macronutrient intake can affect both the adaptive response to exercise training and performance during competition. When timing meals for peak performance, individuals should consider the type of training performed, the time of day of the training, and the rate of desired recovery for the next optimal performance (i.e., practice or competition). Table 9.15 provides a sample meal plan for an athlete who performs two separate training runs (at 7 a.m. and 5:30 p.m.) on a single day. The next several sections provide an overview of the macronutrients for the athlete in training.

Carbohydrate

Carbohydrate stored as glycogen in the skeletal muscle (300-400 g) and liver (75-100 g) is used as the immediate energy source for physical activity and for prolonged exercise above approximately 65% of $\dot{V}O_2$max. It has long been understood that diet can enhance muscle and liver glycogen levels (9). Depending on the athlete and the type of training, skeletal muscle stores can range from less than 100 mmol · kg^{-1} of wet muscle weight to approximately 130 mmol · kg^{-1} for the endurance trained and possibly more than 200 mmol · kg^{-1} when a carbohydrate-rich diet is consumed (also known as *glycogen loading*).

Exercise performed for 120 min at a minimum of 60% to 70% of $\dot{V}O_2$max will lead to near depletion of the stores of carbohydrate in skeletal muscle and the liver. Low levels of carbohydrate stores lead to a reduction in performance; thus, carbohydrate is referred to as a "rate limiting fuel."

Research in the 1960s suggested that glycogen depletion during exercise could be avoided or delayed by the consumption of a high-fat diet followed by glycogen repletion with a diet containing high levels of carbohydrate several days before a race. The following are recommendations for when to ingest carbohydrate under various exercise conditions (41).

- Consume carbohydrate if performing exercise for more than 90 min at intensity levels of 65% to 85% of $\dot{V}O_2$max.

- Eat a general diet of approximately 8 to 10 g · kg^{-1} · d^{-1} of high glycemic index carbohydrate to maintain maximal endogenous glycogen stores.

- Ingesting 1 to 2 g of carbohydrate per kg of body weight 3 to 4 h before practice or competition generally is adequate; however, this can be altered depending on factors such as exercise duration, intensity, and fitness level.

- Because carbohydrate availability during exercise is related to endurance exercise performance, carbohydrate intake during exercise may be beneficial when exercise is performed for more than 60 min at a moderate to high intensity.

- Carbohydrate intake of 30 to 60 g · h^{-1} can maintain blood glucose levels.

- Athletes should consume a 6% to 8% carbohydrate solution in 8 to 16 oz (237-473 mL) every 10 to 15

Table 9.15 Sample Meal Plan for a Runner Performing Two Training Runs on a Single Day

Time	Meal	Diet
7:00 a.m.	Pre-breakfast run	Water
8:00 a.m.	Breakfast	Porridge and banana
10:00-11:00 a.m.	Mid morning	Slow-release carbs (e.g., sweet potatoes)
1:00-2:00 p.m.	Lunch	Lean protein and salad (e.g., eggs, fish)
60-90 m before training	Pretraining	High protein, slow-release carbs (e.g., lentils)
Immediately after training	Posttraining	Protein shake and fruit
8:00-10:00 p.m.	Dinner	Fish or chicken and steam vegetables

The goal is to enhance training performance and recovery (e.g., glycogen replenishment) between sessions.

min of exercise. Ideally, this would begin near the start of a ≥ 60-min exercise session.

- Mixing carbohydrate forms (glucose, fructose, sucrose, maltodextrin) can enhance skeletal muscle carbohydrate use rate from 1 g of carbohydrate per min to 1.2 to 1.75 g of carbohydrate per min. Avoid excessive fructose intake to reduce the chance of gastrointestinal upset.

- Consume carbohydrate within 30 min after an exercise bout to maximize glycogen resynthesis of the skeletal muscles and liver.

- Up to 10 g of carbohydrate $\cdot$ kg^{-1} $\cdot$ day^{-1} is recommended to replenish glycogen stores.

Protein

Protein turnover is defined as the constant process of protein synthesis and breakdown. With respect to skeletal muscle, this concept is important for hypertrophy–**atrophy** balance, particularly for those competing in various athletic events or exercise training. Protein use during exercise typically makes up about 2% to 6% of total energy use. In a glycogen-depleted state, protein is used to a greater degree. This is an important concept for those who fail to adequately replenish glycogen stores between bouts of exercise (training or competition) and those who exercise to a point where glycogen stores are greatly reduced. Recent research suggests that consuming carbohydrate in combination with protein before a resistance training session can enhance the amount of skeletal muscle protein synthesis compared with consuming carbohydrate and protein after exercise (91).

- Ingesting 0.15 to 0.25 g of protein $\cdot$ kg^{-1} of body weight 3 to 4 h before practice or competition generally is adequate; however, this can be altered depending on factors such as exercise duration, intensity, age, and fitness level.

- Consuming protein (or essential amino acids) before exercise has been recommended as a method for enhancing protein synthesis (47) during what is termed the *anabolic window*. However, controversy exists as to whether pre-exercise protein is beneficial, and general sufficient dietary protein intake may be adequate (32).

- The International Society of Sports Nutrition reports that consuming protein and carbohydrate at a ratio of 3 to 4:1 (carbohydrate:protein) during exercise has been shown to increase endurance performance during both acute exercise and subsequent bouts of endurance exercise (41).

- Consuming carbohydrate and protein during exercise may reduce the amount of exercise-induced muscle damage (5).

- Postexercise (immediately to 3 h postexercise) intake of essential amino acids may stimulate protein synthesis.

Fat

Although fat is an important fuel, particularly for endurance performance, fat consumption traditionally has not been thought to have an ergogenic effect on exercise. Because fat enters metabolism in the tricarboxylic acid (Krebs) cycle, it produces ATP at a slower rate than carbohydrate through anaerobic glycolysis. Several studies have assessed exercise performance when individuals consume a high-fat diet versus a lower fat, high-carbohydrate diet (71,88). In general, these studies suggest no benefit or harm from consuming a high-fat diet during submaximal exercise. Despite these findings, some coaches state that increases in blood or muscle levels of free fatty acids or intramuscular triglyceride content along with adaptations of the tricarboxylic acid cycle can aid performance. Part of the argument for a performance effect is the potential sparing of glycogen during events of long distance or duration. This is a plausible theory, and some research evidence supports this hypothesis (77,85). However, high-fat (e.g., paleo, Atkins) or ketogenic diets have not been aggressively studied with respect to exercise performance, and the studies that have been done have reported equivocal findings (77). Therefore, high-fat diets and fat loading currently are not recommended for athletes.

Fluids

It is well established that the loss of 1% to 3% of body weight can impair exercise performance (3). This loss is primarily due to sweating, which can reach rates of 2 to 3 L $\cdot$ h^{-1} in hot and humid environments. A large portion of the fluid volume used for sweat comes from the vascular compartment and therefore decreases plasma volume. This can have effects on cardiac output and the ability to continue to sweat at sufficient rates to control body temperature. Fluid intake during exercise is very common and is used primarily to replace water lost through sweating. Fluids can also be a vehicle for delivering specific nutrients, including the macronutrients carbohydrate and protein and also sodium and other electrolytes. The consumption of these nutrients in liquid (vs. solid) form allows for quicker absorption into the bloodstream. Therefore, consuming various fluids to maintain vascular fluid volume and carbohydrate levels has an ergogenic effect that can occur during both exercise training and competition.

Endurance athletes consuming large amounts of water should note the risk of hyponatremia. This occurs when serum sodium concentration falls below 130 to 135 mEq · L^{-1}. The incidence of hyponatremia in ultraendurance events is reported to be 0.03%, or about 1 person for every 3,000 competitors (65). Symptoms include dizziness, fatigue, headache, and possible mental confusion and nausea.

Rehydration after prolonged exercise is an important ergogenic aid because it prepares for the subsequent bout of exercise. Rehydration typically serves two purposes: to ensure that body water and fluid content return to normal, especially after a bout of exercise associated with heavy sweating, and to deliver the macro- and micronutrients needed to replenish body stores. Fluid retention is enhanced when drinks contain a higher osmolality of carbohydrate and electrolytes compared with the blood.

Supplements

A variety of nutritional supplements are touted as **ergogenic aids**. In the United States, dietary supplements are under the purview of the Food and Drug Administration. However, the regulations are not as strict as those for conventional food and drugs. The Food and Drug Administration's stance is that a substance deemed a dietary supplement is considered safe until proven to be unsafe. A comprehensive review of potential ergogenic supplements is beyond the scope of this chapter. However, table 9.16 provides a brief summary of several common supplements that athletes often use to promote improved performance.

Ergogenic Substances

Since the 1968 Olympics, there has been an ever-changing list of drugs and other substances considered to be illegal ergogenic aids. Table 9.17 lists several of these. In an effort to keep sport "clean," many governing bodies for both amateur and professional sports test for these and other substances. Depending on the sport and the testing period, a sample of either

urine or blood may be assessed. This testing can occur both during competition and during noncompetition periods. Positive drug tests have been reported in 1% to 2% of world-class athletes. The World Anti-Doping Agency (WADA) is an independent international organization that is responsible for developing antidoping policies and monitoring procedures. WADA accredits testing laboratories around the world for this purpose. WADA reports the results to governing organizations, which then use the information to determine whether sanctions (e.g., removal from further competition) is warranted. Samples have been saved for years on cohorts of athletes and tested when a new analysis is developed. For instance, six track and field athletes who competed in the 2005 world championships failed an ergogenic drug test when their samples were reanalyzed in 2013.

Summary

- Exercise performance can be affected by the environment, an individual's genetics and nutrition habits, and ergogenic aids.
- Excessive heat and humidity can adversely affect performance and potentially result in serious medical issues, such as heat illness and stroke.
- Understanding the physiological effects of heat and humidity can help control and reduce adverse effects. Clothing, fluid intake, and acclimation can be modified in an attempt to maintain or enhance performance.
- Cold temperatures can result in skin injury when an individual is not properly clothed, especially when conditions are windy. This environment also may lead to hypothermia.
- When an individual exercises at altitude or underwater, gas exchange at the lungs and diffusion within the body at the tissue level may be adversely affected. In these extreme environments, specialized knowledge is required in order to deal with potential adverse effects.
- Genetics—particularly its effect on stature and sport-specific physiology—can play a role in performance. Despite genetics, the performance of all athletes can be altered at least somewhat through training status, proper nutrition, and coaching.
- Training for optimal performance should be specific to energy systems and skills.

Table 9.16 Legal Nutritional Supplements Used to Enhance Athletic Performance

Supplement	Source	Proposed mechanism	How to use	Precautions	Beneficial?
β-alanine	• Synthesized by the body • Available in powder form (pill or mix in liquid)	Delays fatigue by increasing blood buffering (H+) capacity Enhances ability to train intensely and recover	Approximately 3-6 g · d⁻¹	Generally safe May result in flushing and tingling	Possibly, but research to date is limited
Branched-chain amino acids	• Leucine, isoleucine, and valine typically provided in a concentrated supplement	Affects signals to the brain that cause perception of fatigue (i.e., central fatigue) Enhances protein synthesis	5-g dosages 2-4 times/d often recommended	Not recommended if pregnant or breastfeeding or within 2 wk of surgery due to possible elevation of blood glucose	Possibly; does not appear to enhance performance but enhances recovery
Caffeine	• Naturally occurring and added to certain foods, including energy drinks • Available in pill form	Stimulates the central nervous system to improve alertness, concentration, vigor, and motor unit recruitment Increases cAMP to augment lipolysis and FFA mobilization, sparing glycogen	Consume before exercise Can use daily Usually 3-9 mg · kg⁻¹ doses	Small potential risk of arrhythmia for some Can cause withdrawal symptoms	Yes, but legal only if <12 µg · mL⁻¹ of urine (NCAA allows 15 µg)
Carnitine	• Animal-based food • Available in pill form	Increases FFA (lipid) catabolism Glycogen sparing	Not recommended	Possible side effects: gastrointestinal distress, headache, hypertension	No
Creatine	• Meat, fish • Synthesized by the body • Available in pill form	Increases creatine and CP in skeletal muscle to enhance high-intensity performance May enhance lean body mass	Loading dose: 20 g · d⁻¹ for 5-7 d Maintenance dose: 10 g · d⁻¹	Minor risk of dehydration and muscle cramps but no major risk profile	Yes
Sodium bicarbonate	• Baking soda	Enhances blood buffering capacity Reduces risk of blood acidity (lower pH)	60-90 min before intermittent exercise, allowing for adequate muscle blood flow and H+ buffering 20-30 g with at least 0.5 L of water	Possible gastrointestinal distress Do not use for the first time before an important event	Yes

FFA = free fatty acid; NCAA = National Collegiate Athletic Association; cAMP = cyclic adenosine monophosphate.

Table 9.17 Illegal Ergogenic Substances

Substance	Source	Proposed mechanism	How it's used	Precautions	Beneficial?
Erythropoietin	• Hormone produced by the kidneys • Synthetic available as an injection for those with a reduced ability to produce red blood cells	• Stimulates bone marrow to produce red blood cells • Increases hematocrit	Synthetic form is injectable	Can elevate blood hematocrit to dangerous levels and increase risk of thrombus formation	Yes Has been abused by various athletes (most notably by Tour de France cyclists) to increase blood O_2 carrying capacity.
Anabolic steroids	• Synthetically produced • Similar to those produced by the body	• Increases lean mass • Reduces fat mass	Injected into skeletal muscle or ingested in pill form	Linked to a variety of possible side effects, including mood disturbances, heart disease, high blood pressure, and cancer	Yes, but unknown whether the benefit is linked specifically to acute performance or to the ability to train at higher intensities because recovery is improved
Growth hormone	• Produced by the body for glucose regulation and general growth of bone and muscle • Synthetic form available	Produces skeletal muscle hypertrophy via enhanced protein synthesis and reduced body fat	Synthetic form can be injected	Can result in several diseases or conditions, including diabetes, joint swelling, and pain	Likely-
Amphetamines	A general group of stimulants	• Heightens arousal and alertness • Possibly reduces perception of muscle use discomfort	Typically in pill form	• Can overstimulate the cardiovascular system and result in arrhythmia • Extreme use can lead to addiction	Unknown
Blood doping	Removal of own blood, which is then stored for later use	Enhances hematocrit and red blood cell count to enhance O_2 carrying capacity	• Blood is reinfused after several weeks when the body's red blood cell count has returned to normal • May require 2-3 units	Precautions related to blood storage and infusion	Yes, can double hemoglobin and hematocrit concentrations (erythrocythemia)
Androstenedione	• Precursor to testosterone produced by the body • Available in synthetic form	Theoretically maximizes natural testosterone production, leading to the benefits of increased testosterone	Pill form used daily	Possible reduction in testosterone production with long-term use Possible increase in estrogen production and reduction in high-density lipoprotein cholesterol levels	Unknown

- Macronutrient (carbohydrate, protein, and fat) and fluid intake can affect performance.
- The term *ergogenic aid* refers to anything that can be used to enhance performance. It is often linked to nutrition and supplements. Many ergogenic aids are legal (e.g., caffeine, creatine), but some are illegal (e.g., anabolic steroids, growth hormone, erythropoietin).

Definitions

ad libitum—As much and as often as desired.

acclimation/acclimatization—The process in which an individual organism adjusts to a change in its environment (such as a change in altitude, temperature, humidity, photoperiod, or pH), allowing it to maintain performance across a range of environmental conditions.

ambient—Relating to the immediate surroundings.

ataxia—The loss of full control of bodily movements.

atrophy—Wasting or reducing.

biomechanics—The study of the structure and function of biological systems.

conduction—The transfer of heat between two objects of different temperatures that are in direct contact with one another. Heat always flows from the warmer object to the cooler object (e.g., from the skin surface to the air).

convection—The transfer of heat from one place to another by the motion of a heated substance. An example is air from a fan blowing past the skin surface, where the warmer air (via conduction) is removed and replaced with cooler air.

core body temperature—The internal temperature of the body.

ergogenic aid—External influences that enhance performance during exercise. May include drugs, mechanical devices, supplements, food or nutrition, and psychological aids.

evaporation—The process in which a liquid changes to a gas.

genotype—The genetic makeup of an organism.

glycogenolysis—The breakdown of glycogen to glucose.

kinanthropometry—The study of human size, shape, proportion, composition, maturation, and gross function in order to understand growth, exercise, performance, and nutrition.

phenotype—The observable makeup of an organism that may be influenced by the interaction of the organism's genotype with the environment.

Q10 effect—A measure of enzymatic reaction rate or physiological process for every 10 °C increase.

radiation—The transfer of heat between objects through electromagnetic waves as produced by the sun.

thermodynamic—Concerned with heat and temperature and their relation to energy and work.

thermoneutral—The range of ambient temperatures that do not produce heat production or loss.

validity—A measure of how accurately a measurement technique reflects the actual measure.

Chapter 10

Physical Activity and Exercise for Health and Fitness

More now than ever before, physical activity and exercise are recognized for playing an important role in health maintenance and disease management. If your career aspirations involve helping others through exercise, then you have chosen a path with widening opportunities going forward. The evidence gathered over the past two decades linking physical activity and exercise to both the primary and secondary prevention of disease is overwhelming. (*Primary prevention* refers to the use of a strategy or treatment with the intent of preventing the first occurrence of a disease-related event. *Secondary prevention* means that the treatment is being used to decrease the likelihood that a second event will occur.) And unlike many of the medications used to prevent or treat certain disorders, the benefits of exercise are systemic and apply to numerous organs. In the absence of a safe and effective (and likely very expensive) "polypill" (29) formulated to prevent or treat one or more diseases using a combination of two or more drugs, regular exercise is safe, affordable, and readily available to almost everyone.

In this chapter we focus on the myriad of ways that physical activity lessens the risk of disease and is linked to better health. We also discuss how much exercise is needed to derive its benefits. Such knowledge will serve you well and help you correctly address inquiries from the athletes, clients, or patients you interact with each day as well as friends, family members, and community residents who ask questions regarding exercise and health. Keep in mind that exercise is not only an intervention that can affect the health and well-being of an individual. In the context of public policy, it also becomes a strategy that can help influence public health and health care spending at a societal level.

EPIDEMIOLOGY OF PHYSICAL ACTIVITY, INACTIVITY, AND EXERCISE

Before discussing the frequency (or infrequency) of physical activity and exercise and their relationship to health in the United States and around the world, it is important to define some common terms: *sedentary behavior, physical activity, exercise, physical fitness*, and *health* (86). **Sedentary behavior** includes activities (<1.5 METs, or metabolic equivalents of task; see chapter 3) performed while awake and sitting or reclining that expend low amounts of energy (78). Examples include watching television, reading, and working on a computer. **Physical activity** reflects movement of the body of any kind, including for occupation, housework, leisure time, and transportation. Leisure time physical activity can be subclassified into three types: competitive sport, formal exercise training, and insufficient activity. Insufficient leisure time physical activity represents activities that do not meet current guidelines pertaining to moderate- and vigorous-intensity activities; examples include bowling, fishing, and gardening. **Total daily activity**, typically expressed as kilocalories per day or MET-hours per day, is the sum of sleep, sedentary behavior, and all types of physical activity in a day.

Exercise or exercise training, although often used synonymously with physical activity, is actually physical activity with the expressed purpose of improving or maintaining health or physical fitness. Exercise training is physical activity that is planned, structured, and repeated on a regular basis. Physical fitness has enjoyed many definitions over time. A common one is "the ability to perform work without undue fatigue and with sufficient energy to engage in leisure time interests and emergencies." There are many types of physical fitness. A few examples include cardiorespiratory or aerobic capacity; skeletal muscle strength, endurance, or power; flexibility; balance; agility; and body composition. Success in one type of fitness does not guarantee proficiency in or capacity to perform another type. Also, superior performance in some of these types of fitness does not necessarily confer good health. An athlete might have excellent balance or demonstrate a high level of muscle strength, but these attributes do not necessarily guarantee good health and lower risk for a chronic disease. Moderate and higher levels of cardiorespiratory fitness most often are associated with disease prevention.

Providing a comprehensive definition of health is outside the scope of this chapter. Health is not simply the absence of disease. Instead, it is a state that operates along a continuum, with the ability to enjoy life and respond to changes on one end and medical problems, frailty, and premature death on the other. This chapter provides the evidence that links physical activity and fitness to both improved health and a lower risk for developing a chronic disease.

RELATIONSHIP BETWEEN PHYSICAL ACTIVITY, EXERCISE, FITNESS, AND DISEASE PREVENTION

The burden associated with sedentary behavior, insufficient activity, or poor fitness is immense in terms of reduced quality of life and lives lost as well as economic burden. In a landmark study some 60 yr ago, Morris and Raffle (63) observed higher death rates and heart attacks among the sedentary bus drivers of London compared with the more active conductors on each bus. Since then, much research has documented the independent relationship between physical inactivity and poorer health outcomes. Approximately 2.5 million people die each year in the United States, generally equally split between men and women; 96% of deaths occur in people over 30

yr of age (26). Inadequate physical activity or inactivity is responsible for approximately 190,000 (8%) of these deaths. (Note that cigarette smoking is responsible for approximately 470,000 or 19% of all deaths.) Globally, physical inactivity is estimated to be associated with 3.2 million deaths annually (54).

Using **population attributable risk**, which is a measure that epidemiologists use to estimate the effect of a risk factor on the frequency of new cases of a disease in a population, we can break down the burden of physical inactivity according to disease type. Specifically, insufficient activity or sedentary behavior accounts for approximately 7% of deaths due to coronary heart disease, 8% of deaths due to diabetes, and 12% of deaths due to colon or breast cancer (50). These deaths likely would not have occurred or would have been delayed if the people had exercised regularly.

It is important to emphasize that regular exercise, despite its many benefits, is not a panacea. Instead, regular exercise is but one segment of several important health behaviors that can influence preventable deaths. Other important health behaviors include limiting alcohol consumption to moderate levels, properly managing weight, practicing healthy nutrition and dietary behaviors (e.g., high intake of fruits, vegetables, and omega-3 fatty acids; low intake of sodium and trans fatty acids), and abstaining from using tobacco products and harmful recreational substances (77).

As mentioned previously, insufficient physical activity and sedentary behavior are associated with an economic burden. This burden is not simply direct medical costs but also includes costs of decreased and lost productivity (absenteeism). Health care expenditures associated with inactivity and insufficient activity are estimated to exceed $100 billion annually in the United States (11). In the United Kingdom, the health care costs for physical inactivity were approximately £940 million in 2009-2010 (76). When sorted by type of disease and physical inactivity, costs were £117 million for stroke, £542 million for heart disease, £65 million for colorectal cancer, £54 million for breast cancer, and £158 million for diabetes (8).

Physical Activity and Exercise Guidelines

In 2013, the overall percentage of U.S. adults meeting the Department of Health and Human Services guidelines for both leisure time cardiorespiratory exercise (i.e., at least 150 min · wk^{-1} of moderate exercise or 75 min · wk^{-1} of vigorous exercise) and skeletal muscle-strengthening

Wellness and Disease Management in the Workplace

You've heard the news: Health care costs are too high! But what is responsible for the high costs? Remarkably, treatments associated with managing chronic diseases are the main culprit. Treatments for diabetes, heart disease, hypertension, stroke, and cancer account for 75% of the $2.5 trillion spent on medical care annually in the United States (14).

One does not have to look far to find out who is paying the bill. Employers are the leading provider of health insurance in the United States. The 2014 Employer Health Benefits Survey sponsored by Kaiser Family Foundation reported that over the past 10 yr, the average premium for family coverage has increased 69%. The effect of unhealthy lifestyle choices largely is responsible for the increase in premiums. Productivity loss, absenteeism, and disability are estimated to cost an employer $1,685 per employee per year, or $225.8 billion annually. The rising financial strain and the declining health of Americans make implementing workplace wellness and disease management programs attractive to employers.

Workplace wellness is an organized, employer-sponsored program designed to help employees (and often their spouses) adopt and sustain behaviors (e.g., regular physical activity) that reduce health risks, improve quality of life, enhance self-efficacy, and benefit the organization's bottom line (2). Employees often are incentivized to participate.

Effective programs include primary prevention to prevent disease and secondary prevention to detect a condition early and provide effective interventions. The World Health Organization estimates that 80% of heart disease and stroke, 80% of type 2 diabetes, 90% of obstructive lung disease, and 40% of cancer could be prevented if people would do three basic things: eat a healthy diet, increase levels of physical activity, and stop smoking (89). Successful workplace programs are multidimensional and typically include two common elements: lifestyle management and disease management.

- Lifestyle management programs address modifiable risk behaviors that mitigate longer term health risks. For example, an education program about the importance of healthy eating and increasing physical activity may lessen the potential of developing prediabetes over time.

- Disease management programs address immediate health needs. Programs include identifying the chronically ill population (e.g., hypertension, diabetes), stratifying patients based on their level of risk, identifying gaps in care, implementing evidence-based guidelines to address barriers that impede patients from effectively managing their condition, collaborating with a patient's physician, providing targeted self-management education (face-to-face or telephonic coaching and behavior change using health professionals), and continuously evaluating program outcomes.

One important aspect of most successful workplace wellness programs is use of a health risk appraisal (HRA). The HRA is a comprehensive assessment tool completed by an individual to gather demographic, behavioral risk, and physiological information. Once the HRA is completed, the participant is provided with an individualized risk profile that includes educational resources and strategies for reducing their risks. Obtaining these data allows health educators and health care professionals to develop interventions specific to the needs of a company's population. The HRA can also be used to identify and enroll members in a disease management program.

An often-asked question that is now being answered is "Do workplace wellness programs actually improve health outcomes?" Literature addressing return on investment and behavioral outcomes suggests that wellness programs generally are favorable, but some confusion remains about the optimal timing of these programs. Specifically, after gathering risk data using an HRA, it is important to identify very early on what the program outcomes will be and then quickly put in place evidence-based programs that target the desired outcomes.

Recent data suggest that a comprehensive health program can yield, on average, a $3 return on health care expenditures for every dollar spent over a 2- to 5-yr period (15). Such programs include promoting healthy behaviors (i.e., proper nutrition and regular physical activity), improving employees'

(continued)

Wellness and Disease Management in the Workplace
(continued)

health knowledge and skills through education, reducing barriers for employees and their families to receive necessary health screenings, identifying disease early, and providing immunizations and follow-up care.

For example, PepsiCo's Healthy Living program includes both lifestyle management (87% of participants) and disease management (13% of participants) components. The project involves health risk assessments, on-site wellness events, and a 24-h nurse line. Researchers found that 7 yr of continuous participation was associated with an average reduction of $30 in health care costs per member per month; the disease management component was responsible for the majority of those savings. Employees participating in the disease management program experienced a 30% reduction in hospital admissions (9).

Workplace wellness programs have the potential to reduce health risks and delay or prevent the onset of chronic diseases. Implementing lifestyle management to promote healthy living and disease management to help employees manage an illness can reduce health care costs and improve quality of life.

Source: Contributed by Stephanie R. Keteyian-Stacy, MS, ASR Health Benefits

---

Table 10.1 Summary of Exercise Recommendations for Adults Living in the United States

Type of fitness	Recommendations
Cardiorespiratory or aerobic	• At last 150 min · wk^{-1} of moderate intensity or 75 min · wk^{-1} of vigorous intensity • For additional and more extensive health benefits, increase to 300 min · wk^{-1} of moderate intensity or 150 min · wk^{-1} of vigorous intensity
Muscular	• At least 2d/wk of moderate- to high-intensity resistance training or strengthening activities that engage all major muscle groups
Flexibility or neuromuscular	• 2-3 times/wk engage major muscle-tendon units using dynamic, ballistic, or active or passive stretching exercises; proprioceptive neuromuscular facilitation also effective • 2-3 times/wk engage in exercises that involve balance, agility, coordination, and gait training (e.g., yoga)

Based on Garber et al. (30); U.S. Department of Health and Human Services (87).

exercise (≥2 d/wk) was 20% (lowest = Tennessee, 12%; highest = Colorado, 27%) (18,19). This is slightly higher than the 18% prevalence reported in 2008 (15) when the guidelines were first published and clearly greater than the 14.5% prevalence reported back in 1998 (16). Table 10.1 summarizes the physical activity and exercise guidelines as advanced by the American College of Sports Medicine (ACSM), U.S. Department of Health

and Human Services, and World Health Organization (30,87). These guidelines and recommendations target cardiorespiratory fitness and muscular conditioning as well as joint flexibility and neuromuscular function (i.e., balance).

Conversely, the percentage of people in the United States who were insufficiently active or engaged in no leisure time physical activity was 26% (lowest = Colorado, 18%; highest = Mississippi, 38%) (19). In 2013, the percentage of students in grades 9 to 12 who watched 3 h or more of television every school day was 33% (lowest = Utah, 15%; highest = Mississippi, 40%) (19). Table 10.2 compares the prevalence of sedentary behavior in the United States (25.4%) with that of selected other countries (90). Although most countries report sedentary behavior rates of between 15% and 30%, some rates exceed 50%.

The frequency of U.S. adults exercising at least 150 min · wk^{-1} at a moderate intensity or 75 min · wk^{-1} or more at a vigorous intensity and performing skeletal muscle–strengthening activities 2 d or more/wk varies based on age, race, and gender. The following are a few key points to keep in mind when discussing the demographics associated with participation in physical activity during leisure time in the United States.

- Rates of regular physical activity decrease with age (approximately 32% in people aged 18-24 yr vs. 16% in people aged 65-74 yr).

- Overall, physical activity is lower in women (17%) than in men (25%) (16,18,19).

- More white adults (23%) meet the guidelines for regular physical activity than either African American adults (18%) or Hispanic/Latino adults (17%).

- Regular physical activity levels are consistently lower in people reporting a chronic health problem (e.g., heart failure, arthritis, or emphysema) compared with matched people free of a chronic disease (7,16,25).

Exercise, Activity, and Disease Prevention

Figure 10.1 shows the 10 leading causes of death in 2013 (17,20). In the United States (and many other industrialized nations), most die from one of two causes: cancer or a cardiovascular disorder (i.e., heart disease or stroke). Concerning cardiovascular disease, it is predominately attributable to the underlying disease atherosclerosis (see figure 10.2). Although other disorders occupy the remaining positions in the top 10, cancer (163.2/100,000 people) and cardiovascular-related events (heart disease = 169.8/100,000 people; stroke = 36.2/100,000 people) by far represent the majority of the burden.

Many of the causes of death listed in figure 10.1 are influenced by exercise habits. Specifically, a 2011 epidemiologic study by Wen et al. (88) involving more than 400,000 Taiwanese people helped describe the association between weekly exercise volume and relative risk of total mortality and risk of death due to cardiovascular disease, diabetes, and cancer (88). As shown in figure 10.3, in each of these disorders regular exercise is associated with a reduction in relative risk. In general, more is better. More important, notice that performing just a moderate volume (7.5-16.5 MET-h · wk^{-1}) of exercise or activity provides substantial benefits. In fact, the relative risk of death due to any cause over 8 yr was reduced by approximately 20% for those who regularly engaged in moderate exercise compared with the nonexercising group (<3.75 MET-h · wk^{-1}). For cancer deaths, the reduction in relative risk associated with moderate exercise

Table 10.2 Prevalence of No Leisure Time Physical Activity (Inactivity) Among Men in 2010 in Selected Countries

Country	Prevalence (%)	Country	Prevalence (%)
Australia	20.1	Japan	31.1
Argentina	35.8	Mexico	20.3
Austria	19.2	Netherlands	14.0
Bahamas	29.6	Norway	22.9
Canada	20.3	Poland	13.6
China	22.5	Russian Federation	10.2
Colombia	54.3	Saudi Arabia	53.2
Ecuador	19.6	United Kingdom	32.3
Germany	18.7	United States	25.4
Greece	10.1		

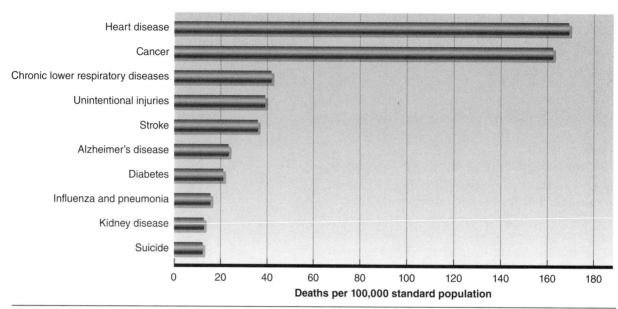

Figure 10.1 Age-adjusted death rates for the 10 leading causes of death in the United States in 2013.

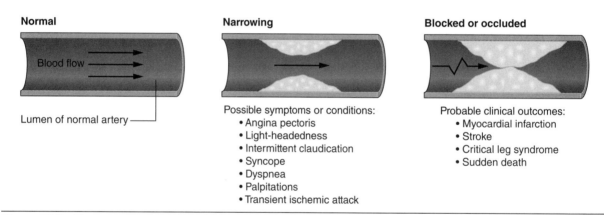

Figure 10.2 The narrowing of the lumen of any artery of the body by atherosclerosis is a progressive disorder. Over time, such narrowing of the arteries involving the heart, brain, kidneys, and skeletal muscles typically leads to many symptoms or conditions that are caused by the associated reduction in blood flow (i.e., ischemia). In an advanced state of disease the blood flow though the artery can be critically reduced or fully blocked, resulting in permanent tissue organ damage or death.

was approximately 15%. For cardiovascular deaths, the reduction was approximately 21%.

Thirty million residents of the United States have diabetes, a disorder associated with an impaired ability of cell membranes to take up glucose into the cell. Individuals with blood glucose levels that remain elevated (i.e., hyperglycemia) are classified as having diabetes (fasting blood glucose ≥ 126 mg · dL^{-1}) or prediabetes (fasting blood glucose 100-125 mg · dL^{-1}). There are two main types of diabetes: type 1 and type 2. Type 1 diabetes, which typically develops due to factors independent of one's lifestyle, is caused by the absence of insulin production

by the pancreas. Type 2 diabetes (present in >28 million people in the United States) is caused when peripheral tissues (i.e., skeletal muscles) resist responding to insulin-stimulated glucose uptake. People with type 2 diabetes usually develop the disorder later in life as a result of sedentary habits and obesity. Regular moderate exercise reduces the relative risk of diabetes-related deaths by 23% (figure 10.3). The message coming through in this discussion is clear and consistent: Exercise is good preventive medicine.

Most of our discussion so far has addressed activity and health relative to aerobic-type exercise or activities

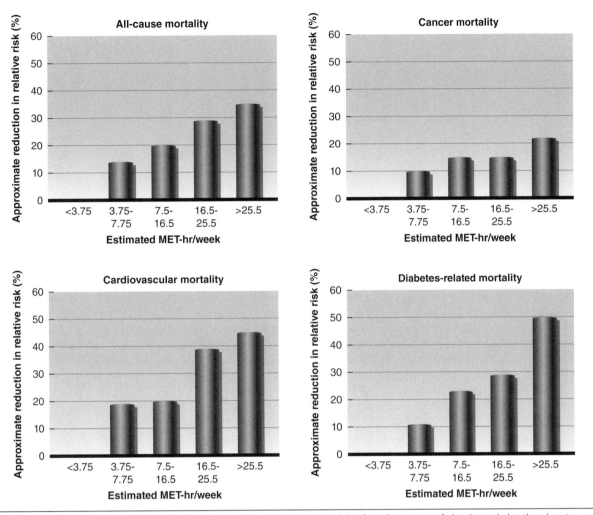

Figure 10.3 Relationship between exercise volume and relative risks for all causes of death and deaths due to cardiovascular disease, cancer, and diabetes. A moderate volume of exercise consistently provides substantial benefits.

Data adapted from Wen et al. (88)

such as brisk walking, jogging, running, and swimming. It is also important to mention that other types of exercise are associated with markers of improved health. For example, although definitive data addressing the relationship between muscular strengthening exercises and survival are lacking, some observational data indicate that reduced levels of muscle strength (e.g., quadriceps strength during knee extension, hand grip strength) are associated with increased all-cause mortality in both men and women (65). Additionally, resistance training lowers the risk for developing metabolic syndrome (40,41) and favorably affects glycemic control and glucose handling in patients with diabetes (12,80). Finally, although no large randomized research studies have investigated the effects of resistance training on bone health in adults, observational studies indicate that such training may blunt the age-related

decline in bone density and lessen the relative risk for bone fracture (47).

Finally, overt sedentary behavior is also an important independent marker for health risk. Over the past 50 yr, occupation-related energy expenditure has decreased by more than 100 kcal · d^{-1} (21,83). Also, computers, automobiles, elevators, escalators, moving sidewalks, garage door openers, washing machines, self-propelled lawnmowers, and a myriad of other labor-saving devices have all made our lives less physically demanding. High levels of sedentary behavior (even in people who exercise 3-4 times/wk) are now a health risk (3,37). In fact, sedentary behavior, independent of formal exercise, is associated with a twofold increase in risk for mortality. Some evidence suggests that among less active adults aged 59 yr and older, replacing 1 h · d^{-1} of inactive behavior with an equal amount of activity (e.g., lawn or garden

---

Performance: Is Too Much Exercise Harmful?

The health benefits of exercise are undeniable. That said, some evidence over the past decade suggests that too much intense endurance-type exercise may result in an overuse injury that involves the heart (49,68,72). As discussed in chapter 6, chronic long-term training is associated with *supranormal* cardiac remodeling, termed *athlete's heart*. The question becomes whether ultraendurance training that has gone on for years or decades can harm the heart. To be clear, we are not referring to people who for 30 yr have been exercising 3 to 5 times/wk for 45 to 60 min each time. Instead, we are addressing people who engage in high-volume exercise that is sufficient to support participation in 10 or more marathons each year or ultraendurance (50 mi, or 80 km) events.

To place the health issues associated with this topic in context, envision a reverse J-shaped relationship between regular exercise and mortality. Specifically, lack of exercise presents the greatest risk for developing a health problem, moderate to higher volumes of regular exercise lower the risk, and extreme endurance exercise possibly increases the risk slightly. Evidence supporting this potential small increase in risk among high-performing endurance athletes comes from both observational studies and some surrogate studies involving cardiac biomarkers (e.g., cardiac troponin) and cardiac imaging suggesting cardiac injury or fibrosis or strain of the right ventricle (68,72). All of this represents a potential substrate for an increased likelihood of atrial and ventricular arrhythmias. However, these findings are not consistently observed in all studies (53,79). Additionally, although some data might suggest that marathon runners develop excessive amounts of coronary atherosclerosis (35), these findings are also inconsistent, suggesting that this premise may also be inconclusive. In fact, much evidence indicates that the coronary arteries of the elite athletes are healthy, with intact and supranormal vasodilatory capacity (53).

To place this in proper context, consider these two points. First, we still know relatively little about how variations in physical activity and exercise dose might affect diseases or conditions involving the heart. Clinicians caring for extreme endurance athletes should be vigilant and observe for symptoms and abnormalities but not assume outright that such exercise is harmful.

Second, for the vast majority of individuals, the current exercise guidelines of 150 min · wk^{-1} or more of moderate-intensity endurance-type exercise or 75 min · wk^{-1} of vigorous exercise are prudent and enormously helpful for inducing a host of important health benefits. More exercise—such as progressively increasing to 300 min · wk^{-1} of moderate-paced exercise or 150 min · wk^{-1} of vigorous-paced exercise—is safe and provides more health benefits.

---

work, household chores, walking) is associated with an approximately 30% reduction in risk of mortality (61). The takeaway message is that in order to optimally reduce the risk for health problems due to insufficient activity and sedentary behavior, both activity throughout the day (at home, work, and school) and formal periods of exercise are needed each week.

Fitness and Health

Although the words *exercise* and *fitness* often are used synonymously, this is not always the case. It is correct to say that regular exercise is the main modifiable behavior that can affect one's fitness. However, one can be moderately fit without engaging in a high amount of regular exercise; genetics does play a role (see chapter 11).

Fitness—in this case, we limit our discussion to predominately cardiorespiratory fitness—which is usually expressed as peak $\dot{V}O_2$ or estimated METs measured during an exercise test. The evidence is fairly clear that the higher one's level of fitness, the better the health outcomes. Figure 10.4 shows the relationships between fitness (as measured by estimated METs) and relative risk of death and the adjusted risk for the development of diabetes or high blood pressure. Compared with 13 METs, less than 6 METs and 6 to 8 METs are associated with a more than 4-fold and almost 2.5-fold increased risk of death over 6 yr, respectively. Achieving at least 12 METs reduces the risk for developing diabetes or hypertension by 62% and 35%, respectively. In summary, a graded relationship exists between level of fitness and health.

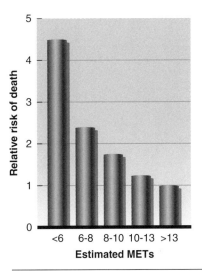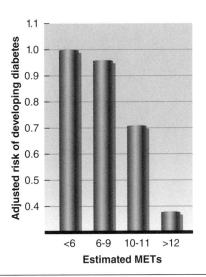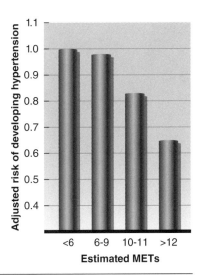

Figure 10.4 The association between *(a)* fitness level (as measured by estimated METs) and relative risk for all-cause mortality and *(b)* adjusted risk for the development of diabetes or *(c)* hypertension (adjusted for age, sex, and race).

Data adapted from Myers et al. (64), Juraschek et al. (38), and Juraschek et al. (39).

Also important is the observation that a change in one's fitness level is related to a change in future health outcomes. This is important to people with a poor level of fitness; in such individuals, changing the level of fitness by starting and sticking with a regular exercise program is indeed beneficial and worth their time and effort. Recent data from our laboratory indicate that changing one's fitness level from unfit (<7 METs) to more fit (>7 METs) is associated with an approximately 48% reduction in relative risk of death over the next 9 yr (6). Although genetics can influence up to 50% of an individual's response in fitness due to exercise training (5), physical activity itself is the only modifiable behavior that can affect cardiorespiratory fitness. This is an important teaching point to convey to clients who are considering an exercise program or who have just started one.

How Does Exercise Improve Health and Lower Risk?

Although it is very important for exercise professionals to be able to communicate the extent to which exercise and regular physical activity help with disease prevention and treatment, it is also important to appreciate how exercise acts at the organ, tissue, and cellular levels to influence health and disease. Such knowledge has been and remains an area of intense scientific investigation. The mechanism

or manner through which exercise improves health or lowers risk is often disease specific. For example, changes in physiology that are responsible for the prevention and treatment of atherosclerosis (e.g., heart disease, stroke) are somewhat different from the changes that develop from regular exercise that help with the primary or secondary prevention of cancer.

Concerning cancer, compared with people with low levels of physical activity, the risk for developing colon, endometrial, and breast cancer is reduced by 12% to 25% in those who are highly physically active (51). Other data suggest that the incidence of other cancers (e.g., prostate, lung, and kidney) also appears to be somewhat favorably influenced by engaging in regular exercise. How exercise works to prevent and treat the abnormal growth of tissue (i.e., neoplasm) is somewhat disease specific because cancer can originate in any organ system, can spread to other organ systems, and can be attributable to abnormalities in many different cellular pathways. Specifically, how exercise influences the development and treatment of breast cancer is mostly different from the mechanisms involved with the prevention and treatment of colon cancer. Therefore, the potential mechanisms or processes through which exercise acts to prevent or treat cancer are many (figure 10.5).

Table 10.3 provides a summary of the main potential mechanisms through which exercise helps in the prevention and treatment of several common diseases. How

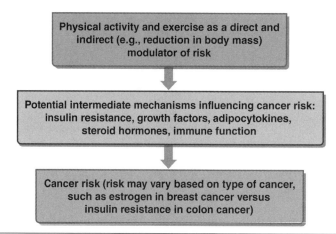

Figure 10.5 Potential mechanisms linking physical activity to cancer. Regular exercise may directly or indirectly affect these mechanisms, with the latter potentially involving the effect of regular exercise on modulating or reducing body mass.

exercise is used in the management of specific diseases and to what extent it helps are discussed later in this chapter.

PROPER SCREENING BEFORE EXERCISE AND THE RISKS ASSOCIATED WITH EXERCISE

Regular exercise is described by some as a double-edged sword. On one hand, the adaptations associated with training can be very helpful and effective in disease management and treatment. On the other hand, very rarely, in select persons an acute bout of exercise can trigger a life-ending event: **sudden cardiac death** (SCD) or a fall or accident. Relative to exercise, SCD is defined as unexpected death due to loss of heart function occurring during or within 1 h of participating in sport. Of the approximately 25 million competitive athletes in the United States, 25 to 125 documented cases of SCD occur per year. A challenge for the exercise professional working with inactive clients (and low-activity clients who wish to exercise more vigorously or to a greater extent) is weighing the need for appropriate screening to ensure client safety against imposing barriers that prevent those who can exercise safely from getting started. For example, if a 48-yr-old premenopausal woman who is sedentary but otherwise healthy wishes to start a moderate intensity exercise program (e.g., brisk walking) to regain stamina and assist with weight maintenance, should the preparticipation screening be the same as that for a sedentary 45-yr-old man who has high blood pressure, complains of palpitations (i.e., perceived abnormal heartbeat), and

wishes to return to playing adult ice hockey? Some obvious differences between these two scenarios can help guide decision making.

The ACSM's Recommendations for Exercise Preparticipation Health Screening identifies three modulators of risk that exercise and health care professionals need to consider when determining someone's risk for an exercise-related event (74). These three variables are as follows:

1. The individual's current level of activity
2. The presence of signs or symptoms and presence of known cardiovascular, metabolic, or renal diseases
3. The individual's desired exercise intensity (light, moderate, or high, with high defined as ≥6 METs)

In the previous example, both an inactive woman and an inactive man want to exercise. However, the man also complains of palpitations and wishes to engage in a vigorous activity. For the woman, we can avoid an unnecessary referral for medical clearance and proceed directly to developing a safe, progressive program of moderate-intensity exercise (e.g., walking) that targets improving her fitness and facilitating calorie management. Conversely, we should counsel the male client to withhold returning to vigorous exercise (i.e., ice hockey) until his symptoms and overall risk are evaluated by a qualified health care professional. The revised ACSM recommendations for exercise preparticipation health screening let the health professional decide the extent and type of evaluation and testing (e.g., interview and examination, exercise stress test, other) needed to ensure a client's safety during moderate to vigorous exercise.

Table 10.3 Mechanisms Through Which Exercise Helps Prevent and Treat Common Select Chronic Diseases

Chronic disease	Actual and potential mechanisms
Heart disease/stroke	• Maintain or increase myocardial oxygen supply through improved coronary artery endothelial function and delayed progression of atherosclerosis (improved lipid profile, decrease in platelet aggregation, and increase in fibrinolysis) • Decrease myocardial oxygen demand due to decrease in heart rate and blood pressure at rest and during exercise; decrease in circulating catecholamines • Improve myocardial function due to improved contractility and stroke volume (and likely reduced afterload)
Cancer (breast, prostate, colon)	Moderate exercise enhances the body's immune system (increase in both number and activity of natural killer cells and enhanced proliferation of T lymphocytes); the role that these favorable responses play in cancer prevention remains an area of study. Disease-specific potential exercise-related mechanisms include the following (10,23,24): • Breast: Increase the likelihood of anovulatory (i.e., oocyte not released from ovaries) menstrual cycles and delay of menarche (age of first menstrual cycle), thus influencing total lifetime exposure to estrogen and progesterone • Colon: Potentially reduce gastrointestinal transit time, decrease insulin growth factors, alter gut flora, and lower level of bile secretions • Endometrial: Prevent or manage estrogen–progesterone imbalance, which is more pronounced among overweight or obese women; likely mediated through less insulin resistance, improved levels of insulin growth factors, and decrease in proinflammatory state (decrease in cytokines)
Diabetes	• Improve and maintain healthy blood glucose as measured by hemoglobin A_{1c}, especially in patients with type 2 diabetes (80) • Improve insulin sensitivity regardless of change in body composition due to increased capillary density and fat oxidation (32) • Increase glucose transporter type 4 (a protein responsible for uptake of glucose into skeletal muscle cells) in muscle (36)
Hypertension	• Vascular remodeling due to increased arterial diameter in trained limbs (e.g., femoral artery) or the increased capillarity that occurs in actively trained muscles (increased capillary-to-muscle fiber ratio) • Downregulate autonomic sympathetic activity to peripheral vasculature • Enhance vasodilatory capacity due to improved endothelial-dependent vasodilation, likely through nitric oxide (endothelial nitric oxide synthase) and possibly non–nitric oxide (e.g., prostacyclin) pathways
Osteoporosis	• Ground reactive force (mechanical stress)–stimulated increase in osteoblast activity and decrease in osteoclast activity, leading to increased bone mineral density or maintenance thereof
Dyslipidemia	• Reduce triglycerides and mildly to moderately increase high-density lipoprotein cholesterol • Increase lipoprotein lipase (LPL) gene expression with eventual increases in skeletal muscle LPL, messenger ribonucleic acid, LPL mass, and LPL enzyme activity • In the absence of weight loss or changes in dietary fat or alcohol consumption, exercise training alone results in no clinically meaningful changes in total cholesterol, low-density lipoprotein cholesterol, or hepatic lipase (28)

Risks Associated With Exercise

Earlier in this chapter, we review how and to what extent exercise prevents or lessens risks associated with developing certain chronic diseases. Later in this chapter, we discuss the extent to which regular exercise specifically helps improve measurable health outcomes in patients with a known, clinically manifest disease. However, despite all the favorable benefits associated with exercise, exercise itself—especially vigorous exercise—can trigger a rare, harmful event.

The idea of dying during exercise is counterintuitive, and it evokes the image of one suddenly collapsing while engaged in vigorous activity. Such an occurrence often receives much media attention and devastates families and communities alike. How can an athlete or an individual who is so healthy and able to perform at an extreme level succumb to such a sudden life-ending event? To answer this question let's first explore the underlying reasons why people die suddenly during exercise. Using autopsy data, we know that SCD during exercise usually results from either coronary artery disease (i.e., atherosclerosis) or nonatherosclerotic causes (46,85). Among persons older than 35 yr of age, atherosclerosis of the coronary arteries is the underlying cause of SCD during exertion approximately 75% to 80% of the time. In most of these cases, the SCD usually is due to either an acute disruption (i.e., rupture, erosion) of an existing atherosclerotic plaque, leading to acute thrombotic occlusion of an artery, or myocardial ischemia that results in a lethal irregular heart rhythm (i.e., arrhythmia) (85).

The second cause of SCD during exercise (i.e., those deaths that are not attributable to coronary atherosclerosis) is much more common (>85%) in persons under 40 yr of age. In these people, SCD is related to abnormalities that were either present at birth or acquired over time and independent of a patient's lifestyle habits. The more common examples include the following:

- Abnormal left ventricular chamber size or wall thickness (hypertrophic cardiomyopathy)

- Rupture of the ascending aorta

- Stenosis of the aortic valve

- Abnormalities in how the cardiac cell (myocyte) transports sodium and potassium ions across its membrane (channelopathies or long QT syndrome)

- Anomalies of the location and direction from which coronary arteries arise or originate from the aorta

- Commotio cordis (i.e., the sudden, forceful impact of a ball or object to the chest wall during ventricular repolarization)

One prominent study analyzed 134 SCDs that occurred in young competitive athletes. Among the deceased, the median age was 17 yr (range = 12-40 yr), 90% were male, 52% were white, and 86% had a standard preparticipation medical evaluation. Of the SCDs, 90% occurred during or immediately after training or competition, 35% involved basketball, and 34% involved football (58). Over the past decade, several studies have evaluated the role of completing a standard resting electrocardiogram (ECG) in athletes as a way to identify those who might be at increased risk for one of the previously mentioned abnormalities. These conditions are rare, and many do not cause abnormalities that are manifest in the resting ECG. Several countries and organizations do perform resting ECGs on all of their athletes (56,82). Such a practice has not received universal endorsement in the United States (59).

The overall frequency of exercise-related deaths (cardiac and noncardiac) during exercise is very low. This is true for young athletes, generally healthy people, and patients with known heart disease alike. In high school and collegiate athletes, the occurrence of SCD is approximately 1 event per every 150,000 persons (59). In apparently healthy people, such as individuals who work out at a commercial health or fitness center, occurrence is approximately 1 event per 100,000 persons. Finally, in patients with known cardiovascular disease, the frequency of deaths during exercise is approximately 1 per every 125,000 or more patient-hours of exercise (52).

Consistent with the ACSM recommendations for exercise preparticipation health screening, it is important to place the issue of death during exercise in perspective. As discussed, in rare instances, exercise (especially vigorous exercise) can trigger a harmful sudden event. Not surprisingly, this transient and finite increase in risk for SCD during vigorous exercise is lower in people who are regularly active than in those who are routinely sedentary and then engage in vigorous exercise.

Athletes, clients, and patients who inquire about the safety of exercise should first be told that the overall health and disease prevention and treatment benefits associated with exercise far outweigh the small likelihood of experiencing an exercise-related event. Among those individuals in which an event unfortunately occurs, it is more often linked to vigorous exercise and more likely to occur in those who are sedentary compared with those who are regularly active and engage in vigorous exercise. This latter point emphasizes the importance of developing an exercise prescription that progressively increases the amount of exercise and intensity of effort, especially in patients who previously were sedentary.

HOW MUCH EXERCISE IS ENOUGH?

We've all taken medication to treat an illness or infection. In each case, the medication was prescribed based on research that demonstrated that the dose was effective, safe, and associated with few unwanted or harmful side effects. The same concerns apply when developing an exercise prescription to help an individual improve fitness and health. The type and volume (dose) of exercise prescribed must be based on research that shows that the exercise being recommended is safe, effective, and associated with few side effects (e.g., an injury).

In chapters 6 and 7 we discuss the two main principles—specificity of training and progressive overload—associated with developing training regimens designed to improve aerobic and anaerobic performance, respectively. Table 10.4 reviews each of these factors in the context of how much exercise is needed to accumulate the general health benefits associated with regular cardiorespiratory-type exercise. The acronym FITT-VP (frequency, intensity, time, type, volume, progression) is used to help guide

the development of a safe, effective exercise prescription that induces the proper volume of exercise and uses the correct types of exercises. Concerning exercise intensity, table 10.5 displays typical exercise training heart rates calculated at 60% and 80% of heart rate reserve.

Over the past 20 yr, dozens of articles and documents published by expert panels and professional organizations have detailed how much exercise is needed to improve health and prevent disease. In the United States, one landmark document, published in 1996, was the Surgeon General's Report on Physical Activity and Health (86). The main recommendation for exercise and better health in this report was as follows: "All children and adults should accumulate a minimum of 30 min of moderate physical activity on most and preferably all days of the week."

The statement adopted a public health tone that focused on physical activity, health, and behavior. Note the emphasis placed on minutes per week rather than per session and on the level of effort (moderate intensity). Also, the statement specifically supports accumulating minutes of activity throughout a day, highlighting the adoption of behaviors that allow people to include shorter duration (10 min) bouts as a way to progress activity in a manner that might be more favorable for those who are currently sedentary. Regarding the recommendation of accumulating 150 min · wk^{-1} from physical activities and exercise, such an amount is equivalent to approximately 750 to 1,000 kcal · wk^{-1} or about 10 to 15 MET-h · wk^{-1}

Table 10.4 Summary of Guidelines for Prescribing Exercise to Improve Cardiorespiratory Fitness Using the FITT-VP Method

Item	Training principle	Guideline
Frequency	Progressive overload	≥ 5 d/wk moderate or ≥3 d/wk vigorous
Intensity	Progressive overload	Moderate = 45%-60% of heart rate reserve Vigorous = 60%-<90% of heart rate reserve
Time or duration	Progressive overload	30-60 min/d of moderate 20-60 min/d of vigorous
Type	Specificity of training	Gross body exercises that involve the major muscle groups and are continuous or rhythmic (aerobic-type exercises; e.g., walking, jogging, and Nordic skiing)
Volume	Progressive overload	8-16 MET-h · wk^{-1}
Progression	Progressive overload	To facilitate compliance and lessen the risk of orthopedic injury or untoward event, adjust the frequency, time, and intensity of effort in a progressive manner that facilitates achievement of an individual's personal goals.

Starting an exercise regimen at levels below the stated recommendations may be more proper and effective in previously sedentary or deconditioned individuals.

Based on American College of Sports Medicine, 2017, *ACSM's guidelines for exercise testing and prescription*, 10th ed. (Philadelphia, PA: Lippincott, Williams, and Wilkins).

Table 10.5 Exercise Training Heart Rate (HR) by Age for Healthy Persons

	Age (yr)				
	21-30	31-40	41-50	51-60	61-70
Measured or estimated peak HR	195	185	175	165	155
TRAINING HR AT 60%					
$0.60\ (HR_{peak} - 75) + 75$	147	141	135	129	123
TRAINING HR AT 80%					
$0.80\ (HR_{peak} - 75) + 75$	171	163	155	147	139

Exercise training heart rate is computed using the heart rate reserve method. It assumes a resting heart rate of 75 beats · min⁻¹ and a peak heart rate either measured during a symptom limited exercise test or estimated as 220 − age.

(87)—a volume of exercise recognized to be associated with inducing many of the initial health gains (improved fitness, lower rate of cardiovascular death) that result from being more physically active. Interestingly, recent data suggests that performing all of the recommended exercise (i.e., > 75 min of vigorous intensity or > 150 min of moderate intensity) in just 1 or 2 sessions per week (so-called "weekend-warrior") may be sufficient to substantially reduce future risk for all-cause, cardiovascular or cancer mortality (67).

As shown in figure 10.6, the lower one's initial level of physical activity or fitness, the more gains and improvement one makes upon becoming more active. Conversely, people who are already active and fit likely need to engage in relatively more activity in order to make further health gains, and the magnitude of the gains will likely be relatively less for the same amount of time or volume of exercise. This is not meant to suggest that exercise is not helpful. Rather, given a person's higher level of fitness, more (i.e., a greater stimulus) is needed to induce similar incremental benefits.

Over the past 20 yr the guidelines for exercise and health have evolved. Table 10.1 summarizes the current guidelines for health professionals who counsel and prescribe exercise to apparently healthy individuals. These guidelines for exercise, as advanced by the ACSM (30) and the U.S. Department of Health and Human Services (87), are also summarized below:

Most adults engage in moderate-intensity cardiorespiratory exercise training for ≥30 min · d⁻¹ on ≥5 d · wk⁻¹ for a total of 150 min · wk⁻¹, vigorous-intensity cardiorespiratory exercise training for ≥20 min · d⁻¹ on ≥3 d · wk⁻¹ (≥75 min · wk⁻¹), or a combination of moderate- and vigorous-intensity exercise to achieve a total energy expenditure of ≥500

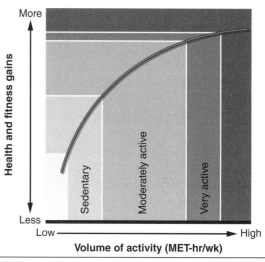

Figure 10.6 The relationship between volume of activity and derived health and fitness benefits. A moderate volume of activity markedly improves health, and more activity provides additional benefits, which likely will be developed at a somewhat slower rate in people who are already engaged in moderate-high to high levels of activity.

to 1,000 MET min · wk⁻¹ (8-17 MET hr · wk⁻¹). On 2 to 3 d · wk⁻¹, adults should also perform resistance exercises for each of the major muscle groups and neuromotor exercise involving balance, agility, and coordination. Crucial to maintaining joint range of movement, completing a series of flexibility exercises for each of the major muscle-tendon groups (a total of 60 s per exercise) on ≥2 d · wk⁻¹ is recommended.

Compared with the 1996 surgeon general's report, the current guideline statement highlights several important differences. First, more components of personal fitness (i.e., muscular strength, flexibility) are now included. Second, the recommendation includes volume of aerobic-type exercise (i.e., 500-1,000 MET-min · wk⁻¹), which allows for including the contribution of exercise intensity. Finally, although the word *accumulated* is not included in the main wording of the recommendation, the full report does acknowledge that bouts of exercise as short as 10 min may indeed be helpful. More recently, bouts of moderate to vigorous exercise that are less than 10 min have also been shown to be associated with favorable health outcomes (55). Table 10.6 lists a variety of activities and the associated recommended volume based on a specified duration of time and intensity of effort.

The public health messages from the previous discussion that are relevant to professionals prescribing exercise to individuals are as follows.

• Strive to achieve 500 to 1,000 MET-min · wk⁻¹ of moderate- or vigorous-intensity cardiorespiratory exercise.

• For those who are currently sedentary, the health benefits gained may initially be larger. The volume of exercise should be gradually progressed in a manner that modifies duration, frequency, and intensity (minimum threshold intensity around 40%-50% of heart rate reserve) to achieve the stated goals. More fit people and individuals who wish to engage in more exercise will derive further health benefits, although likely in a manner that is somewhat consistent with diminishing return.

• The pattern of training likely involves continuous exercise. However, bouts of 10 min (or even shorter) may prove helpful.

• Incorporate exercises that improve muscular strength, joint range of motion, and neuromotor function at least 2 d/wk. Gradually progress the duration, frequency, and intensity of the exercises to develop these types of fitness.

EXERCISE IN THE TREATMENT OF COMMON NONCOMMUNICABLE CHRONIC DISEASES

In this chapter we have discussed the role of exercise and physical activity in the primary prevention of all-cause

Table 10.6 Examples of Activities or Exercises That Can Be Combined to Achieve the Recommended Volume of Activity Each Week*

Activity	Duration (min)	METs	Volume (MET-min)
Washing and waxing a car	45	2.0	90
Vacuuming (moderate)	30	3.4	102
Playing noncompetitive volleyball	45	3.0	135
Gardening	45	3.0	135
Walking 1.75 mi (2.8 km) at 3 mph (4.8 kph)	35	3.5	123
Shooting baskets in basketball	30	4.5	135
Leisurely bicycling 5 mi (8 km) at 10 mph (16.1 kph)	30	6.0	180
Pushing a stroller 1.5 mi (2.4 km)	30	4.0	120
Raking leaves	30	3.8	114
Walking 2 mi (3.2 km) at 4 mph (6.4 kph)	30	5.0	150
Water aerobics	30	5.3	160
Mowing the lawn with a power mower	30	4.8	144
Golf, pulling clubs	60	5.0	300
Square dancing	25	6.0	150
Playing a game of basketball	20	4.5	90
Bicycling 4 mi (6.4 km) at 16 mph (25.7 kph)	15	10.0	150
Shoveling snow by hand (moderate effort)	15	5.3	80
Stair walking (slow pace)	15	4.0	60

*A recommended goal is 500 to 1,000 MET-min · wk⁻¹.

and disease-specific (e.g., heart disease) mortality. Table 10.7 presents the effects of exercise on various primary clinical outcome measures (e.g., mortality, hospitalization) in patients with common chronic health disorders. For each disorder, a cardiorespiratory-type exercise usually is linked with improved clinical outcomes. This does not mean that resistance or neuromotor exercises are not important or helpful. In fact, resistance training has been shown to improve important intermediate outcomes such as muscle strength in cancer patients or blood glucose in diabetics. However, relative to key clinical outcomes such as survival or level of pain, aerobic-type exercise has received most of the attention in clinical research to date.

Table 10.7 Benefits of Regular Cardiorespiratory or Aerobic Exercise on Important Clinical Outcomes in Patients With Known Common Diseases

Disease/disorder	Programming dose	Key clinical outcomes	Other comments
Coronary artery disease	F: 3-6 d/wk I: 60%-80% HRR T: 20-60 min	• Mortality reduced 5%-20% • Reduction in hospitalizations	• Progression of atherosclerosis slowed • Intermediate markers (endothelial function, blood pressure, glucose handling, automatic balance) improved • Dose–response relationship suggests that more exercise is associated with fewer clinical events.
Heart failure with reduced ejection fraction	F: 3-5 d/wk I: 60%-80% HRR, or rating of perceived exertion level of 11-14 T: 30-60 min	• All-cause mortality reduced 10%-15% • Rehospitalizations reduced 20%-30%	• Target volume = 3-5 MET-h · wk^{-1} • Long-term adherence a challenge
Cancers: breast, colon, prostate, ovarian, lung	F: 3-5 d/wk I: 40%-60% up to 60%-80% HRR, or rating of perceived exertion level of 12-15 T: 30-50 min	Population-based evidence suggests improved disease-free survival and decrease in all-cause mortality.	• Most data currently exist for breast cancer; several trials now ongoing for other cancers • Progress volume to 10-20 MET-h · wk^{-1} • Large randomized trials needed
Peripheral artery disease	F: 3 d/wk I: Exercise to grade 3-4 on a 1-5 claudication pain scale. Rest until pain becomes 1 on pain scale, or resolves, then resume. Progress speed and grade as tolerated. T: 60 min	Supervised exercise improves total walking time more than stent revascularization.	1-2 times/wk incorporate intermittent use of stationary biking to achieve a higher work rate and heart rate that is not limited by claudication; set training intensity at 60%-80% of HRR.
Diabetes	F: 3-7 d/wk I: 40%-70% HRR T: 20-50 min	Caloric restriction and exercise shown to have no significant effect on cardiovascular events in overweight and obese adults with type 2 diabetes.	• Markers of glucose handling (insulin sensitivity, blood glucose, glycolated hemoglobin) improved • Long-term adherence a challenge • Energy expenditure goal of 1,000-2,000 kcal · wk^{-1}

F = frequency; I = intensity; T = time; HRR = heart rate reserve.

*Although the primary focus of this table is cardiorespiratory or aerobic exercise, resistance training can also help improve intermediate makers in select patients (e.g., glucose handling in patients with diabetes). Cardiorespiratory or aerobic exercise involves the large muscle groups and is rhythmic in nature (e.g., walking, cycling).

Aging, Exercise, and Fall Prevention

Falls are a major threat to the health and independence of people aged 65 yr and older. Each year about one third of all older adults experience a fall, and falls are the leading cause of unintentional injury-related deaths in this population, accounting for approximately 25,000 deaths annually (20). About 20% of all falls result in a serious injury such as a hip fracture or head injury, both of which are associated with much pain and require hospitalization. Falls carry an enormous economic burden as well; total direct costs for fatal and nonfatal falls exceed $30 billion annually.

The reasons for falls are many. There are three types of risk factors (13):

Biological

- Poor mobility due to muscle weakness or balance problems
- Health disorder such as arthritis or stroke
- Poor nutrition (vitamin D deficiency)
- Sensory deficits such as hearing or vision loss

Behavioral

- Sedentary behavior
- Drug side effects or interactions
- Use of alcohol or recreational drugs

Environmental

- Hazards such as floor clutter, poor lighting, or uneven steps
- Absence or incorrect sizing of assistance devices (e.g., canes, walkers)
- Poor design of space

Typically, two or more of these factors are involved in or lead to a fall. Any individual who has had one prior fall is at a markedly increased risk of having another. Most important, many falls are preventable.

Many nonexercise- and exercise-related strategies can be used to attenuate one's risk of falling. A very important nonexercise factor involves modifying the individual's environments. This could include removing trip hazards (e.g., wires, newspapers, shoes, and throw rugs) from the floors. In addition, proper lighting in each room is essential along with use of support rails and antislip shoes and surfaces. Finally, steps should be wide, even, easy to access and navigate (no sharp turns), and clearly visible.

Concerning exercise, participating in group- and home-based exercise programs reduces the risk of falling by 15% to 30% (31). Modern forms of tai chi that emphasize balance, proprioception, slow movements, and physical and mental health have been shown to reduce the risk of falls to a similar extent. Although muscle weakness is a risk factor for falling, simply strengthening the lower limbs with resistance training does not appear to be sufficient to meaningfully reduce one's risk of falling. Implementing a simple walking program also seems to be insufficient.

Current exercise programming for fall prevention emphasizes a multicomponent model that involves two or more of the following: gait, balance, and functional training (e.g., balance platform, agility); strength training; flexibility (e.g., yoga); three-dimensional exercises (e.g., tai chi); improved general daily activity (e.g., walking); and aerobic endurance (e.g., cycling, treadmill walking) (75). Gait and balance training is a key component to include in any fall prevention initiative. Current guidelines regarding a fall prevention exercise regimen recommend 2 d/wk or more for a total of 60 min or more (30). Grabiner et al. (33) suggest that fall prevention should also include task-specific perturbation training. This involves developing the motor skills needed to stop or interrupt a fall by placing individuals in conditions that mimic a fall—for example, learning to step laterally for sideways-directed falls or adjusting the step response to tripping or slipping. Although the ability to learn motor skills aimed at interrupting a fall erodes with age, older individuals do retain the ability to relearn simple and complex motor skills. However, more time may be required to acquire the skill.

Rehabilitation of Patients With Cardiovascular Diseases

Up until the late 1960s and early 1970s, the average length of hospital stay for a patient who survived a myocardial infarction was approximately 21 d; the stay was even longer if the event was associated with complications such as heart failure or arrhythmia. The hospital stay consisted mainly of bed rest, which was intended to allow sufficient time for the damaged cardiac muscle cells to heal by scarring over.

However, many patients complained of marked exercise intolerance and weakness after being discharged from the hospital. Questions were raised about whether the activity restrictions imposed on the patients while in the hospital (and often continued at home) actually contributed to their deconditioning and fatigue. Fortunately, we have learned much over the past 40 yr, and now patients who experience an uncomplicated myocardial infarction are discharged from the hospital after only 4 to 5 d! While in the hospital and after initial stabilization, the activities patients are allowed to engage in are progressively increased such that most patients freely walk around their ward before going home.

Current treatment guidelines (81) indicate that eligible patients with cardiovascular disease should be referred to an outpatient cardiac rehabilitation program before being discharged. Eligible patients include patients who have

- experienced an acute coronary syndrome (myocardial infarction, unstable angina) or heart failure,
- undergone coronary revascularization (coronary artery bypass surgery, percutaneous coronary intervention such as angioplasty or stent deployment), or
- undergone heart surgery for heart valve replacement or repair.

Such programs often are called *early* or *phase II cardiac rehabilitation*. (Phase I cardiac rehabilitation usually consists of the education and supervised exercise therapy that patients receive while in the hospital. This service has been modified over time to accommodate the shortened length of hospital stays.) Whenever possible, the rehabilitation program is started several days to 1 to 2 wk after discharge from the hospital. These programs typically involve 3 sessions/wk for up to 12 wk and usually are supervised by exercise physiologists, registered nurses, or physical therapists. In the United States, most programs involve (at least for several sessions) having the patient undergo continuous ECG telemetry monitoring during exercise to ensure that the patient is training at a level that is safe for his or her medical condition. The volume of exercise completed each session and the total number of prescribed sessions is individualized based on a variety of factors, including clinical condition, pre-existing comorbidities, functional status, extent of insurance coverage and patient copayment, and return-to-work status.

In addition to exercise, patients are expected to receive individualized treatment programs that include structured educational counseling delivered in a group or one-to-one setting. All such education involves behavioral strategies aimed at reducing future risk through symptom recognition and management, medical compliance, proper long-term eating and exercise habits, smoking cessation, and weight management. As indicated in table 10.7, in patients with heart disease suffering a myocardial infarction or undergoing coronary revascularization, regular exercise and cardiac rehabilitation reduces all-cause mortality between 5% and 20% and reduces the risk for rehospitalization (84). Among patients with chronic heart failure, all-cause mortality is reduced 10% to 15% and rehospitalization is reduced 20% or more (66). Numerous other important intermediate effects are observed as well, including improved functional capacity or fitness, lowered blood pressure, improved glucose handing and endothelial function, and improved mood.

Rehabilitation of Patients With Cancer

The rehabilitation of patients being treated for or surviving cancer stands today where the rehabilitation of patients with cardiovascular disease stood in the mid-1970s. That said, the integration of regular exercise into the care of patients with cancer will progress much more quickly than it did for patients with cardiovascular disease because much of the research over the past 40 yr detailing how best to improve function, decrease risk, and change health behaviors in patients with other health disorders can be applied to patients with cancer. Also, the now-known benefits and acceptance of exercise in the treatment of patients with diabetes and cardiovascular disease has alerted health care professionals to incorporate exercise into the care of patients with cancer. Finally, with much preliminary evidence detailing the benefits of exercise in patients with cancer (42), large randomized clinical trials are now investigating the effects of exercise on disease-free survival.

The Patient Athlete

Ever since the pioneering work of Dr. Woldemar Gerschler and Dr. Herbert Reindel in the 1950s and 1960s, athletes of all ages and at all levels of competition have used higher intensity intermittent work (higher-intensity interval training; HIIT) to improve their performance. As the name implies, HIIT is a series of repeated bouts of higher intensity work intervals alternated with periods of relief (light or mild-intensity exercise). Although rigorous, this approach to training athletes leads to improvements in aerobic-type athletic events due to improvement in the energy capacity of the skeletal muscles (ATP production via both glycolysis and oxidative phosphorylation), improvements in cardiac stroke volume, and other favorable adaptations, all of which lead to less fatigue and greater improvements in cardiorespiratory fitness (as measured by peak $\dot{V}O_2$).

Patients with stable coronary heart disease were the first to have moderate-intensity exercise training formally incorporated into their routine care; this began in the late 1960s and early 1970s. From that time and up until today, exercise was prescribed at approximately 60% to 80% of heart rate reserve, 20 to 60 min/session, 3 to 4 sessions/wk. This regimen typifies what was prescribed in many cardiac rehabilitation programs in North America and Europe. One reason for its long-term use is that such training improves exercise capacity and quality of life and lowers the risk that a subsequent clinical event will occur. However, over the past 10 to 15 yr more and more emphasis has been placed on incorporating HIIT training into routine care for patients with heart disease as well as patients diagnosed with diabetes, cancer, heart failure, or hypertension (44,73).

In HIIT, the shorter the work interval, the longer the relief interval. Work intervals of 1 min might be set at an intensity of 95% of heart rate reserve and involve 2 min or more of recovery (relief interval). Conversely, work intervals of 4 min typically are set at 85% to 90% of heart rate reserve and involve 3 min of recovery. All of this is repeated across a 30-min bout of exercise 1 to 2 times/wk. Prior work in our lab showed that 10 wk of interval training in patients with heart disease led to a 3.6 mL · kg^{-1} · min^{-1} increase in cardiorespiratory fitness compared with an increase of 1.7 mL · kg^{-1} · min^{-1} in patients randomized to a traditional moderate-intensity training regimen that is typical of what is used in cardiac rehabilitation today (44).

The application of HIIT now extends far beyond individuals with heart disease. Compared with moderate continuous training, patients with other chronic diseases who practice HIIT have demonstrated similar or greater improvements in cardiorespiratory fitness (often twofold greater), vascular function, metabolic markers such as blood glucose and blood pressure, cardiac function, and skeletal muscle oxidative function (43,73). However, several questions remain about optimal HIIT work rate intervals, rate of progression, long-term application and safety, and influence on clinical events in patients with known disease. Concerning the latter, because a person's aerobic fitness is linked so closely to risk for future clinical events and because HIIT appears to induce greater improvements in fitness than does moderate-intensity training, further research is needed to determine whether HIIT also induces even greater reductions in risk.

Before closing on the topic of interval training, we introduce the novel observations of Kusy and Zieliński (48). They advance the premise that although much attention over the years has focused on younger distance runners who continue to run through adulthood and, as a result, enjoy healthy aging, proportionately little is discussed about the exercise habits of sprinters as they age. Without any exercise, all former young athletes are susceptible to the health perils of aging, including loss of aerobic capacity and submaximal endurance, decrease in lean body mass, and challenges with neuromuscular function. However, these researchers suggest that sprinters who continue to engage in higher intensity or sprint-type activities as they age experience the associated health benefits (48). Although the endurance training model is the classic model for healthy exercise habits across the life span, current research also suggests that continued participation in sprint-training or sprint-type activities with increasing age may be a viable option for ensuring healthy aging.

Although similarities exist between the exercise training of patients with cancer and those with heart disease, differences exist as well. A big difference is that unlike patients with cardiovascular disease, a formally detailed program involving supervision and monitoring by health professionals (and that is covered by health insurance) currently does not exist for patients with cancer. That said, most oncology rehabilitation that is provided today (usually through research projects or options not covered by insurance) involves aerobic-type training and resistance training and results in improved function, less fatigue, and improved mood (42). Concerning the development of improved cardiorespiratory fitness, many exercise training programs involve starting cancer patients at an exercise volume of 6 to 10 MET-h · wk^{-1} and then progressively increasing weekly volume. Such a program likely incorporates an exercise training intensity of 60% to 80% of heart rate reserve performed daily for 30 min or more. It is most common to start exercise training in these patients 2 mo after surgery or after chemotherapy or radiation therapy treatments are all completed. However, several studies have used exercise between chemotherapy or radiation therapy visits as a way to help attenuate the fatigue-inducing effects of the therapy.

Hypertension and Exercise

Hypertension, or elevated blood pressure, is one of the most prevalent chronic diseases in the United States, and treatment costs exceed $130 million annually. Hypertension requiring medical treatment is defined as a blood pressure of 140/90 mmHg (systolic/diastolic) or greater in adults aged 18 to 59 yr and 160/90 mmHg or greater in people aged 60 yr of age and older (57). More than 78 million adults (1 in every 3) have the disorder. The prevalence is similar between sexes, and the highest prevalence (approximately 42%) is observed in African American adults.

Although heredity plays a role in the development of hypertension, the remaining primary risk factors are potentially modifiable lifestyle traits such as obesity, excessive alcohol intake, and sedentary behavior. Persons with hypertension usually have the disorder without symptoms (asymptomatic) and are at increased risk for stroke, heart failure, heart disease, and kidney disorders.

The nonpharmacological strategies used to treat known hypertension include stopping smoking; losing weight as needed; limiting alcohol intake; reducing dietary sodium intake to less than 2,400 mg · d^{-1}; and consuming a diet that is rich in fruits, vegetables, and whole grains and that includes low-fat dairy, poultry, fish, and legumes (27). Aerobic-type exercise is also a very effective treatment for hypertension, In patients with mild to moderate hypertension, regular exercise training can reduce systolic and diastolic blood pressures by up to 6 mmHg (70). The effectiveness of resistance training in the treatment of hypertension is less defined. Table 10.3 reviews the potential mechanisms responsible for exercise-induced reductions in blood pressure.

Fatigue

Fatigue or exercise intolerance is a chief complaint that is nonspecific, which means it is a symptom that is associated with or results from many chronic disorders rather than being specific to one. In most instances (including heart failure, emphysema, and chronic kidney disease), fatigue is caused by abnormalities attributable to the disease. However, in certain disorders fatigue can also be due to the therapies (e.g., medications) used to treat the disease. This often occurs in patients being treated for cancer. Regardless of the cause or underlying health problem (e.g., heart disease, heart failure, cancer, kidney failure), if a patient complains of fatigue or exercise intolerance, regular exercise should be considered as a way to attenuate or reverse this disorder.

The exercise professional needs to first determine whether the exercise intolerance is due to decreased cardiorespiratory fitness, loss of muscle strength or endurance, or both. Once determined, proper exercise programming for patients with most of the chronic diseases mentioned can help reverse the exercise intolerance or fatigue. Often, cardiorespiratory fitness is improved 15% to 30% and muscular strength and endurance are improved 50% or more. Figure 10.7 shows an approximately 15% improvement in exercise capacity, as measured by peak $\dot{V}O_2$, across a 24-wk randomized trial involving supervised exercise 3 times/wk in patients with chronic heart failure (compared with patients not involved in formal exercise) (45). Notice the very low peak $\dot{V}O_2$ at baseline in both the treatment and control groups (16.0 and 14.7 mL · kg^{-1} · min^{-1}, respectively) and how most of the improvement (+2.2 mL · kg^{-1} · min^{-1}) in the exercise group is achieved with just 12 wk of training. The message to remember is that regular exercise allowed patients who initially showed marked exercise intolerance or fatigue to engage in mild to moderately higher levels of activity as well as their routine daily activities (e.g., climb a flight of stairs, shop for groceries) with less associated fatigue and other symptoms.

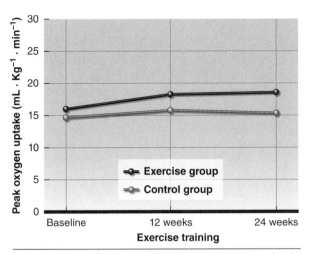

Figure 10.7 Changes in peak oxygen uptake in response to 24 wk of exercise training in patients with stable heart failure compared with a no-exercise control.

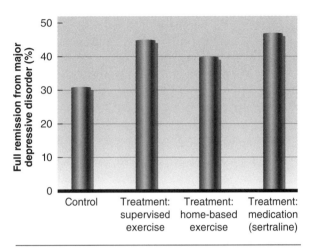

Figure 10.8 Effect of 16 wk of regular supervised or home-based exercise on remission from major depression compared with nontreated control patients and those undergoing medical care (sertraline). Rates of remission for those engaged in supervised exercise or home-based exercise were similar to those for patients in the drug treatment group.

Adapted from Blumenthal et al. (4).

Exercise and Depression and Health-Related Quality of Life

In addition to improved physical function, two other general adaptations often accompany participation in a regular exercise programs. Specifically, both depression and health-related quality of life typically improve. Depression, which often coexists with many chronic diseases, is a condition typically associated with diminished pleasure, a change in weight or appetite, poor sleep habits, fatigue or loss of energy, and a diminished ability to concentrate. Aerobic-type exercise has been shown to be an effective treatment for depression (22). In fact, one study reported in 2007 that involved 202 adults with major depression compared supervised exercise 3 times/wk, home-based exercise, and a standard drug treatment (sertraline) with a no-exercise control. After 16 wk, 45% of the patients in the supervised-exercise group and 40% of the patients in the home-based exercise group were no longer suffering from major depression; these findings were significantly better than those for the patients randomized to the no-exercise control. The rate of remission in those undergoing supervised exercise was very similar to that of patients in the drug treatment group (figure 10.8) (4).

Most data indicate that exercise has a favorable effect on quality of life. However, some lingering issues surround how quality of life is measured, the tools used to measure it, its multidimensional components (i.e., social, spiritual, physical, and emotional), and its variable nature across individuals and settings. Two decades ago, Mondin et al. (62) conducted a simple, provocative study in which they asked 10 healthy volunteer subjects (4 women, 6 men) who regularly exercised 6 to 7 d/wk to stop exercising! After just 3 d of no exercise, follow-up testing for changes in quality of life revealed that the subjects complained of mood disturbances, increased stress, and tension. These findings were reversed with the resumption of regular exercise.

CHALLENGES OF EXERCISE ADHERENCE AND FACILITATING BEHAVIOR CHANGE

Although the benefits of regular exercise are well documented, myriad clinical, behavioral, and psychosocial factors affect compliance. Common reasons for not exercising include lack of time, personal safety, sedentary family or social environment, confounding medical problems, perceptions of body weight or image, and self-efficacy (34,60,69). Depending on the population being referenced, current estimates are that as low as 20% and as high as 60% of people who start to exercise stop within 12 mo. Given what some may say is a bleak outlook, it is important to discuss the various strategies available to help people maintain regular exercise habits in order to derive the long-term health and fitness benefits.

Many theories have been advanced detailing how people best modify or adopt new behaviors. One effective model that has been applied to physical activity as well as smoking cessation and weight management is the transtheoretical model (71). This model conceptualizes five stages of behavioral change that, at any given time for each specific person, operate along a continuum. The five stages, adapted to physical activity, are as follows.

1. *Precontemplation.* The individual is not thinking about exercise or physical activity. If they have thought about it, they have no desire to start such a program.

2. *Contemplation.* The individual knows he or she needs to be more active and is considering starting such a program in the next 6 mo but may not have the skills, knowledge, or incentive to do so.

3. *Preparation.* The individual undertakes planning and behavioral steps to start an exercise or activity program within the next 30 d. Counseling should focus on identifying barriers to becoming active and developing concrete strategies for successfully resolving or eliminating each barrier.

4. *Action.* Individuals in this stage follow through and are engaged in practices that involve exercise or increased activity. The action stage takes place within 6 mo of starting a program. Type and volume of activity may or may not be sufficient to promote health benefits. Additional counseling and support may be required to facilitate achievement of proper exercise dose and long-term compliance.

5. *Maintenance.* Individuals demonstrate regular exercise habits for more than 6 mo. They require periodic contact from exercise professionals to ensure safe and proper implementation of their program and for help progressing and adapting the program as preferences develop.

Although one can envision movement through the five stages as being linear—for example, a person starts at the precontemplation phase and progresses through the subsequent stages until achieving maintenance—such is not always the case. People often and easily move between stages as life events, health conditions, and personal motivators shift over time. Many people do not succeed on their first attempt to stick with exercise (i.e., they relapse), which itself can provide useful information to draw upon when they again consider or resume exercise.

For people in the action and maintenance phases, the exercise professional can use two important strategies to help facilitate long-term compliance: self-monitoring and social support. Self-monitoring includes keeping records (e.g., exercise diaries, logs, or personal calendars) to document frequency and duration of exercise. Heart rate, symptoms, general feelings of enjoyment or satisfaction, and perceived exertion can also be measured. Social support involves both perceived and actual support from the individual's family or social network. Perceived support means that the person trying to become more active receives verbal and nonverbal cues that others indeed support his or her intentions. Perceived support often helps people who are contemplating or just starting an exercise program. Actual support means that the actions and behaviors of other people support the individual's desire to exercise. For example, friends or partners might actually join the person when he or she goes for a walk. Group exercise classes can provide actual support.

One should not overlook the importance of educating patients about implementing a safe and effective program along with what to expect as a result of one's efforts. How to properly warm up and cool down, body positioning and mechanics when performing an activity, warning signs and symptoms, proper clothing, accommodating and adapting to environmental extremes, and exercise progression are all important topics to discuss and can facilitate compliance.

Finally, always keep in mind that it is much easier to develop a safe, individualized exercise prescription for someone than it is to implement it. Your skills as an exercise professional should include helping those who are already active and wish to be more active as well as embracing those who are interested or just getting started—and, even more difficult, engaging people who have little to no interest in being active even though they know doing so will be helpful. When prescribing exercise and counseling others, it may behoove you to think a little less about developing the exact exercise prescription and instead begin with simply getting people started by identifying solutions to the barriers that impede mild to moderate activity. Then you can titrate exercise volume, as tolerated, and as their exercise habits favorably progress.

Summary

- Approximately one fourth of U.S. adults are sedentary.
- The burden associated with inactivity is immense in terms of health risks as well as lost income and dollars spent for the care of a chronic disease.
- Current guidelines recommend 150 min · wk⁻¹ of moderate-intensity exercise or 75 min · wk⁻¹ of vigorous aerobic-type exercise along with 1 to 2 d of resistance training and regular activities targeting neuromuscular coordination.
- The competent exercise professional should be able to prescribe exercise that is safe and effective in healthy people as well as in those with a clinically manifest chronic condition.
- The competent exercise professional should appreciate and convey to others the expected health benefits associated with regular exercise.
- The competent exercise professional should counsel both currently active and inactive people in a manner that strives to maintain long-term compliance.

Definitions

exercise or exercise training—Physical activity with the expressed purpose of improving or maintaining health or physical fitness.

physical activity—Movement of the body of any kind (i.e., occupation, household, leisure time, and transportation). Leisure time physical activity can be subclassified into three types: competitive sport, formal exercise training, and insufficient activity.

population attributable risk—A measure that epidemiologists use to estimate the effect of a risk factor on the frequency of new cases of a disease in a population.

sedentary behavior—A state of low energy expenditure; involves activities (<1.5 METs) performed while awake and sitting or reclining.

sudden cardiac death—Unexpected death due to sudden loss of heart function (often due to arrhythmia).

total daily activity—The sum of sleep, sedentary behavior, and all types of physical activity in a day; typically expressed as kilocalories per day or MET-hours per day.

11

Emerging Concepts: Exercise Pharmacology and Exercise Genomics

Like all other fields of scientific study, exercise physiology is dynamic. To that end, this chapter discusses two distinct key topics in the field. The first section is quite clinical and addresses the interaction between pharmacology and exercise. The second section examines the rapidly expanding field of exercise genomics. Both topics make important contributions to the learning of the graduate student interested in exercise physiology, and both give credence to the fact that exercise physiology is a field that continues to evolve.

EXERCISE PHARMACOLOGY

The use of drugs and **nutraceuticals** (e.g., herbal supplements, vitamins, and minerals) for health and wellness is pervasive throughout the United States. Nearly half of all Americans report taking at least one prescription medication, and approximately 80% of Americans take over-the-counter drugs or nutraceuticals. Throughout history, athletes have attempted to use drugs and nutraceuticals to improve performance and enhance recovery. This dates back to the ancient Greek athletes, who would use various substances such as sesame seeds and hallucinogenic mushrooms (24). Similar to today, it was considered unethical for ancient Greek athletes to seek an unfair advantage by these means, and perpetrators would be sold into slavery (24). A more recent example of the use of

drugs and nutraceuticals is the designer steroid scandals of the late 1990s and early 2000s. The Bay Area Laboratory Co-Operative (BALCO) was a clandestine sport supplement company that supplied banned substances such as testosterone cream, erythropoietin, and human growth hormone to high-profile professional athletes (11). The trial and subsequent convictions surrounding BALCO led to improved drug testing and drug enforcement rules by organizations such as Major League Baseball to deter the use of anabolic steroids and other banned substances.

Athletic performance notwithstanding, more adults are engaging in exercise than in the previous two decades, including many individuals who take prescription medications for chronic health problems (31). Although many types of drug–drug interactions are known, relatively less is known regarding the drug–exercise interaction. Having knowledge of both exercise physiology and pharmacology can provide better insight into the increasingly important relationship between drugs and exercise.

Basics of Pharmacokinetics

The term *pharmacokinetics* simply means how the body affects a drug. This includes how a drug is absorbed, metabolized, distributed, and finally disposed of in the body. How a drug enters the body is known as the **route of administration** (table 11.1).

Table 11.1 Routes of Drug Administration

Route of administration	Example	First-pass metabolism?	Advantage(s)
Oral (PO)	Ibuprofen	Yes	• Easiest way to administer • Reversible
Sublingual (SL) or buccal	Testosterone	No	• Rapid absorption • Avoids first-pass metabolism
Transdermal	Nicotine	No	• Easy to administer • Avoids first-pass metabolism
Rectal	Acetaminophen	No	• Alternative route when gastrointestinal tract is upset (e.g., vomiting)
Intravenous (IV)	Morphine	No	• Rapid absorption • More precise control of drug levels • Avoids first-pass metabolism
Intramuscular (IM)	Influenza vaccine	No	• More precise control of smaller amounts of drugs • Avoids first-pass metabolism
Subcutaneous (SC)	Insulin	No	• Allows slower release of drug • Avoids first-pass metabolism

Drug Absorption

More than 80% of all prescription drugs are taken orally. However, the oral route is actually the most complicated because much of the drug that is ingested often does not make it to systemic circulation. This is partially due to physical (i.e., cell membrane) and chemical (i.e., low pH in the stomach) barriers in the gastrointestinal tract but largely due to first-pass metabolism, which occurs mainly in the liver and to a lesser extent in the gastrointestinal tract. Drugs that enter the stomach filter through the hepatic (i.e., liver) or portal circulation before reaching the systemic circulation. The chemical composition of a drug is changed when enzymes in the liver act on the agent, sometimes transforming a drug to its inactive form. Because of this, many drugs (e.g., insulin, testosterone, and nitroglycerine) are administered using alternative methods (e.g., intramuscular, sublingual). The percentage of a drug that successfully absorbs into the systemic circulation and thus becomes available for use is known as its **bioavailability**.

Drug Distribution

How a drug is distributed throughout the body influences how long it stays in the body and thus its potential for both therapeutic and adverse effects. In addition to the systemic bloodstream (i.e., blood plasma), drugs can redistribute into the interstitial or intracellular fluid compartments, or both (table 11.2). Drugs that stay localized in the plasma have greater bioavailability, meaning that higher concentrations of a drug can exert its effect on activation sites. However, they are also more readily eliminated. Conversely, drugs that are distributed into all three fluid compartments typically stay in the body longer and are said to have a larger **volume of distribution**. An example of this and of how exercise can influence pharmacokinetics is the heart-failure drug digoxin. During exercise, plasma concentration of digoxin is reduced because increased blood flow to muscles redistributes the drug, which has a strong affinity to bind to muscle. Therefore, digoxin's volume of distribution increases with exercise.

In general, lipophilic (i.e., fat-soluble) drugs of low molecular weight have a large volume of distribution spreading throughout all three fluid compartments, whereas large drugs that contain a charge (+ or −) remain within the plasma. This is because lipophilic drugs can pass through cell membranes easier, and drugs with smaller molecules can pass through spaces between capillary walls known as *slit junctions*.

To determine the volume of distribution (V_d) for a drug, you need to know the amount of total drug in the body and the plasma concentration of that drug for a given sample. For example, if 10 mg of a drug is administered intravenously (thus, the bioavailability is 100% because it

Table 11.2 Fluid Compartments of the Body and Characteristics of Drugs Found in Them

Fluid compartment	Drug characteristic	Ability to move into other compartments?
Plasma	High molecular weight, hydrophilic, protein binding	No
Interstitial	Low molecular weight, hydrophilic	Yes, plasma
Intracellular	Low molecular weight, lipophilic	Yes, interstitial and plasma

does not first go through portal circulation) and a blood sample shows a concentration of 1 mg/L, then using equation 11.1 you can calculate the V_d by dividing 10 mg by 1 mg/L; the result is 10 L. Of course, this is a simplified example because it does not take into account how much of a drug is being eliminated.

$$V_d = D/C_D \qquad \text{(Equation 11.1)}$$

Where

V_d = volume of distribution (L)

D = total amount of the drug in the body (mg)

C_D = plasma concentration of the drug (mg/L)

Another important factor that influences drug distribution is protein binding. Drugs that bind to blood plasma proteins are inert, meaning that they cannot have a therapeutic effect while bound. Although this can delay drugs from binding to activation sites, it also can create an immediate reservoir by compartmentalizing the drug within the plasma and helps maintain steady concentrations of the free drug in the blood. Albumin is the most common plasma protein that drugs bind to.

Many drug-to-drug interactions occur when two or more drugs have a high affinity for albumin. An example of this is the interaction of warfarin (an anticoagulant agent) with fluoxetine (an antidepressant). When fluoxetine competes with warfarin for albumin, this allows for greater concentrations of warfarin, which in turn increases the risk of serious bleeding complications.

Drug Metabolism and Elimination

The vast majority of drugs are eliminated through the kidneys (i.e., urine). Other sources of elimination include feces, expired air, sweat, tears, and breast milk. Although some medications (e.g., atenolol) are eliminated relatively unchanged, lipid-soluble drugs must first be metabolized before they can be eliminated by the kidneys. This is because lipid-soluble drugs readily

diffuse back into the bloodstream when passing through the kidney. Therefore, before a drug is eliminated, it must first undergo drug metabolism in which it is oxidized or reduced (causing it to become polar—i.e., positively or negatively charged) and then must be made water soluble through a conjugation reaction (e.g., addition of an amino acid group). Once a drug becomes polarized and water soluble, it can no longer reabsorb into the bloodstream and essentially becomes trapped in the kidneys and thus eliminated.

Most drug metabolism occurs in the liver, although metabolism can also occur elsewhere, such as in the plasma and intestines. The **cytochrome P450** (C-P450) system is responsible for catalyzing most drug metabolic reactions. C-P450 is found throughout the body where metabolism occurs. There are several isoforms of C-P450, which are influenced by the genetic makeup of an individual. This genetic variation can explain individual and ethnic differences in metabolizing certain drugs. The C-P450 system is also a reason for potential drug interactions. For example, grapefruit can interfere with a C-P450 isoform that is responsible for metabolizing caffeine. Thus, plasma concentrations of caffeine are higher when an individual concomitantly consumes grapefruit!

An important index that measures how long it takes the body to eliminate a drug is known as the **half-life**. Knowledge of a drug's half-life helps determine how frequently a drug needs to be dosed in order to maintain therapeutic plasma levels in the body (figure 11.1). As the name suggests, the half-life indicates the time required for 50% of a drug concentration to be eliminated from the body. The half-life is directly related to the volume of distribution and inversely related to the rate of clearance, which is the amount of drug eliminated per hour. In other words, the greater the volume of distribution or the slower the clearance, the longer the half-life.

Students often think that the amount of drug given (i.e., dose) affects the half-life, but generally this is

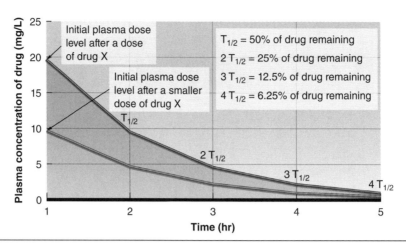

Figure 11.1 Example of drug half-life using different doses of the same drug. Despite different initial doses, the percentage of the same drug removed from the plasma per hour is the same. This demonstrates how the rate of drug metabolism increases proportionally to the amount of drug in the plasma.

not the case. As the concentration of a drug increases, more enzymes associated with drug metabolism become involved, thus increasing the rate of drug metabolism proportionally to drug concentrations. This results in the same half-life regardless of the initial dose. The exception to this rule would be if the amount of drug in the body exceeded the enzymes available to metabolize it. This is known as **zero-order** drug metabolism.

To better appreciate the difference between first-order drug metabolism (i.e., regular drug metabolism) and zero-order drug metabolism, think about checkout lines at a grocery store. If a store has 10 open registers, customers can immediately purchase items from any cashier as long as 10 or fewer customers are checking out at the same time (first-order metabolism). However, lines begin to form when the number of customers wanting to check out exceeds the number of open registers. In this example, the customers (and their items) are the drugs and the cashier is the metabolic catalyst (i.e., enzyme). Aspirin is a good example of first-order versus zero-order drug metabolism. When taking the standard doses of aspirin (i.e., 81-325 mg/d) associated with the secondary prevention of cardiovascular events, the body metabolizes these at the same rate regardless of the dose. However, when the dose exceeds four times the standard dose amount, the half-life increases from 3 h to 15 h!

Pharmacodynamics

Pharmacodynamics describes how a drug affects the body. It is important to remember that drugs do not create novel physiologic responses but instead enhance, diminish, or mimic endogenous processes of the body. This often is accomplished through drugs that mimic endogenous ligands and bind to cell receptors, resulting in an intracellular response. Most drugs bind to receptors on the surface of cell membranes (e.g., beta receptors); however, some lipid-soluble drugs cross through the cell membrane and bind to intracellular receptors. For example, hormone analogues, such as exogenous testosterone, translocate into the cell nucleus and bind to transcriptional factors that regulate gene expression that increases muscle protein synthesis and other anabolic responses. Depending on the site of activation (cell membrane vs. intracellular) and receptor type, it can take seconds to hours before the effects from the drug begin.

Drugs that bind to a receptor and cause an increased cell response are known as **agonists**. Conversely, drugs that bind to receptors and cause attenuation, a given biologic response, are known as **antagonists**. Table 11.3 shows some common agonists and antagonists.

Therapeutic Index

The **therapeutic index** is a measure of drug safety. Certain drugs are more known for their adverse effects (e.g., warfarin

Table 11.3   Common Drug Agonists and Antagonists

Drug classification	Generic example (trade name example)	Receptor type	Biological effect
Beta blockers (antagonist)	Metoprolol (Toprol-XL)	Beta-adrenergic receptor	Decreases heart rate, myocardial contractility, and cardiac output
Antihistamines (antagonist)	Cetirizine (Zyrtec)	H1 receptor	Reduces nasal and bronchial mucus, prevents bronchial constriction, and blocks histamine-associated itching
Opioids (agonist)	Oxycodone (OxyContin)	μ receptor	Inhibits the release of neurotransmitters responsible for pain
Insulin (agonist)	Insulin aspart (NovoLog)	Insulin receptor	Activates the translocation of glucose transporter receptors to absorb glucose into various cells throughout the body
Selective estrogen receptor modulators (antagonist)	Tamoxifen (Nolvadex)	Estrogen receptor	Blocks estrogen from binding to certain types of breast cancer cells, preventing their proliferation

and risk of bleeding); however, it is important to understand that all drugs have the potential for toxicity. Even a benign substance such as water can cause death due to hyponatremia if consummated in extremely large amounts.

The therapeutic index is a ratio of the dose that produces a toxic effect to the clinically effective dose (equation 11.2). The higher the ratio, the safer the drug. An example of a toxic drug with a low therapeutic index is one that has a 1-mg effective dose but an only 2-mg toxic dose. If this hypothetical medication were given in 1-mg tablets and someone accidently took two tablets, then adverse effects would be likely.

$$\text{Therapeutic index (TI)} = TD_{50}/ED_{50}$$
(Equation 11.2)

Where

TD_{50} = toxic dose in 50% of the population

ED_{50} = effective dose in 50% of the population

Putting It Together: The Interaction Between Exercise and Drugs

Research is limited regarding interactions between exercise and many specific drugs. However, knowledge of how a drug is absorbed, metabolized, distributed, and

eliminated can provide clues about how exercise may interact. The following sections describe some common exercise–drug interactions and the mechanisms associated with each.

Exercise and Drug Absorption

Blood redistribution during exercise is significant: The proportion of blood flow to the metabolically more active skeletal muscles shifts from approximately 20% of cardiac output to 70% or more depending on the intensity of exercise (14). This can result in a shunting of blood away from important drug absorption sites in the gastrointestinal system. As described in chapter 2, exercise intensities at 70% $\dot{V}O_2$ peak or greater result in decreased gastric emptying; thus, the absorption of oral drugs may be slowed. Also, because the majority of oral drugs are absorbed through the small intestines, decreases in gut transit time (due to blood redistribution during exercise) may, in theory, be a source of an exercise–drug interaction, although evidence for this is still lacking.

Cardiac output (i.e., blood flow) is redistributed to the metabolically more active skeletal muscles as well as to the skin to assist with maintenance of body temperature. This is especially important for individuals with transdermal medications, such as the nicotine patch, because increased blood flow to the skin allows for quicker drug absorption. Several studies have shown that compared with rest, exercise increases the amount of drug available in plasma as much as twofold when the drug is administered by transdermal methods (14). This increased transdermal absorption associated with exercise has been shown to increase drug-related side effects (e.g., nausea and palpitations) for those using a nicotine patch (14).

Another common example of how exercise can interact with drug administration and potentially lead to adverse reactions is the administration of insulin. Insulin is given subcutaneously, often in the thigh or abdomen (19). Several studies have shown that injecting insulin into the thigh before exercise on a stationary cycle increased the absorption of insulin compared with the subcutaneous administration at an abdominal site (14). Thus, increasing exogenous insulin levels in the body can lead to hypoglycemia. Interestingly, in a study that investigated the long-acting insulin glargine, which is known to exert its effects in the body for more than 20 h, exercise was shown not to effect insulin absorption (22).

Exercise and Drug Metabolism and Distribution

As mentioned previously, the liver is the main organ responsible for the metabolism of most drugs. Reductions in hepatic blood flow, caused by acute exercise, might therefore slow down drug metabolism and elimination, resulting in greater concentrations in the blood (19). A study by van Baak et al. (30) showed that during a single bout of exhaustive exercise at 70% of $\dot{V}O_2$ peak, concentrations of the beta blocker propranolol temporarily increased before returning to levels similar to those of the resting control group. Because propranolol has a high hepatic extraction rate, it was conjectured that the increased concentration was the result of reduced hepatic blood flow. However, although an acute bout of exercise may increase concentrations of hepatic-dependent drugs, regular exercise training may have the opposite effect on drug metabolism. Some studies have found greater hepatic blood flow and increased liver enzymes associated with habitual exercise (30).

One well-known exercise–drug interaction involves the medication warfarin (common brand name = Coumadin), an anticoagulant used to prevent the formation or progression of thrombosis. As mentioned earlier, warfarin has a strong affinity for the plasma protein albumin. While bound to albumin, warfarin is essentially inactive. Exercise has been shown to increase albumin synthesis by 51% for up to 22 h after exercise (18). Thus, more plasma protein albumin is available to bind to warfarin, consequently decreasing its effectiveness and

increasing the amount of drug that must be given. Studies have confirmed an association between increased levels of physical activity and higher required doses of warfarin (14). Interestingly, increased exercise also has been shown to decrease the incidence of bleeding while taking warfarin (26).

Exercise and Drug Elimination

As mentioned previously, the primary method of drug elimination is renal excretion. Therefore, it stands to reason that any effect that exercise may have on the kidney will in all likelihood affect drug elimination. This points to the effects exercise has on blood redistribution because acute exercise has been shown to reduce renal blood flow by up to 50% (14).

One common medication used in pharmacokinetic studies done during exercise is atenolol (14,15). Atenolol is a beta blocker that is mostly unchanged in the body and is highly reliant on renal elimination (14). Renal elimination of atenolol has been shown to decrease during exercise, leading to subsequent increases of plasma concentrations (4,15,28). However, these effects seem to be transient and return to baseline shortly after an exercise bout. Because of this temporary effect on the kidneys, any effect exercise may have on atenolol and other drugs that are dependent on renal elimination may be limited to exercise bouts of longer duration.

Looking at drugs that are eliminated from the body through other means, such as perspiration and respiration, a plausible case can also be made for the effects of exercise on those drugs. In fact, studies have shown that drugs that typically are eliminated by the lungs (e.g., the catecholamines norepinephrine and dopamine) have a faster rate of elimination during exercise, which may simply be due to the increase in blood flow to the lungs during exercise (2).

Drug–Exercise Pharmacodynamics

When examining the relationship between drugs and exercise, it is also important to consider the pharmacodynamic effects of drugs (i.e., how a drug affects the exercise response). Some drugs are taken for purported enhancements during exercise and sport (see table 11.4); however, many drug–exercise interactions can impair exercise performance and blunt training adaptations. A common example of a drug–exercise interaction involves beta-adrenergic blocking agents (so-called beta blockers)

The Pharmacology of Anabolic-Androgenic Steroids

Anabolic-androgenic steroids (AAS) are synthetically derived compounds that are similar to testosterone, which is endogenously produced by the body in the gonads and adrenal glands (12,24,32). Due to first-pass metabolism, testosterone itself cannot be administrated orally. However, pharmaceutical companies alter testosterone and other similar AAS molecules through the process of alkylation, thus improving its bioavailability when taken orally (12). Other routes of administration for AAS include buccal, transdermal (e.g., patch, cream, gel), and intramuscular injection (32).

Synthetically manufactured AAS act as antagonists, binding to androgen receptors throughout the body and resulting in androgenic (i.e., masculinizing) and anabolic (i.e., tissue building) changes as well as other steroid-related effects (e.g., sodium retention, anti-inflammatory effects) (12). The few well-controlled studies that have looked at AAS showed a strong dose–response relationship, with higher doses leading to greater gains in strength and muscle mass (24).

Some AAS are approved and used for clinical reasons (e.g., attenuate or reverse sarcopenia or muscle wasting in patients with acquired immunodeficiency syndrome). However, AAS have become synonymous with abuse among both athletes and nonathletes. Prior to 1974, AAS abuse had become widespread in sporting events and included government-supported use in some Eastern bloc countries (24). Because most types of AAS are excreted in the urine, drug testing mainly involves urine samples. However, detecting AAS use has become more of a challenge with the advent of designer steroids. In fact, in order to circumvent standard drug tests, athletes involved in the BALCO scandal were also given tetrahydrogestrinone to mimic the testosterone/epitestosterone ratio found in urine samples (11). Currently, the World Anti-Doping Agency monitors and oversees drug testing for collegiate, professional, and Olympic athletes.

Over the past 50 years, substantial evidence has grown demonstrating considerable adverse effects associated with AAS. Some documented adverse effects include male pattern baldness, hirsutism, hypogonadism, mood disorders, premature epiphyseal closure, and cardiomyopathy. Additionally, the use of oral alkylated AAS drugs (to avoid first-pass metabolism in the liver) has been linked to liver abnormalities (12). A likely contributing factor to some adverse effects is the supraphysiologic blood levels of testosterone achieved by illicit users. Individuals who abuse AAS typically take amounts that are 15 to 30 times higher than a standard clinical dose (12). Further, many sources of illegal AAS come through clandestine Internet sites as supplements, which at best supply ineffective AAS and at worst (because many new designer steroids do not undergo any type of testing for quality) place the buyer at extreme risk with an unknown, unregulated substance. However, thanks to efforts to increase awareness of these known side effects, use among high school students decreased from 5% in 2001 to 3.2% in 2013 (13).

and their effect on attenuating heart rate and contractility and, as a result, possibly peak $\dot{V}O_2$ (i.e., $\dot{V}O_2$ = stroke volume × heart rate × A-$\dot{V}O_2$ difference) (29). Interestingly, individuals taking a beta blocker typically show a compensatory augmentation of stroke volume during exercise, and although peak $\dot{V}O_2$ may still be slightly reduced (compared with no beta blocker), the increase in stroke volume negates some of the negative chronotropic effects of the drug (29).

Additionally, exercise can exacerbate specific drug side effects because of parallel physiological pathways leading to similar outcomes. Examples of these include antihypertensive agents and several of the drugs used to treat diabetes. Exercise can concomitantly lower blood pressure after exercise and blood glucose during exercise, leading to a synergetic effect that can cause adverse responses when taken in concert.

EXERCISE GENOMICS

Contribution by Mark A. Sarzynski, PhD,
University of South Carolina

The rapidly expanding field of exercise genomics advances our understanding of the preventive and

Table 11.4 Select Performance-Enhancing Drugs Banned by the World Anti-Doping Agency

Drug classification	Clinical indications	Sport/event used in	Purported benefit	Adverse effects
Beta blockers (e.g., atenolol)	Antihypertensive Prophylactic in patients who suffered a heart attack Migraine prophylactic Stable angina	Archery, shooting, golf, billiards	Reduces heart rate and blood pressure, which may reduce tremors and help with sports requiring high accuracy	Dizziness, fatigue, arrhythmias
Diuretics (e.g., furosemide)	Edema Heart failure Renal impairment	Bodybuilding, boxing, wrestling	Used for rapid weight loss in sports with weight classes; also taken with other performance-enhancing drugs as a masking agent	Dizziness, muscle cramps, electrolyte imbalances
Stimulants (e.g., methylphenidate)	Attention-deficit/ hyperactivity disorder Narcolepsy	Football, hockey, weightlifting, rowing	May improve reaction time, increase acceleration, and reduce fatigue	Hypertension, stroke, myocardial infarction, rhabdomyolysis, cardiac arrhythmias, anxiety, headache, nausea
Beta-agonists (e.g., albuterol)	Asthma Bronchitis	Speed skating, swimming, triathlon, weightlifting	Oral: may increase skeletal muscle and inhibit muscle breakdown; inhaled: open bronchioles, thus decreasing airway resistance	Cardiac arrhythmias, anxiety, hypertension, hyperglycemia

therapeutic properties of exercise by increasing our knowledge about the physiology of exercise and human behavior. Although understanding the effects of various factors (e.g., environmental) on exercise-related **phenotypes** has long been of interest, understanding of the role of genetic factors (i.e., exercise genomics) is still in its infancy. Most of the major developments have occurred over the past 50 years. This section of the chapter does not have the capacity to fully review the field of genetics; therefore, the following sections introduce the basic concepts of genetics, deoxyribonucleic acid (DNA), and molecular biology relative to exercise physiology. A more thorough review on the topic can be found in *Genetics Primer for Exercise Science and Health* by Stephen M. Roth (2007).

Basics of Human Genetics

Genetics is the study of genes, heredity, and genetic variation in living organisms. The **human genome**, which contains all of the genetic material in human cells, consists of about 3 billion DNA base pairs. DNA encodes the instructions for the development, function, and reproduction of the entire human being. DNA is located in both the nucleus of the cell and the mitochondria, the majority of which is located in the nucleus. The nuclear genome comprises 22 pairs of autosomes (non-sex-specific chromosomes) and one pair of sex chromosomes for a total of 46 chromosomes. Humans inherit half of their genome from their mother (22 autosomes and an X chromosome) and half from their father (22 autosomes and either an X or Y chromosome). Women carry two copies of the X chromosome, whereas men carry one X chromosome and one Y chromosome. The presence or absence of the Y chromosome (responsible for testis development) determines the sex of the offspring.

Each DNA molecule consists of two complementary strands that each comprise four nucleotide bases—adenine (A), cytosine (C), guanine (G), and thymine (T)—that form a double helix. The binding between the two DNA strands is complementary: A always binds to T and C with G, creating complementary base pairs. The order of these nucleotide bases determines the meaning of the biological information encoded in that part of the DNA molecule. Specific sequences or regions of DNA that provide instructions to build ribonucleic acid (RNA) and proteins are called **genes**. Genes are the basic physical and functional units of heredity. The human genome contains about 20,500 genes. DNA found in the mitochondria is a circular double helix and contains 37 genes. The typical structure of a gene consists of many regions, including the coding region (exons), noncoding region (introns), promoter region (5'-end; upstream region; beginning of gene), and 3'-untranslated and terminator regions (3'-end; downstream region; end of gene; figure 11.2). The promoter region contains elements that direct the transcription of a gene (i.e., turn the gene on or off), whereas the terminator region contains the DNA sequence responsible for stopping the transcription of a gene.

A central tenet of molecular biology describes the flow of information from DNA to RNA to proteins (figure 11.3). During the process of transcription, or gene expression, the nuclear DNA sequence found in a gene is read and copied onto a complementary strand of RNA called *premature messenger RNA* (mRNA) by the enzyme RNA polymerase. RNA is single stranded and contains the nucleotides A, G, C, and uracil (U), which replaces T and similarly binds only to A. Before the premature mRNA can be used to build protein, the noncoding regions of introns are removed, resulting in mature mRNA that consists of the coding exonic regions only. The mature mRNA then travels out of the nucleus and into the cell's cytoplasm, where it undergoes the process of translation. During translation, the genetic code is read by special organelles in the cell called *ribosomes,* which pair the mRNA to transfer RNA and eventually attach a corresponding amino acid to the polypeptide chain. Repetition of this step assembles a protein one amino acid at a time (figure 11.4). With just the four nucleotides used in three-letter combinations, the 20 standard amino acids are assembled.

Heritability and Genetic Variation

More than 99.9% of the DNA sequence in humans is similar. The 0.1% that differs is what makes each individual unique. Because family members share an even greater proportion of DNA sequence compared with unrelated individuals, analysis of familial aggregation (i.e., resemblance across family members) often is used to determine whether genetic variation contributes to a trait. For a given trait, familial aggregation is tested by comparing the variance between families with the variance within family members. For example, the HERITAGE Family Study examined the familial resemblance for maximal oxygen uptake ($\dot{V}O_2max$), the gold standard measure of cardiorespiratory fitness, in 426 sedentary subjects from 86 nuclear families. The authors found that achieved levels of $\dot{V}O_2max$ clearly aggregate in families; there was 2.7 times more variance between different

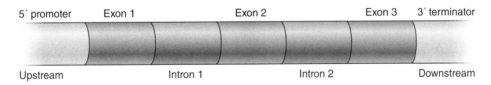

Figure 11.2   The basic structure of a gene, including the upstream promoter region, the coding exons separated by introns, and the downstream terminator region at the end of the gene.

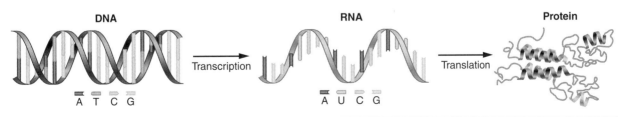

Figure 11.3   The central principle of molecular biology, whereby DNA codes for RNA, which codes for proteins.

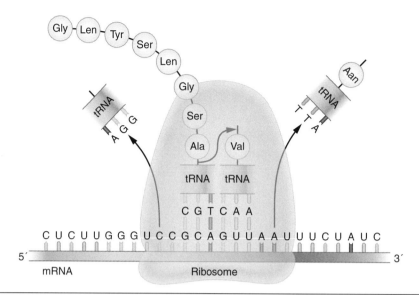

Figure 11.4 Overview of the translation of mRNA into protein. The ribosome binds to mRNA and moves one codon at a time downstream, where transfer RNA drops off an amino acid that links with the previous amino acid, thereby forming a polypeptide chain that continues until the ribosome reaches a stop codon.

families than within families for adjusted $\dot{V}O_2$max (4). Figure 11.5 shows that some families tended to have below-average $\dot{V}O_2$max values, whereas others had above-average values. Because families share genetic factors as well as similar environmental factors (e.g., diet, education, physical activity, and sedentary behaviors), further analyses are needed to determine whether the familial resemblance in $\dot{V}O_2$max is due to genetic or environmental factors.

Heritability is an estimate of how much of the variation in a trait is due to genetic factors alone. Heritability estimates are quantified by comparing trait similarities between pairs of relatives with different degrees of relatedness. Thus, twin, adoption, and family studies commonly are used to distinguish between the contribution of genetic and environmental factors on the familial aggregation of a trait. For example, a pattern of significant correlations among siblings and between parents and siblings but not between spouses would imply that the familial resemblance primarily is due to genetic factors, whereas significant correlations between spouses and lesser correlations between siblings or between parents and siblings would suggest a stronger influence of shared environmental factors such as diet and exercise habits. In the HERITAGE Family Study, the authors estimated that the genetic heritability of $\dot{V}O_2$max was 51% (4). In addition to $\dot{V}O_2$max, numerous exercise-related traits have been shown to have a significant genetic component, as evidenced by

their estimated heritability levels from twin and family studies (table 11.5).

After familial aggregation and heritability have been estimated for a trait, the next step is to determine which specific genetic factors contribute to the trait. Here, the term *genetic factors* refers to genetic variation or differences in the DNA sequence that may influence the information contained in a gene, thus resulting in trait variability between individuals. A variant nucleotide found at a **mutation** or **polymorphism** is known as an **allele**. Typically, two alleles are found at a polymorphism in the genome (e.g., A and G). Because genes come in pairs (one inherited from the mother and the other from the father), every individual has a pair of alleles for each mutation or polymorphism. The combination of alleles at any mutation or polymorphism is known as a **genotype** (i.e., that part of the genetic makeup that determines specific characteristics or phenotype). Thus, an individual may have either two copies of the same allele or one copy each of the variant allele. For example, if the nucleotides A and G are found at a particular polymorphism, an individual could have one of the following three genotypes: A/A, A/G, or G/G. An individual is a homozygote if they carry two copies of the same allele (e.g., A/A or G/G), whereas a heterozygote carries one of each allele (e.g., A/G). There are several forms of genetic variation, the most common of which is the single nucleotide polymorphism (SNP). An SNP is a single position in the DNA sequence where more than one nucleotide base is found in greater than

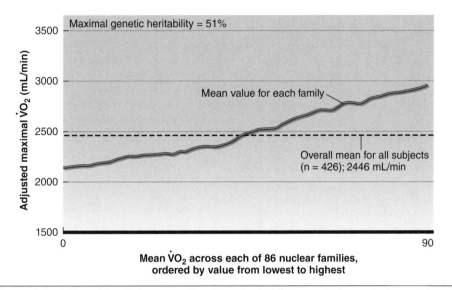

Figure 11.5 General depiction of familial aggregation for adjusted $\dot{V}O_2$max in the sedentary state; concept derived from the HERITAGE Family Study (4). Average $\dot{V}O_2$max (solid line) for each nuclear family is ordered from lowest to highest along the *x*-axis. The higher average $\dot{V}O_2$max for families to the right of the graph is due to family members with a $\dot{V}O_2$max above the average value for all subjects, which was 2,446 mL · min^{-1} (dashed horizontal line). $\dot{V}O_2$max values are adjusted for age, gender, height, weight, fat mass, and fat-free mass.

Table 11.5 Heritability Estimates for Intrinsic Levels of Exercise-Related Traits

Trait	Study design	*N*	Heritability
Physical activity	Nuclear families	9,500	0.23
	Twins	26,724	Males: 0.57 Females: 0.50
Exercise participation	Twins	85,198	0.27-0.71
% time inactive	Nuclear families	696	0.25
$\dot{V}O_2$max (adjusted for body weight)	Nuclear families	429	0.20-0.52
	Twins	6 studies	0.40-0.94
Submaximal aerobic performance (e.g., $\dot{V}O_2$ measured at 50% and 80% of maximum)	Nuclear families	483	0.29-0.70
	Twins	2 studies	0.38-0.55
Static strength	Nuclear families	5 studies	0.27-0.58
	Twins	15 studies	0.14-0.83
Dynamic strength (isokinetic, concentric, eccentric)	Nuclear families	6 studies	0.22-0.85
	Twins	3 studies	0.29-0.90
Muscular endurance	Nuclear families	2 studies	0.40-0.74
	Twins	2 studies	0.45-0.75
Flexibility (sit and reach)	Nuclear families	1,264	0.64
	Twins and parents	391	Males: 0.72 Females: 0.51

Table created using data from Blangero and Kent (1); Bray, Fulton, Kaluapahana, and Lightfoot (7); Perusse (20).

1% of the population. There are more than 10 million SNPs in the human genome. A mutation differs from a polymorphism in that the rare allele is found in less than 1% of the population.

Genetic Variation, Cardiorespiratory Fitness, and Muscular Strength

The field of exercise genomics continues to expand. As a result, we cannot summarize all of the genetics-related studies addressing all exercise-related traits. Instead, we focus on cardiorespiratory fitness and muscular strength and endurance traits in humans that are based on several key studies.

As previously shown in table 11.5, cardiorespiratory fitness and muscular strength and endurance traits are influenced by a substantial genetic component, with heritability estimates ranging from 29% to 94% for maximal and submaximal cardiorespiratory fitness traits and from 14% to 90% for muscular strength and endurance traits. This has led many researchers to search for the genetic factors that explain the heritability of these traits. Several strategies for identifying the genetic factors associated with complex exercise-related traits have been used, the most common of which are linkage analysis and association studies. Briefly, linkage analysis attempts to identify gene regions (i.e., **loci**) across the entire genome that constitute the genetic component of a trait. The loci are known as quantitative trait loci (QTL), and special mapping procedures can be used to search for the specific genes and variants contributing to each QTL. Genetic association studies typically examine whether mean quantitative trait values (e.g., $\dot{V}O_2$max) differ between genotype groups or whether allele or genotype frequency differs between cases and controls for a dichotomous trait (e.g., elite athletes vs. sedentary controls).

In 2000, a group of researchers began publishing an annual review of the published findings for genetic associations with exercise performance and health-related fitness phenotypes. In its final update in 2006-2007, 214 candidate autosomal genes and QTL and 7 loci on the X chromosome had been identified in relation to these phenotypes (8). The 2006-2007 human gene map for performance and fitness phenotypes identified 24 candidate genes and 19 QTLs associated with cardiorespiratory endurance–related phenotypes (i.e., particularly $\dot{V}O_2$max) and 19 candidate genes and 8 QTLs associated

with muscular strength and anaerobic traits (8). Variation in six candidate genes was associated with both endurance and muscular strength and power traits.

An emerging approach in genetic association studies is the genome-wide association study (GWAS). A GWAS differs from a candidate gene study in that it analyzes genetic markers across the entire genome in an unbiased fashion. As such, GWASs require much larger sample sizes in order to detect significant associations. In 2009, a meta-analysis of GWAS results from 1,644 Dutch and 978 American-Caucasian adults evaluated participation in leisure-time physical activity using questionnaires and classified those that exercised based on achieving more or less than 4 metabolic equivalent hours (MET-h) per week (9). A strong association was found at the rs10887741 SNP in the *PAPSS2* gene on chromosome 10. The odds of being a regular exerciser increased by 32% for each T allele carried at this SNP.

Genetics of Adaptation to Regular Exercise

Two untrained individuals who have a similar fitness level at baseline and then undertake the exact same exercise regimen (e.g., each jog 10 MET-h/wk) will in all likelihood show differences or variability relative to the magnitude of the increase in $\dot{V}O_2$max. Specifically, one individual might experience an 18% improvement in fitness, whereas the other might experience a 28% increase. In all instances, a portion of the differences between individuals is attributable in part to genetic factors. Thus, better explaining human variation in the adaptation to regular exercise is of great interest to exercise physiologists. The most extensive data on individual differences in trainability come from the HERITAGE Family Study, in which 742 healthy but sedentary men and women from 204 families completed a highly standardized, controlled, laboratory-based endurance training program for 20 wk. As expected, on average the training program resulted in significant changes in $\dot{V}O_2$max and other exercise- and health-related phenotypes or characteristics. However, there was large heterogeneity in the responsiveness of these traits across individuals (5). For example, the mean increase in $\dot{V}O_2$max in HERITAGE was 384 mL of O_2, and the standard deviation was 202 mL of O_2. The training responses for $\dot{V}O_2$max ranged from no change to increases of more than 1,000 mL of $O_2 \cdot min^{-1}$.

Results from twin and family exercise training studies have shown that the heterogeneity of trait responses to regular exercise is characterized by significant familial aggregation (25). For example, three separate endurance training studies involving identical (monozygotic) twins demonstrated that there is about nine times more variance in the response of $\dot{V}O_2$max to a standardized exercise program when the data from twin pairs from different families (across genotypes) was compared with the observed variability within genotypes (within a pair of twins from the same family) (25). In 481 Caucasian-sedentary subjects from 98 families of the HERTAGE Family Study, the heritability of $\dot{V}O_2$max response to training was estimated at 47% (3). Although this heritability estimate of 47% for $\dot{V}O_2$max in response to training is similar to the heritability contribution of 51% for measured $\dot{V}O_2$max in a sedentary state at baseline (i.e., before starting an exercise program; table 11.6), in HERITAGE the genetic factors underlying the two phenotypes likely are different. This is because no association was observed between $\dot{V}O_2$max measured at baseline and change or response of $\dot{V}O_2$max to training.

The heritability of training-induced changes for several other phenotypes (e.g., submaximal oxygen uptake, heart rate, stroke volume, cardiac output, exercise blood pressure, and blood lipids) was also estimated in the HERITAGE Family Study (table 11.6). Interestingly, the magnitude of the heritability for each of these other variables is less than what was observed for $\dot{V}O_2$max.

Finally, the HERITAGE study also showed significant familial aggregation in the response of skeletal muscle enzyme activities to exercise training (25), particularly for enzymes in the glycolytic and oxidative energy production pathways (e.g., creatine kinase, hexokinase, citrate synthase).

Candidate Gene Studies

The 2006-2007 human gene map for performance and fitness phenotypes identified 11 candidate genes and 12 QTLs associated with the response of cardiorespiratory fitness–related phenotypes to exercise training as well as 5 genes associated with muscle strength–related phenotypes (8). Additionally, 49 candidate genes were associated with exercise training–induced changes in hemodynamics, body composition, insulin and glucose metabolism, blood lipid, and hemostatic phenotypes. However, most of the candidate genes associated with exercise-response traits were based on only one study with positive findings. One candidate gene (identified as *ACE*) was investigated in two independent studies that involved 140 British Army recruits who underwent 10 wk of basic military training and found that the increases in left ventricular mass after 10 wk of physical training were 2.7 times or more greater with the *ACE* D/D or *ACE* I/D genotype than with the *ACE* I/I gene (16,17). A more recent study involving more than 600 young adults exposed to unilateral upper arm strength training found that elbow flexor strength increased only in *ACE* I-allele

Table 11.6 Estimated Heritabilities of Training Response in Exercise- and Health-Related Traits From the HERITAGE Family Study

Trait	Heritability
$\dot{V}O_2$max	0.47
Submaximal $\dot{V}O_2$ (50 watts, 60%, 80%)	0.23-0.57
Submaximal SV and Q (50 watts, 60%)	0.24-0.38
Submaximal HR (50 watts, 60%)	0.29-0.34
Submaximal SBP (50 watts)	0.22
Blood lipids (TC, TG, HDL-C)	0.26-0.32

SV = stroke volume; Q = cardiac output; HR = heart rate; SBP = systolic blood pressure; TC = total cholesterol; TG = triglycerides; HDL-C = high-density lipoprotein cholesterol.

Submaximal traits were measured at the heart rate associated with 50 watts or 60% or 80% of maximum, as determined from a maximal exercise test.

Created using data from Sarzynski, Rankien, and Bouchard (25).

carriers; no change was found in individuals with the D/D genotype (21). Given the number of genes potentially influencing the training response and the overall paucity of data, the need to explore such associations through innovative research remains important.

Genome-Wide Approaches

Advancements in high-throughput technologies, including the sequencing of the human genome in 2001, have allowed for hypothesis-free genome-wide analyses of exercise-response phenotypes. Several genome-wide linkage and GWAS reports related to exercise training phenotypes have been published. The HERITAGE Family Study used genome-wide linkage analysis to find genes related to several exercise training response phenotypes, including $\dot{V}O_2$max, submaximal exercise stroke volume and cardiac output, and insulin and glucose phenotypes. For example, QTLs for training-induced changes in stroke volume and heart rate during steady-state submaximal exercise at 50 watts were found on chromosomes 10p11 and 2q33.3-q34, respectively (23,27).

The first GWAS for an exercise-response trait was published in 2011 by Bouchard et al. (6), where the authors examined the association of more than 324,000 SNPs with exercise-induced changes in $\dot{V}O_2$max in 473 Caucasian subjects. The authors found that none of the SNPs reached genome-wide significance, although 39 SNPs were associated with change in $\dot{V}O_2$max. Further analysis of these 39 SNPs found that the top 21 SNPs explained 49% of the variance in change in $\dot{V}O_2$max, a value very similar to the heritability estimate of 47% previously reported (3).

Future Directions and Outlook

The field of exercise genomics has rapidly expanded over the past two decades with the advent of next-generation sequencing technologies, allowing for higher throughput and less expensive sequencing of common and rare variants. However, the field is mired by a general reliance on the candidate gene approach and observational studies with small sample sizes. Furthermore, lack of suitable replication studies or available DNA samples has made the replication of findings a difficult task, particularly for genome-wide studies. Thus, moving forward, exercise genomics needs to utilize a systems biology approach that integrates data from multiple technologies, including genomics, transcriptomics, metabolomics, proteomics, and epigenomics, among others. This will require large, well-designed, well-

powered collaborative studies with replication from multiple sources along with the development of computational and bioinformatics tools and expertise within the field. Ultimately, identifying the genetic factors underlying the variability in health- and fitness-related traits due to regular exercise would significantly contribute to the study of the biology of adaptation to exercise and the development of an exercise component of personalized preventive and therapeutic medicine.

Summary

- The important role of genetic factors (i.e., exercise genomics) in determining one's fitness level and response to an exercise training regimen has received increased attention over the past 25 years.
- Heritability estimates suggest that 51% of $\dot{V}O_2$max is due to genetic heritability; estimates for muscular strength and endurance range from 14% to 90%.
- Response to a regular exercise regimen is also influenced by genetics. In the Heritage Family Study, 742 healthy but sedentary men and women from 204 families completed a 20 wk endurance training program. The heritability estimate for change in $\dot{V}O_2$max in response to the endurance training was 47%.

Definitions

agonist—A drug or substance that binds to cell receptors, causing an increased biological response.

allele—A variant form of a gene in which a mutation or polymorphism is found.

antagonist—A drug or substance that binds to cell receptors, causing a decreased biological response.

bioavailability—The percentage of a drug that successfully absorbs into the systemic circulation to produce a therapeutic effect.

cytochrome P450—A ubiquitous class of enzymes responsible for the metabolism of many drugs.

gene—The functional unit of heredity. Comprising DNA that ranges from base pairs to more than a million, genes provide the instructions to build RNA.

genotype—An individual's collection of genes; an individual's genetic makeup that determines a characteristic or trait.

half-life—An index describing how long it takes a drug to be eliminated by the body.

heritability—The amount of the variation in an observed phenotypic trait that is attributable to genetic factors alone.

human genome—The complete set of genetic material in human cells (approximately 3 billion DNA base pairs encoded in the 23 chromosome pairs).

loci—Genetic location or region on a chromosome that constitutes the genetic component responsible for a trait.

mutation—A permanent alteration or variant of the nucleotide sequence of the genome in an organism.

nutraceutical—A consumable product (e.g., food, vitamin, or dietary supplement) containing a substance purported to have a medicinal benefit.

pharmacodynamics—The study of how a drug affects the body or organism.

pharmacokinetics—The study of how a body or organism affects a drug with respect to absorption, metabolism, bioavailability, distribution, and elimination.

phenotype—Observed, expressed traits (e.g., hair color, blood type) in an individual.

polymorphism—A genetic mutation that yields a different, alternate, or morph form of a phenotype.

route of administration—How a drug is taken into the body.

therapeutic index—A measure of drug safety; the ratio of an amount of drug causing toxic effects divided by the amount of drug causing therapeutic effects.

volume of distribution—The displacement of a drug throughout the body into one or more of the body's fluid compartments (i.e., plasma, interstitial fluid, intracellular fluid).

zero order—A constant rate of drug metabolism, which occurs when the amount of metabolizing enzymes become saturated.

Appendix A: Calculations for Oxygen Consumption and Carbon Dioxide Production

Oxygen uptake or consumption ($\dot{V}O_2$) and carbon dioxide production are fundamental exercise physiology metrics that are used for several indications in both athletes and clinical populations. Early methods of gas analysis during exercise were cumbersome and required the use of a sealed bag known as a Douglas bag, which collected expired air and sampled it for oxygen and carbon dioxide. Although modern advances in technology make obtaining volume and gas measures much faster and easier, the principles behind calculating these metabolic gases essentially are the same.

Step 1: Determine the Volume of Inspired or Expired Air

When measuring inspired or expired air, it must first be converted from atmospheric temperature and pressure saturated (ATPS) to standard temperature and pressure dry (STPD). This is because the gas laws state that the volume of a gas can vary depending on temperature (Charles' law), pressure (Boyle's law), and water vapor. Therefore, regardless of whether physiological measurements are taken in Denver, Miami, or Detroit, the gas samples are all standardized and thus comparable.

$$\text{STPD (L)} = \text{ATPS (L)} \times 273\,°K / (273\,°K + TA) \times (PB - PH_2O)/760\,\text{mmHg}$$
$$\text{(Equation A.1)}$$

273 = standard temperature (absolute temperature kelvin)

760 mmHg = standard pressure

TA = temperature of room

PB = barometric pressure

PH_2O = water vapor pressure (see table A.1)

Step 2: Find the Volume of Inspired Oxygen

Simply put, $\dot{V}O_2$ is the difference between inspired oxygen and expired oxygen. In step 1 we calculated the volume of inspired air. To determine the amount of oxygen, we simply multiply the volume of inspired air by the percentage of oxygen in it. Oxygen remains relatively constant at 20.93% or 210,000 parts per million.

Known Gas Concentrations in Room Air

O_2 = 20.93%

CO_2 = 0.03%

N_2 = 79.04%

$$\dot{V}O_2 = (\dot{V}_I \times F_IO_2) - (\dot{V}_E \times F_EO_2)$$
$$\dot{V}_IO_2 = \dot{V}_I \times F_IO_2 \qquad \text{(Equation A.2)}$$

$\dot{V}O_2$ = volume of oxygen consumed

$\dot{V}_IO_2$ = volume of inspired oxygen

$\dot{V}_EO_2$ = volume of expired oxygen

$\dot{V}_I$ = volume of inspired air (measured in the lab by the dry gas meter)

$\dot{V}_E$ = volume of expired air (unknown; this is the key to finding $\dot{V}O_2$)

F_IO_2 = fraction of inspired oxygen (assumed to be 0.2093)

F_IO_2 = fraction of expired oxygen (measured by the O_2 analyzer)

Step 3: Find the Volume of Expired Air via the Nitrogen Equation (i.e., the Haldane Transformation)

Nitrogen is neither consumed nor produced. Therefore, we can assume that the volume of nitrogen inspired ($\dot{V}_I N_2$) is equal to the volume of nitrogen expired ($\dot{V}_E N_2$).

$$\dot{V}_I N_2 = \dot{V}_E N_2$$

$$\left(\dot{V}_I \times F_I N_2\right) = \left(\dot{V}_E \times F_E N_2\right) \qquad \text{(Equation A.3)}$$

Through algebra, we can restate the equation as follows:

$$\dot{V}_E = \left(\dot{V}_I \times F_I N_2\right) / F_E N_2 \qquad \text{(Equation A.4)}$$

$F_E N_2$ can be solved using the equation $F_E N_2 = 1 - (F_E CO_2 + F_E O_2)$, which is why we measure expired carbon dioxide.

$\dot{V}_I N_2$ = volume of inspired nitrogen

$\dot{V}_E N_2$ = volume of expired nitrogen

$F_I N_2$ = fraction of inspired nitrogen (known air concentration of 0.7904)

$F_I N_2$ = fraction of expired nitrogen

$F_E CO_2$ = fraction of expired carbon dioxide

Step 4: Solve for $\dot{V}O_2$

$$\dot{V}O_2 = \left(\dot{V}_I O_2\right) - \left(\dot{V}_E \times F_E O_2\right) \qquad \text{(Equation A.5)}$$

Note that $\dot{V}O_2$ should be reported in liters per minute.

Step 5: Respiratory Exchange Ratio

Respiratory exchange ratio (RER) is an important measurement in terms of exercise intensity. As the ratio of carbon dioxide produced to oxygen consumed increases, our bodies shift from aerobic to anaerobic sources of energy. This shift increases our production of metabolic waste products, including carbon dioxide.

$$RER = \left(V_E \times F_E CO_2\right) / \dot{V}O_2$$

$$RER = CO_2 / \dot{V}O_2$$

$$C_6H_{12}O_6 + 6\,O_2 \rightarrow 6\,CO_2 + 6\,H_2O \quad 6/6 = 1.000$$

$$C_6H_{32}O_2 + 23\,O_2 \rightarrow 16\,CO_2 + 16\,H_2O \quad 16/23 = 0.696$$

$$\text{(Equation A.6)}$$

Step 6: Determine the Kilocalories

Depending on the RER, the kilocalories per liter of oxygen varies between 4.686 and 5.047. Thus:

$$1\text{ L of }O_2 = 5\text{ kcal} \qquad \text{(Equation A.7)}$$

Table A.1 Water Vapor Pressure (PH_2O) at Selected Gas Collection (Ambient) Temperatures

Ambient temperature (°C)	PH$_2$O (mmHg)	Ambient temperature (°C)	PH$_2$O (mmHg)
18	15.5	28	28.3
19	16.5	29	30.0
20	17.5	30	31.8
21	18.7	31	33.7
22	19.8	32	35.7
23	21.1	33	37.7
24	22.4	34	39.9
25	23.8	35	42.2
26	25.2	36	44.6
27	26.7	37	47.1

For computer-programming purposes, the following equation may be used: $PH_2O \approx 13.955 - 0.6584T + 0.0419T2$, where T = ambient temperature.

Reprinted, by permission, from M.L. Foss, S.J. Keteyian, and E.L. Fox, 1998, *Fox's physiological basis for exercise and sport*, 6th ed. (Pittsburg, PA: William C. Brown). By permission of the authors.

Appendix B: Efficiency and Energy Expenditure

Efficiency refers to the ratio of work output to work input.

Work output = kilocalories or watts produced

Work input = kilocalorie expenditure

$$\% \text{ efficiency} = \frac{\text{work output}}{\text{work input} \times 100}$$

(Equation B.1)

Devices called *ergometers* can measure work output. An important concept in the use of ergometers is calibration, which is the process of comparing measurement values of an ergometer against a known standard. Instructions on how to calibrate a piece of equipment should be available in the user manual for that specific item. The goal is to have as accurate a measure as possible. This is important because it allows the tester (e.g., exercise physiologist, personal trainer) to be sure that the work rate that is imposed on the person being tested is exactly as desired and accurately measured as the work output.

When measuring efficiency, the work, power, or energy units can be used to compute efficiency. The steps for measuring efficiency are listed as follows and can be applied to all types of ergometers (e.g., cycle, treadmill, arm, stepping).

MEASURING EFFICIENCY ON A CYCLE ERGOMETER

Figure B.1 shows a mechanically braked cycle ergometer. A belt runs around the rim of the flywheel and can be adjusted to increase or decrease the tension and thus the amount of friction applied to the flywheel. The greater the amount of friction, the greater the resistance when pedaling. To determine work rate or output, we first must know some specifics about the ergometer. In the case of this cycle ergometer, the needed information is as follows:

- The flywheel travels 6 m for each complete turn of the pedals.
- The pedal rate should be constant and known; 50 rpm is common.
- The scale on the cycle depicting the tension is in kiloponds (kp); 1 kp is the force acting on the mass of 1 kg at the normal acceleration of gravity.
- The force (in kiloponds) multiplied by the distance pedaled (in meters) provides the work value in kilopond-meters (kpm).
- If the distance traveled is expressed per unit of time, then the unit is a power unit and can be expressed as kpm per minute (kpm/min). This can also be converted to other power units such as watts (1 watt = 6.1 kpm/min).

The relationship among the various work units at 50 rpm is as follows:

1 kp = 300 kpm/min or 300 kilogram-meters per min

$$(\text{kgm/min}) = \text{approximately 50 watts}$$

(Equation B.2)

We also know the following:

1 kcal = 462.9 kgm/min   (Equation B.3)

Other types of cycle ergometers (e.g., electrically braked, air braked) compensate (i.e., increase and decrease resistance) for variations in pedal frequency to maintain the desired work rate. Although calibration is as important to accuracy in these types of ergometers as in the mechanically braked type, it requires a more sophisticated process (1).

The following is a sample calculation of work output in an individual who cycled for 10 min at a resistance of 3 kp at 50 rpm.

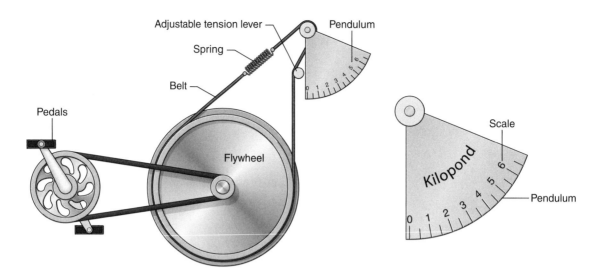

Figure B.1 A mechanically braked cycle ergometer.

Work Output

To determine the work output, we must use what is known. The primary information regarding resistance, pedaling rate, and duration has been provided. We must also know, as stated previously, that this cycle ergometer moves the flywheel 6 m for each pedal revolution.

Work (W) Output

$$W = \text{force} \times \text{distance}$$

$$W = 3 \text{ kp} \times \left(\frac{50 \text{ rpm} \times 10 \text{ min} \times 6 \text{ m/}}{\text{revolution of the flywheel}} \right)$$

$$W = 3 \text{ kp} \times 3{,}000 \text{ m}$$

$$W = 9{,}000 \text{ kpm} = 9{,}000 \text{ kgm}$$

$$\text{kcal} = 9{,}000 \text{ kgm} / 426.9 \text{ kgm/kcal}$$

$$\text{kcal} = 21.1 \text{ kcal} \qquad \text{(Equation B.4)}$$

Work Input

To determine work input, we must know the respiratory exchange ratio (RER) to determine the kilocalorie equivalent of 1 L of oxygen (O_2) consumed. It is important to note that RER can be used only when determined under steady-state exercise conditions. Thus, the work output must be submaximal, or below a person's anaerobic (i.e., lactic acid) threshold.

$$\text{RER} = \dot{V}CO_2 / \dot{V}O_2 \text{ (Note: The RER represents an estimate of the respiratory quotient, which reflects}$$

CO_2 production and O_2 consumption at the cellular level)

Assume that RER = 0.85 and $\dot{V}O_2 = 2.0 \text{ L} \cdot \text{min}^{-1}$

We know that 1 L of O_2 consumed at an RER of 0.85 = 4.87 kcal expended

Total $\dot{V}O_2 = 2.0 \text{ L} \cdot \text{min}^{-1} \times 10 \text{ min} = 20 \text{ L of } O_2$ consumed

Total kcal = 20 L of $O_2 \times 4.87 \text{ kcal} \cdot \text{min}^{-1} = 97.3$ kcal expended in 10 min

Efficiency

Efficiency is the measure of the amount of work input required to produce a given amount of work output. Using the information for work output and work input computed above, efficiency for our example can be computed as follows:

$$\text{Efficiency} = \frac{\text{work output}}{\text{work input}} = \frac{21.1 \text{ kcal}}{97.3 \text{ kcal}} =$$

$$0.2168 \times 100 = 21.7\%$$

$$\text{(Equation B.5)}$$

In summary, the calibration of ergometers is very important for the accurate measurement of work output. Although the example provided here uses a mechanically braked cycle ergometer, any piece of exercise equipment from which an assessment of output can be determined (e.g., treadmill, arm ergometer) can be used to assess efficiency.

Appendix C: Metabolic Equivalent of Task Values of Common Activities

The acronym MET stands for metabolic equivalent of task. Understanding the caloric energy expenditure (in kilocalories) of a physical activity is important for several tasks, including estimating caloric energy expenditure, prescribing safe activities for those with a limited functional capacity, and quantifying total activity as it relates to affecting disease risk. (Activity is quantified in MET-hours or MET-minutes, which are expressed by multiplying the average MET level of a physical activity by the duration the activity is performed.)

An MET is defined as the supine or seated resting metabolic rate. It is equivalent to approximately 3.5 mL of $O_2 \cdot$ kg of body weight$^{-1} \cdot$ min^{-1} or an energy expenditure of 1 kcal $\cdot$ kg$^{-1} \cdot$ h^{-1}. This definition of a MET may not be precise for certain populations. For instance, obese individuals may have a lower resting metabolic rate. Table C.1 provides MET values for some common physical activities and is based on the calculation 1 MET = 3.5 mL$\cdot$kg$^{-1} \cdot$ min^{-1}. These examples are taken from the comprehensive list of MET levels for more than 800 activities that has been published in a compendium of physical activities [1]. Approximately two thirds of these MET levels were determined by field and laboratory tests that measured expired gases to determine the oxygen consumption for the activity.

Understanding the relationship of METs to kilocalories expended allows for a rough calculation of caloric expenditure:

Because 1 MET = 1 kcal $\cdot$ kg$^{-1} \cdot$ h^{-1}

Then kcal = MET × body weight (kg) × duration (h)

Example: 3.8-MET activity performed by a 70-kg person for 15 min

3.8 × 70 × 0.25 = 66.5 kcal expended

The entire Compendium of Physical Activities can be accessed at https://sites.google.com/site/compendiumofphysicalactivities. An interesting item on this website is a discussion regarding a "corrected" MET. This discussion is based on a report that the standard 1 MET value of 3.5 mL of O_2 per kg underestimates the true measured MET level more than 80% of the time [4]. Here is an example using the standard (3.5 mL $\cdot$ kg$^{-1} \cdot$ min^{-1}) and a measured MET:

Standard MET: 20 mL $\cdot$ kg$^{-1} \cdot$ min^{-1}/3.5 mL $\cdot$ kg$^{-1} \cdot$ min^{-1}
= 5.5 METs (moderate intensity)
Measured MET: 20 mL $\cdot$ kg$^{-1} \cdot$ min^{-1}/2.5 mL $\cdot$ kg$^{-1} \cdot$ min^{-1}
= 8.0 METs (vigorous intensity)

(Equation C.1)

Several proposals have suggested appropriate methods for correcting MET values. These include measurement of resting metabolic rate (RMR), which is costly and time consuming. Another suggestion [2,4] is to divide the standard MET (3.5 mL$\cdot$kg$^{-1}\cdot$min^{-1}) by the predicted RMR from the Harris-Benedict equation [3]. A full discussion of this topic can be found at the aforementioned Compendium of Physical Activities website.

Table C.1 Selected Activities and Their Respective MET Values

Activity	MET value	Activity	MET value
REST AND SELF-CARE ACTIVITIES			
Rest (supine)	1.0	Showering	2.0
Sitting	1.5	General grooming, standing	2.0
Eating	1.5	Dressing or undressing, standing	2.5
Bathing	1.5		
HOME ACTIVITIES			
Knitting or hand sewing, light effort	1.3	Vacuuming (general, moderate effort)	3.3
Washing dishes	1.8	Making beds, changing linens	3.3
Ironing	1.8	Cleaning (scrubbing floor, washing car, washing windows)	3.5
Laundry, folding or hanging clothes	2.0-2.3	Sweeping, moderate effort	3.8
Cooking or food preparation	2.0-3.5	Moving furniture, carrying boxes	5.8
Machine sewing	2.8	Scrubbing floors on hands and knees, vigorous effort	6.5
OCCUPATIONAL			
Sitting tasks, office work, working at a computer	1.5	Construction (outside)	4.0
Driving a delivery truck, taxi, school bus, and so on	2.0	Hotel housekeeper	4.0
Cook, chef	2.5	Yard work	4.0
Standing tasks, light to moderate effort	3.0-4.5	Manual or unskilled labor	2.8-6.5
Custodial work	2.5-4.0	Farming, light to vigorous effort	2.0-7.8
Carpentry (general, light to moderate effort)	2.5-4.3	Firefighter on the job	6.8-9.0
PHYSICAL CONDITIONING			
Walking			
2.5 mph, level	3.0	4.5 mph, level	7.0
3.5 mph, level	4.3	5.0 mph, level	8.3
4.0 mph, level	5.0	5.0 mph, 3% grade	9.8
Jogging or running on level surface			
4.0 mph	6.0	10.0 mph	14.5
6.0 mph	9.8	12.0 mph	19.0
8.0 mph	11.8	14.0 mph	23.0
Swimming			
Freestyle, vigorous effort	9.8	Breaststroke, recreational or training and competition	5.3-10.3
Freestyle, slow to moderate	5.8	Sidestroke, general	7.0
Backstroke, recreational or training and competition	4.8-9.5		

Activity	MET value	Activity	MET value
PHYSICAL CONDITIONING (CONTINUED)			
Cycling			
Leisure, 5.5 mph	3.5	Leisure, 14.0-15.9 mph (vigorous effort)	10.0
Leisure, 10.0-11.9 mph (slow, light effort)	6.8	Racing, 16.0-19.0 mph (vigorous effort)	12.0
Leisure, 12.0-13.9 mph (moderate effort)	8.0	Racing, >20 mph (vigorous effort)	15.8
RECREATIONAL ACTIVITIES			
Aerobic dance	5.0-7.3	General resistance training	3.5-6.0
Video game activities	2.3-6.0	Rowing machines	4.8-12.0
Stationary cycle ergometer	3.5-14.0	Water aerobics	5.3
Circuit training	4.3-8.0	Video exercise workouts, light to vigorous	2.3-6.0
SPORT ACTIVITIES			
Archery	4.3	Rock or mountain climbing	5.0-8.0
Badminton	5.5-7.0	Roller skating	7.0
Basketball	6.0-9.3	Rugby	6.3-8.3
Bowling, lawn bowling	3.0-3.8	Skateboarding	5.0-6.0
Football, flag or touch	4.0-8.0	Soccer	7.0-10.0
Golf	4.8	Softball	5.0-6.0
Handball	12.0	Squash	7.3-12.0
Hockey, field	7.8	Table tennis (ping-pong)	4.0
Hockey, ice	8.0-10.0	Tennis, singles	7.3-8.0
Horseback riding	5.8-7.3	Tennis, doubles	4.5-6.0
Lacrosse	8.0	Volleyball	3.0-4.0
Orienteering	9.0	Volleyball, competitive	8.0
Racquetball	7.0-10.0	Volleyball, competitive beach	6.0

Data from Ainsworth et al. Healthy Lifestyles Research Center, College of Nursing and Health Innovation, Arizona State University. Available: http://sites.google.com/site/compendiumofphysicalactivities.

Appendix D: Professionalization of the Exercise Professional

The rich history of exercise physiology reaches back to before the AD calendar began (5) and has been comprehensively reviewed elsewhere. The history of exercise physiology as a profession, on the other hand, is much shorter. Only over the past 50 years has there been steady formation and growth in the exercise professions. Exercise professionals are individuals who possess a level of training required to work with individuals or groups in the development and implementation of an exercise training or physical activity program. The training of the exercise professional can be quite varied and range from a single weekend or online course (e.g., personal trainer) to an advanced college degree (e.g., PhD prepared exercise physiologist).

More recently, the exercise profession has experienced unprecedented growth, largely because of the increase in published research that indicates that all individuals should engage in regular exercise and physical activity as a means to reduce the risk of many adverse health outcomes. We know that some exercise and physical activity is more beneficial than sedentary living; we also know that more benefits are derived at higher levels of intensity, higher frequency, and longer duration (6). Some might ask why qualified exercise professionals are needed or required for others to exercise or might view such a requirement as an added barrier to an individual's performance of exercise or physical activity. Although it is true that the presence of fewer barriers is associated with an increased likelihood that an individual will begin an exercise program, it is also true that many individuals lack the understanding and knowledge to implement a safe and effective exercise program, may require individual or group instruction in order to maintain adequate adherence, and have an underlying chronic disease (either diagnosed or undiagnosed) that may require expertise in the design and implementation of an exercise training regimen. These individuals might benefit from the intervention of a trained and experienced exercise professional when either starting out or performing a regular exercise routine.

The next logical question is "What are the qualifications of an exercise professional?" Several different types of exercise professionals exist, and each may require different levels of education, experience, and certification to practice their craft.

Professionalism, also called *credentialism*, is the social process by which any trade or occupation transforms itself into a true profession of the highest integrity and competence. The term *professionalization* is used to describe the ongoing process of improving the quality of training as well as the identification of the minimum acceptable knowledge and skills required to perform the duties of a profession. To date, most of the professionalization efforts relative to exercise have focused on the education, training programs, internships, and certification programs available in an effort to regulate those who use or deliver their knowledge of exercise physiology on the job. Because this textbook is designed for graduate-level exercise science students, you likely have already decided the direction of your career path in exercise. In case you have not, the following information may aid in your decision making. Additionally, understanding the profession of exercise at all levels of practice helps set boundaries within the profession. This is important because the profession of exercise is ever evolving.

The exercise profession lies on a continuum ranging from service sector employees to allied health care professionals. Although many exercise and fitness certifications are available to those who have a high school degree rather than a degree in higher education, many require a college degree. A college degree may not be necessary for some of the jobs listed in "Common Exercise Professionals" (e.g., personal trainer, aerobics/group exercise instructor, strength and conditioning specialist/coach, sport coach). However, an individual's performance of each of these jobs might be advanced if he or she has formal academic coursework in exercise physiology.

Conversely, several jobs clearly require formal knowledge and training in exercise physiology. These include exercise physiologist, athletic trainer, clinical exercise physiologist, and kinesiotherapist, all of which require the completion of at least one exercise physiology course. The principles of exercise physiology often are incorporated in the performance of other professions listed (e.g., physician, chiropractor, dietitian), but formal education for these jobs typically does not require an exercise physiology course. "2012 U.S. Median Pay for Various Exercise Professionals" lists the approximate hourly wage of select jobs listed by the U.S. Bureau of Labor Statistics.

Common Exercise Professionals

Personal trainer

Aerobics/group exercise instructor

Athletic trainer

Exercise physiologist

Clinical exercise physiologist

Biomechanist

Strength and conditioning specialist/coach

Sport coach

Physician (e.g., sports medicine specialist)

Physical or occupational therapist

Sport dietitian/nutritionist

Physical education teacher

Researcher

Chiropractor

Kinesiotherapist

2012 U.S. Median Pay for Various Exercise Professionals

Profession	Hourly pay rate
Personal trainer or instructor	$15.25 (AU$26.90 Australia)
Massage therapist	$17.29
Athletic trainer	$20.25
Exercise physiologist	$20.25 (AU$26.41 Australia)
Dietitian/nutritionist	$26.56
Chiropractor	$31.80
Occupational therapist	$36.25
Physical therapist	$38.39
Physician	$90.00+

Source: U.S. Bureau of Labor Statistics (http://www.bls.gov/); www.payscale.com

Outlook

Gathering information about the outlook of the various exercise professions is not an easy task due to the constantly changing employment environment. The following sections look at two important professionals in the exercise field: personal trainers and clinical exercise physiologists.

Personal Trainer

The profession of personal training is growing rapidly both in the United States and throughout the world. In 2015, both the American Fitness Professionals & Associates and the U.S. Bureau of Labor Statistics estimated that there are more than 250,000 personal trainers in the United States (1); that number is expected to increase 13% between 2012 and 2022. In Australia, there are an estimated 28,000 personal trainers, 32% of which are employed full time. Growth over the past 10 yr has been dramatic, and the number of personal trainers in Australia is expected to grow an additional 10% over the next 5 yr (2). The Canadian government does not specifically address the job title *personal trainer* (4). Instead, it lists personal trainer with several other titles (e.g., exercise physiologist, exercise therapist, and kinesiologist) under the category "recreation, sports, and fitness program supervisors and consultants."

Several professional organizations define the duties of a personal trainer and include a scope of practice statement—a description that defines a legal and moral range of responsibilities for a given job. For the exercise professional, the scope of practice describes the setting in which the job can be performed, the particular population to which the job should be confined, and the parameters that define the range of advice and actions that can be provided. Because many exercise professions exist, there are many scopes of practice, and several overlap from one specific profession to another. For similar professions, the scope of practice definition might differ based on education, practical training (e.g., internship), certifications or licenses required, and any specific regional or state governance issues.

Let's look at personal training for an example of a scope of practice that may differ depending on the aforementioned criteria. The example in table D.1, which comes from the IDEA Health & Fitness Association, begins with a general scope of practice statement. Because many credentialing organizations exist, there are many scopes of practice for personal trainers. The example in figure D.1 comes from the American Council on Exercise (ACE), a well-respected certifying organization that developed

Table D.1   IDEA Personal Fitness Trainers' Scope of Practice

Fitness professionals DO NOT	Fitness professionals DO
Diagnose	• Receive exercise, health, or nutrition guidelines from a physician, physical therapist, or registered dietitian • Follow national consensus guidelines for exercise programming for medical disorders • Screen for exercise limitations • Identify potential risk factors through screening • Refer clients to an appropriate allied health professional or medical practitioner
Prescribe	• Design exercise programs • Refer clients to an appropriate allied health professional or medical practitioner for an exercise prescription
Prescribe diets or recommend specific supplements	• Provide general information on healthy eating according to the MyPlate Food Guidance System • Refer clients to a dietitian or nutritionist for a specific diet plan
Treat injury or disease	• Refer clients to an appropriate allied health professional or medical practitioner for treatment • Use exercise to help improve overall health • Help clients follow physician or therapist advice
Monitor progress for medically referred clients	• Document progress • Report progress to an appropriate allied health professional or medical practitioner • Follow physician, therapist, or dietitian recommendations
Rehabilitate	• Design an exercise program once a client has been released from rehabilitation
Counsel	• Coach • Provide general information • Refer clients to a qualified counselor or therapist
Work with patients	• Work with clients

IDEA Health & Fitness Association's Opinion Statement: Benefits of a working relationship between medical and allied health practitioners and personal fitness trainers. IDEA Health and Fitness Source, September 2001. www.ideafit.com. Reprinted with permission.

a scope of practice specific to personal trainers who are credentialed through their organization.

Clinical Exercise Physiologist

Clinical exercise physiology is a subspecialty of exercise physiology that addresses exercise testing and exercise training in patients with a clinically manifest disease. It requires knowledge of the relationship between exercise and chronic disease; the mechanisms and adaptations by which exercise influences the disease process; and the role and importance of exercise testing and training in the prevention, evaluation, and treatment of chronic diseases. This information can be used to develop and implement

exercise training programs. Therefore, clinical exercise physiologists must be knowledgeable about the broad range of exercise responses, including those that occur both within a disease class and across several different chronic diseases (e.g., cancer, heart disease, diabetes). The sidebar presents the definition of a clinical exercise physiologist from two leading professional organizations.

It is well accepted that exercise training prevents and delays disability and improves outcomes for many diseases and conditions, including cardiovascular, skeletal, muscle, and pulmonary diseases. For these conditions, exercise training often is part of a comprehensive treatment plan. Additionally, some population groups (e.g.,

ACE-Certified Personal Trainer Scope of Practice

The ACE-certified personal trainer is a fitness professional who has met all requirements of the ACE to develop and implement fitness programs for individuals who have no apparent physical limitations or special medical needs. The ACE-certified personal trainer realizes that personal training is a service industry focused on helping people enhance fitness and modify risk factors for disease to improve health. As members of the allied health care continuum with a primary focus on prevention, ACE-certified personal trainers have a scope of practice that includes the following:

- Developing and implementing exercise programs that are safe, effective, and appropriate for individuals who are apparently healthy or have medical clearance to exercise
- Conducting health-history interviews and stratifying risk for cardiovascular disease with clients in order to determine the need for referral and identify contraindications for exercise
- Administering appropriate fitness assessments based on the client's health history, current fitness, lifestyle factors, and goals using research-proven and published protocols
- Assisting clients in setting and achieving realistic fitness goals
- Teaching correct exercise methods and progressions through demonstration, explanation, and proper cueing and spotting techniques
- Empowering individuals to begin and adhere to their exercise programs using guidance, support, motivation, lapse-prevention strategies, and effective feedback
- Designing structured exercise programs for one-on-one and small-group personal training
- Educating clients about fitness- and health-related topics to help them adopt healthful behaviors that facilitate exercise program success
- Protecting client confidentiality according to the Health Insurance Portability and Accountability Act and related regional and national laws
- Always acting with professionalism, respect, and integrity
- Recognizing what is within the scope of practice and always referring clients to other health care professionals when appropriate
- Being prepared for emergency situations and responding appropriately when they occur

Although there does not appear to be a major university that offers a specific degree for personal or group exercise training, an exercise science or physiology degree from many universities will prepare a student very well for work in these fields. Additionally, at the time of this writing, the National Strength and Conditioning Association (NSCA) recognizes more than 30 university-based accredited academic institutions that meet the guidelines of the NSCA Education Recognition Program (www.nsca.com/Programs/Education-Recognition-Program/Recognized-ERP-Schools). The NSCA also lists more than 75 academic institutions that meet the requirement for the Education Recognition Program strength and conditioning program. Each of these programs requires at least one semester of general exercise physiology as part of the curriculum.

Figure D.1 ACE-certified personal trainer scope of practice.

Reprinted, by permission, from American Council on Exercise, 2010, *ACE personal trainer manual*, 4th ed., edited by C.X. Bryant and D.J. Green (San Diego, CA: ACE Fitness), 9. For more information, visit acefitness.org.

women, children, certain races and ethnicities) may be at greater risk than others for developing a chronic disease or disability because of physical inactivity. This risk also increases in all persons with age. More than ever before, clinical exercise physiology is at the forefront of advances in clinical care and public policy directed at improving health and lowering the risk of future disease through regular exercise training and increased physical activity.

Definitions of a Clinical Exercise Physiologist from Two Leading Professional Organizations

Clinical Exercise Physiology Association

A clinical exercise physiologist is an allied health care professional who is trained to work with patients with chronic diseases where exercise training has been shown to be of therapeutic benefit, including but not limited to cardiovascular disease, pulmonary disease, and metabolic disorders.

Clinical exercise physiologists work primarily in a medically supervised environment that provides a program or service that is directed by a licensed physician. A clinical exercise physiologist holds a minimum of a master's degree in exercise physiology, exercise or movement science, or kinesiology and either is licensed under state law or holds a professional certification from a national organization that is functionally equivalent to either the American College of Sports Medicine's (ACSM) Certified Clinical Exercise Specialist or the ACSM's Registered Clinical Exercise Physiologist credentials. An individual with a bachelor's degree in exercise physiology, exercise or movement science, or kinesiology who is an ACSM Certified Clinical Exercise Specialist is also considered qualified to perform exercise physiology services.

American College of Sports Medicine

The ACSM Certified Clinical Exercise Physiologist (CEP) is an allied health professional with a minimum of a bachelor's degree in exercise science. The CEP works with patients and clients who are challenged with cardiovascular, pulmonary, and metabolic diseases and disorders as well as with apparently healthy populations in cooperation with other health care professionals to enhance quality of life, manage health risk, and promote lasting health behavior change. The CEP conducts preparticipation health screenings, maximal and submaximal graded exercise tests, and strength, flexibility, and body composition tests. The CEP develops and administers programs designed to enhance cardiorespiratory fitness, muscular strength and endurance, balance, and range of motion. The CEP educates their clients about testing, exercise program components, and clinical and lifestyle self-care for control of chronic disease and health conditions.

The ACSM Registered Clinical Exercise Physiologist (RCEP) is an allied health professional with a minimum of a master's degree in exercise science who works in the application of physical activity and behavioral interventions for those clinical conditions where they have been shown to provide therapeutic and/or functional benefit. Persons for whom RCEP services are appropriate may include, but are not limited to, individuals with cardiovascular, pulmonary, metabolic, orthopedic, musculoskeletal, neuromuscular, neoplastic, immunologic, or hematologic disease. The RCEP provides primary and secondary prevention strategies designed to improve fitness and health in populations ranging from children to older adults.

The RCEP provides exercise screening, exercise and fitness testing, exercise prescriptions, exercise and physical activity counseling, exercise supervision, exercise and health education and promotion, and measurement and evaluation of outcome measures related to exercise and physical activity. The RCEP works individually or as part of an interdisciplinary team in a clinical, community, or public health setting. The practice and supervision of the RCEP is guided by published professional guidelines, standards, and applicable state and federal laws and regulations.

Since the surgeon general's report on physical activity and health was published in 1996, practice guidelines, position statements, and ongoing research have led to the evolution of exercise-related services. These services have progressed from simple exercise training regimens to programs that represent sophisticated exercise and behavioral management programs administered by multidisciplinary teams that include clinical exercise physiologists. Cardiac and pulmonary rehabilitation programs, bariatric and weight management programs, exercise oncology programs, and diabetes management programs are now commonplace due to the clinical and research contributions of clinical exercise physiologists.

The development of employment opportunities for those with clinical exercise physiology credentials continues to grow. Traditionally, most clinical exercise physiologists worked in the cardiac rehabilitation setting. However, employment opportunities for the CEP continue to expand and now include formal program offerings for patients with obesity, peripheral arterial disease, diabetes, and cancer. Evidence has also recently emerged on the favorable effect of physical activity and exercise on those who are intellectually or physically disabled. Additionally, clinical exercise professionals increasingly are being employed to work in clinical research trials that involve exercise and noninvasive exercise testing, including cardiopulmonary exercise testing studies. Finally, the emerging importance of physical activity and health and an increase in both professional and public awareness of the skills of clinical exercise physiologists have led to increasing employment in nonclinical settings. These include, but are not limited to, personal training, corporate fitness programs, medically affiliated fitness centers, schools and communities, professional and amateur sport consulting, and general weight management programs.

A survey by the Clinical Exercise Physiology Association estimated there are more than 6,000 individuals in the United States who consider themselves to be a clinical exercise physiologist. Of these, about 3,600 are ACSM-certified clinical exercise specialists and 850 are ACSM-registered clinical exercise physiologists.

In summary, the involvement of exercise physiologists in a variety of settings has grown dramatically over the past 30 yr. Historically, most were engaged in human performance–related research or academic instruction, but many now provide professional advice and services in the clinical, preventive, community, or corporate settings. Clinical exercise physiologists pro-

vide a number of important services, including fitness assessments or screenings, exercise testing and outcome assessments, exercise prescriptions or recommendations, exercise leadership, and exercise supervision. To date in the United States, only the state of Louisiana has opted to license exercise physiologists as a means to govern the delivery of specified and defined services to consumers. However, efforts have occurred or are already underway in other states, including Massachusetts, Minnesota, California, and Kentucky. The next section presents information about licensure of the exercise physiologist.

Certification and Licensure of Exercise Professionals

Licensure—the most restrictive form of credentialing—refers to a process developed by the local, state, or federal government that places requirements or criteria (and thus limitations) on who can practice in a given profession. In health care, there are licenses for physicians, nurses, and registered dietitians. The desired goal of licensure is to protect the health, safety, and welfare of the general public.

Certification confers some level of assurance that an individual has the minimum knowledge, skills, and abilities necessary to perform in a certain profession. The level of education and experience required for certification as an exercise professional varies greatly. Some certifications require only an online or weekend course, with or without a qualifying examination, whereas others require a combined level of education and experience. Certifying organizations may provide self-administered examinations or examinations that are accredited by an external independent organization. For instance, two common accrediting organizations are the National Commission for Certifying Agencies and the Commission on Accreditation of Allied Health Education Programs. Although some exercise-related professions such as athletic training and physical and occupational therapy are regulated at the state level, many are not and therefore do not require a certification or registration.

There have been attempts to regulate the exercise profession industry. For instance, in Washington, DC, a bill was passed in 2014 that required all personal trainers within city boundaries to register with the mayor's office and pay a fee. This essentially was an attempt at licensure.

This bill was rescinded by the end of 2015, and once again there is no licensure of personal trainers in the United States. An issue with these types of attempts at regulation is the lack of an industry standard for the education and training required to be a personal trainer. Competing organizations and professions, each guarding their own domain, have lined up against one another and have made the various attempts at passing licensure bills difficult. In 2008, an attempt by the New Jersey state senate to require licensure for personal trainers was defeated, partially because the International Health, Racquet & Sportsclub Association IHRSA mobilized its members to voice their opinion that the bill would not include many working in the personal training industry who have certifications from a certifying body or organization other than the one that the bill required. They also argued that the education requirements were not fair and would add a layer of expense and thus artificially limit those able to practice as a personal trainer.

The U.S. Registry of Exercise Professionals (USREPS) was developed recently with a stated mission "to secure recognition of registered exercise professionals for their distinct roles in medical, health, fitness, and sports performance fields." As of early 2016, they are allied with seven member organizations. In addition to a variety of resources available for exercise professionals and the public, the U.S. Registry of Exercise Professionals has a searchable registry that provides an individual's name, credential (from any organization) and credential number, date of credential expiration, and location.

Member Organizations of the U.S. Registry of Exercise Professionals

ACE	American Council on Exercise
ACSM	American College of Sports Medicine
NCSF	National Council on Strength & Fitness
NSCA	National Strength and Conditioning Association
NETA	National Exercise Trainers Association
The Cooper Institute	The Cooper Institute
PMA	Pilates Method Alliance

Examples of Certifying Organizations for Various Exercise Physiology–Related Professions

Profession	Certifying Organizations
Exercise physiologist	ACSM, ASEP, ESSA
Clinical exercise physiologist	ACE, ACSM, ESSA
Personal trainer	ACE, ACSM, CrossFit, CSEP, NSCA
Aerobics/group exercise instructor	ACE, ACSM
Athletic trainer	NATA
Strength and conditioning specialist or coach	NSCA
Sport coach	USSA, NAYS
Health coach	ACE, Wellcoaches
Physician (i.e., sports medicine specialist)	ABFM, ACSM
Physical or occupational therapist	ABPTS, APTA
Sports dietitian or nutritionist	AASDN, AND
Kinesiotherapist	AKTA, COPS-KT

AASDN = American Academy of Sport Dietitians and Nutritionists; ABFM = American Board of Family Medicine; ABPTS = American Board of Physical Therapy Specialties; ACE = American Council on Exercise; ACSM = American College of Sports Medicine; AKTA = American Kinesiotherapy Association; AND = Academy of Nutrition and Dietetics; APTA = American Physical Therapy Association; ASEP = American Society of Exercise Physiologists; CSEP = Canadian Society for Exercise Physiology; COPS-KT = Council on Professional Standards for Kinesiotherapy; ESSA = Exercise and Sports Science Australia; NATA = National Athletic Trainers' Association; NAYS = National Youth Sports Coaches Association; NSCA = National Strength and Conditioning Association; USSA = United States Sports Academy.

Cardiopulmonary Resuscitation and Automatic External Defibrillator Certification

Although the risk of cardiac arrest during exercise is low in the general population and in patients with cardiovascular disease who attend cardiac rehabilitation, the risk of cardiac arrest is very slightly elevated in everyone during exercise as opposed to rest. When persons certified in cardiopulmonary resuscitation (CPR) and the use of an automatic external defibrillator (AED) are involved in the resuscitation of an individual suffering a cardiac arrest, the success—and thus survival—rate is greatly enhanced. For every minute that passes during a cardiac arrest, the chances of survival decrease by 7% to 10% if CPR and defibrillation are not provided (3). Therefore, CPR and AED knowledge and skills as provided and certified by the American Heart Association, the National Safety Council, the American or Australian Red Cross, and other organizations are recommended for the practicing exercise professional.

Advanced cardiac life support (ACLS) training and certification identifies those individuals who have a higher level of knowledge, and possibly skill, in the face of cardiac arrest. Training and testing for ACLS focuses on responding to life-threatening situations in which the level of care delivered is higher than that provided during CPR or AED alone. Because ACLS involves the administration of medications, the provision of ACLS requires a physician or a midlevel provider (i.e., nurse practitioner or physician assistant). Therefore, many exercise professionals will not be in a position to perform or assist. Thus, ACLS certification is not routinely recommended for most exercise professionals. However, clinical exercise physiologists who work in the cardiac setting (e.g., cardiac rehabilitation or noninvasive testing) may benefit from ACLS instruction and training and may be able to assist in the emergency care of patients when supervised by a qualified health care provider.

Ideal Personal Attributes

Although education and experience are very important for the exercise professional to acquire, a variety of personal attributes also allow an individual to practice their profession at an effective level. The ability to communicate effectively (to both individuals and groups in both words and writing), attention to detail, a high level of initiative, and moral integrity are just a few of the key attributes required by an exercise professional. Note that these attributes focus on the relationship between an exercise professional and his or her client or patient and on the ability to effectively deliver a contemporary and effective exercise program. Although these attributes do not naturally occur in everyone, with effort they certainly can be learned and developed.

Attributes of an Effective Exercise Professional

- Self-motivated
- Outgoing/friendly
- Problem solver
- Positive and charismatic attitude
- Compassionate/empathetic
- Able to "walk the walk"
- Passionate
- Effective listener
- Always willing to learn and adapt
- Organized
- Patient
- Energetic
- Optimistic
- Visionary

Appendix E: Common Scientific Abbreviations and Units

Note: SI units (Le Système International d'Unités) are the preferred units of measurement in exercise and sport physiology. In this text, alternate units in common use are often provided as well.

BASIC SI UNITS

Time: seconds (s)
Quantity of a substance: mole (mol)
Length: meter (m)
Mass: kilogram (kg)

From these 4 basic units, other standard units can be derived:

Force: newton (N)
Energy or heat: joule (J)
Power: watt (W)
Velocity: meters per second ($m \cdot s^{-1}$)
Torque: Newton-meter (N-m)
Acceleration: meters per second per second ($m \cdot s^2$)
Angle: radian (rad)
Pressure: Pascal (Pa)
Volume: liter (L)

PREFIXES FOR MULTIPLES OF UNITS COMMON TO EXERCISE AND SPORT PHYSIOLOGY:

10^{-1}: deci- (d)
10^{-2}: centi- (c)
10^{-3}: milli- (m)
10^{-6}: micro- (μ)
10^{-9}: nano- (n)
10^{-12}: pico- (p)
10^3: kilo- (k)
10^6: mega- (M)

OTHER ABBREVIATIONS

Length (Distance)

in. = inch
ft = foot
yd = yard
mi = mile

Velocity

mph (or $mi \cdot h^{-1}$) = miles per hour

Electrical Potential Difference

V = volt

Mass and Weight

g = gram
lb = pound
oz = ounce
kp = kilopond

Energy

kcal = kilocalorie (sometimes written as Calorie)
cal = calorie
BTU = British Thermal Unit
N-m = Newton-meter
ft-lb = foot-pound
kp-m = kilopond-meter
kg-m = kilogram-meter
W-h = watt-hour

Power

hp = horsepower
$kp\text{-}m \cdot min^{-1}$ = kilopond-meter per minute
$kcal \cdot min^{-1}$ = kilocalories per minute

Pressure

atm = atmosphere
mmHg = millimeters of mercury = torr
psi = pounds per square inch
mbar = millibar

Temperature

°C = degrees Celsius or centigrade
°F = degrees Fahrenheit

Time

h = hour
min = minute
s = second

Volume

mL = milliliter = cubic centimeter (cc or cm^3)
gal = gallon
qt = quart
oz = ounce
c = cup
tbsp = tablespoon
tsp = teaspoon

UNIT CONVERSIONS

Quantity of a Substance

1 mol of a gas = 22.4 L (standard conditions) = 6.022 × 10^{23} molecules (Avogadro's number)
1 L of gas (standard conditions) = 44.6 mmol
[mol = mass (g)/molecular mass]
(molality of a solution = mol of substance/kg of solvent)
(molarity of a solution = mol of substance/L of solvent)

Length (Distance)

1 m = 39.370 in. = 3.281 ft =1.094 yd
1 km = 0.621 mi
1 cm = 0.394 in.
1 in. = 2.540 cm
1 ft = 12 in. = 30.480 cm
1 yd = 3 ft = 0.914 m
1 mi = 5,280 ft = 1,760 yd = 1,609.344 m

Velocity

1 mi · h^{-1} (mph) = 26.822 m · min^{-1} = 1.467 ft · s^{-1} = 0.447 m · s^{-1}
1 m · s^{-1} = 2.237 mi · h^{-1} (mph) = 3.281 ft · s^{-1}

Force

1 N = 0.225 lb of force = 0.102 kg of force
1 lb force = 4.448 N
1 kg force = 9.81 N

Torque

1 N-m = 0.738 ft-lb
1 ft-lb = 1.356 N

Mass and Weight

1 kg = 2.205 lb
1 g = 0.035 oz
1 lb = 16 oz = 0.454 kg
1 oz = 28.350 g
(1 L of water weighs 1 kg)

Energy

1 kcal = 4.186 kJ = 426.935 kg-m = 1.163 W-h
1 BTU = 0.252 kcal = 1.055 kJ = 107.586 kg-m
1 J = 1 N-m = 0.102 kg-m = 0.239 cal
1 kg-m = 1 kp-m = 9.807 J = 2.342 cal
[1 L of oxygen consumed = 5.05 kcal = 21.143 kJ (at an RQ of 1.00)]

Power

1 W = 1 J · s^{-1} = 6.118 kg-m · min^{-1} = 0.860 kcal · h^{-1} = 0.00134 hp
1 kg-m· min^{-1} = 1 kp-m · min^{-1} = 0.163 W = 0.141 kcal· h^{-1}
1 kcal · min^{-1} = 69.780 W

Pressure

1 atm = 760 mmHg = 101.325 kPa = 14.696 psi
1 mmHg = 1 torr = 0.0193 psi = 133.322 Pa = 0.00132 atm
1 kPa = 0.01 mbar

Temperature

°C = 0.555 [(°F) – 32]
°F = 1.8 (°C) + 32

Volume

1 L = 1.057 qt
1 qt = 0.946 L = 2 pt = 32 oz
1 (U.S.) gal = 4 qt = 128 oz = 3.785 L
1 c = 8 oz = 0.237 L
1 oz = 2 tbsp = 6 tsp = 29.574 mL
1 tbsp = 3 tsp = 14.787 mL
1 tsp = 4.929 mL

References

CHAPTER 1

1. Alberts B, Johnson A, Lewis J, Raff M, Roberts K, Walter P. *Molecular biology of the cell.* 4th ed. New York: Garland Science; 2002.

2. Baker DJ, Greenhaff PL, MacInnes A, Timmons JA. The experimental type 2 diabetes therapy glycogen phosphorylase inhibition can impair aerobic muscle function during prolonged contraction. *Diabetes.* 2006;55(6):1855-61.

3. Bangsbo J, Johansen L, Quistorff B, Saltin B. NMR and analytic biochemical evaluation of CrP and nucleotides in the human calf during muscle contraction. *J Appl Physiol.* 1993;74(4):2034-9.

4. Betik AC, Baker DJ, Krause DJ, McConkey MJ, Hepple RT. Exercise training in late middle-aged male Fischer 344 x Brown Norway F1-hybrid rats improves skeletal muscle aerobic function. *Exp Physiol.* 2008;93(7):863-71.

5. Blaak EE, Glatz JF, Saris WH. Increase in skeletal muscle fatty acid binding protein (FABPC) content is directly related to weight loss and to changes in fat oxidation following a very low calorie diet. *Diabetologia.* 2001;44(11):2013-7.

6. Bogdanis GC, Nevill ME, Boobis LH, Lakomy HK. Contribution of phosphocreatine and aerobic metabolism to energy supply during repeated sprint exercise. *J Appl Physiol.* 1996;80(3):876-84.

7. Boon H, Jonkers RA, Koopman R, Blaak EE, Saris WH, Wagenmakers AJ, et al. Substrate source use in older, trained males after decades of endurance training. *Med Sci Sports Exerc.* 2007;39(12):2160-70.

8. Bricker DK, Taylor EB, Schell JC, Orsak T, Boutron A, Chen YC, et al. A mitochondrial pyruvate carrier required for pyruvate uptake in yeast, *Drosophila*, and humans. *Science.* 2012;337(6090):96-100.

9. Brooks GA, Dubouchaud H, Brown M, Sicurello JP, Butz CE. Role of mitochondrial lactate dehydrogenase and lactate oxidation in the intracellular lactate shuttle. *Proc Natl Acad Sci.* 1999;96(3):1129-34.

10. Brooks GA, Mercier J. Balance of carbohydrate and lipid utilization during exercise: the "crossover" concept. *J Appl Physiol.* 1994;76(6):2253-61.11.

11. Broskey NT, Greggio C, Boss A, Boutant M, Dwyer A, Schlueter L, et al. Skeletal muscle mitochondria in the elderly: effects of physical fitness and exercise training. *J Clin Endocrinol Metab.* 2014;99(5):1852-61.

12. Cartee GD. Aging skeletal muscle: response to exercise. Exerc Sport Sci Rev. 1994;22:91-120.

13. Carter H, Pringle JS, Boobis L, Jones AM, Doust JH. Muscle glycogen depletion alters oxygen uptake kinetics during heavy exercise. *Med Sci Sports Exerc.* 2004;36(6):965-72.

14. Chabowski A, Gorski J, Luiken JJ, Glatz JF, Bonen A. Evidence for concerted action of FAT/CD36 and FABPpm to increase fatty acid transport across the plasma membrane. *Prostaglandins Leukot Essent Fatty Acids.* 2007;77(5-6):345-53.

15. Chaussain M, Camus F, Defoligny C, Eymard B, Fardeau M. Exercise intolerance in patients with McArdle's disease or mitochondrial myopathies. *Eur J Med.* 1992;1(8):457-63.

16. Cheetham ME, Boobis LH, Brooks S and Williams C. Human muscle metabolism during sprint running. *Journal of applied physiology* (Bethesda, Md : 1985). 1986; 61: 54-60.

17. Coggan AR, Abduljalil AM, Swanson SC, Earle MS, Farris JW, Mendenhall LA, et al. Muscle metabolism during exercise in young and older untrained and endurance-trained men. *J Appl Physiol.* 1993;75(5):2125-33.

18. Colberg SR, Sigal RJ, Fernhall B, Regensteiner JG, Blissmer BJ, Rubin RR, et al. Exercise and type 2 diabetes: the American College of Sports Medicine and the American Diabetes Association: joint position statement executive summary. *Diabetes Care.* 2010;33(12):2692-6.

19. Dawson B, Fitzsimons M, Green S, Goodman C, Carey M, Cole K. Changes in performance, muscle metabolites, enzymes and fibre types after short sprint training. *European J Appl Physiol Respir Environ Exerc Physiol Occup Physiol.* 1998;78(2):163-9.

20. Dzeja P, Terzic A. Adenylate kinase and AMP signaling networks: metabolic monitoring, signal

communication and body energy sensing. *Int J Mol Sci.* 2009;10(4):1729-72.

21. Ferrara CM, Goldberg AP, Ortmeyer HK, Ryan AS. Effects of aerobic and resistive exercise training on glucose disposal and skeletal muscle metabolism in older men. J Gerontol A Biol Sci Med Sci. 2006;61(5):480-7.

22. Fournier M, Ricci J, Taylor AW, Ferguson RJ, Montpetit RR, Chaitman BR. Skeletal muscle adaptation in adolescent boys: sprint and endurance training and detraining. *Med Sci Sports Exerc.* 1982;14(6):453-6.

23. Garcia CK, Goldstein JL, Pathak RK, Anderson RG, Brown MS. Molecular characterization of a membrane transporter for lactate, pyruvate, and other monocarboxylates: implications for the Cori cycle. *Cell.* 1994;76(5):865-73.

24. Garland PB, Randle PJ, Newsholme EA. Citrate as an intermediary in the inhibition of phosphofructokinase in rat heart muscle by fatty acids, ketone bodies, pyruvate, diabetes, and starvation. *Nature.* 1963;200:169-70.

25. Gastin PB. Energy system interaction and relative contribution during maximal exercise. *Sports Med.* 2001;31(10):725-41.

26. Glatz JF, Schaap FG, Binas B, Bonen A, van der Vusse GJ, Luiken JJ. Cytoplasmic fatty acid-binding protein facilitates fatty acid utilization by skeletal muscle. *Acta Physiol Scand.* 2003;178(4):367-71.

27. Gollnick PD, Piehl K, Saltin B. Selective glycogen depletion pattern in human muscle fibres after exercise of varying intensity and at varying pedalling rates. *J Physiol.* 1974;241(1):45-57.

28. Guenard D, Morange M, Buc H. Comparative study of the effect of 5' AMP and its analogs on rabbit glycogen phosphorylase b isoenzymes. *Eur J Biochem.* 1977;76(2):447-52.

29. Hirvonen J, Nummela A, Rusko H, Rehunen S, Harkonen M. Fatigue and changes of ATP, creatine phosphate, and lactate during the 400-m sprint. *Can J Sport Sci.* 1992;17(2):141-4.

30. Holloszy JO, Booth FW. Biochemical adaptations to endurance exercise in muscle. *Annu Rev Physiol.* 1976;38:273-91.

31. Hue L, Taegtmeyer H. The Randle cycle revisited: a new head for an old hat. *Am J Physiol Endocrinol Metab.* 2009;297(3):E578-91.

32. Hussey SE, McGee SL, Garnham A, McConell GK and Hargreaves M. Exercise increases skeletal muscle GLUT4 gene expression in patients with type 2 diabetes. *Diabetes, obesity & metabolism.* 2012; 14: 768-71.

33. Ivy JL, Young JC, McLane JA, Fell RD, Holloszy JO. Exercise training and glucose uptake by skeletal muscle in rats. *J Appl Physiol Respir Environ Exerc Physiol.* 1983;55(5):1393-6.

34. Jacobs I, Tesch PA, Bar-Or O, Karlsson J, Dotan R. Lactate in human skeletal muscle after 10 and 30 s of supramaximal exercise. *J Appl Physiol Respir Environ Exerc Physiol.* 1983;55(2):365-7.

35. Karatzaferi C, de Haan A, Ferguson RA, van Mechelen W, Sargeant AJ. Phosphocreatine and ATP content in human single muscle fibres before and after maximum dynamic exercise. *Pflugers Archiv.* 2001;442(3):467-74.

36. Katz A, Sahlin K and Henriksson J. Muscle ATP turnover rate during isometric contraction in humans. *Journal of applied physiology* (Bethesda, Md : 1985). 1986; 60: 1839-42.

37. Kiens B, Richter EA. Utilization of skeletal muscle triacylglycerol during postexercise recovery in humans. *Am J Physiol.* 1998;275(2 Pt 1):E332-7.

38. Krause U, Wegener G. Control of adenine nucleotide metabolism and glycolysis in vertebrate skeletal muscle during exercise. *Experientia.* 1996;52(5):396-403.

39. Kushmerick MJ, Moerland TS, Wiseman RW. Mammalian skeletal muscle fibers distinguished by contents of phosphocreatine, ATP, and Pi. *Proc Natl Acad Sci.* 1992;89(16):7521-5.

40. Little JP, Gillen JB, Percival ME, Safdar A, Tarnopolsky MA, Punthakee Z, et al. Low-volume high-intensity interval training reduces hyperglycemia and increases muscle mitochondrial capacity in patients with type 2 diabetes. *J Appl Physiol.* 2011;111(6):1554-60.

41. MacDougall JD, Hicks AL, MacDonald JR, McKelvie RS, Green HJ, Smith KM. Muscle performance and enzymatic adaptations to sprint interval training. *J Appl Physiol.* 1998;84(6):2138-42.

42. Medbo JI, Tabata I. Relative importance of aerobic and anaerobic energy release during short-lasting exhausting bicycle exercise. *J Appl Physiol.* 1989;67(5):1881-6.

43. Mitchell P. Chemiosmotic coupling in oxidative and photosynthetic phosphorylation. *Biol Rev Camb Philos Soc.* 1966;41(3):445-502.

44. Mitchell P, Moyle J. Evidence discriminating between the chemical and the chemiosmotic mechanisms of electron transport phosphorylation. *Nature.* 1965;208(5016):1205-6.

45. Moerland TS, Wolf NG, Kushmerick MJ. Administration of a creatine analogue induces isomyosin

transitions in muscle. *Am J Physiol.* 1989;257(4 Pt 1):C810-6.

46. Mourtzakis M, Gonzalez-Alonso J, Graham TE, Saltin B. Hemodynamics and O$_2$ uptake during maximal knee extensor exercise in untrained and trained human quadriceps muscle: effects of hyperoxia. *J Appl Physiol.* 2004;97(5):1796-802.

47. Nelson DC, Lehninger M. *Principles of biochemistry.* 5th ed. New York: Freeman; 2008.

48. Peeters RA, Groen MA, Veerkamp JH. The fatty acid-binding protein from human skeletal muscle. *Arch Biochem Biophys.* 1989;274(2):556-63.

49. Pittelli M, Formentini L, Faraco G, Lapucci A, Rapizzi E, Cialdai F, et al. Inhibition of nicotinamide phosphoribosyltransferase: cellular bioenergetics reveals a mitochondrial insensitive NAD pool. *J Biol Chem.* 2010;285(44):34106-14.

50. Randle PJ, Garland PB, Hales CN, Newsholme EA. The glucose fatty-acid cycle. Its role in insulin sensitivity and the metabolic disturbances of diabetes mellitus. *Lancet.* 1963;1(7285):785-9.

51. Rasmussen UF, Rasmussen HN, Krustrup P, Quistorff B, Saltin B, Bangsbo J. Aerobic metabolism of human quadriceps muscle: in vivo data parallel measurements on isolated mitochondria. *Am J Physiol Endocrinol Metab.* 2001;280(2):E301-7.

52. Richardson RS, Noyszewski EA, Leigh JS, Wagner PD. Lactate efflux from exercising human skeletal muscle: role of intracellular PO$_2$. *J Appl Physiol.* 1998;85(2):627-34.

53. Romijn JA, Coyle EF, Sidossis LS, Gastaldelli A, Horowitz JF, Endert E, et al. Regulation of endogenous fat and carbohydrate metabolism in relation to exercise intensity and duration. *Am J Physiol.* 1993;265(3 Pt 1):E380-91.

54. Rosing J and Slater EC. The value of G degrees for the hydrolysis of ATP. *Biochimica et biophysica acta.* 1972; 267: 275-90.

55. Ross A, Leveritt M. Long-term metabolic and skeletal muscle adaptations to short-sprint training: implications for sprint training and tapering. *Sports Med.* 2001;31(15):1063-82.

56. Roth DA, Brooks GA. Lactate transport is mediated by a membrane-bound carrier in rat skeletal muscle sarcolemmal vesicles. *Arch Biochem Biophys.* 1990;279(2):377-85.

57. Sahlin K, Areskog NH, Haller RG, Henriksson KG, Jorfeldt L, Lewis SF. Impaired oxidative metabolism increases adenine nucleotide breakdown in McArdle's disease. *J Appl Physiol.* 1990;69(4):1231-5.

58. Schantz PG, Henriksson J. Enzyme levels of the NADH shuttle systems: measurements in isolated muscle fibres from humans of differing physical activity. *Acta Physiol Scand.* 1987;129(4):505-15.

59. Schantz PG, Sjoberg B, Svedenhag J. Malate-aspartate and alpha-glycerophosphate shuttle enzyme levels in human skeletal muscle: methodological considerations and effect of endurance training. *Acta Physiol Scand.* 1986;128(3):397-407.

60. Schenk S, Horowitz JF. Coimmunoprecipitation of FAT/CD36 and CPT I in skeletal muscle increases proportionally with fat oxidation after endurance exercise training. *Am J Physiol Endocrinol Metab.* 2006;291(2):E254-60.

61. Spriet LL. Anaerobic ATP provision, glycogenolysis and glycolysis in rat slow-twitch muscle during tetanic contractions. *Pflugers Arch.* 1990;417(3):278-84.

62. Spriet LL. Phosphofructokinase activity and acidosis during short-term tetanic contractions. *Canadian J Physiol Pharmacol.* 1991;69(2):298-304.

63. Spriet LL. Anaerobic metabolism in human skeletal muscle during short-term, intense activity. *Canadian J Physiol Pharmacol.* 1992;70(1):157-65.

64. Spriet LL, Howlett RA, Heigenhauser GJ. An enzymatic approach to lactate production in human skeletal muscle during exercise. *Med Sci Sports Exerc.* 2000;32(4):756-63.

65. Spriet LL, Soderlund K, Bergstrom M, Hultman E. Anaerobic energy release in skeletal muscle during electrical stimulation in men. *J Appl Physiol.* 1987;62(2):611-5.

66. Stanford KI, Goodyear LJ. Exercise and type 2 diabetes: molecular mechanisms regulating glucose uptake in skeletal muscle. *Adv Physiol Educ.* 2014;38(4):308-14.

67. Talanian JL, Holloway GP, Snook LA, Heigenhauser GJ, Bonen A, Spriet LL. Exercise training increases sarcolemmal and mitochondrial fatty acid transport proteins in human skeletal muscle. *Am J Physiol Endocrinol Metab.* 2010;299(2):E180-8.

68. Thorstensson A, Sjodin B, Karlsson J. Enzyme activities and muscle strength after "sprint training" in man. *Acta Physiol Scand.* 1975;94(3):313-8.

69. Torres NV, Mateo F, Riol-Cimas JM, Melendez-Hevia E. Control of glycolysis in rat liver by glucokinase and phosphofructokinase: influence of glucose concentration. *Mol Cell Biochem.* 1990;93(1):21-6.

70. van Loon LJ, Greenhaff PL, Constantin-Teodosiu D, Saris WH, Wagenmakers AJ. The effects of increasing

exercise intensity on muscle fuel utilisation in humans. *J Physiol.* 2001;536(Pt 1):295-304.

71. Villar-Palasi C, Wei SH. Conversion of glycogen phosphorylase b to a by non-activated phosphorylase b kinase: an in vitro model of the mechanism of increase in phosphorylase a activity with muscle contraction. *Proc Natl Acad Sci.* 1970;67(1):345-50.

72. Walter G, Vandenborne K, Elliott M, Leigh JS. In vivo ATP synthesis rates in single human muscles during high intensity exercise. *J Physiol.* 1999;519(Pt 3):901-10.

73. Westerblad H, Bruton JD and Lannergren J. The effect of intracellular pH on contractile function of intact, single fibres of mouse muscle declines with increasing temperature. *The Journal of physiology.* 1997; 500 (Pt 1): 193-204.

74. Wojtaszewski JF, Higaki Y, Hirshman MF, Michael MD, Dufresne SD, Kahn CR, et al. Exercise modulates postreceptor insulin signaling and glucose transport in muscle-specific insulin receptor knockout mice. *J Clin Invest.* 1999;104(9):1257-64.

75. Yoshida Y, Holloway GP, Ljubicic V, Hatta H, Spriet LL, Hood DA, et al. Negligible direct lactate oxidation in subsarcolemmal and intermyofibrillar mitochondria obtained from red and white rat skeletal muscle. *J Physiol.* 2007;582(Pt 3):1317-35.

76. Zinker BA, Britz K, Brooks GA. Effects of a 36-hour fast on human endurance and substrate utilization. *J Appl Physiol.* 1990;69(5):1849-55.

CHAPTER 2

1. International Association of Athletics Federations. 12th IAAF World Championships in Athletics: 100 meters men final results 2009 [Internet]. Available from: http://www.iaaf.org/competitions/iaaf-world-championships/12th-iaaf-world-championships-in-athletics-3658/results/men/100-metres/final/result

2. Experiments and observations on the different modes in which death is produced by certain vegetable poisons by Sir Benjamin Collins Brodie. *Int Anesthesiol Clin.* 1968;6(2):423-3.3.

3. Antonio J, Gonyea WJ. Role of muscle fiber hypertrophy and hyperplasia in intermittently stretched avian muscle. *J Appl Physiol.* 1993;74(4):1893-8.

4. Antonio J, Gonyea WJ. Skeletal muscle fiber hyperplasia. *Med Sci Sports Exerc.* 1993;25(12):1333-45.

5. Baker LB, Rollo I, Stein KW, Jeukendrup AE. Acute effects of carbohydrate supplementation on intermittent sports performance. *Nutrients.* 2015;7(7):5733-63.

6. Barany M, Close RI. The transformation of myosin in cross-innervated rat muscles. *J Physiol.* 1971;213(2):455-74.

7. Batters C, Veigel C, Homsher E, Sellers JR. To understand muscle you must take it apart. *Front Physiol.* 2014;5:90.

8. Bawa P, Murnaghan C. Motor unit rotation in a variety of human muscles. *J Neurophysiol.* 2009;102(4):2265-72.

9. Berchtold MW, Brinkmeier H, Muntener M. Calcium ion in skeletal muscle: its crucial role for muscle function, plasticity, and disease. *Physiol Rev.* 2000;80(3):1215-65.

10. Bowman WC. Neuromuscular block. *Br J Pharmacol.* 2006;147(Suppl 1):S277-86.

11. Boyle T, Keegel T, Bull F, Heyworth J, Fritschi L. Physical activity and risks of proximal and distal colon cancers: a systematic review and meta-analysis. *J Natl Cancer Inst.* 2012;104(20):1548-61.

12. Chalmers GR, Row BS. Common errors in textbook descriptions of muscle fiber size in nontrained humans. *Sports Biomech.* 2011;10(3):254-68.

13. Cheng LK. Slow wave conduction patterns in the stomach: from Waller's foundations to current challenges. *Acta Physiol* (Oxf). 2015;213(2):384-93.

14. Clark CS, Kraus BB, Sinclair J, Castell DO. Gastroesophageal reflux induced by exercise in healthy volunteers. *J Am Med Assoc.* 1989;261(24):3599-601.

15. Coggan AR, Spina RJ, Rogers MA, King DS, Brown M, Nemeth PM, et al. Histochemical and enzymatic characteristics of skeletal muscle in master athletes. *J Appl Physiol.* 1990;68(5):1896-901.

16. Costill DL, Daniels J, Evans W, Fink W, Krahenbuhl G, Saltin B. Skeletal muscle enzymes and fiber composition in male and female track athletes. *J Appl Physiol.* 1976;40:149-54.

17. Costill DL, Saltin B. Factors limiting gastric emptying during rest and exercise. *J Appl Physiol.* 1974;37(5):679-83.

18. Courneya KS, Booth CM, Gill S, O'Brien P, Vardy J, Friedenreich CM, et al. The Colon Health and Life-Long Exercise Change trial: a randomized trial of the National Cancer Institute of Canada Clinical Trials Group. *Curr Oncol.* 2008;15(6):279-85.

19. Dayanidhi S, Lieber RL. Skeletal muscle satellite cells: mediators of muscle growth during development and implications for developmental disorders. *Muscle Nerve.* 2014;50(5):723-32.

20. de Oliveira EP, Burini RC. Carbohydrate-dependent, exercise-induced gastrointestinal distress. *Nutrients.* 2014;6(10):4191-9.

21. de Oliveira EP, Burini RC, Jeukendrup AE. Gastrointestinal complaints during exercise: prevalence, etiology, and nutritional recommendations. *Sports Med.* 2014;44(Suppl 1):S79-85.

22. Delbono O, O'Rourke KS, Ettinger WH. Excitation-calcium release uncoupling in aged single human skeletal muscle fibers. *J Membr Biol.* 1995;148(3):211-22.

23. Eken T, Gundersen K. Electrical stimulation resembling normal motor-unit activity: effects on denervated fast and slow rat muscles. *J Physiol.* 1988;402:651-69.

24. Enoka RM. Eccentric contractions require unique activation strategies by the nervous system. *J Appl Physiol.* 1996;81(6):2339-46.

25. Fitts RH, Trappe SW, Costill DL, Gallagher PM, Creer AC, Colloton PA, et al. Prolonged space flight-induced alterations in the structure and function of human skeletal muscle fibres. *J Physiol.* 2010;588(18):3567-92.

26. Fukutani A, Kurihara T, Isaka T. Factors of force potentiation induced by stretch-shortening cycle in plantar flexors. *PLoS One.* 2015;10(6):e0120579.

27. Gamu D, Bombardier E, Smith IC, Fajardo VA, Tupling AR. Sarcolipin provides a novel muscle-based mechanism for adaptive thermogenesis. *Exerc Sport Sci Rev.* 2014;42(3):136-42.

28. Gardiner P. *Advanced neuromuscular exercise physiology.* Champain, IL: Human Kinetics; 2011.

29. Gonzalez-Freire M, de Cabo R, Studenski SA, Ferrucci L. The neuromuscular junction: aging at the crossroad between nerves and muscle. *Front Aging Neurosci.* 2014;6:208.

30. Gordon AM, Huxley AF, Julian FJ. The variation in isometric tension with sarcomere length in vertebrate muscle fibres. *J Physiol.* 1966;184(1):170-92.

31. Hanson J, Huxley HE. Structural basis of the cross-striations in muscle. Nature. 1953;172(4377):530-2.

32. Harris AJ, Duxson MJ, Butler JE, Hodges PW, Taylor JL, Gandevia SC. Muscle fiber and motor unit behavior in the longest human skeletal muscle. *J Neurosci.* 2005;25(37):8528-33.

33. Henneman E, Somjen G, Carpenter DO. Excitability and inhibitability of motoneurons of different sizes. *J Neurophysiol.* 1965;28(3):599-620.

34. Herzog W, Powers K, Johnston K, Duvall M. A new paradigm for muscle contraction. *Front Physiol.* 2015;6:174.

35. Huxley AF, Niedergerke R. Structural changes in muscle during contraction; interference microscopy of living muscle fibres. *Nature.* 1954;173(4412):971-3.

36. Jackson JR, Mula J, Kirby TJ, Fry CS, Lee JD, Ubele MF, et al. Satellite cell depletion does not inhibit adult skeletal muscle regrowth following unloading-induced atrophy. *Am J Physiol Cell Physiol.* 2012;303:C854-61.

37. Kaczkowski W, Montgomery DL, Taylor AW, Klissouras V. The relationship between muscle fiber composition and maximal anaerobic power and capacity. *J Sports Med Phys Fitness.* 1982;22(4):407-13.

38. Keynes R. J.Z. and the discovery of squid giant nerve fibres. *J Exp Biol.* 2005;208(2):179-80.

39. Kirkendall DT, Garrett WE, Jr. The effects of aging and training on skeletal muscle. *Am J Sports Med.* 1998;26(4):598-602.

40. Koffler KH, Menkes A, Redmond RA, Whitehead WE, Pratley RE, Hurley BF. Strength training accelerates gastrointestinal transit in middle-aged and older men. *Med Sci Sports Exerc.* 1992;24(4):415-9.

41. Lammers WJ, van der Vusse GJ. Introduction to "electrical propagation in smooth muscle organs." *Acta Physiol* (Oxf). 2015;213(2):347-8.

42. Lexell J. Human aging, muscle mass, and fiber type composition. *J Gerontol A Biol Sci Med Sci.* 1995;50:11-6.

43. Lowey S, Waller GS, Trybus KM. Skeletal muscle myosin light chains are essential for physiological speeds of shortening. *Nature.* 1993;365(6445):454-6.

44. McGavock JM, Hastings JL, Snell PG, McGuire DK, Pacini EL, Levine BD, et al. A forty-year follow-up of the Dallas Bed Rest and Training study: the effect of age on the cardiovascular response to exercise in men. *J Gerontol A Biol Sci Med Sci.* 2009;64(2):293-9.

45. Morrow D. Development of a continuum mechanics model of passive skeletal muscle. Dissertation. Michigan Technological University; 2011.

46. Ortenblad N, Lunde PK, Levin K, Andersen JL, Pedersen PK. Enhanced sarcoplasmic reticulum Ca(2+) release following intermittent sprint training. *Am J Physiol Regul Integr Comp Physiol.* 2000;279(1):R152-60.

47. Pallafacchina G, Blaauw B, Schiaffino S. Role of satellite cells in muscle growth and maintenance of muscle mass. *Nutr Metab Cardiovasc Dis.* 2013;23(Suppl 1):S12-8.

48. Pandolfino JE, Bianchi LK, Lee TJ, Hirano I, Kahrilas PJ. Esophagogastric junction morphology predicts susceptibility to exercise-induced reflux. *Am J Gastroenterol.* 2004;99(8):1430-6.

49. Perko MJ, Nielsen HB, Skak C, Clemmesen JO, Schroeder TV, Secher NH. Mesenteric, coeliac and splanchnic blood flow in humans during exercise. *J Physiol.* 1998;513(Pt 3):907-13.

50. Pfeiffer B, Stellingwerff T, Hodgson AB, Randell R, Pottgen K, Res P, et al. Nutritional intake and gastrointestinal problems during competitive endurance events. *Med Sci Sports Exerc.* 2012;44(2):344-51.

51. Rehrer NJ, Beckers EJ, Brouns F, ten Hoor F, Saris WH. Effects of dehydration on gastric emptying and gastrointestinal distress while running. *Med Sci Sports Exerc.* 1990;22(6):790-5.

52. Sanchez LD, Tracy JA, Berkoff D, Pedrosa I. Ischemic colitis in marathon runners: a case-based review. *J Emerg Med.* 2006;30(3):321-6.

53. Scott W, Stevens J, Binder-Macleod SA. Human skeletal muscle fiber type classifications. *Phys Ther.* 2001;81(11):1810-6.

54. Sola OM, Christensen DL, Martin AW. Hypertrophy and hyperplasia of adult chicken anterior latissimus dorsi muscles following stretch with and without denervation. *Exp Neurol.* 1973;41(1):76-100.

55. Spillane J, Beeson DJ, Kullmann DM. Myasthenia and related disorders of the neuromuscular junction. *J Neurol Neurosurg Psychiatry.* 2010;81(8):850-7.

56. Susuki K. Myelin: a specialized membrane for cell communication. Nat Ed. 2010;3(9):59.

57. van Nieuwenhoven MA, Brouns F, Brummer RJ. The effect of physical exercise on parameters of gastrointestinal function. *Neurogastroenterol Motil.* 1999;11(6):431-9.

58. van Wijck K, Lenaerts K, Grootjans J, Wijnands KA, Poeze M, van Loon LJ, et al. Physiology and pathophysiology of splanchnic hypoperfusion and intestinal injury during exercise: strategies for evaluation and prevention. *Am J Physiol Gastrointest Liver Physiol.* 2012;303(2):G155-68.

59. Widmaier E, Raff H, Strang K. *Vander's human physiology.* 10th ed. New York: McGraw-Hill; 2006.

60. Wilson JM, Loenneke JP, Jo E, Wilson GJ, Zourdos MC, Kim JS. The effects of endurance, strength, and power training on muscle fiber type shifting. *J Strength Cond Res.* 2012;26(6):1724-9.

61. Wilson PB. Multiple transportable carbohydrates during exercise: current limitations and directions for future research. *J Strength Cond Res.* 2015;29(7):2056-70.

CHAPTER 3

1. Armstrong RB, Delp MD, Goljan EF, Laughlin MH. Distribution of blood flow in muscles of miniature swine during exercise. *J App Physiol.* 1987:62;1285-98.

2. Astrand P, Cuddy T, Saltin B, Stenberg J. Cardiac output during submaximal and maximal work. *J App Physiol.* 1964;19:268-74.

3. Bevegard S, Freyschuss U, Strandell T. Circulatory adaptation to arm and leg exercise in supine and sitting position. *J App Physiol.* 1966;21:37-46.

4. Brubaker PH, Kitzman DW. Chronotropic incompetence causes, consequences, and management. *Circulation.* 2011;123:1010-20.

5. Chapman CB, Fisher JN, Sproule BJ. Behavior of stroke volume at rest and during exercise in human beings. *J Clin Invest.* 1960;39:1208-13.

6. Cole CR, Blackstone EH, Pashkow FJ, Snader CE, Lauer MS. Heart-rate recovery immediately after exercise as a predictor of mortality. *N Engl J Med.* 1999;341:1351-57.

7. Coyle EF, Gonzalez-Alonso J. Cardiovascular drift during prolonged exercise. New perspectives. *Exer Sports Sci Rev.* 2001;29:86-92.

8. Davies PF. Flow-mediated endothelial mechanotransduction. *Physiol Rev.* 1995;75:519-60.

9. Egashira K, Inou T, Hirooka Y, Kai H, Sugimachi M, Suzuki S, et al. Effects of endothelian-dependent vasodilatation of resistance coronary vessels by acetycholine. *Circulation.* 1993;88:77-81.

10. Ekblom B, Astrand P, Saltin B, Stenberg J, Wallstrom B. Effect of training on circulatory response to exercise. *J Appl Physiol.* 1968;24:518-28.

11. Ekblom B, Hermansen L. Cardiac output in athletes. *J App Physiol.* 1968;25:619-25.

12. Fleg JL, Morrell CH, Bos AG, Brant LJ, Talbot LA, Wright JG, et al. Accelerated longitudinal decline of aerobic capacity in healthy older adults. *Circulation.* 2005;112:674-82.

13. Freedson P, Katch VL, Sady S, Weltman A. Cardiac output differences in males and females during mild

cycle ergometer exercise. *Med Sci Sports.* 1979;11:16-19.

14. Freund BJ, Shizuru EM, Hashiro GM, Claybaugh JR. Hormonal, electrolyte, and renal responses to exercise are intensity dependent. *J Appl Physiol.* 1991;70:900-6.

15. Gellish RL, Goslin BR, Olson RE, McDonald A, Russi GD, Moudgil VK. Longitudinal modeling of the relationship between age and maximal heart rate. *Med Sci Sports Exerc.* 2007;39:822-29.

16. Gertz EW, Wisneski JA, Stanley WC, Neese RA. Myocardial substrate utilization during exercise in humans. Dual carbon-labeled carbohydrate isotope experiments. *J Clin Invest.* 1988;82:2017-25.

17. Gorlin R, Cohen L, Elliott W, Klein M, Lane F. Effect of supine exercise on left ventricular volumes and oxygen consumption in man. *Circulation.* 1965;32:361.

18. Hartley L, Grimby G, Kilbom A, Nilsson N, Astrand I, Ekblom B, et al. Physical training in sedentary middle-aged and older men. III: Cardiac output and gas exchange at submaximal and maximal exercise. *Scand and J Clin Lab Invest.* 1969;24:335-44.

19. Hasking GJ, Esler MD, Jennings GL, Burton D, Johns JA, Korner PI. Norepinephrine spillover to plasma in patients with congestive heart failure: evidence of increased overall and cardiorenal sympathetic nervous activity. *Circulation.* 1986;73:615-21.

20. Janicki JS, Sheriff DD, Robotham JL, Wise RA. Cardiac output during exercise: contributions of the cardiac, circulatory, and respiratory systems. In: Rowell LB, Shephard, JT, editors. *Handbook of physiology.* New York: American Physiology Society; 1996. p 649-704.

21. Kasch FW. Thirty-three years of aerobic exercise adherence. *Quest.* 2001;53: 362-65.

22. Keteyian SJ, Marks CRC, Brawner CA, Levine AB, Kataoka T, Levine TB. Responses to arm exercise in patients with compensated heart failure. *J Cardiopulmon Rehabil.* 1996;16:366-71.

23. Keteyian SJ, Marks CRC, Levine AB, Kataoka T, Fedel F, Levine TB. Cardiovascular responses to submaximal arm and leg exercise in cardiac transplants. *Med Sci Sports Exerc.* 1994;26: 420-24.

24. Klocke FJ. Coronary blood flow in man. *Prog Cardiovasc Dis.* 1976;19:117.

25. Mach C, Foster C, Brice G, Mikat RP, Porcari JP. Effect of exercise duration on postexercise hypotension. *J Cardiopulmon Rehabil.* 2005;25:366-69.

26. Massie BM, Schwartz GG, Garcia J, Wisneski JA, Weiner MW, Owens T. Myocardial metabolism during increased work states in the porcine left ventricle in vivo. *Circulation Res.* 1994;74:64-73.

27. McGuire DK, Levine BD, Williamson JW, Snell PG, Blomqvist G, Saltin B, et al. A 30-year follow-up of the Dallas bed rest and training study. *Circulation.* 2001;104:1350-57.

28. McIlveen SA, Hayes SG, Kaufman MP. Both central command and exercise pressor reflex reset carotid sinus baroreflex. *Am J Physiol.* 2001;280:H1454-63.

29. Mithieux SM, Weiss AS. Elastin. *Adv Protein Chem.* 2005:70;437-61.

30. O'Leary DS, Mueller PJ, Sala-Mercado JA. The cardiovascular system: Design and control. In: Farrell PA, Joyner MJ, Caiozszo VJ, editors. *ACSM's advanced exercise physiology.* 2nd ed. Baltimore: Lippincott, Williams & Wilkins; 2012. pp 297-312.

31. O'Leary DS, Seamans DP. Effect of exercise on autonomic mechanisms of baroreflex control of heart rate. *J Appl Physiol.* 1993;75:2251-7.

32. Ogoh S, Fadel PJ, Nissen P, Jans Ø, Selmer C, Secher NH, et al. Baroreflex-mediated changes in cardiac output and vascular conductance in response to alterations in carotid sinus pressure during exercise in humans. *J Physiol* (Lond). 2003;550:317-24.

33. Poortmans JR. Exercise and renal function. *Sports Med.* 1984;1:125-53.

34. Potts JT, Li J. Interaction between carotid baroreflex and exercise pressor reflex depends on baroreceptor afferent input. *Am J Physiol.* 1998;274:H1841-7.

35. Potts JT, Shi XR, Raven PB. Carotid baroreflex responsiveness during dynamic exercise in humans. *Am J Physiol.* 1993;265:H1928-38.

36. Rowell L, O'Leary DS, Kellog DL, Jr. Integration of cardiovascular control systems in dynamic exercise. In: Rowell LB, Shephard, JT, editors. *Handbook of physiology.* New York: American Physiology Society; 1996. pp. 770-840.

37. Sarnoff SJ, Braunwald E, Welch GH Jr, Case RB, Stainsby WN, Macruz R. Hemodynamic determinates of oxygen consumption of the heart with special reference to the tension-time index. *Am J Physiol.* 1958;192:148-156.

38. Schairer JR, Stein PD, Keteyian SJ, Fedel F, Ehrman J, Alam M., et al. Left ventricular response to submaximal exercise in endurance-trained athletes and sedentary adults. *Am J Cardiol.* 1992;70:930-33.

39. Scott JC. Physical activity and the coronary circulation. *Can Med Assoc J.* 1967;96:853-61.

40. Seals DR, Hagberg JM, Hurley BF, Ehsani AA, Holloszy JO. Endurance training in older men and women I: Cardiovascular responses to exercise. *J Appl Physiol.* 1984;57:1024-9.

41. Simon GH, Dickhuth H, Starger J, Essig C, Kindermann W, Keul J. The value of echocardiography during physical exercise. *Int Sport Sci.* 1980;1(11):900.

42. Spina RJ, Ogawa T, Kohrt WM, Martin WH III, Holloszy JO, Ehsani AA. Differences in cardiovascular adaptations to endurance exercise training between older men and women. *J Appl Physiol.* 1993;75:849-55.

43. Stenberg J, Astrand PO, Ekblom B, Royce J, Saltin B. Hemodynamic response to work with different muscle groups, sitting and supine. *J Appl Physiol.* 1967;22:61-70.

44. Tipton CM. Exercise, training, and hypertension: An update. *Exer Sport Sci Rev.* 1991;18:447-505.

45. Warren BJ, Nieman DC, Dotson, RG, Atkins CH, O'Donnell KA, Haddock BL, et al. Cardiorespiratory responses to exercise training in septuagenarian women. *Int J Sports Med.* 1993;14:60-5.

46. Wilmore JH, Stanforth PR, Gagnon J, Rice T, Mandel S, Leon AS, et al. Cardiac output and stroke volume changes with endurance training: the HERITAGE Family Study. *Med Sci Sports Exer.* 2001;33:99-106.

47. Younis LT, Melin JA, Schoevaerdts JC, Van Dyck M, Robert AS, Chalant C, et al. Left ventricular systolic function and diastolic filling at rest and during upright exercise after orthotopic heart transplantation: comparison with young and aged normal subjects. *J Heart Transplant.* 1990;9:683-92.

CHAPTER 4

1. Beaver W, Wasserman K, Whipp B. A new method for detecting anaerobic threshold by gas exchange. *J Appl Physiol.* 1986;60(6):2020-7.

2. Bianchi R, Gigliotti F, Romagnoli I, Lanini B, Castellani C, Binazzi B, Stendardi L, Bruni GI, Scano G. Impact of a rehabilitation program on dyspnea intensity and quality in patients with chronic obstructive pulmonary disease. Respiration. 2011;81(3):186-95.

3. Brooks GA. Anaerobic threshold: review of the concept and directions for future research. Med Sci Sports Exerc. 1985;17:22-31.

4. Burtsher M. Exercise limitations by the oxygen delivery and utilization systems in aging and disease: coordinated adaptation and deadaptation of the lung-heart muscle axis—a mini-review. Gerontology. 2013;59:289-96.

5. Caiozzeo VJ, Davis JA, Ellis JF, Azus JL, Vandagriff R, Prietto CA, McMaster WC. A comparison of gas exchange coincides used to detect the anaerobic threshold. J Appl Phyiol. 1982;53(5):1184-9.

6. Chen H-I, Kuo C-S. Relationship between respiratory muscle function and age, sex, and other factors. J Appl Physiol. 1989;66:943-8.

7. Cote CG, Celli BR. Pulmonary rehabilitation and the BODE index in COPD. Eur Respir J. 2005;26:630-6.

8. Davis JA. Anaerobic threshold: review of the concept and directions for future research. Med Sci Sports Exerc. 1985;17:6-18.

9. Dempsey JA, Mitchell G, Smith C. Exercise-induced arterial hypoxemia in healthy persons at sea level. J Physiol. 1984;355:161-75.

10. Du Bois RM, Weycker D, Albera C, Bradford WZ, Costabel U, Kartashov A, Lancaster L, Noble PW, Sahn SA, Szwarcberg J, Thomeer M, Valeyre D, King TE Jr. Six-minute-walk test in idiopathic pulmonary fibrosis: test validation and minimal clinically important difference. Am J Respir Crit Care Med. 2011;183(9):1231-7.

11. Eldridge FL, Waldrop TG. Neural control of breathing during exercise. In: Whipp B, Wasserman K, editors. Exercise: pulmonary physiology and pathophysiology. New York: Dekker; 1991. pp. 309-70.

12. Habedank D, Reindl I, Viezke G, Bauer U, Sperfeld A, Gläser S, Wernecke KD, Kleber FX. Ventilatory efficiency and exercise tolerance in 101 healthy volunteers. Eur J Appl Physiol Occup Physiol. 1998;77:421-6.

13. Hagberg JM, Yerg JE, Seals DR. Pulmonary function in young and older athletes and untrained men. J Appl Physiol. 1988;65:101-5.

14. HajGhanbari B, Ymabayashi C, Buna TR, Coelho JD, Freedman KD, Morton TA, Palmer SA, Toy MA, Walsh C, Sheel AW, Reid WD. Effects of respiratory muscle training on performance in athletes: a systematic review with meta-analyses. J Strength Cond Res. 2013;27:1643-63.

15. Hankinson JL, Odencrantz JR, Fedan KB. Spirometric reference values from a sample of the general U.S. population. Am J Respir Crit Care Med. 1999;159:179-87.

16. Hudson LD, Tyler ML, Petty TL. Hospitalization needs during an outpatient rehabilitation program for severe chronic airway obstruction. Chest. 1976;70:606-10.

17. Illi SK, Held U, Frank I, Spengler CM. Effect of respiratory muscle training on exercise performance in healthy individuals: a systematic review and meta-analysis. Sports Med. 2012;42:707-24.

18. Janaudis-Ferreira T, Hill K, Goldstein RS, Robles-Ribeiro P, Beauchamp MK, Dolmage TE, Wadell K, Brooks D. Resistance arm training in patients with COPD: a randomized controlled trial. Chest. 2011;139:151-8.

19. Jensen D, Ofir D, O'Donnell DE. Effects of pregnancy, obesity and aging on the intensity of perceived breathlessness during exercise in healthy humans. Resp Physiol Neurobiol. 2009;167:87-100.

20. Johnson BD, Dempsey JA. Demand vs. capacity in the aging pulmonary system. Exerc Sport Sci Rev. 1991;19:171-210.

21. Katz A, Sahlin K. Regulation of lactic acid production during exercise. J Appl Physiol. 1988;65:509-18.

22. Kilding AE, Brown S, McConnell AK. Inspiratory muscle training improves 100 and 200 m swimming performance. Eur J Appl Physiol. 2010;108(3):505-11.

23. Lomax M, Thomaidis SP, Iggleden C, Toubekis AG, Tiligadas G, Tokmakidis SP, Oliveira RC, Costa AM. The impact of swimming speed on respiratory muscle fatigue during front crawl swimming: a role for critical velocity? Int J Swim Kinet. 2013;2:3-12.

24. Maltais F, LeBlanc P, Simard C, Jobin J, Bérubé C, Bruneau J, Carrier L, Belleau R. Skeletal muscle adaptation to endurance training in patients with chronic obstructive pulmonary disease. Am J Respir Crit Care Med. 1996;154:442-7.

25. McCarthy B, Casey D, Devane D, Murphy K, Murphy E, Lacasse Y. Pulmonary rehabilitation for chronic obstructive pulmonary disease. Cochrane Database Syst Rev. 2015;Feb 23(2):CD00379.

26. Neder JA, Andreoni S, Castelo-Filho A, Nery LE. Reference values for lung function tests. I. Static volumes. Braz J Med Biol Res. 1999;32:703-17.

27. Neder JA, Andreoni S, Lerario MC, Nery LE. Reference values for lung function tests. II. Maximal respiratory pressures and voluntary ventilation. Braz J Med Biol Res. 1999;32:719-27.

28. Pelkonen M, Notkola IL, Lakka T, Tukiainen HO, Kivinen P, Nissinen A. Delaying decline in pulmonary function with physical activity: a 25-year follow-up. Am J Respir Crit Care Med. 2003;168:494-9.

29. Powers SK, Dodd S, Lawler J, Landry G, Kirtley M, McKnight T, Grinton S. Incidence of exercise induced hypoxemia in the elite endurance athlete at sea level. Eur J Apply Physiol. 1988;58:298-302.

30. Pruthi N, Multani NK. Influence of age on lung function tests. J Exer Sci Physiother. 2012;8:1-6.

31. Robergs RA, Ghiasvand F, Parker D. Biochemistry of exercise-induced metabolic acidosis. Am J Physiol Regul Integr Comp Physiol. 2004;287:R502-16.

32. Schneider DA, Berwick JP. VE and $\dot{V}CO_2$ remain tightly coupled during incremental cycling performed after a bout of high-intensity exercise. Eur J Appl Physiol Occup Physiol. 1998;77(1-2):72-6.

33. Sharma G, Goodwin J. Effect of aging on respiratory system physiology and immunology. Clin Interven Aging. 2006;1(3):253-60.

34. Shephard R. The oxygen cost of breathing during vigorous exercise. Q J Exp Physiol. 1966;51:336-50.

35. Sue DY, Wasserman K, Moricca RB, Casaburi R. Metabolic acidosis during exercise in patients with chronic obstructive pulmonary disease. Use of the V-slope method for anaerobic threshold determination. Chest. 1988;94(5):931-8.

36. Taylor BJ, Johnson BD. The pulmonary circulation and exercise responses in the elderly. Semin Respir Crit Care Med. 2010;31:528-38.

37. The Global Asthma Report 2014. Auckland, New Zealand: Global Asthma Network; 2014.

38. Waldrop TG, Eldridge FL, Iwamoto GA, Mitchell JH. Central neural control of respiration and circulation during exercise. In: Rowell LB, Shepherd JT, editors. Exercise: regulation and integration of multiple systems. New York: Oxford University Press; 1996. pp. 333-80.

39. Ware JH, Dockery DW, Louis TA, Xu XP, Ferris BG Jr, Speizer FE. Longitudinal and cross-sectional estimates of pulmonary function decline in never-smoking adults. Am J Epidemiol. 1990;132:685-700.

40. Wasserman K, Whipp BJ, Koyal SN, Beaver WL. Anaerobic threshold and respiratory gas exchange during exercise. J Appl Physiol. 1973;35(2):236-43.

CHAPTER 5

1. Ahlborg B, Ahlborg G. Exercise leukocytosis with and without beta-adrenergic blockade. *Acta Med Scand*. 1970;187(4):241-6.

2. Ali S, Ullah F, Jan R. Effects of intensity and duration of exercise on differential leucocyte count. *J Ayub Med Coll Abbottabad*. 2003;15(1):35-7.

3. Arner P, Kriegholm E, Engfeldt P, Bolinder J. Adrenergic regulation of lipolysis in situ at rest and during exercise. *J Clin Invest*. 1990;85(3):893-8.

4. Baj Z, Kantorski J, Majewska E, Zeman K, Pokoca L, Fornalczyk E, et al. Immunological status of competitive cyclists before and after the training season. *Int J Sports Med*. 1994;15(6):319-24.

5. Barnard RJ, Ngo TH, Leung PS, Aronson WJ, Golding LA. A low-fat diet and/or strenuous exercise alters the IGF axis in vivo and reduces prostate tumor cell growth in vitro. *Prostate*. 2003;56(3):201-6.

6. Beavers KM, Brinkley TE and Nicklas BJ. Effect of exercise training on chronic inflammation. *Clinica chimica acta; international journal of clinical chemistry*. 2010; 411: 785-93.

7. Booth FW, Roberts CK and Laye MJ. Lack of exercise is a major cause of chronic diseases. *Compr Physiol*. 2012; 2: 1143-211.

8. Borer KT. *Exercise endocrinology*. Champaign, IL: Human Kinetics; 2003.

9. Campbell JE, Fediuc S, Hawke TJ, Riddell MC. Endurance exercise training increases adipose tissue glucocorticoid exposure: adaptations that facilitate lipolysis. *Metabolism*. 2009;58(5):651-60.

10. Cannon WB, De La Paz D. Emotional stimulation of adrenal secretion. *Am J Physiol*. 1911;28:64-70.

11. Chiang LM, Chen YJ, Chiang J, Lai LY, Chen YY, Liao HF. Modulation of dendritic cells by endurance training. *Int J Sports Med*. 2007;28(9):798-803.

12. Colberg SR, Albright AL, Blissmer BJ, Braun B, Chasan-Taber L, Fernhall B, et al. Exercise and type 2 diabetes: American College of Sports Medicine and the American Diabetes Association: joint position statement. *Med Sci Sports Exerc*. 2010;42(12):2282-303.

13. Convertino VA, Brock PJ, Keil LC, Bernauer EM, Greenleaf JE. Exercise training-induced hypervolemia: role of plasma albumin, renin, and vasopressin. *J Appl Physiol Respir Environ Exerc Physiol*. 1980;48(4):665-9.

14. Convertino VA, Keil LC, Greenleaf JE. Plasma volume, renin, and vasopressin responses to graded exercise after training. *J Appl Physiol Respir Environ Exerc Physiol*. 1983;54(2):508-14.

15. Deuster PA, Chrousos GP, Luger A, DeBolt JE, Bernier LL, Trostmann UH, et al. Hormonal and metabolic responses of untrained, moderately trained, and highly trained men to three exercise intensities. *Metabolism*. 1989;38(2):141-8.

16. Douketis JD, Macie C, Thabane L, Williamson DF. Systematic review of long-term weight loss studies in obese adults: clinical significance and applicability to clinical practice. *Int J Obes* (Lond). 2005;29(10):1153-67.

17. Downey GP, Worthen GS. Neutrophil retention in model capillaries: deformability, geometry, and hydrodynamic forces. *J Appl Physiol*. 1988;65(4):1861-71.

18. Duncan GE, Perri MG, Theriaque DW, Hutson AD, Eckel RH, Stacpoole PW. Exercise training, without weight loss, increases insulin sensitivity and postheparin plasma lipase activity in previously sedentary adults. *Diabet Care*. 2003;26(3):557-62.

19. Fatouros I, Chatzinikolaou A, Paltoglou G, Petridou A, Avloniti A, Jamurtas A, et al. Acute resistance exercise results in catecholaminergic rather than hypothalamic-pituitary-adrenal axis stimulation during exercise in young men. *Stress*. 2010;13(6):461-8.

20. French DN, Kraemer WJ, Volek JS, Spiering BA, Judelson DA, Hoffman JR, et al. Anticipatory responses of catecholamines on muscle force production. *J Appl Physiol*. 2007;102(1):94-102.

21. Fry RW, Morton AR, Crawford GP, Keast D. Cell numbers and in vitro responses of leucocytes and lymphocyte subpopulations following maximal exercise and interval training sessions of different intensities. *Eur J Appl Physiol Occup Physiol*. 1992;64(3):218-27.

22. Gleeson M. Immune function in sport and exercise. *J Appl Physiol*. 2007;103(2):693-9.

23. Gleeson M, Williams C. Intense exercise training and immune function. *Nestle Nutr Inst Workshop Ser*. 2013;76:39-50.

24. Goldhammer E, Tanchilevitch A, Maor I, Beniamini Y, Rosenschein U, Sagiv M. Exercise training modulates cytokines activity in coronary heart disease patients. *Int J Cardiol*. 2005;100(1):93-9.

25. Hackney AC. Stress and the neuroendocrine system: the role of exercise as a stressor and modifier of stress. *Expert Rev Endocrinol Metab.* 2006;1(6):783-92.

26. Haff G, Triplett NT, National Strength and Conditioning Association. *Essentials of strength training and conditioning.* Fourth edition. Champaign, IL: Human Kinetics; 2016.

27. Hakkinen K, Pakarinen A, Alen M, Kauhanen H, Komi PV. Neuromuscular and hormonal adaptations in athletes to strength training in two years. *J Appl Physiol.* 1988;65(6):2406-12.

28. Horn PL, Pyne DB, Hopkins WG, Barnes CJ. Lower white blood cell counts in elite athletes training for highly aerobic sports. *Eur J Appl Physiol.* 2010;110(5):925-32.

29. Kadoglou NP, Iliadis F, Angelopoulou N, Perrea D, Ampatzidis G, Liapis CD, et al. The anti-inflammatory effects of exercise training in patients with type 2 diabetes mellitus. *Eur J Cardiovasc Prev Rehabil.* 2007;14(6):837-43.

30. Kasapis C, Thompson PD. The effects of physical activity on serum C-reactive protein and inflammatory markers: a systematic review. *J Am Coll Cardiol.* 2005;45(10):1563-9.

31. King DE, Carek P, Mainous AG III, Pearson WS. Inflammatory markers and exercise: differences related to exercise type. *Med Sci Sports Exerc.* 2003;35(4):575-81.

32. Klein S, Burke LE, Bray GA, Blair S, Allison DB, Pi-Sunyer X, et al. Clinical implications of obesity with specific focus on cardiovascular disease: a statement for professionals from the American Heart Association Council on Nutrition, Physical Activity, and Metabolism: endorsed by the American College of Cardiology Foundation. *Circulation.* 2004;110(18):2952-67.

33. Kraemer WJ, Marchitelli L, Gordon SE, Harman E, Dziados JE, Mello R, et al. Hormonal and growth factor responses to heavy resistance exercise protocols. *J Appl Physiol.* 1990;69(4):1442-50.

34. Kruijsen-Jaarsma M, Revesz D, Bierings MB, Buffart LM, Takken T. Effects of exercise on immune function in patients with cancer: a systematic review. *Exerc Immunol Rev.* 2013;19:120-43.

35. Lakier Smith L. Overtraining, excessive exercise, and altered immunity: is this a T helper-1 versus T helper-2 lymphocyte response? *Sports Med.* 2003;33(5):347-64.

36. Lehmann M, Keul J, Huber G, Da Prada M. Plasma catecholamines in trained and untrained volunteers during graduated exercise. *Int J Sports Med.* 1981;2(3):143-7.

37. Luckey AE, Parsa CJ. Fluid and electrolytes in the aged. *Arch Surg.* 2003;138(10):1055-60.

38. Lund S, Holman GD, Schmitz O, Pedersen O. Contraction stimulates translocation of glucose transporter GLUT4 in skeletal muscle through a mechanism distinct from that of insulin. Proc Natl Acad Sci USA. 1995;92(13):5817-21.

39. Madssen E, Moholdt T, Videm V, Wisloff U, Hegbom K, Wiseth R. Coronary atheroma regression and plaque characteristics assessed by grayscale and radiofrequency intravascular ultrasound after aerobic exercise. *Am J Cardiol.* 2014;114(10):1504-11.

40. McCarthy DA, Dale MM. The leucocytosis of exercise. A review and model. *Sports Med.* 1988;6(6):333-63.

41. McCarthy DA, Grant M, Marbut M, Watling M, Wade AJ, Macdonald I, et al. Brief exercise induces an immediate and a delayed leucocytosis. *Br J Sports Med.* 1991;25(4):191-5.

42. Mikines KJ, Sonne B, Farrell PA, Tronier B, Galbo H. Effect of physical exercise on sensitivity and responsiveness to insulin in humans. *Am J Physiol.* 1988;254(3 Pt 1):E248-59.

43. Nehlsen-Cannarella SL, Nieman DC, Jessen J, Chang L, Gusewitch G, Blix GG, et al. The effects of acute moderate exercise on lymphocyte function and serum immunoglobulin levels. *Int J Sports Med.* 1991;12(4):391-8.

44. Nelson RK, Horowitz JF. Acute exercise ameliorates differences in insulin resistance between physically active and sedentary overweight adults. *Appl Physiol Nutr Metab.* 2014;39(7):811-8.

45. Nieman DC, Miller AR, Henson DA, Warren BJ, Gusewitch G, Johnson RL, et al. Effect of high- versus moderate-intensity exercise on lymphocyte subpopulations and proliferative response. *Int J Sports Med.* 1994;15(4):199-206.

46. Okutsu M, Suzuki K, Ishijima T, Peake J, Higuchi M. The effects of acute exercise-induced cortisol on CCR2 expression on human monocytes. *Brain Behav Immun.* 2008;22(7):1066-71.

47. Ortega E, Collazos ME, Maynar M, Barriga C, De la Fuente M. Stimulation of the phagocytic function of neutrophils in sedentary men after acute

moderate exercise. *Eur J Appl Physiol Occup Physiol.* 1993;66(1):60-4.

48. Ostrowski K, Hermann C, Bangash A, Schjerling P, Nielsen JN, Pedersen BK. A trauma-like elevation of plasma cytokines in humans in response to treadmill running. *J Physiol.* 1998;513(Pt 3):889-94.

49. Peake J, Suzuki K. Neutrophil activation, antioxidant supplements and exercise-induced oxidative stress. *Exerc Immunol Rev.* 2004;10:129-41.

50. Peake JM, Suzuki K, Hordern M, Wilson G, Nosaka K, Coombes JS. Plasma cytokine changes in relation to exercise intensity and muscle damage. *Eur J Appl Physiol.* 2005;95(5-6):514-21.

51. Pedersen BK. The anti-inflammatory effect of exercise: its role in diabetes and cardiovascular disease control. *Essays Biochem.* 2006;42:105-17.

52. Pedersen BK, Toft AD. Effects of exercise on lymphocytes and cytokines. *Br J Sports Med.* 2000;34(4):246-51.

53. Re R, Bryan SE. Functional intracellular renin-angiotensin systems may exist in multiple tissues. *Clin Exp Hypertens A.* 1984;6(10-11):1739-42.

54. Rowbottom DG, Green KJ. Acute exercise effects on the immune system. *Med Sci Sports Exerc.* 2000;32(7 Suppl):S396-405.

55. Sato H, Suzuki K, Nakaji S, Sugawara K, Totsuka M, Sato K. Effects of acute endurance exercise and 8 week training on the production of reactive oxygen species from neutrophils in untrained men. *Nihon Eiseigaku Zasshi.* 1998;53(2):431-40.

56. Sato K, Iemitsu M. Exercise and sex steroid hormones in skeletal muscle. *J Steroid Biochem Mol Biol.* 2015;145:200-5.

57. Shoemaker JK, Green HJ, Ball-Burnett M, Grant S. Relationships between fluid and electrolyte hormones and plasma volume during exercise with training and detraining. *Med Sci Sports Exerc.* 1998;30(4):497-505.

58. Sim YJ, Yu S, Yoon KJ, Loiacono CM, Kohut ML. Chronic exercise reduces illness severity, decreases viral load, and results in greater anti-inflammatory effects than acute exercise during influenza infection. *J Infect Dis.* 2009;200(9):1434-42.

59. Spence L, Brown WJ, Pyne DB, Nissen MD, Sloots TP, McCormack JG, et al. Incidence, etiology, and symptomatology of upper respiratory illness in elite athletes. *Med Sci Sports Exerc.* 2007;39(4):577-86.

60. Spielmann G, Bollard CM, Bigley AB, Hanley PJ, Blaney JW, LaVoy EC, et al. The effects of age and latent cytomegalovirus infection on the redeployment of CD8[+] T cell subsets in response to acute exercise in humans. *Brain Behav Immun.* 2014;39:142-51.

61. Starkie R, Ostrowski SR, Jauffred S, Febbraio M, Pedersen BK. Exercise and IL-6 infusion inhibit endotoxin-induced TNF-alpha production in humans. Faseb J. 2003;17(8):884-6.

62. Starkie RL, Rolland J, Angus DJ, Anderson MJ, Febbraio MA. Circulating monocytes are not the source of elevations in plasma IL-6 and TNF-alpha levels after prolonged running. *Am J Physiol Cell Physiol.* 2001;280(4):C769-74.

63. Steensberg A, Fischer CP, Keller C, Moller K, Pedersen BK. IL-6 enhances plasma IL-1ra, IL-10, and cortisol in humans. *Am J Physiol Endocrinol Metab.* 2003;285(2):E433-7.

64. Stokes K. Growth hormone responses to submaximal and sprint exercise. *Growth Horm IGF Res.* 2003;13(5):225-38.

65. Szivak TK, Hooper DR, Dunn-Lewis C, Comstock BA, Kupchak BR, Apicella JM, et al. Adrenal cortical responses to high-intensity, short rest, resistance exercise in men and women. *J Strength Cond Res.* 2013;27(3):748-60.

66. Tank AW, Lee Wong D. Peripheral and central effects of circulating catecholamines. *Compr Physiol.* 2015;5(1):1-15.

67. Tidball JG, Villalta SA. Regulatory interactions between muscle and the immune system during muscle regeneration. *Am J Physiol Regul Integr Comp Physiol.* 2010;298(5):R1173-87.

68. Trappe TA, White F, Lambert CP, Cesar D, Hellerstein M, Evans WJ. Effect of ibuprofen and acetaminophen on postexercise muscle protein synthesis. *Am J Physiol Endocrinol Metab.* 2002;282(3):E551-6.

69. Tvede N, Kappel M, Klarlund K, Duhn S, Halkjaer-Kristensen J, Kjaer M, et al. Evidence that the effect of bicycle exercise on blood mononuclear cell proliferative responses and subsets is mediated by epinephrine. *Int J Sports Med.* 1994;15(2):100-4.

70. Vingren JL, Kraemer WJ, Hatfield DL, Volek JS, Ratamess NA, Anderson JM, et al. Effect of resistance exercise on muscle steroid receptor protein content in strength-trained men and women. *Steroids.* 2009;74(13-14):1033-9.

71. Volek JS. Influence of nutrition on responses to resistance training. *Med Sci Sports Exerc.* 2004;36(4):689-96.

72. Walsh NP, Gleeson M, Shephard RJ, Gleeson M, Woods JA, Bishop NC, et al. Position statement. Part

one: Immune function and exercise. *Exerc Immunol Rev.* 2011;17:6-63.

73. Weltman A, Weltman JY, Womack CJ, Davis SE, Blumer JL, Gaesser GA, et al. Exercise training decreases the growth hormone (GH) response to acute constant-load exercise. *Med Sci Sports Exerc.* 1997;29(5):669-76.

74. Woods JA, Wilund KR, Martin SA, Kistler BM. Exercise, inflammation and aging. *Aging Dis.* 2012;3(1):130-40.

75. Zuhl M, Dokladny K, Mermier C, Schneider S, Salgado R, Moseley P. The effects of acute oral glutamine supplementation on exercise-induced gastrointestinal permeability and heat shock protein expression in peripheral blood mononuclear cells. *Cell Stress Chaperones.* 2015;20(1):85-93.

CHAPTER 6

1. American College of Sports Medicine. *ACSM's guidelines for exercise testing and prescription,* 2nd ed. Philadelphia: Lea & Febiger; 1980

2. American College of Sports Medicine. *ACSM's guidelines for exercise testing and prescription,* 10th ed. Baltimore: Lippincott, Williams & Wilkins; 2017.

3. Adhihetty PJ, Hood DA. Mechanisms of apoptosis in skeletal muscle. *Basic Appl Myol.* 2003;13:171-9.

4. Ahlskog JE, Geda YE, Graff-Radford NR, Petersen RC. Physical exercise as a preventive or disease-modifying treatment of dementia and brain aging. *Mayo Clin Proc.* 2011;86:876-84.

5. Ainsworth BE, Haskell WL, Herrmann SD, Meckes N, Bassett DR Jr, Tudor-Locke C, et al. 2011 Compendium of Physical Activities: a second update of codes and MET values. *Med Sci Sports Exerc.* 2011;43:1575-81.

6. Al-Mallah MH, Keteyian SJ, Brawner CA, Whelton S, Blaha MJ. Rationale and design of the Henry Ford Exercise Testing Project (the FIT project). *Clin Cardiol.* 2014;37:456-61.

7. Armstrong LE, VanHeest JL. The unknown mechanism of the overtraining syndrome: clues from depression and psychoneuroimmunology. *Sports Med.* 2002;32:185-209.

8. Asmussen E, Boje O. Body temperature and capacity for work. *Acta Physiol Scand.* 1945;10:1-22.

9. Baggish AL, Wood MJ. Athlete's heart and cardiovascular care of the athlete: scientific and clinical update. *Circulation.* 2011;123:2723-35.

10. Balady GJ, Arena R, Sietsema K, Myers J, Coke L, Fletcher GF, et al. Clinician's guide to cardiopulmo-

nary exercise testing in adults: a scientific statement from the American Heart Association. *Circulation.* 2010;122:191-225.

11. Benjamin M, Toumi H, Ralphs JR, Bydder G, Best TM, Milz S. Where tendons and ligaments meet bone: attachment sites ("entheses") in relation to exercise and/or mechanical load. *J Anat.* 2006;208:471-90.

12. Bergh U, Ekblom B. Physical performance and peak aerobic power at different body temperatures. *J Appl Physiol.* 1979;46(5):885-9.

13. Bevegard S, Holmgren A, Jonsson B. Circulatory studies in well-trained athletes at rest and during heavy exercise, with special reference to stroke volume and the influence of body position. *Acta Physiol Scand.* 1963;57:26-50.

14. Bi L, Triadafilopoulos G. Exercise and gastrointestinal function and disease: an evidence-based review of risks and benefits. *Clin Gastroenterol Hepatol.* 2003;1:345-55.

15. Borg G. Subjective effort in relation to physical performance and working capacity. In: Pick HJ, Liebewitz HW, Singer JE, Steinschneider A, Stevenson H, editors. *Psychology: from research to practice.* New York: Plenum; 1978. p. 333-61.

16. Boushel R, Lundby C, Qvortrup K, Sahlin K. Mitochondrial plasticity with exercise training and extreme environments. *Exerc Sport Sci Rev.* 2014;42:169-74.

17. Brawner CA, Lewis B, Schairer JR, Ehrman JK, Kerrigan DJ, Keteyian SJ. Association between estimated MET intensity during cardiac rehabilitation and prognosis among patients with heart disease. *Am J Cardiol* 2016;117:1236-41.

18. Brodal P, Inger F, Hermansen L. Capillary supply of skeletal muscle fibers in untrained and endurance-trained men. *Am J Physiol.* 1977;232:H705-12.

19. Carlile F. Effects of preliminary passive warming-up on swimming performance. *Res Q.* 1956;27:143-51.

20. Convertino VA. Blood volume: Its adaptation to endurance training. *Med Sci Sports Exerc.* 1991;23:1338-48.

21. Costill DL, Thomason H, Roberts E. Fractional utilization of the aerobic capacity during distance running. *Med Sci Sports.* 1973;5(4):248-52.

22. Coyle EF. Integration of the physiological factors determining endurance performance ability. In: Hollosy JO, editor. *Exercise and sports sciences reviews.* Vol. 23. Baltimore: Williams & Wilkins; 1995. p. 25-63.

23. Davis C, Barnes C, Godfrey S. Body composition and maximal exercise performance in children. *Human Biol.* 1972;44:195-215.

24. Degens H, Rittweger J, Parviainen T, Timonen KL, Suominen H, Heinonen A, et al. Diffusion capacity of the lung in young and old endurance athletes. *Int J Sports Med.* 2013;34:1051-7.

25. Diedrich A, Paranjape SY, Robertson D. Plasma and blood volume in space. *Am J Med Sci.* 2007;334:80-5.

26. Donnelly JE, Blair SN, Jakicic JM, Manore MM, Rankin JW, Smith BK, et al. American College of Sports Medicine Position Stand. Appropriate physical activity intervention strategies for weight loss and prevention of weight regain for adults. *Med Sci Sports Exerc.* 2009;41(2):459-71.

27. Drexler H, Riede U, Munzel T, Konig H, Funke E, Just H. Alterations of skeletal muscle in chronic heart failure. *Circulation.* 1992;85:1751-9.

28. Drinkwater B, Horvath S, Wells C. Aerobic power of females, ages 10 to 68. *J Gerontology.* 1973;30:385-94.

29. Eckstein F, Hudelmaier M, Putz R. The effects of exercise on human articular cartilage. *J Anat.* 2006;208:491-512.

30. Ehsani AA, Hagberg JM, Hickson RC. Rapid changes in left ventricular dimensions and mass in response to physical conditioning and deconditioning. *Am J Cardiol.* 1978;43:52-6.

31. Ekblom B, Astrand P, Saltin B, Stenberg J, Wallstrom B. Effects of training on circulatory response to exercise. *J Appl Physiol.* 1968;24:518-28.

32. Ekblom B, Hermansen L. Cardiac output in athletes. *J App Physiol.* 1968;25:619-25.

33. Fleg JL, Morrell CH, Bos AG, Brant LJ, Talbot LA, Wright JG, et al. Accelerated longitudinal decline of aerobic capacity in healthy older adults. *Circulation.* 2005;112:674-82.

34. Fleg JL, O'Connor F, Gerstenblith G, Becker LC, Clulow J, Schulman SP, et al. Impact of age on the cardiovascular response to dynamic upright exercise in healthy men and women. *J Appl Physiol.* 1995;78:890-900.

35. Fox E, McKenzie D, Cohen K. Specificity of training: metabolic and circulatory responses. *Med Sci Sports.* 1975;7:83.

36. Fox EL. Differences in metabolic alterations with sprint versus interval training programs. In: Howald H, Poortmans J, editors. *Metabolic adaptation to prolonged physical exercise.* Basel, Switzerland: Birkhauser Verlag; 1975. p. 119-26.

37. Fox SM, Naughton JP, Haskell WL. Physical activity and the prevention of coronary heart disease. *Am Clin Res.* 1971;3:404-32.

38. Frick M, Konttinen A, Sarajas S. Effects of physical training on circulation at rest and during exercise. *Am J Cardiol.* 1963;12:142-7.

39. Garber CE, Blissmer B, Deschenes MR, Franklin BA, Lamonte MJ, Lee IM, et al. American College of Sports Medicine position stand. Quantity and quality of exercise for developing and maintaining cardiorespiratory, musculoskeletal, and neuromotor fitness in apparently healthy adults: guidance for prescribing exercise. *Med Sci Sports Exerc.* 2011;43:1334-59.

40. Gellish RL, Goslin BR, Olson RE, McDonald A, Russi GD, Moudgil VK. Longitudinal modeling of the relationship between age and maximal heart rate. *Med Sci Sports Exerc.* 2007;39:822-9.

41. Gormley SE, Swain DP, High R, Spina RJ, Dowling EA, Kotipalli US, et al. Effect of intensity of aerobic training on $\dot{V}O_2$max. *Med Sci Sports Exerc.* 2008;40:1336-43.

42. Green H, Jones LL, Painter DC. Effects of short-term training on cardiac function during prolonged exercise. *Med Sci Sports Exerc.* 1990;22:488-93.

43. Grimby G, Haggendal E, Saltin B. Local xenon 133 clearance from the quadriceps muscle during exercise in man. *J Appl Physiol.* 1967;22:305-10.

44. Häggmark T, Eriksson E, Jansson E. Muscle fiber type changes in human skeletal muscle after injuries and immobilization. *Orthopedics.* 1986;9:181-5.

45. Hambrecht R, Fiehn E, Yu J, Niebauer J, Weigl C, Hilbrich L, et al. Effects of endurance training on mitochondrial ultrastructure and fiber type distribution in skeletal muscle of patients with stable chronic heart failure. *J Am Coll Cardiol.* 1997;29:1067-73.

46. Hamdy RC, Anderson JS, Whalen KE, Harvill LM. Regional differences in bone density of young men involved in different exercises. *Med Sci Sports Exerc.* 1994;26(7):884-8.

47. Heath GW, Hagberg JM, Ehsani AA, Holloszy JO. A physiological comparison of young and older endurance athletes. *J Appl Physiol Respir Environ Exerc Physiol.* 1981;51:634-40.

48. Henricksson J, Reitman JS. Time course of changes in human skeletal muscle succinate dehydrogenase and cytochrome oxidase activities and maximal

oxygen uptake with physical activity and inactivity. *Acta Physiol Scand.* 1977;99:91-7.

49. Hickson RC. Skeletal muscle cytochrome c and myoglobin endurance, and frequency of training. *J Appl Physiol.* 1981;51:746-9.

50. Hooper SL, MacKinnon LT, Howard A, Gordon RD, Bachmann AW. Markers for monitoring overtraining and recovery. *Med Sci Sports Exerc.* 1995;27(1):106-12.

51. Houmard JA. Impact of reduced training on performance endurance athletes. *Sports Med.* 1991;12:380-93.

52. Houmard JA, Scott BK, Justice CL, Chenier TC. The effects of taper on performance in distance runners. *Med Sci Sports Exerc.* 1994;26:624-31.

53. Inger F, Stromme SB. Effects of active, passive or no warm up on the physiological response to heavy exercise. *Eur J Appl Physiol.* 1979;40:273-82.

54. Janicki JS, Sheriff DD, Robotham JL, Wise RA. Cardiac output during exercise: contributions of the cardiac, circulatory, and respiratory systems. In: Rowell LB, Shephard JT, editors. *Handbook of physiology.* New York: American Physiology Society; 1996. p. 649-704.

55. Johns RA, Houmard JA, Kobe RW, et al. Effects of taper on swim power, stroke distance and performance. *Med Sci Sports Exerc.* 1992;24:1141-6.

56. Joyner MJ. Physiological limiting factors and distance running: influence of gender and age on record performances. In: Holloszy JO, editor. *Exercise and sports sciences reviews.* Baltimore: Williams & Wilkins; 1993. p. 103-33.

57. Karpovich P. Hale C. Effect of warming-up upon physical performance. *J Am Med Assoc.* 1956;162:1117-9.

58. Kaufmann D, Swenson E, Fencl J, Lucas A. Pulmonary function of marathon runners. *Med Sci Sports.* 1974;6(2):114-7.

59. Keteyian SJ, Duscha BD, Brawner CA, Green HJ, Marks CR, Schachat FH, et al. Differential effects of exercise training in men and women with chronic heart failure. *Am Heart J.* 2003;145:912-8.

60. Kielberg S, Rudhe U, Sjostrand T. Increase of the amount of hemoglobin and blood volume in connection with physical training. *Acta Physiol Scand.* 1949;19:146-51.

61. Kielberg S, Rudhe U, Sjostrand T. The amount of hemoglobin and the blood volume in relation to the pulse rate and cardiac volume during rest. *Acta Physiol Scan.* 1949;19:136-45.

62. Klassen G, Andrew G, Becklake M. Effect of training on total and regional blood flow and metabolism in paddlers. *J Appl Physiol.* 1970;28(4):397-406.

63. Konopka AR, Harber MP. Skeletal muscle hypertrophy after aerobic exercise training. *Exerc Sport Sci Rev.* 2014;42:53-61.

64. Konopka AR, Sreekumaran Nair K. Mitochondrial and skeletal muscle health with advancing age. Mol Cell Endocrinol. 2013;379:19-29.

65. Kreher JB, Schwartz JB. Overtraining syndrome: a practical guide. *Sports Health.* 2012;4:128-38.

66. Larsson L, Ansved T. Effects of long-term physical training and detraining on enzyme histochemical and functional skeletal muscle characteristic in man. *Muscle Nerve.* 1985;8:714-22.

67. Lee EJ, Long KA, Risser WL, Poindexter HBW, Gibbons WE, Goldzieher J. Variations in bone status contralateral and regional sites in young athletic women. *Med Sci Sports Exerc.* 1995;27(10):1354-61.

68. Levine BD, Lane LD, Watenpaugh DE, Gaffney FA, Buckey JC, Blomqvist CG. Maximal exercise performance after adaptation to microgravity. *J Appl Physiol.* 1996;81:686-94.

69. Lusk, G. *Science of nutrition.* 4th edition. Philadelphia: Saunders; 1928.

70. Magel J, Anderson K. Pulmonary diffusing capacity and cardiac output in young trained Norwegian swimmers and untrained subjects. *Med Sci Sports.* 1969;1(3):131-9.

71. Maron BJ. Structural features of the athlete heart as defined by echocardiography. *J Am Coll Cardiol.* 1986;7:190-203.

72. Maron BJ, Pelliccia A. The heart of trained athletes: cardiac remodeling and the risks of sports, including sudden death. *Circulation.* 2006;114:1633-44.

73. Martin BJ, Robinson S, Wiegman DL, Aulick LH. Effect of warm up on metabolic responses to strenuous exercise. *Med Sci Sports.* 1975;7(2):146-9.

74. Matthews D, Snyder H. Effect of warm up on the 440-yard dash. *Res Q.* 1959;30:446-51.

75. Molé P, Oscai L, Holloszy JO. Adaptation of muscle to exercise: increase in levels of palmityl CoA synthase, carnitine palmityltransferase, and palmityl CoA dehydrogenase, and in the capacity to oxidize fatty acid. *J Clin Invest.* 1971;50:2323-30.

76. Myers J, Buchanan N, Walsh D, Kraemer M, McAuley P, Hamilton-Wessler M, et al. Comparison of the ramp versus standard exercise protocols. *J Am Coll Cardiol*. 1991;17:1334-42.

77. Myers J, Prakash M, Froelicher V, Do D, Partington S, Atwood JE. Exercise capacity and mortality among men referred for exercise testing. *N Engl J Med*. 2002;346:793-801.

78. Newman F, Smalley B, Thompson M. A comparison between body size and lung function of swimmers and normal school children. *J Physiol*. (London). 1961;156:9.

79. Pacheco B. Improvement in jumping performance due to preliminary exercise. *Res Q*. 1957;28:55-63.

80. Powers SK, Martin D, Dodd S. Exercise-induced hypoxaemia in elite endurance athletes: Incidence, causes and impact on VO$_2$max. *Sports Med*. 1993;16:14-22.

81. Proctor DN, Miller JD, Dietz NM, Minson CT, Joyner MJ. Reduced submaximal leg blood flow after high-intensity aerobic training. *J Appl Physiol*. 2001;91:2619-27.

82. Rico-Sanz J, Rankinen T, Joanisse DR, Leon AS, Skinner JS, Wilmore JH, et al. Familial resemblance for muscle phenotypes in the HERITAGE Family Study. *Med Sci Sports Exerc*. 2003;35:1360-6.

83. Rosenkilde M, Reichkendler MH, Auerbach P, Bonne TC, Sjödin A, Ploug T, et al. Changes in peak fat oxidation in response to different doses of endurance training. *Scand J Med Sci Sports*. 2015;25:41-52.

84. Saltin B, Blomqvist G, Mitchell JH, Johnson RL Jr., Wildenthal K, Chapman CB. Response to exercise after bed rest and after training. *Circulation*. 1968;38(5 Suppl):1-78.

85. Saltin B, Rowell LB. Functional adaptations to physical activity and inactivity. Federation Proceedings. 1980;39:1506-13.

86. Savage PD, Toth MJ, Ades PA. A re-examination of the metabolic equivalent concept in individuals with coronary heart disease. *J Cardiopulmon Rehab Prev*. 2007;27:143-8.

87. Schlosser Covell GE, Hoffman-Snyder CR, Wellik KE, Woodruff BK, Geda YE, Caselli RJ, et al. Physical activity level and future risk of mild cognitive impairment or dementia: a critically appraised topic. *Neurologist*. 2015;19:89-91.

88. Shepley B, MacDougall JD, Cipriano N, Sutton JR, Tarnopolsky MA, Coates G. Physiological effects of tapering in highly trained athletes. *J Appl Physiol*. 1984;57:1668-73.

89. Skubic V, Hodgkins J. Effect of warm-up activities on speed, strength and accuracy. *Res Q*. 1957;28:147-52.

90. Smith ML, Hudson DL, Graitzer HM, Raven PB. Exercise training bradycardia: the role of autonomic balance. *Med Sci Sports Exerc*. 1989;21:40-4.

91. Stromme SB, Ingjer F, Meen ID. Assessment of maximal aerobic power in specifically trained athletes. *J Appl Physiol*. 1977;42:833-7.

92. Sullivan MJ, Green HJ, Cobb FR. Skeletal muscle biochemistry and histology in ambulatory patients with long-term heart failure. *Circulation*. 1990;81:518-27.

93. Sullivan MJ, Higginbotham MB, Cobb FR. Exercise training in patients with severe left ventricular dysfunction. Hemodynamic and metabolic effects. *Circulation*. 1988;78:506-15.

94. Swain DP, Franklin BA. V̇O$_2$ reserve and the minimal intensity for improving cardiorespiratory fitness. *Med Sci Sports Exerc*. 2002;34:152-7.

95. Swain DP, Leutholtz BC. Heart rate reserve is equivalent to %V̇O$_2$ reserve, not to %V̇O$_2$max. *Med Sci Sports Exerc*. 1997;29:410-4.

96. Tanaka, H. Effects of cross training. *Sports Med*. 1994;18:330-9.

97. Tanaka H, Monahan KD, Seals DR. Age-predicted maximal heart rate revisited. *J Am Coll Cardiol*. 2001;37:153-6.98.

98. Trappe SW, Costill DL, Fink WJ, Pearson DR. Skeletal muscle characteristics among distance runners: a 20-yr follow-up study. *J Appl Physiol*. 1995;78:823-9.

99. Trappe SW, Costill DL, Gallagher P, Creer A, Peters JR, Evans H, et al. Exercise in space: human skeletal muscle after 6 months aboard the International Space Station. *J Appl Physiol*. 2009;106:1159-68.

100. Trappe SW, Costill DL, Vukovich MD, Jones J, Melham T. Aging among elite distance runners: a 22-yr longitudinal study. *J Appl Physiol*. 1996;80:285-90.

101. Waburton DE, Haykowsky MJ, Quinney HA, Blackmore D, Teo KK, Taylor DA, et al. Blood volume expansion and cardiorespiratory function: Effects of training modality. *Med Sci Sports Exerc*. 2004;36:991-1000.

102. Williams PT. Lower risk of Alzheimer's disease mortality with exercise, statin, and fruit intake. *J Alzheimers Dis*. 2015;44:1121-9.

103. Winder, WW, Hagberg JM, Hickson RC, Ehsani AA, McLane JA. Time course of sympathoadrenal adaptation to endurance exercise training in man. *J Appl Physiol.* 1978;45:370-4.

104. Woltmann ML, Foster C, Porcari JP, Camic CL, Dodge C, Haible S, et al. Evidence that the talk test can be used to regulate exercise intensity. *J Strength Cond Res.* 2015;29:1248-54.

CHAPTER 7

1. Acevedo RJ, Rivera-Vega A, Miranda G, Micheo W. Anterior cruciate ligament injury: identification of risk factors and prevention strategies. Curr Sports Med Rep. 2014;13(3):186-91.

2. Agbuga B, Konukman F, Yilmaz I. Prediction of upper body strength by using grip strength test in division II American college football players' grip strength. Hacettepe J Sport Sci. 2009;20:16-23.

3. Almuzaini KS, Fleck SJ. Modification of the standing long jump test enhances ability to predict anaerobic performance. J Strength Cond Res. 2008; 22(4):1265-72.

4. Alvehus M, Boman N, Soderlund K, Svensson MB, Buren J. Metabolic adaptations in skeletal muscle, adipose tissue, and whole-body oxidative capacity in response to resistance training. Eur J Appl Physiol. 2014;114:1463-71.

5. American College of Sports Medicine, Chodzko-Zajko WJ, Proctor DN, Fiatarone Singh MA, Minson CT, Nigg CR, Salem GJ, Skinner JS. Exercise and physical activity for older adults. Med Sci Sports Exerc. 2009;41:1510-30.

6. American College of Sports Medicine, Garber CE, Blissmer B, Deschenes MR, Franklin BA, Lamonte MJ, Lee IM, Nieman DC, Swain DP. Quantity and quality of exercise for developing and maintaining cardiorespiratory, musculoskeletal, and neuromotor fitness in apparently healthy adults: guidance for prescribing exercise. Med Sci Sports Exerc. 2011;43:1334-59.

7. Babault N, Basine W, Deley G, Paizis C, Lattier G. Direct relation of acute effects of static stretching on isokinetic torque production are directly related to the initial flexibility level. Int J Sports Physiol Perform. 2014;2[Epub ahead of print].

8. Bachle TR, Earle RW, editors. Essentials of strength training and conditioning. Third edition. Champaign, IL: Human Kinetics; 2008.

9. Baeza-Velasco C, Gély-Nargeot MC, Pailhez G, Vilarrasa AB. Joint hypermobility and sport: a review of advantages and disadvantages. Curr Sports Med Reports. 2013;12(5):291-5.

10. Bar-Or O. The Wingate anaerobic test: an update on methodology, reliability and validity. Sports Med. 1987;4:381-94.

11. Baumgart C, Gokeler A, Donath L, Hoppe MW, Freiwald J. Effects of static stretching and playing soccer on knee laxity. Clin J Sport Med. 2015;25(6):541-5.

12. Beighton P, Horan F. Orthopaedic aspects of the Ehlers-Danlos syndrome. J Bone Joint Surg Br. 1969;51:444-53.

13. Benjaminse A, Gokeler A, Fleisig GS, Sell TC, Otten B. What is the true evidence for gender-related differences during plant and cut maneuvers? A systematic review. Knee Surg Sports Traumatol Arthrosc. 2011;19:42-54.

14. Boone DC, Azen SP. Normal range of motion of joints in male subjects. J Bone Joint Surg Am. 1979;61(5):756-9.

15. Buttelli AC, Pinto SS, Schoenell MC, Almada BP, Camargo LK, de Oliveira Conceição M, Kruel LF. Effects of single vs. multiple sets water-based resistance training on maximal dynamic strength in young men. J Hum Kinet. 2015;47:169-77.

16. Cahill BR, Misner JE, Boileau RA. The clinical importance of the anaerobic energy system and its assessment in human performance. Am J Sports Med. 1997;25:863-72.

17. Carter C, Wilkinson J. Persistent joint laxity and congenital dislocation of the hip. J Bone Joint Surg Br. 1964;46:40-5.

18. Cheema BS, Kilbreath SL, Fahey PP, Delaney GP, Atlantis E. Safety and efficacy of progressive resistance training in breast cancer: a systematic review and meta-analysis. Breast Cancer Res Treat. 2014;148:249-68.

19. Costa PB, Herda TJ, Herda AA, Cramer JT. Effects of dynamic stretching on strength, muscle imbalance, and muscle activation. Med Sci Sports Exerc. 2014;46:586-93.

20. Dawson B, Fitzsimons M, Green S, Goodman C, Carey M, Cole K. Changes in performance, muscle metabolites, enzymes and fibre types after short sprint training. Eur J Appl Physiol Occup Physiol. 1998;78:163-9.

21. Decoster LC, Cleland J, Altieri C, Russell P. The effects of hamstring stretching on range of motion: a systematic literature review. J Orthop Sports Phys Ther. 2005;35:377-87.

22. Drinkwater EJ, Lawton TW, McKenna MJ, Lindsell RP, Hunt PH, Pyne DB. Increased number of forced repetitions does not enhance strength development with resistance training. J Strength Cond Res. 2007;21:841-7.

23. Enwemeka CS. Radiographic verification of knee goniometry. Scand J Rehabil Med. 1986;18:47-9.

24. Fournier M, Ricci J, Taylor AW, Ferguson RJ, Montpetit RR, Chaitman BR. Skeletal muscle adaptation in adolescent boys: sprint and endurance training and detraining. Med Sci Sports Exerc. 1982;14:453-6.

25. Fowles JR, Sale DG, MacDougall JD. Reduced strength after passive stretch of the human plantarflexors. J Appl Physiol. 2000;89:1179-88.

26. García-López J, Morante JC, Ogueta-Alday A, Rodríguez-Marroyo JA. The type of mat (contact vs. photocell) affects vertical jump height estimated from flight time. J Strength Cond Res. 2013;27(4):1162-7.

27. Gardinier ES, Manal K, Buchanan TS, Snyder-Mackler L. Gait and neuromuscular asymmetries after acute anterior cruciate ligament rupture. Med Sci Sports Exerc. 2012;44:1490-6.

28. Garrett WE, Almedinders L, Seaber AV. Biomechanics of muscle tears and stretching injuries. Trans Orthop Res Soc. 1984;9:384.

29. Gollnick PD, Piehl K, Saltin B. Selective glycogen depletion pattern in human muscle fibers after exercise of varying intensity and at varying pedaling rates. J Physiol. 1974;241:45-57.

30. Goss FL, Robertson RJ, Gallagher M, Piroli A, Nagle EF. Response normalized omni rating of perceived exertion at the ventilatory breakpoint in division I football players. Percept Mot Skills. 2011;112:539-48.

31. Harvey L, Herbert R, Crosbie J. Does stretching induce lasting increases in joint ROM? A systematic review. Physiotherapy Res Int. 2002;7(1):1-13.

32. Heitkamp HC, Holdt M, Scheib K. The reproducibility of the 4 mmol/l lactate threshold in trained and untrained women. Int J Sports Med. 1991;12(4):363-8.

33. Herbert RD, de Noronha M. Stretching to prevent or reduce muscle soreness after exercise. Cochrane Database Syst Rev. 2011;17:CD004577.

34. Hetzler RK, Vogelpohl RE, Stickley CD, Kuramoto AN, Delaura MR, Kimura MF. Development of a modified Margaria-Kalamen anaerobic power test for American football athletes. J Strength Cond Res. 2010;24(4):978-84.

35. Hori N, Newton RU, Andrews WA, Kawamori N, McGuigan MR, Nosaka K. Does performance of the hang power clean differentiate performance of jumping, sprinting, and changing of direction? J Strength Cond Res. 2008;22(2):412-8.

36. Howarth SJ, Glisic D, Lee JG, Beach TA. Does prolonged seated deskwork alter the lumbar flexion relaxation phenomenon? J Electromyogr Kinesiol. 2013;23:587-93.

37. Janssen I, Heymsfield SB, Wang Z, Ross R. Skeletal muscle mass and distribution in 468 men and women aged 18-88 yr. J Appl Physiol. 2000.;89:81-8.

38. Kenny WL, Wilmore JH, Costill DL. Physiology of sport and exercise. Fifth edition. Champaign, IL: Human Kinetics; 2011.

39. Kraemer WJ, Ratamess NA. Fundamentals of resistance training: progression and exercise prescription. Med Sci Sports Exerc. 2004;36:674-88.

40. MacDougall JD, Hicks AL, MacDonald JR, McKelvie RS, Green HJ, Smith KM. Muscle performance and enzymatic adaptations to sprint interval training. J Appl Physiol. 1998;84:2138-42.

41. Macdougall JD, Wenger HA, Green HJ. The purpose of physiological testing. In: Macdougall JD, Wenger HS, Green HJ, editors. Physiological testing of the high-performance athlete. Champaign, IL: Human Kinetics; 1991. pp. 1-6.

42. Maglischo EW. Swimming even faster. Mountain View, CA: Mayfield; 1993.

43. Margaria R, Aghemo P, Rovelli E. Measurement of muscular power (anaerobic) in man. J Appl Physiol. 1966;21:1662-4.

44. Mitchell JH, Haskell WL, Raven PB. Classification of sports 26th Bethesda Conference. J Am Coll Cardiol. 1994;24:866.

45. Monod H, Scherrer J. The work capacity of a synergic muscular group. Ergonomics. 1965;8:329-38.

46. Moritani T, Nagata A, deVries HA, Muro M. Critical power as measure of physical work capacity and anaerobic threshold. Ergonomics. 1981;24:339-50.

47. Norkin CC, White DJ. Measurement of joint motion: a guide to goniometry. Fourth edition. Philadelphia: Davis; 2009.

48. Nuzzo JL, Anning JH, Scharfenberg JM. The reliability of three devices used for measuring vertical jump height. J Strength Cond Res. 2011;25(9):2580-90.

49. Nuzzo JL, McBride JM, Cormie P, McCaulley G. Relationship between countermovement jump performance and multijoint isometric and dynamic tests of strength. J Strength Cond Res. 2008;22(3):699-707.

50. Pescatello L. ACSM's guidelines for exercise testing and prescription. Ninth edition. Philadelphia: Wolters Kluwer/Lippincott, Williams & Wilkins Health; 2014.

51. Rantanen T, Era P, Heikkinen E. Maximal isometric muscle strength and socio-economic status, health, and physical activity in 75-year-old persons. J Aging Phys Act. 1994;2:206-20.

52. Rantanen T, Volpato S, Ferrucci L, Heikkinen E, Fried LP, Guralnik JM. Handgrip strength and cause-specific total mortality in older disabled women: exploring the mechanism. J Am Ger Soc. 2003;51:636-41.

53. Ratamess NA, Alvar BA, Evetoch TK, Housh TJ, Kible WB, Kraemer WJ, Triplett NT. Progression models in resistance training for healthy adults. Med Sci Sports Exerc. 2009;41:687-708.

54. Roach KE, Miles TP. Normal hip and knee active range of motion: the relationship to age. Phys Ther. 1991;71:656-65.

55. Rodriguez FA, Mader A. Energy systems in swimming. In: L. Seifert, D. Chollet, I. Mujika, editors. World book of swimming: from science to performance. Location: Nova Science Publishers, Inc., Hauppauge NY; 2010.

56. Roig M, O'Brien K, Kirk G, Murray R, McKinnon P, Shadgan B, Reid WD. The effects of eccentric versus concentric resistance training on muscle strength and mass in healthy adults: a systematic review with meta-analysis. Br J Sports Med. 2009;43:556-68.

57. Ross A, Leveritt M. Long-term metabolic and skeletal muscle adaptations to short-sprint training: implications for sprint training and tapering. Sports Med. 2001;31:1063-82.

58. Rothstein JM, Miller PJ, Roettger RF. Goniometric reliability in a clinical setting. Elbow and knee measurements. Phys Ther. 1983;63:1611-15.

59. Ryan ED, Everett KL, Smith DB, Pollner C, Thompson BJ, Sobolewski EJ, Fiddler RE. Acute effects of different volumes of dynamic stretching on vertical jump performance, flexibility and muscular endurance. Clin Physiol Funct Imaging. 2014;34:485-92.

60. Scherr J, Wolfarth B, Christle JW, Pressler A, Wagenpfeil S, Halle M. Associations between Borg's rating of perceived exertion and physiological measure of exercise intensity. Eur J Appl Physiol. 2013;113:147-55.

61. Senechal M, Mcgavock JM, Church TS, Lee DC, Earnest CP, Sui X, Blair SN. Cut points of muscle strength associated with metabolic syndrome in men. Med Sci Sports Exerc. 2014;46:1475-81.

62. Serpell BG, Scarvell JM, Ball NB, Smith PN. Mechanisms and risk factors for noncontact ACL injury in age mature athletes who engage in field or court sports: a summary of the literature since 1980. J Strength Cond Res. 2012;26(11):3160-76.

63. Serra AJ, Silva JA, Marcolongo AA, Manchini MT, Oliveira JV, Santos LF, Rica RL, Bocalini DS. Experience in resistance training does not prevent reduction in muscle strength evoked by passive static stretching. J Strength Cond Res. 2013;27(8):2304-8.

64. Tanaka H, Swensen T. Impact of resistance training on endurance performance. A new form of cross-training? Sports Med. 1998;25:191-200.

65. Tang JE, Hartman JW, Phillips SM. Increased muscle oxidative potential following resistance training induced fibre hypertrophy in young men. Appl Physiol Nutr Metab. 2006;31:495-501.

66. Taylor DC, Dalton JD, Seaber AV, Garrett WE. Viscoelastic properties of muscle-tendon units. The biomechanical effects of stretching. Am J Sports Med. 1990;18:300-9.

67. Thorstensson A, Sjodin B, Karlsson J. Enzyme activities and muscle strength after "sprint training" in man. Acta Physiol Scand. 1975;94:313-8.

68. Tomlin DL, Wenger HA. The relationship between aerobic fitness and recovery from high intensity intermittent exercise. Sports Med. 2001;31:1-11.

69. Vollestad NK, Blom PCS. Effect of varying exercise intensity on glycogen depletion in human muscle fibers. Acta Physiol Scand. 1985;125:395-405.

70. Walker S, Peltonen H, Sautel J, Scaramella C, Kraemer WJ, Avela J, Häkkinen K. Neuromuscular adaptations to constant vs. variable resistance training in older men. Int J Sports Med. 2014;35:69-74.

71. Weppler CH, Magnussion SP. Increasing muscle extensibility: a matter of increasing length or modifying sensation? Phys Ther. 2010;90:438-49.

72. Westcott WL, Winett RA, Anderson ES, Wojcik JR, Loud RL, Cleggett E, Glover S. Effects of regular and slow speed resistance training on muscle strength. J Sports Med Phys Fitness. 2001;41:154-8.

73. Whitmer TD, Fry AC, Forsythe CM, Andre MJ, Lane MT, Hudy A, Honnold DE. Accuracy of a vertical jump contact map for determining jump height and flight time. J Strength Cond Res. 2015;29(4):877-81.

74. Witvrouw E, Mahieu N, Danneels L, McNair P. Stretching and injury prevention: an obscure relationship. Sports Med. 2004;34(7):443-9.

75. Zipes DP, Libby P, Bonow RO. Braunwald's heart disease. A textbook of cardiovascular medicine. Philadelphia: Saunders; 2007.

CHAPTER 8

1. Cardiovascular effects of intensive lifestyle intervention in type 2 diabetes. *New Engl J Med.* 2013;369(2):145-54.

2. Reduction in the incidence of type 2 diabetes with lifestyle intervention or metformin. *New Engl J Med.* 2002;346(6):393-403.

3. The female athlete triad. *Med Sci Sports Exerc.* 2007;39(10):1867-82.4.

4. Behnke A, Feen B, Welham W. The specific gravity of healthy men. *J Am Med Assoc.* 1942;118:495-8.

5. Berggren JR, Boyle KE, Chapman WH, Houmard JA. Skeletal muscle lipid oxidation and obesity: influence of weight loss and exercise. *Am J Physiol Endocrinol Metab.* 2008;294(4):E726-32.

6. Brozek J, Grande F, Anderson J, Keys A. Densitometric analysis of body composition: revision of some quantitative assumptions. *Ann NY Acad Sci.* 1963;110:113-40.

7. Buch E, Bradfield J, Larson T, Horwich T. Effect of bioimpedance body composition analysis on function of implanted cardiac devices. *Pacing Clin Electrophysiol.* 2012;35(6):681-4.

8. Carels RA, Darby LA, Rydin S, Douglass OM, Cacciapaglia HM, O'Brien WH. The relationship between self-monitoring, outcome expectancies, difficulties with eating and exercise, and physical activity and weight loss treatment outcomes. *Ann Behav Med.* 2005;30(3):182-90.

9. Casazza K, Fontaine KR, Astrup A, Birch LL, Brown AW, Bohan Brown MM, et al. Myths, presumptions, and facts about obesity. *New Engl J Med.* 2013;368(5):446-54.

10. Church TS, Thomas DM, Tudor-Locke C, Katzmarzyk PT, Earnest CP, Rodarte RQ, et al. Trends over 5 decades in US occupation-related physical activity and their associations with obesity. *PLoS One.* 2011;6(5):e19657.

11. Clarys JP, Martin AD, Drinkwater DT. Gross tissue weights in the human body. *Hum Biol.* 1984;56:459-473.

12. Collins M, Millard-Stafford M, Sparling P, Snow T, Rosskopf L, Webb S, et al. Evaluation of bod pod for assessing body fat in collegiate football players. *Med Sci Sports Exerc.* 1999;31:1350-6.

13. Colquitt JL, Pickett K, Loveman E, Frampton GK. Surgery for weight loss in adults. *Cochrane Database Syst Rev.* 2014;8:CD003641.

14. Cook CM, Edwards C. Success habits of long-term gastric bypass patients. *Obes Surg.* 1999;9(1):80-2.

15. Cooper Z, Fairburn CG. A new cognitive behavioral approach to the treatment of obesity. *Behav Res Ther.* 2001;39(5):499-511.

16. Courcoulas AP, Goodpaster BH, Eagleton JK, Belle SH, Kalarchian MA, Lang W, et al. Surgical vs. medical treatments for type 2 diabetes mellitus: a randomized clinical trial. *J Am Med Assoc Surg.* 2014;149(7):707-15.

17. Das SK, Roberts SB, Kehayias JJ, Wang J, Hsu LKG, Shikora SA, et al. Body composition assessment in extreme obesity and after massive weight loss induced by gastric bypass surgery. *Am J Physiol Endocrinol Metab.* 2003;284(6):E1080-8.

18. Deitel M, Gawdat K, Melissas J. Reporting weight loss 2007. *Obes Surg.* 2007;17:565-8.

19. Elfhag K, Rössner S. Who succeeds in maintaining weight loss? A conceptual review of factors associated with weight loss maintenance and weight regain. *Obesity Rev.* 2005;6(1):67-85.

20. Eng DS, Lee JM, Gebremariam A, Meeker JD, Peterson K, Padmanabhan V. Bisphenol A and chronic disease risk factors in US children. *Pediatrics.* 2013;132(3):e637-45.

21. Fawcett KA, Barroso I. The genetics of obesity: FTO leads the way. *Trends Genet.* 2010;26(6):266-74.

22. Fields DA, Goran MI, McCrory MA. Body-composition assessment via air-displacement plethysmography in adults and children: a review. *Am J Clin Nutr.* 2002;75(3):453-67.

23. Fields DA, Wilson GD, Gladden LB, Hunter GR, Pascoe DD, Goran MI. Comparison of the BOD POD with the four-compartment model in adult females. *Med Sci Sports Exerc.* 2001;33(9):1605-10.

24. Flegal KM, Carroll MD, Kit BK, Ogden CL. Prevalence of obesity and trends in the distribution of body

mass index among US adults 1999-2010. *J Am Med Assoc.* 2012;307(5):491-7.

25. Flegal KM, Carroll MD, Ogden CL, Curtin LR. Prevalence and trends in obesity among US adults, 1999-2008. *J Am Med Assoc.* 2010;303(3):235-41.

26. Flegal KM, Kit BK, Orpana H, Graubard BI. Association of all-cause mortality with overweight and obesity using standard body mass index categories: a systematic review and meta-analysis. *J Am Med Assoc.* 2013;309(1):71-82.

27. Forbes R, Cooper A, Mitchell H. The composition of the human body as determined by chemical analysis. *J Biol Chem.* 1953;203:359-66.

28. Forbes R, Mitchell H, Cooper A. Further studies on the gross composition and mineral elements of the adult human body. *J BiolChem.* 1956;223:969-75.

29. Gonzalez MC, Pastore CA, Orlandi SP, Heymsfield SB. Obesity paradox in cancer: new insights provided by body composition. *Am J Clin Nutr.* 2014;99(5):999-1005.

30. Goodpaster BH. Measuring body fat distribution and content in humans. Curr Opin Clin Nutr *Metab Care.* 2002;5(5):481-7.

31. Guare JC, Wing RR, Marcus MD, Epstein LH, Burton LR, Gooding WE. Analysis of changes in eating behavior and weight loss in type II diabetic patients. *Diabet Care.* 1989;12:500-3.

32. Hall KD, Heymsfield SB, Kemnitz JW, Klein S, Schoeller DA, Speakman JR. Energy balance and its components: implications for body weight regulation. *Am J Clin Nutr.* 2012;95(4):989-94.

33. Hall KD, Sacks G, Chandramohan D, Chow CC, Wang YC, Gortmaker SL, et al. Quantification of the effect of energy imbalance on bodyweight. *Lancet.* 2011;378(9793):826-37.

34. Hangartner TN, Warner S, Braillon P, Jankowski L, Shepherd J. The official positions of the International Society for Clinical Densitometry: acquisition of dual-energy X-ray absorptiometry body composition and considerations regarding analysis and repeatability of measures. *J Clin Densitom.* 2013;16(4):520-36.

35. Heymsfield SB, van Mierlo CA, van der Knaap HC, Heo M, Frier HI. Weight management using a meal replacement strategy: meta and pooling analysis from six studies. *Int J Obes Relat Metab Disord.* 2003;27(5):537-49.

36. Heyward V, Wagner D. *Applied body composition assessment.* 2nd ed. Champaign, IL: Human Kinetics; 2004.

37. Hill JO, Wyatt HR, Peters JC. Energy balance and obesity. *Circulation.* 2012;126(1):126-32.

38. Hu HH, Kan HE. Quantitative proton MR techniques for measuring fat. *NMR Biomed.* 2013;26(12):1609-29.

39. Jackson AS, Pollock ML. Generalized equations for predicting body density of men. *Br J Nutr.* 1978;40:497-504.

40. Jackson AS, Pollock ML. A generalized equation for predicting body density of women. *Br J Nutr.* 1980;12:175-81.

41. Jacquelin-Ravel N, Pichard C. Clinical nutrition, body composition and oncology: A critical literature review of the synergies. *Crit Rev Oncol/Hematol.* 2012;84:37-46.

42. Jeffery RW. How can health behavior therapy be made more useful for intervention research? *Int J Behav Nutr Phys Activ.* 2004;1(10):1-10.

43. Johansson K, Neovius M, Hemmingsson E. Effects of anti-obesity drugs, diet, and exercise on weight-loss maintenance after a very-low-calorie diet or low-calorie diet: a systematic review and meta-analysis of randomized controlled trials. *Am J Clin Nutr.* 2014;99(1):14-23.

44. Klein S, Allison DB, Heymsfield SB, Kelley DE, Leibel RL, Nonas C, et al. Waist circumference and cardiometabolic risk: a consensus statement from Shaping America's Health: Association for Weight Management and Obesity Prevention; NAASO, The Obesity Society; the American Society for Nutrition; and the American Diabetes Association. *Am J Clin Nutr.* 2007;85(5):1197-202.

45. Kyle UG, Bosaeus I, De Lorenzo AD, Deurenberg P, Elia M, Gomez JM, et al. Bioelectrical impedance analysis–part I: review of principles and methods. *Clin Nutr.* 2004;23:1226-43.

46. Lee SY, Gallagher D. Assessment methods in human body composition. *Curr Opin Clin Nutr Metab Care.* 2008;11(5):566-72.

47. Lockner DW, Heyward VH, Baumgartner RN, Jenkins KA. Comparison of air-displacement plethysmography, hydrodensitometry, and dual X-ray absorptiometry for assessing body composition of children 10 to 18 years of age. *Ann NY Acad Sci.* 2000;904:72-8.

48. Lukaski HC, Johnson PE, Bolonchuk WW, Lykken GI. Assessment of fat-free mass using bioelectrical impedance measurements of the human body. *Am J Clin Nutr* 1985;41:810-7.

49. Mason C, Xiao L, Imayama I, Duggan CR, Campbell KL, Kong A, et al. The effects of separate and combined dietary weight loss and exercise on fast-

ing ghrelin concentrations in overweight and obese women: a randomized controlled trial. *Clin. Endocrinol.* 2015;82:369-76.

50. Mayer J, Roy P, Mitra KP. Relation between caloric intake, body weight, and physical work: studies in an industrial male population in West Bengal. *Am J Clin Nutr.* 1956;4(2):169-75.

51. McCullough PA, Gallagher MJ, Dejong AT, Sandberg KR, Trivax JE, Alexander D, et al. Cardiorespiratory fitness and short-term complications after bariatric surgery. *Chest.* 2006;130(2):517-25.

52. National Collegiate Athletic Association. Weight management education [Internet]. Indianapolis: National Collegiate Athletic Association; year. Available from: http://www.ncaa.com/news/wrestling/2002-01-01/weight-management-education2012

53. Nevill AM, Metsios GS, Jackson AS, Wang J, Thornton J, Gallagher D. Can we use the Jackson and Pollock equations to predict body density/fat of obese individuals in the 21st century? *Int J Body Compos Res.* 2008;6(3):114-21.

54. Ng M, Fleming T, Robinson M, Thomson B, Graetz N, Margono C, et al. Global, regional, and national prevalence of overweight and obesity in children and adults during 1980-2013: a systematic analysis for the Global Burden of Disease Study 2013. *Lancet.* 2014;384(9945):766-81.

55. O'Connor DP, Bray MS, Mcfarlin BK, Sailors MH, Ellis KJ, Jackson AS. Generalized equations for estimating DXA percent fat of diverse young women and men: The TIGER Study. *Med Sci Sports Exerc.* 2010;42(10):1959-65.

56. O'Neil PM. Assessing dietary intake in the management of obesity. *Obesity Res.* 2001;9(Suppl 5):S361-6.

57. Pierson R, Wang J, Thornton J. Body composition comes of age: A modest proposal for the next generation. *Ann NY Acad Sci.* 2000;904:1-11.

58. Pietrobelli A, Wang Z, Formica C, Heymsfield SB. Dual-energy X-ray absorptiometry: fat estimation errors due to variation in soft tissue hydration. *Am J Physiol.* 1998;274(5 Pt 1):E808-16.

59. Ross R. Advances in the application of imaging methods in applied and clinical physiology. *Acta Diabet.* 2003;40(1):S45-50.

60. Rueda-Clausen CF, Padwal RS. Pharmacotherapy for weight loss. *Br Med J.* 2014;348:g3526.

61. Sato T, Ida T, Nakamura Y, Shiimura Y, Kangawa K, Kojima M. Physiological roles of ghrelin on obesity. *Obes Res Clin Pract.* 2014;8(5):e405-13.

62. Siri W. The gross composition of the body. In: Tobias C, Lawrence J, editors. *Advances in biological and medical physics.* New York: Academic Press; 1956. p. 239-280.

63. Stunkard AJ, Sorensen TI, Hanis C, Teasdale TW, Chakraborty R, Schull WJ, et al. An adoption study of human obesity. *N Engl J Med.* 1986;314(4):193-8.

64. Sun G, French CR, Martin GR, Younghusband B, Green RC, Xie YG, et al. Comparison of multifrequency bioelectrical impedance analysis with dual-energy X-ray absorptiometry for assessment of percentage body fat in a large, healthy population. *Am J Clin Nutr.* 2005;81(1):74-8.

65. Thomas JG, Bond DS, Phelan S, Hill JO, Wing RR. Weight-loss maintenance for 10 years in the National Weight Control Registry. *Am J Prev Med.* 2014;46(1):17-23.

66. Thomson R, Brinkworth GD, Buckley JD, Noakes M, Clifton PM. Good agreement between bioelectrical impedance and dual-energy X-ray absorptiometry for estimating changes in body composition during weight loss in overweight young women. *Clin Nutr.* 2007;26(6):771-7.

67. Thorland WG, Tipton CM, Lohman TG, Bowers RW, Housh TG, Johnson GO, et al. Midwest wrestling study: Prediction of minimal weight for high school wrestlers. *Med Sci Sports Exerc.* 1991;23:1102-10.

68. Tsai AG, Wadden TA. The evolution of very-low-calorie diets: an update and meta-analysis. *Obesity.* 2006;14(8):1283-93.

69. Wang J, Thornton JC, Bari S, Williamson B, Gallagher D, Heymsfield SB, et al. Comparisons of waist circumferences measured at 4 sites. *Am J Clin Nutr.* 2003; 77(2):379-384.

70. Wang ZM, Heshka S, Pierson RN Jr., Heymsfield SB. Systematic organization of body-composition methodology: an overview with emphasis on component-based methods. *Am J Clin Nutr.* 1995;61(3):457-65.

71. Webster S, Rutt R. Physiological effects of a weight loss regimen practiced by college wrestlers. *Med Sci Sports Exerc.* 1990;22(2):229-34.

72. Widdowson E, McCance R, Spray C. The chemical composition of the human body. *Clin Sci.* 1951;10:113-25.

73. Wing RR, Hill JO. Successful weight loss maintenance. *Annu Rev Nutr.* 2001;21:323-41.

74. Withers RT, Laforgia J, Heymsfield SB. Critical appraisal of the estimation of body composition via two-, three-, and four-compartment models. *Am J Hum Biol.* 1999;11(2):175-85.

75. Xia Q, Grant SFA. The genetics of human obesity. *Ann NY Acad Sci.* 2013;1281(1):178-90.

76. Xu L, Cheng X, Wang J, Cao Q, Sato T, Wang M, et al. Comparisons of body-composition prediction accuracy: a study of 2 bioelectric impedance consumer devices in healthy Chinese persons using DXA and MRI as criteria methods. *J Clin Densitom.* 2011;14(4):458-64.

CHAPTER 9

1. Ackland T, Lohman T, Sundgot-Borgen J, Maughan R, Meyer N, Stewart A, Muller W. Current assessment of body composition in sport: review and position statement on behalf of the ad hoc research working group on body composition health and performance, under the auspices of the I.O.C. Medical Commission. Sports Med. 2012;42(3):227-49.

2. Adams WC, Bernauer EM, Dill DB, Bomar JB. Effects of equivalent sea-level and altitude training on $\dot{V}O_2$max and running performance. J Appl Physiol. 1975;39(2):262-6.

3. Armstrong LE, Casa DJ, Millard-Stafford M, Moran DS, Pyne SW, Roberts WO. Exertional heat illness during training and competition. Med Sci Sports Exerc. 2007;39(3):556-72.

4. Arngrimsson SA, Petitt DS, Stueck MG, Jorgensen DK, Cureton KJ. Cooling vest worn during active warm-up improves 5-km run performance in the heat. J Appl Physiol. 2004;96:1867-74.

5. Baty JJ, Hwang H, Ding Z, Bernard JR, Wang B, Kwon B, Ivy JL. The effect of a carbohydrate and protein supplement on resistance exercise performance, hormonal response, and muscle damage. J Strength Cond Res. 2007;21(2):321-9.

6. Beattie K, Kenny IC, Lyons M, Carson BP. The effect of strength training on performance in endurance athletes. Sports Med. 2014;44(6):845-65.

7. Bergeron MF. Heat cramps: fluid and electrolyte challenges during tennis in the heat. J Sci Med Sport. 2003;6:19-27.

8. Bergh U, Ekblom B. Physical performance and peak power at different body temperatures. J Appl Physiol. 1979;45(5):885-9.

9. Bergsrom JL, Hermansen E, Hutlman E, Saltin B. Diet, muscle glycogen and physical performance. Nature. 1967;210:309-10.

10. Bergstrom J, Hermansen L, Hultman E, Saltin B. Diet, muscle glycogen and physical performance. Acta Physiol Scand. 1968;71:140-50.

11. Blomstrand E, Kaijser L, Martinsson A, Bergh U, Ekblom B. Temperature-induced changes in metabolic and hormonal responses to intensive dynamic exercise. Acta Physiol Scand. 1986;127:477-84.

12. Bray MS, Hagberg JM, Perusse L, Rankinen T, Roth SM, Wolfarth B, Bouchard C. The human gene map for performance and health-related fitness phenotypes: the 2006-07 update. Med Sci Sports Exerc. 2009;41(1):35-73.

13. Brooks GA, Hittelman KJ, Faulkner JA, Beyer RE. Temperature, skeletal muscle mitochondrial functions, and oxygen debt. Am J Physiol. 1971;220:1053-9.

14. Case DJ, Becker SM, Ganio MS, Brown CM, Yeargin SW, Roti MW, Siegler J, Blowers JA, Glaviano NR, Huggins RA, Armstrong LE, Maresh CM. Validity of devices that assess body temperature during outdoor exercise in the heat. J Athl Train. 2007;42(3):333-42.

15. Chapman RF, Stray-Gundersen J, Levine BD. Individual variation in response to altitude training. J Appl Physiol. 1998;85:1448-56.

16. Cockburn E, Fortune A, Briggs M, Rumbold P. Nutrition knowledge of UK coaches. Nutrients. 2014;6(4):1442-53.

17. Cohen JS, Gisolfi CV. Effects of interval training on work-heat tolerance of young women. Med Sci Sports Exerc. 1982;14:46-52.

18. Convertino VA. Exercise and adaptation to microgravity environments. In: Fregley MJ, Blatteis CM, editors. Handbook of physiology, section 4: environmental physiology. New York: Oxford University Press; 1996, pp. 815-43.

19. Convertino VA. Physiological adaptations to weightlessness: effects on exercise and work performance. Exerc Sport Sci Rev. 1990;18:119-66.

20. Coris EF, Ramirez AM, Van Durme JD. 2004. Heat illness in athletes: the dangerous combination of heat, humidity, and exercise. Sports Med. 34(1):9-16.

21. Costill DL, Branam G, Eddy D, Sparks K. Determinants of marathon running success. Int Z Angew Physiol. 1971;29:249-54.

22. Coyle EF, Gonzalez-Alonso J. Cardiovascular drift during prolonged exercise: new perspectives. Exerc Sports Sci Rev. 2001;29(2):88-92.

23. Daniels J, Oldridge N. The effects of alternate exposure to altitude and sea level on world-class middle distance runners. Med Sci Sports. 1970;2(3):107-12.

24. Davis JA, Vodak P, Wilmore JH, Vodak J, Kurtz P. Anaerobic threshold and maximal aerobic power for three modes of exercise. J Appl Physiol. 1976;41(4):544-50.

25. Dawson, B. Exercise training in sweat clothing in cool conditions to improve heat tolerance. Sports Med. 1994;17:233-44.

26. Eynon N, Ruiz JR, Femia P, Pushkarev VP, Cieszczyk P, Maciejewska-Karlowska A, Sawczuk M, Dyatlov DA, Lekontsev EV, Kulikov LM, Birk R, Bishop DJ. The ACTN3 R577X polymorphism across three groups of elite male European athletes. PLoS One. 2012;7:e43132.

27. Foster C, Hector LL, Welsh M, Schrager M, Green MA, Snyder AC. Effects of specific versus cross-training on running performance. Eur J Appl Physiol. 1995;70:367-72.

28. Fulco CS, Rock PB, Cymerman A. Maximal and submaximal exercise performance at altitude. Aviat Space Environ Med. 1998;69:793-801.

29. Ganio MS, Brown CM, Casa DJ, Becker SM, Yeargin SW, McDermott BP, Boots LM, Boyd PW, Armstrong LE, Maresh CM. Validity and reliability of devices that assess body temperature during indoor exercise in the heat. J Athl Train. 2009;44(2):124-35.

30. Gary D, Steffes GD, Megura AE, Adams J, Claytor RP, Ward RM, Horn TS, Potteiger JA. Prevalence of metabolic syndrome risk factors in high school and NCAA division I football players. J Strength Cond Res. 2013;27(7):1749-57.

31. Gisolfi CV. Work-heat tolerance derived from interval training. J Appl Physiol. 1971;35:349-54.

32. Glynn EL, Fry CS, Timmerman KL, Drummond MJ, Volpi E, Rasmussen BB. Addition of carbohydrate or alanine to an essential amino acid mixture does not enhance human skeletal muscle protein anabolism. J Nutr. 2013;143:307-14.

33. Guth LM, Roth SM. Genetic influence on athletic performance. Curr Opin Pediatr. 2013;26(6):653-8.

34. Harrison MH, Edwards RJ, Leitch DR. Effect of exercise and thermal stress on plasma volume. J Appl Physiol. 1975;39(6):925-31.

35. Hochachka PW, Beatty CL, Burelle Y, Trump ME, McKenzie DC, Matheson GO. The lactate paradox in human high-altitude physiological performance. Physiology. 2002;17(3):122-6.

36. Höpfl G, Ogunshola O, Gassman M. Hypoxia and high altitude. The molecular response. Adv Exp Med Biol. 2003;543:89-115.

37. Houmard JA, Johns RA. Effects of taper on swim performance. Practical implications. Sports Med. 1994;17(4):224-32.

38. Houtkooper L. Body composition. In: Manore M, Thompson J, editors. Sport nutrition for health and performance. Champaign, IL: Human Kinetics; 2000.

39. Hultman EH, Sahlin K. Acid-base balance during exercise. Exerc Sport Sci Rev. 1980;7:41-128.

40. Kearney, JT. Sport performance enhancement: design and analysis of research. Med Sci Sports Exerc. 1999;31:755-6.

41. Kerksick C, Harvey T, Stout J, Campbell B, Wilborn C, Kreider R, Kalman D, Ziegenfuss T, Lopez H, Landis J, Ivy JL, Antonio J. International Society of Sports Nutrition position stand: nutrient timing. J Int Soc Sports Nutr. 2008;5:17.

42. Kirby CR, Convertino VA. Plasma aldosterone and sweat sodium concentrations after exercise and heat acclimation. J Appl Physiol. 1986;61(3):967-70.

43. Klausen T, Ghisler U, Mohr T, Fogh-Andersen N. Erythropoietin, 2,3 diphosphoglycerate and plasma volume during moderate-altitude training. Scand J Med Sci Sports. 1992;2(1):16-20.

44. Klissouras V. Heritability of adaptive variation. J Appl Physiol. 1971;31(3):338-44.

45. Klissouras V, Pirnay F, Petit J. Adaptation to maximal effort: genetics and age. J Appl Physiol. 1973;35(2):288-93.

46. Kraemer WJ, Torine JC, Silvestre R, French DN, Ratamess NA, Spiering BA, Hatfield DL, Vingren JL, Volek JS. Body size and composition of national football league players. J Strength Cond Res. 2005;19(3):485-9.

47. Lemon PW, Berardi JM, Noreen EE. The role of protein and amino acid supplements in the athlete's diet: does type or timing of ingestion matter? Curr Sports Med Rep. 2002;1(4):214-21.

48. Levine BD, Stray-Gundersen J. "Living high-training low": effect of moderate-altitude acclimatization with low-altitude training on performance. J Appl Physiol. 1997;83:102-12.

49. Li X, Wang H, Yang Y, Qi C, Wang F, Jin M. Effect of height on motor coordination in college students participating in a dancesport program. Med Probl Perform Art. 2015;30:20-5.

50. Lorenzo S, Halliwill JR, Sawka MN, Minson CT. Heat acclimation improves exercise performance. J Appl Physiol. 2010;109(4):1140-7.

51. Lundby C, Millet GP, Calbet JA, Bartsch P, Subudhi AW. Does "altitude training" increase exercise performance in elite athletes? Br J Sports Med. 2012;46:792-5.

52. Machado M, Pereira R, Sampaio-Jorge F, Kinfis F, Hackney A. Creatine supplementation: effects on blood creatine kinase activity to resistance exercise and creatine kinase activity measurement. Braz J Pharmaceut Sci. 2009;45(4):751-7.

53. Mack GW, Nadel ER. Body fluid balance during heat stress in humans. In: Fregly MJ, Blatteis CM, editors. Environmental physiology. New York: Oxford University Press; 1996. pp. 187-214.

54. Maher JT, Jones LG, Hartley LH. Effects of high-altitude exposure on submaximal endurance capacity of men. J Appl Physiol. 1974; 37:895-8.

55. Mazzeo RS, Bender PR, Brooks GA, Butterfield GE, Groves BM, Sutton JR, Wolfel EE, Reeves JT. Arterial catecholamine responses during exercise with acute and chronic high-altitude exposure. Am J Physiol. 1991;261(4):E419-24.

56. Mazzeo RS, Fulco CS. Physiological systems and their responses to conditions of hypoxia. In: Tipton CM, editor. ACSM's advanced exercise physiology. Baltimore: Lippincott, Williams & Wilkins; 2006. pp. 564-80.

57. Mcardle WD, Magel JR, Lesmes GR, Pechar GS. Metabolic and cardiovascular adjustment to work in air and water at 18, 25, and 33 degrees C. J Appl Physiol. 1976;40:85-90.

58. Mevaloo SF, Shahpar FM. Talent identification programmes [Internet]. Available from: http://www.fina.org/

59. Mujika I, Padilla S, Geyssant A, Chatard JC. Hematological response to training and taper in competitive swimmers: relationships with performance. Arch Physiol Biochem. 1998;105(4):379-85.

60. Murach KA, Bagley JR. Less is more: the physiological basis for tapering in endurance, strength, and power athletes. Sports. 2015;3:209-18.

61. Myerson S, Hemingway H, Budget R, Martin J, Humphries S, Montgomery H. Human angiotensin I-converting enzyme gene and endurance performance. J Appl Physiol. 1999;87(4):1313-6.

62. Nadel ER. Temperature regulation during exercise. In: Houdus Y, Guieu JD, editors. New trends in thermal physiology. Paris: Masson; 1978. pp. 143-53.

63. Neary JP, McKenzie DC, Bhambhani YN. Muscle oxygenation trends after tapering in trained cyclists. Dyn Med. 2005;24(4):1-9.

64. Nielsen B, Hales JRS, Strange S, Juel Christensen N, Warberg J, Saltin B. Human circulatory and thermoregulatory adaptations with heat acclimation and exercise in a hot, dry environment. J Physiol. 1993;460:467-85.

65. Noakes TD. Hyponatremia during endurance running. A physiological and clinical interpretation. Med Sci Sports Exerc. 1992;24(4):403-5.

66. Nybo L, Nielsen B. Hyperthermia and central fatigue during prolonged exercise in humans. J Appl Physiol. 2001;91:1055-60.

67. Pandolf KB. Effects of physical training and cardiorespiratory physical fitness on exercise-heat tolerance: recent observations. Med Sci Sports. 1979;11(1):60-5.

68. Pandolf KB. Time course of heat acclimation and its decay. Int J Sports Med. 1998;19(Suppl 2):S157-60.

69. Pendergast DR. The effects of body cooling on oxygen transport during exercise. Med Sci Sports Exerc. 1988;20:S171-6.

70. Pion JA, Fransen J, Depres DN, Segers VI, Vaeyens R, Philippaerts RM, Lenoir M. Stature and jumping height are required in female volleyball, but motor coordination is a key factor for future elite success. J Strength Cond Res. 2015;29(6):1480-5.

71. Pogliaghi S, Veicsteinas A. Influence of low and high dietary fat on physical performance in untrained males. Med Sci Sports Exerc. 1999;31(1):149-55.

72. Qvist J, Hurford WE, Park YS, Radermacher P, Falke KJ, Ahn DW, et al. Arterial blood gas tensions during breath-hold diving in the Korean ama. J Appl Physiol. 1993;75(1):285-93.

73. Reeves JT, Groves BM, Sutton JR, Wagner PD, Cymerman A, Malconian MK, et al. Operation

Everest II: preservation of cardiac function at extreme altitude. J Appl Physiol. 1987;63:531-9.

74. Riebe D. The American College of Sports Medicine guidelines for exercise testing and prescription. Baltimore: Lippincott, Williams & Wilkins; 2017.

75. Roberts MF, Wenger CB, Stolwijk JAJ, Nadel ER. Skin blood flow and sweating changes following exercise training and heat acclimation. J Appl Physiol. 1977;43:133-7.

76. Rowell LB. Human circulation: regulation during physical stress. New York: Oxford University Press; 1986.

77. Rowlands DS, Hopkins WG. Effect of high-fat, high-carbohydrate, and high-protein meals on metabolism and performance during endurance cycling. Int J Sport Nutr Exerc Metab. 2002;12(3):318-35.

78. Sawka MN, Pandolf KB, Avellini BA, Shapiro Y. Does heat acclimation lower the rate of metabolism elicited by muscular exercise? Aviat Space Environ Med. 1983;54:27-31.

79. Sawka MN, Young AJ. Physiological systems and their responses to conditions of heat and cold. In: Tipton CM, editor. ACSM's advanced exercise physiology. Baltimore: Lippincott, Williams & Wilkins; 2006. pp. 535-563.

80. Schvartz E, Shapiro Y, Magazanik A, Meroz A, Birnfeld H, Mechtinger A, Shibolet S. Heat acclimation, physical fitness, and responses to exercise in temperate and hot environments. J Appl Physiol. 1977;43(4):678-83.

81. Schwellnus MP, Derman EW, Noakes TD. Aetiology of skeletal muscle "cramps" during exercise: a novel hypothesis. J Sports Sci. 1997;15:277-85.

82. Scoon GS, Hopkins W, Mayhew S, Cotter JD. Effect of post-exercise sauna bathing on the endurance performance of competitive male runners. J Sci Med Sport. 2007;10:239-62.

83. Senay LC. Changes in plasma volume and protein content during exposures of working men to various temperatures before and after acclimatization to heat: separation of the roles of cutaneous and skeletal muscle circulation. J Physiol. 1972;224(1):61-81.

84. Shapiro Y, Magazanik A, Vdassin R, Ben-Baruch GM, Shvartz E, Shoenfeld Y. Heat intolerance in former heatstroke patients. Ann Intern Med. 1979;90(6):913-6.

85. Sherman WM, Leenders N. Fat loading: the next magic bullet? Int J Sport Nutr. 1995;5(Suppl):S1-12.

86. Shields C, Whitney FE, Zomar VD. Exercise performance of professional football players. Am J Sports Med. 1984;12(6):455-9.

87. Stanley J, Halliday A, D'Auria S, Buchheit M, Leicht AS. Effect of sauna-based heat acclimation on plasma volume and heart rate variability. Eur J Appl Physiol. 2015;115:785-94.

88. Staudacher HM, Carey AL, Cummings NK, Hawley JA, Burke LM. Short-term high-fat diet alters substrate utilization during exercise but not glucose tolerance in highly trained athletes. Int J Sport Nutr Exerc Metab. 2001;11(3):273-86.

89. Steffes GD, Megura AE, Adams J, Claytor RP, Ward RM, Horn TS, Potteiger JA. Prevalence of metabolic syndrome risk factors in high school and NCAA division 1 football players. J Strength Cond Res. 2013;27(7):1749-57.

90. Terrados N, Melichna J, Sylvén C, Jansson E, Kaijser L. Effects of training at simulated altitude on performance and muscle metabolic capacity in competitive road cyclists. Eur J Appl Physiol. 1988;57(2):203-9.

91. Tipton KD, Rasmussen BB, Miller SL, Wolf SE, Owens-Stovall SK, Petrini BE, Wolfe RR.. Timing of amino acid-carbohydrate ingestion alters anabolic response of muscle to resistance exercise. Am J Physiol Endocrinol Metab. 2001;281:E197-206.

92. Torres-McGehee TM, Pritchett KL, Zippel D, Minton DM, Cellamare A, Sibilia M. Sports nutrition knowledge among collegiate athletes, coaches, athletic trainers, and strength and conditioning specialists. J Athletic Train. 2012;47:205-11.

93. Tsianos G, Sanders J, Dhamrait S, Humphries S, Grant S, Montgomery H. The ACE gene insertion/deletion polymorphism and elite endurance swimming. Eur J Appl Physiol. 2004;92(3):360-2.

94. Vallerand AL, Frim J, Kavanagh MF. Plasma glucose and insulin responses to oral and intravenous glucose in cold-exposed humans. J Appl Physiol. 1988;65(6):2395-9.

95. Wagner PD. Reduced maximal cardiac output at altitude—mechanisms and significance. Resp Physiol. 2000;120(1):1-11.

96. Woods D, Hickman M, Jamshidi Y, Brull D, Vassiliou V, Jones A, Humphries S, Montgomery H. Elite swimmers and the D allele of the ACE I/D polymorphism. Hum Genet. 2001;108(3):230-2.

97. Woods D, Onambele G, Woledge R, Skelton D, Bruce S, Humphries SE, Montgomery H. Angiotensin-I converting enzyme genotype-dependent benefit from hormone replacement therapy in isometric muscle strength and bone mineral density. J Clin Endocrinol Metab. 2001;86(5):2200-4.

98. Wyndham CH, Rogers GG, Senay LC, Mitchell D. Acclimatization in a hot, humid environ-

ment: cardiovascular adjustments. J Appl Physiol. 1976;40(5):779-85.

99. Yegerov AD, Itsekhovsky OG, Polyakova AP, Turchanimova VF, Alferova IV, Savelyeva VG, et al. Results of studies of hemodynamics and phase structure of the cardiac cycle during functional test with graded exercise during 140 days aboard the Salyut-6 station. Kosm Biol Aviakosm Med. 1981;15(3):18-22.

100. Young AJ. Energy substrate utilization during exercise in extreme environments. Exerc Sport Sci Rev. 1990;18:65-117.

101. Young AJ, Sawka MN, Levine L, Cadarette BS, Pandolf KB. Skeletal muscle metabolism during exercise is influenced by heat acclimation. J Appl Physiol. 1985;59:1929-35.

102. Zarkadas PC, Carter JB, Banister EW. Modelling the effect of taper on performance, maximal oxygen uptake, and the anaerobic threshold in endurance triathletes. Adv Exp Med Biol. 1995;393:179-86.

CHAPTER 10

1. American College of Sports Medicine. *ACSM's guidelines for exercise testing and prescription, 10th ed.* Baltimore: Lippincott, Williams & Wilkins; 2017.

2. Berry L, Mirabito A, Baun W. What's the hard return on employee wellness programs? *Harvard Bus Rev.* 2010;89:104.

3. Biswas A, Oh PI, Faulkner GE, Bajaj RR, Silver MA, Mitchell MS, et al. Sedentary time and its association with risk for disease incidence, mortality, and hospitalization in adults: a systematic review and meta-analysis. *Ann Intern Med.* 2015;162:123-32.

4. Blumenthal JA, Babyak MA, Doraiswamy PM, Watkins L, Hoffman BM, Barbour KA, et al. Exercise and pharmacotherapy in the treatment of major depressive disorder. *Psychosom Med.* 2007;69:587-96.

5. Bouchard C, Daw EW, Rice T, Pérusse L, Gagnon J, Province MA, et al. Familial resemblance for $\dot{V}O_2$max in the sedentary state: the HERITAGE Family Study. *Med Sci Sports Exerc.* 1998;30:252-8.

6. Brawner CA, Al-Mallah MH, Ehrman JK, Qureshi WT, Blaha MJ, Keteyian SJ. Change in cardiorespiratory fitness is inversely related to mortality among men and women. *Mayo Clin Proc.* 2017;92:383-390.

7. Brawner CA, Churilla JR, Keteyian SJ. Prevalence of Physical Activity is Lower among Individuals with Chronic Disease. *Med Sci Sports Exerc.* 2016;48:1062-7.

8. British Health Foundation National Centre. Economic costs of physical inactivity [Internet]. Loughborough, UK: British Health Foundation National Centre; 2014. Available from: http://www.bhfactive.org.uk/resources-and-publications-item/40/420/index.html

9. Caloyeras JP, Liu H, Exum E, Broderick M, Mattke S. Managing manifest diseases, but not health risks, saved PepsiCo money over seven years. *Health Aff.* 2014;33:124-31.

10. Carlson MJ, Thiel KW, Yang S, Leslie KK. Catch it before it kills: progesterone, obesity, and the prevention of endometrial cancer. *Discov Med.* 2012;14:215-22.

11. Carlson SA, Fulton JE, Pratt M, Yang Z, Adams EK. Inadequate physical activity and health care expenditures in the United States. *Prog Cardiovasc Dis.* 2015;57:315-23.

12. Castaneda C, Layne JE, Munoz-Orians L, Gordon PL, Walsmith J, Foldvari M, et al. A randomized controlled trial of resistance exercise training to improve glycemic control in older adults with type 2 diabetes. *Diabet Care.* 2002;25:2335-41.

13. Centers for Disease Control and Prevention. Preventing falls: how to develop community-based fall prevention programs for older adults [Internet]. Atlanta: Centers for Disease Control and Prevention; 2008. Available from: http://www.cdc.gov/HomeandRecreationalSafety/images/CDC_Guide-a.pdf

14. Centers for Disease Control and Prevention. The power of prevention: chronic disease . . . the public health challenge of the 21st century [Internet]. Atlanta: Centers for Disease Control and Prevention; 2009. Available from: https://www.cdc.gov/chronicdisease/pdf/2009-power-of-prevention.pdf

15. Centers for Disease Control and Prevention. Comprehensive workplace health programs to address physical activity, nutrition, and tobacco use in the workplace [Internet]. Atlanta: Centers for Disease Control and Prevention; 2012. Available from: http://www.cdc.gov/workplacehealthpromotion/nhwp/index.html

16. Centers for Disease Control and Prevention. Table 63. Participation in leisure-time aerobic and muscle-strengthening activities that meet federal 2008 Physical Activity Guidelines for Americans among adults 18 and over, by selected characteristics: United States selected years 1998-2013 [Internet]. Atlanta: Centers for Disease Control and Prevention; 2014. Available from: https://www.cdc.gov/nchs/data/hus/2014/063/pdf.

17. Centers for Disease Control and Prevention. Mortality in the United States, 2013 [Internet]. Atlanta: Centers for Disease Control and Prevention; 2014.

Available from: http://www.cdc.gov/nchs/data/databriefs/db178.htm

18. Centers for Disease Control and Prevention. Percentage of adults aged ≥18 years who met national guidelines for aerobic activity and muscle strengthening, by age group—National Health Interview Survey, United States, 2008-2013. *Morbid Mortal Weekly Rep.* 2015;64(23):655.

19. Centers for Disease Control and Prevention. Nutrition, Physical Activity and Obesity: Data, Trends and Maps [Internet]. Atlanta: Centers for Disease Control and Prevention; 2017. Available from: https://nccd.cdc.gov/NPAO_DTM/IndicatorSummary.aspx?category=71&indicator=33

20. Centers for Disease Control and Prevention. Leading Causes of Death Reports, National and Regional, 1999-2015 [Internet]. Atlanta: Centers for Disease Control and Prevention; 2017. Available from: https://webappa.cdc.gov/sasweb/ncipc/leadcaus10_us.html

21. Church TS, Thomas DM, Tudor-Locke C, Katzmarzyk PT, Earnest CP, Rodarte RQ, et al. Trends over 5 decades in U.S. occupation-related physical activity and their associations with obesity. *PLoS One.* 2011;6:e19657.

22. Cooney GM, Dwan K, Greig CA, Lawlor DA, Rimer J, Waugh FR, et al. Exercise for Depression. *Cochrane Database Syst Rev.* 2013;9:CD004366.

23. Courneya KS. Physical activity and cancer survivorship: a simple framework for a complex field. *Exerc Sport Sci Rev.* 2014;42:102-9.

24. Courneya KS, Booth CM, Gill S, O'Brien P, Vardy J, Friedenreich CM, et al. The Colon Health and Life-Long Exercise Change trial: a randomized trial of the National Cancer Institute of Canada Clinical Trials Group. *Curr Oncol.* 2008;15:279-85.

25. Crespo CJ, Keteyian SJ, Snelling A, Smit E, Anderson RE. Prevalence of no leisure-time physical activity in persons with chronic disease. *Clin Exerc Physiol.* 1999;1:6.

26. Danaei G, Ding EL, Mozaffarian D, Taylor B, Rehm J, Murray CJ, et al. The preventable causes of death in the United States: comparative risk assessment of dietary, lifestyle, and metabolic risk factors. *PLoS Med.* 2009;6:e1000058.

27. Eckel RH, Jakicic JM, Ard JD, de Jesus JM, Houston Miller N, Hubbard VS, et al. 2013 AHA/ACC guideline on lifestyle management to reduce cardiovascular risk: a report of the American College of Cardiology/American Heart Association Task Force on Practice Guidelines. *Circulation.* 2014;129(25 Suppl 2):S76-99.

28. Franklin BA, Durstine JL, Roberts CK. Impact of diet and exercise on lipid management in the modern era. *Clin Endocrinol Metab.* 2014;28:405-21.

29. Franklin BA, Kahn JK, Gordon NF, Bonow RO. A cardioprotective "polypill"? Independent and additive benefits of lifestyle modification. *Am J Cardiol.* 2004;94:162-6.

30. Garber CE, Blissmer B, Deschenes MR, Franklin BA, Lamonte MJ, Lee IM, et al. American College of Sports Medicine position stand. Quantity and quality of exercise for developing and maintaining cardiorespiratory, musculoskeletal, and neuromotor fitness in apparently healthy adults: guidance for prescribing exercise. *Med Sci Sports Exerc.* 2011;43:1334-59.

31. Gillespie LD, Robertson MC, Gillespie WJ, Sherrington C, Gates S, Clemson LM, et al. Interventions for preventing falls in older people living in the community. *Cochrane Database Syst Rev.* 2012;9:CD007146

32. Goodpaster BH, Katsiaras A, Kelley DE. Enhanced fat oxidation through physical activity is associated with improvements in insulin sensitivity in obesity. *Diabetes.* 2003;52:2191-7.

33. Grabiner MD, Crenshaw JR, Hurt CP, Rosenblatt NJ, Troy KL. Exercise-based fall prevention: can you be a bit more specific? *Exerc Sport Sci Rev.* 2014;42:161-8.

34. Harris MB. Feeling fat: motivation, knowledge, and attitudes of overweight women and men. *Psych Reports.* 1990;67;1191-1202.

35. Haskell WL, Sims C, Myll J, Bortz WM, St. Goar FG, Alderman EL. Coronary artery size and dilating capacity in ultradistance runners. *Circulation.* 1993;87:1076-82.

36. Holten MK, Zacho M, Gaster M, Juel C, Wojtaszewski JF, Dela F. Strength training increases insulin-mediated glucose uptake, GLUT4 content, and insulin signaling in skeletal muscle in patients with type 2 diabetes. *Diabetes.* 2004;53:294-305.

37. Hu FB, Li TY, Colditz GA, Willett WC, Manson JE. Television watching and other sedentary behaviors in relation to risk of obesity and type 2 diabetes mellitus in women. *J Am Med Assoc.* 2003;289:1785-91.

38. Juraschek SP, Blaha MJ, Blumenthal RS, Brawner C, Qureshi W, Keteyian SJ, et al. Cardiorespiratory fitness and incident diabetes: the FIT (Henry Ford Exercise Testing) project. *Diabet Care.* 2015;38:1075-81.

39. Juraschek SP, Blaha MJ, Whelton SP, Blumenthal R, Jones SR, Keteyian SJ, et al. Physical fitness and hypertension in a population at risk for cardiovascular disease: the Henry Ford Exercise Testing (FIT) project. *J Am Heart Assoc.* 2014;3:e001268.

40. Jurca R, Lamonte MJ, Barlow CE, Kampert JB, Church TS, Blair SN. Association of muscular strength with incidence of metabolic syndrome in men. *Med Sci Sports Exerc.* 2005;37:1849-55.

41. Jurca R, Lamonte MJ, Church TS, Earnest CP, Fitzgerald SJ, Barlow CE, et al. Associations of muscle strength and fitness with metabolic syndrome in men. *Med Sci Sports Exerc.* 2004;36:1301-7.

42. Kerrigan DJ, Schairer JR, Jones LW. Cancer. In: Ehrman JK, Gordon PM, Visich PS, Keteyian S, editors. *Clinical exercise physiology, 3rd ed.* Champaign, IL: Human Kinetics; 2013. pp. 379-396.

43. Keteyian S. High intensity interval training in patients with cardiovascular disease: a brief review of physiologic adaptations and suggestions for future research. *J Clin Exerc Physiol.* 2013;2:13-19.

44. Keteyian SJ, Hibner BA, Bronsteen K, Kerrigan D, Aldred HA, Reasons LM, et al. Greater improvement in cardiorespiratory fitness using higher-intensity interval training in the standard cardiac rehabilitation setting. *J Cardiopulm Rehab Prev.* 2014;34:98-105.

45. Keteyian SJ, Levine AB, Brawner CA, Kataoka T, Rogers FJ, Schairer JR, et al. Exercise training in patients with heart failure. A randomized, controlled trial. *Ann Intern Med.* 1996;124:1051-7.

46. Kohl HW, Powell KE, Gordon NF, Blair SN, Paffenbarger RS Jr. Physical activity, physical fitness, and sudden cardiac death. *Epidemiol Rev.* 1992;14:37-58.

47. Kohrt WM, Bloomfield SA, Little KD, Nelson ME, Yingling VR, American College of Sports Medicine. American College of Sports Medicine position stand: physical activity and bone health. *Med Sci Sports Exerc.* 2004;36:1985-96.

48. Kusy K, Zieliński J. Sprinters versus long-distance runners: how to grow old healthy. *Exerc Sport Sci Rev.* 2015;43:57-64.

49. La Gerche A, Heidbuchel H. Can intensive exercise harm the heart? You can get too much of a good thing. *Circulation.* 2014;130(12):992-1002.

50. Lee IM, Shiroma EJ, Lobelo F, Puska P, Blair SN, Katzmarzyk PT, Lancet Physical Activity Series Working Group. Effect of physical inactivity on major non-communicable diseases worldwide: an analysis of burden of disease and life expectancy. *Lancet.* 2012;380:219-29.

51. Leitzmann M, Powers H, Anderson AS, Scoccianti C, Berrino F, Boutron-Ruault MC, et al. European Code against cancer 4th edition: Physical activity and cancer. *Cancer Epidemiol.* 2015;39(Suppl 1):S46-55.

52. Leon AS, Franklin BA, Costa F, Balady GJ, Berra KA, Stewart KJ, et al. Cardiac rehabilitation and secondary prevention of coronary heart disease: an American Heart Association scientific statement from the Council on Clinical Cardiology (Subcommittee on Exercise, Cardiac Rehabilitation, and Prevention) and the Council on Nutrition, Physical Activity, and Metabolism (Subcommittee on Physical Activity), in collaboration with the American Association of Cardiovascular and Pulmonary Rehabilitation. *Circulation.* 2005;111:369-76.

53. Levine BD. Can intensive exercise harm the heart? The benefits of competitive endurance training for cardiovascular structure and function. *Circulation.* 2014;130:987-91.

54. Li L. The financial burden of physical inactivity. *J Sport Health Sci.* 2014;3:56-9.

55. Loprinzi PD. Accumulated short bouts of physical activity are associated with reduced premature all-cause mortality: implications for physician promotion of physical activity and revision of current U.S. Government physical activity guidelines. *Mayo Clin Proc.* 2015;90:1168-9.

56. Magalski A, McCoy M, Zabel M, Magee LM, Goeke J, Main ML, et al. Cardiovascular screening with electrocardiography and echocardiography in collegiate athletes. *Am J Med.* 2011;124:511-8.

57. Mahvan TD, Mlodinow SG. JNC 8: what's covered, what's not, and what else to consider. *J Fam Pract.* 2014;63:574-84.

58. Maron BJ, Shirani J, Poliac LC, Mathenge R, Roberts WC, Mueller FO. Sudden death in young competitive athletes. Clinical, demographic, and pathological profiles. *J Am Med Assoc.* 1996;276:199-204.

59. Maron BJ, Thompson PD, Ackerman MJ, Balady G, Berger S, Cohen D, et al. Recommendations and considerations related to preparticipation screening for cardiovascular abnormalities in competitive athletes: 2007 update: a scientific statement from the American Heart Association Council on Nutrition, Physical Activity, and Metabolism: endorsed by the American College of Cardiology Foundation. *Circulation.* 2007;115:1643-55.

60. Martin JE, Dubbert PM. Exercise applications and promotion in behavioral medicine: current

status and future directions. *Consult Clin Psychol.* 1982;50:1004-17.

61. Matthews CE, Moore SC, Sampson J, Blair A, Xiao Q, Keadle SK. Mortality benefits for replacing sitting time with different physical activities. *Med Sci Sports Exerc.* 2015;47:1833-40.

62. Mondin GW, Morgan WP, Piering PN, Stegner AJ, Stotesbery CL, Trine MR, et al. Psychological consequences of exercise deprivation in habitual exercisers. *Med Sci Sports Exerc.* 1996;28:1199-203.

63. Morris JN, Raffle PA. Coronary heart disease in transport workers: a progress report. *Br J Indust Med.* 1954;11:260-4.

64. Myers J, Prakash M, Froelicher V, Do D, Partington S, Atwood JE. Exercise capacity and mortality among men referred for exercise testing. *N Engl J Med.* 2002;346:793-801.

65. Newman AB, Kupelian V, Visser M, Simonsick EM, Goodpaster BH, Kritchevsky SB, et al. Strength, but not muscle mass, is associated with mortality in the health, aging and body composition study cohort. *J Gerontol A Biol Sci Med Sci.* 2006;61:72-7.

66. O'Connor CM, Whellan DJ, Lee KL, Keteyian SJ, Cooper LS, Ellis SJ, et al. Efficacy and safety of exercise training in patients with chronic heart failure: HF-ACTION randomized controlled trial. *J Am Med Assoc.* 2009;301:1439-50.

67. O'Donovan G, Lee IM, Hamer M, Stamatakis E. Association of "Weekend Warrior" and Other Leisure Time Physical Activity Patterns with Risks for All Cause, Cardiovascular Disease, and Cancer Mortality. *JAMA Intern Med.* 2017;177:335-42.

68. O'Keefe JH, Patil HR, Lavie CJ, Magalski A, Vogel RA, McCullough PA. Potential adverse cardiovascular effects from excessive endurance exercise. *Mayo Clin Proc.* 2012;87:587-95.

69. Pate RR, Pratt M, Blair SN, Haskell WL, Macera CA, Bouchard C, et al. Physical activity and public health. A recommendation from the Centers for Disease Control and Prevention and the American College of Sports Medicine. *J Am Med Assoc.* 1995;273:402-7.

70. Pescatello LS, MacDonald HV, Ash GI, Lamberti LM, Farquhar WB, Arena R, et al. Assessing the existing professional exercise recommendations for hypertension: a review and recommendations for future research priorities. *Mayo Clin Proc.* 2015;90:801-12.

71. Prochaska JO, DiClemente CC. Stages and processes of self-change of smoking: toward an integrative model of change. *J Consult Clin Psychol.* 1983;51:390-5.

72. Rahman I, Bellavia A, Wolk A, Orsini N. Physical activity and heart failure risk in a prospective study of men. *J Am Coll Cardiol Heart Fail.* 2015;3:681-7.

73. Ramos JS, Dalleck LC, Tjonna AE, Beetham KS, Coombes JS. The impact of high-intensity interval training versus moderate-intensity continuous training on vascular function: a systematic review and meta-analysis. *Sports Med.* 2015;45:679-92.

74. Riebe D, Franklin BA, Thompson PD, Garber CE, Whitfield GP, Maga, M, et al. Updating ACSM's recommendations for exercise preparticipation health screening. *Med Sci Sports Exerc.* 2015;47:2473-9.

75. Rose DJ. The role of exercise in preventing falls among older adults. *ACSM Health Fit J.* 2015;19:23-9.

76. Scarborough P, Bhatnagar P, Wickramasinghe KK, Allender S, Foster C, Rayner M. The economic burden of ill health due to diet, physical inactivity, smoking, alcohol and obesity in the UK: an update to 2006-07 NHS costs. *J Public Health (Oxf).* 2011;33:527-35.

77. Schroeder SA. We can do better—improving the health of the American people. *N Engl J Med.* 2007;357:1221-8.

78. Sedentary Behavior Research Network. Letter to the editor: standardized use of the terms "sedentary" and "sedentary behavior." *Appl Physiol Nutr Metab.* 2012;37:540-2.

79. Sharma S, Merghani A, Mont L. Exercise and the heart: the good, the bad, and the ugly. *Eur Heart J.* 2015;36:1445-53.

80. Sigal RJ, Kenny GP, Boulé NG, Wells GA, Prud'homme D, Fortier M, et al. Effects of aerobic training, resistance training, or both on glycemic control in type 2 diabetes: a randomized trial. *Ann Intern Med.* 2007;147:357-69.

81. Smith SC Jr, Benjamin EJ, Bonow RO, Braun LT, Creager MA, Franklin BA, et al. AHA/ACCF secondary prevention and risk reduction therapy for patients with coronary and other atherosclerotic vascular disease: 2011 update: a guideline from the American Heart Association and American College of Cardiology Foundation. *Circulation.* 2011;124(22):2458-73.

82. Steinvil A, Chundadze T, Zeltser D, Rogowski O, Halkin A, Galily Y, et al. Mandatory electrocardiographic screening of athletes to reduce their risk for sudden death: proven fact or wishful thinking? *J Am Coll Cardiol*. 2011;57:1291-6.

83. Swank A. A first step to health: Just stand up and move. ACSM Health Fit J. 2015;19:34-6.

84. Taylor RS, Sagar VA, Davies EJ, Briscoe S, Coats AJ, Dalal H, et al. Exercise-based rehabilitation for heart failure. *Cochrane Database Syst Rev.* 2014;4:CD003331.

85. Thompson PD, Franklin BA, Balady GJ, Blair SN, Corrado D, Estes NA III, et al. Exercise and acute cardiovascular events. Placing the risks into perspective: a scientific statement from the American Heart Association Council on Nutrition, Physical Activity, and Metabolism and the Council on Clinical Cardiology. *Circulation*. 2007;115:2358-68.

86. U.S. Department of Health and Human Services. Physical activity and health: a report of the surgeon general. Atlanta, GA: U.S. Department of Health and Human Services; 1996.

87. U.S. Department of Health and Human Services. Physical activity guidelines for Americans [Internet]. Atlanta, GA: U.S. Department of Health and Human Services; 2008. Available from: http://health.gov/paguidelines/guidelines

88. Wen CP, Wai JPM, Tsai MK, Cheng TYD, Lee MC, Chan HT, et al. Minimum amount of physical activity for reduced mortality and extended life expectancy: a prospective cohort study. *Lancet*. 2011;378:1244-53.

89. World Health Organization. Preventing chronic diseases: a vital investment [Internet]. Geneva, Switzerland: World Health Organization; 2005. Available from: http://www.who.int/chp/chronic_disease_report/full_report.pdf

90. World Health Organization. Insufficient physical activity, 2010. Prevalence of insufficient physical activity among adults ages 18+ (age standardized estimates): Male [Internet]. Geneva, Switzerland: World Health Organization; 2017. Available from: http://www.who.int/gho/ncd/risk_factors/physical_activity/en/

CHAPTER 11

1. Blangero J, Kent JW Jr. Characterizing the extent of human genetic variation for performance-related traits. In: Bouchard C, Hoffman EP, editors. Genetic and molecular aspects of sport performance. West Sussex, UK: Wiley-Blackwell; 2011. pp. 33-45.

2. Boer F. Drug handling by the lungs. Br J Anaesthes. 2003;91:50-60.

3. Bouchard C, An P, Rice T, Skinner JS, Wilmore JH, Gagnon J, et al. Familial aggregation of $\dot{V}O_2$max response to exercise training: results from the HERITAGE Family Study. J Appl Physiol. 1999;87:1003-8.

4. Bouchard C, Daw EW, Rice T, Perusse L, Gagnon J, Province MA, et al. Familial resemblance for $\dot{V}O_2$max in the sedentary state: the HERITAGE family study. Med Sci Sports Exerc. 1998;30:252-8.

5. Bouchard C, Rankinen T. Individual differences in response to regular physical activity. Med Sci Sports Exerc. 2001;33(6 Suppl):S446-51.

6. Bouchard C, Sarzynski MA, Rice TK, Kraus WE, Church TS, Sung YJ, et al. Genomic predictors of the maximal O_2 uptake response to standardized exercise training programs. J Appl Physiol. 2011;110:1160-70.

7. Bray MS, Fulton JE, Kalupahana NS, Lightfoot JT. Genetic epidemiology, physical activity, and inactivity. In: Bouchard C, Hoffman EP, editors. Genetic and molecular aspects of sport performance. West Sussex, UK: Wiley-Blackwell; 2011. pp 81-9.

8. Bray MS, Hagberg JM, Perusse L, Rankinen T, Roth SM, Wolfarth B, et al. The human gene map for performance and health-related fitness phenotypes: the 2006-2007 update. Med Sci Sports Exerc. 2009;41:35-73.

9. De Moor MH, Liu YJ, Boomsma DI, Li J, Hamilton JJ, Hottenga JJ, et al. Genome-wide association study of exercise behavior in Dutch and American adults. Med Sci Sports Exerc. 2009;41:1887-95.

10. Frid A, Ostman J, Linde B. Hypoglycemia risk during exercise after intramuscular injection in the thigh in IDDM. Diabet Care. 1990;13:473-7.

11. Kazlauskas R. Designer steroids. In: James E. Barrett, editor. Handbook of experimental pharmacology. London, UK, Springer; 2010. pp. 155-85.

12. Kersey RD, Elliot DL, Goldberg L, Kanayama G, Leone JE, Pavlovich M, et al. National Athletic Trainers' Association position statement: anabolic-androgenic steroids. J Athlet Train. 2012;47:567-88.

13. LaBotz M, Griesemer BA, Brenner JS, LaBella CR, Brooks MA, Diamond A, et al. Use of performance-enhancing substances. Pediatrics. 2016;doi:10.1542/peds.2016-1300.

14. Lenz TL. The effects of high physical activity on pharmacokinetic drug interactions. Expert Opin Drug Metab Toxicol. 2011;7:257-66.

15. Mason WD, Kopchak G, Winer N, Cohen I. Effect of exercise on the renal clearance of atenolol. J Pharm Sci. 1980;69:344-50.

16. Montgomery HE, Clarkson P, Dollery CM, Prasad K, Losi MA, Hemingway H, et al. Association of angiotensin-converting enzyme gene I/D polymorphism with change in left ventricular mass in response to physical training. Circulation. 1997;93:741-7.

17. Myerson SG, Montgomery HE, Whittingham M, Jubb M, World MJ, Humphries SE, et al. Left ventricular hypertrophy with exercise and ACE gene insertion/deletion polymorphism: a randomized controlled trial with losartan. Circulation. 2001;103:226-30.

18. Nagashima K, Cline GW, Mack GW, Shulman GI, Nadel ER. Intense exercise stimulates albumin synthesis in the upright posture. J Appl Physiol. 2000;88:41-6.

19. Persky AM, Eddington ND, Derendorf H. A review of the effects of chronic and physical fitness level on resting pharmacokinetics. Int J Clin Pharmacol Therapeut. 2002;41:504-16.

20. Perusse, L. Role of genetic factors in sport performance: Evidence from family studies. In: Bouchard C, Hoffman EP, editors. Genetic and molecular aspects of sport performance. West Sussex, UK: Wiley-Blackwell; 2011. pp. 90-100.

21. Pescatello LS, Kostek MA, Gordish-Dressman H, Thompson PD, Seip RL, Price TB, et al. ACE ID genotype and the muscle strength and size response to unilateral resistance training. Med Sci Sports Exerc. 2006;38:1074-81.

22. Peter R, Luzio SD, Dunseath G. Effects of exercise on the absorption of insulin glargine in patients with type I diabetes. Diabet Care. 2005;28:560-5.

23. Rankinen T, An P, Perusse L, Rice T, Chagnon YC, Gagnon J, et al. Genome-wide linkage scan for exercise stroke volume and cardiac output in the HERITAGE Family Study. Physiol Genom. 2002;10:57-62.

24. Reardon CL, Creado S. Drug abuse in athletes. Substance Abuse Rehab. 2014;5:95-105.

25. Sarzynski MA, Rankinen T, Bouchard C. Twin and family studies of training responses. In: Bouchard C, Hoffman EP, editors. Genetic and molecular aspects of sport performance. West Sussex, UK: Wiley-Blackwell; 2011. pp. 110-20.

26. Shendre A, Beasley TM, Brown TM, Hill CE, Arnett DK, Limdi NA. Influence of regular physical activity on warfarin dose and risk of hemorrhagic complications. Pharmacotherapy. 2014;34:545-54.

27. Spielmann N, Leon AS, Rao DC, Rice T, Skinner JS, Rankinen T, et al. Genome-wide linkage scan for submaximal exercise heart rate in the HERITAGE family study. Am J Physiol Heart Circ Physiol. 2007;293:H3366-71.

28. Stoschitzky K, Linder W, Klein W. Stereoselective release of S-atenolol from adrenergic nerve endings at exercise. Lancet. 1992;340:696-700.

29. Tesch PA. Exercise performance and beta-blockade. Sports Med. 1985;2:389-412.

30. van Baak MA, Mooij JM, Schiffers PM. Exercise and the pharmacokinetics of propranolol, verapamil and atenolol. Eur J Clin Pharmacol. 1992;43:547-50.

31. Ward BW, Clarke TC, Freeman G, Schiller JS. Early release of selected estimates based on data from the 2014 National Health Interview Survey [Internet]. Atlanta: National Center for Health Statistics; 2015. Available from: http:///www.cdc.gov/nchs/nhis.htm

32. Yesalis CE, Bahrke MS. Anabolic-androgenic steroids and related substances. Curr Sports Med Rep. 2002;1:246-52.

APPENDIX B

1. Maxwell B.F., Withers R.T., Ilsley A.H., Waking M.J., Woods G.F., Day L. Dynamic calibration of mechanically, air- and electromagnetically braked cycle ergometers. *Eur J Appl Physiol Occup Physiol.* 1998;78(4):346-52.

APPENDIX C

1. Ainsworth BE, Haskell WL, Herman SD, Meckes N, Bassett DR, Jr., Tudor-Locke C, et al. Compendium of Physical Activities: a second update of codes and MET values. Med Sci Sports Exerc 2011;43(8):1575-81.

2. Kozey S, Lyden K, Staudenmayer J, Freedson P. Errors in MET estimates of physical activities using $3.5 \, ml \times kf^{-1} \times min^{-1}$ as the baseline oxygen consumption. J Phys Act Health 2010;4(12):3703.

3. Byrne NM, Hills AP, Hunter GR, Weinsier RL, Schutz Y. Metabolic equivalent: one size does not fit all. J Appl Physiol (1985) 2005;99(3):1112-9.

4. Harris JA, Benedict FG. A Biometric Study of Human Basal Metabolism. Proc Natl Acad Sci U S A 1918;4(12):3703.

APPENDIX D

1. American Fitness Professionals and Associates. A career in fitness: personal training statistics and facts [Internet]. Ship Bottom, NJ: American Fitness Professionals and Associates; 2014. Available from: http://www.afpafitness.com/blog/personal-training-statistics-2015Australia Department of Employment.

2. Job outlook: an Australian government initiative [Internet]. Canberra, ACT: Australia Department of Employment; 2012. Available from: http://joboutlook.gov.au/occupation.aspx?code=4521&search=&Tab=prospects

3. Larsen MP, Eisenberg MS, Cummins RO, Hallstrom AP. Predicting survival from out-of-hospital cardiac arrest: a graphic model. *Ann Emerg Med.* 1993;22:1652-8.

4. Service Canada. Job Futures Quebec [Internet]. Location: Service Canada; 2017. Available from: http://www.jobbank.gc.ca/content_pieces-eng.do?cid=10424&lang=eng

5. Tipton CM. *History of exercise physiology.* Champaign, IL: Human Kinetics; 2014.

6. U.S. Department of Health and Human Services. 2008 physical activity guidelines for Americans: be active, healthy and happy [Internet]. Washington, DC: U.S. Department of Health and Human Services; 2008. Available from: www.health.gov/paguidelines/pdf/paguide.pdf

Index

Note: Page numbers followed by an italicized *f* or *t* refer to the figure or table on that page, respectively.

About the Authors

Jonathan K. Ehrman, PhD, FACSM, FAACVPR, is the associate program director of preventive cardiology and of the Exercise Physiology Core Laboratory and the director of the clinical weight management program at Henry Ford Hospital in Detroit, Michigan. He specializes in cardiac rehabilitation and preventive cardiology. He is also a clinical assistant professor in the exercise science program at Oakland University. He is a fellow of the American Association of Cardiovascular and Pulmonary Rehabilitation (AACVPR) and the American College of Sports Medicine (ACSM), and he is certified by the ACSM as a clinical exercise physiologist. In addition to the ACSM and AACVPR, he is also an active member of the Clinical Exercise Physiology Association (CEPA), the American Heart Association, and the American College of Cardiology. Dr. Ehrman earned his PhD in clinical exercise physiology from The Ohio State University.

Dr. Ehrman is a coeditor of *Clinical Exercise Physiology, Fourth Edition*, and served as a section editor of the 10th edition of *ACSM's Guidelines for Exercise Testing and Prescription*. In addition, he has published more than 100 research manuscripts, abstracts, and book chapters. He is also the current editor of the *Journal of Clinical Exercise Physiology*, which is the official journal of the CEPA.

Dennis J. Kerrigan, PhD, FACSM, is a senior exercise physiologist in preventive cardiology in the Division of Cardiovascular Medicine and the William Clay Ford Center for Athletic Medicine at Henry Ford Hospital in Detroit, Michigan. His current role is as the director of outpatient exercise programs in preventive cardiology, and he oversees exercise programs for individuals with chronic health conditions. In addition to his clinical duties, Dr. Kerrigan also conducts research in patients with heart disease, cancer, and obesity. He was the lead author in a randomized trial that showed improvements in fitness and quality of life in patients with left ventricular assist devices who participated in cardiac rehabilitation. In 2017, Dr. Kerrigan was elected president of the Clinical Exercise Physiology Association (CEPA). He earned his PhD in clinical exercise physiology from The Ohio State University.

Dr. Kerrigan coauthored a book chapter in Guidelines for Cardiac Rehabilitation and Secondary Prevention Programs, Fifth Edition, and has published scholarly articles in peer-reviewed journals, including the Journal of Cardiopulmonary Rehabilitation and Prevention, American Journal of Cardiology, Journal of Cardiac Failure, and Medicine and Science in Sports and Exercise. In 2012, he received the Midwest ACSM Clinical Exercise Professional of the Year award.

Steven J. Keteyian, PhD, FACSM, is the director of preventive cardiology in the Division of Cardiovascular Medicine and director of the Exercise Physiology Core Laboratory, both at Henry Ford Hospital in Detroit, Michigan. He specializes in clinical exercise physiology and preventive cardiology. He also serves as a clinical professor in the exercise science program at Oakland University and as an adjunct professor in the physiology department at Wayne State University, where he earned his PhD. He is an active member of the American Heart Association and the American Association of Cardiovascular and Pulmonary Rehabilitation (AACVPR).

Dr. Keteyian has published more than 200 research manuscripts and book chapters and has coauthored or coedited four textbooks, including *Clinical Exercise Physiology, Third Edition*. He has also served as an editor-in-chief, associate editor, or editorial board member for several academic journals. Dr. Keteyian received the President's Award from the AACVPR in 2013 and the Established Investigator Award from the AACVPR in 2009.